Introduction to Data Systems

In family life, love is the oil that eases friction, the cement that binds closer together, and the music that brings harmony. [Nietzsche]

To Suzanne, your enduring love and support inspire and sustain me. To Amanda and Ben, I love you deeply and am so proud of you. And, to the best of all time—the family of George and Leigh.

−Tom

Preface

The sources and forms of data, along with the demand for acquiring, processing, and analyzing that data are increasing at a prodigious rate. From a curricular perspective, majors, programs, and courses are being designed to enable interested students to understand these topics and prepare themselves for the practice of these associated data skills. As the sources and forms of data increase, with a commensurate increase in the data providers and interfaces for accessing data, our interpretation of what constitutes a data system must keep pace. An understanding of these forms and sources of data is exactly what is needed in a broad introduction to data systems.

Data systems encompasses the study of forms and sources of data, an increasingly important topic for both computer scientists and data scientists. This book covers data acquisition, wrangling, normalization, and curation, requiring only basic prior exposure to Python. The book includes a detailed treatment of tidy data, relational data, and hierarchical data, laying a conceptual basis for the structure, operations, and constraints of each data model, while simultaneously providing hands-on skills in Python, SQL, and XPath. The sources of data studied encompass local files, text applications and regular expressions, database servers, HTTP requests, REST APIs, and web scraping.

Who Is This Book for?

As university curricula expand to include content on data systems, the student audience for such curricular efforts is broadening as well. We identify three student constituencies at the undergraduate level:

- *Computer Science*: students recognizing the value of being facile while working with data and seeing the synergy with the systems and algorithmic problem-solving aspects of the computer science discipline.

- *Data Science/Data Analytics*: students pursuing the emerging undergraduate majors, who desire to build algorithmic and practical skills with an eye toward enabling data exploration, visualization, statistical analysis, and application of machine learning techniques for use in coursework as well as in practicums and internships.
- *Multidisciplinary/domain-centric*: students whose focus is on practical aspects of acquiring and transforming data for use in their own disciplines. These include students in social sciences as well as natural sciences, and often as a prelude to a "methods" course specific to their discipline.

Within computer science education, the union of these audiences has been termed the "data-centric" audience. We have carefully identified a set of essential data-related topics, and a desired level of understanding, that these student constituencies will need to be successful in their future endeavors.

This book is ideal for first- or second-year undergraduate students, either for a classroom setting or for self-directed learning, and does not require prerequisites of data structures, algorithms, or other courses beyond a first course in Python programming. This book equips students with understanding and skills that can be applied in computer science, data science/data analytics, and information technology programs as well as for internships and research experiences. By drawing together content normally spread across a set of upper-level computer science courses, it offers a single source providing the essentials for data science practitioners, an accessible and foundational second course for computer science majors, a potential second course for data science majors, and content that can supplement introductory courses seeking more exposure to real-world data. In our increasingly data-centric world, students from all domains both want and need the *data-aptitude* built by the material in this book.

Philosophy of This Book

We see two primary dimensions to data-aptitude, corresponding to the *sources* and *forms* of data. Data sources range from local files, downloaded or given to the student or analyst, to a relational database system, to a wide assortment of network-based data providers. Data forms/formats include comma-separated value or other delimited flat files, tables within a relational database, XML and JSON used for data interchange, as well as unstructured data and HTML from which data can be extracted. Data models give a framework for understanding the structure, operations, and constraints for these various forms of data. A third dimension of data-aptitude relates to the protection and privacy of the data. Some data is explicitly open and freely available. Other data may be limited to use by particular applications. Still other data may be comprised of protected resources of one or more resource owners, and the acquisition and use of such data must be appropriately authorized.

An Introduction to Data Systems, as advanced by this book, takes the perspective that students can, early in their academic careers, learn these dimensions of data-aptitude. Students need, in a structured way, to learn about the forms and sources of data in modern data systems, defined broadly. As *users* of data systems, students need to be able to build applications to acquire data and, given the form of the data acquired, be able to extract and transform the data into a normalized form, where subsequent statistics and analysis may be easily performed on the data.

This book uses the framework of a set of data models, from a simple tabular model with appropriate constraints, to the relational data model, and to a hierarchical data model, giving structure to the various data forms seen in practice. In the data sources dimension of data-aptitude, the course follows a progression of data sources, starting the students with local files, moving on to MySQL database server and SQLite databases as representative of relational databases, and then covering more advanced client–server interaction over HTTP using provider APIs, and also extracting data from HTML. In this way, we feel we can cover the topics of data-aptitude in sufficient depth, while keeping material broad and applicable to the wide range of data-centric students.

Since specific programming languages and packages are bound to change over the course of the reader's lifetime, we stress a conceptual understanding over an exhaustive coverage of packages. For each topic, we begin with a high-level discussion, situating the topic within its disciplinary context. We then discuss *how* to work with the topic, with an emphasis on problem solving and algorithms. Lastly, we illustrate using tools such as Python, `pandas`, SQL, XPath, `curl`, etc. In this way, we ensure that readers understand the principles underlying these tools, rather than only how to use them as a black box. We select tools that are powerful enough to illustrate the concepts, and simultaneously as easy as possible for readers to learn. Readers are free to focus their energy on the essential content, rather than struggling to learn the tools. We include detailed references that can guide interested readers to further exploration.

To avoid overwhelming readers with too many different data situations, we center our exposition on three data sets—based on economic data, sociological data, and educational data—that we carry with us throughout the book. These illustrate our foundational material, the three data models, and the process of data acquisition. This allows us to center our discussion on compelling real-world problems, rather than on the tools used to solve these problems. This approach has been shown to be effective for a diverse array of student learners, and to increase retention in the discipline. When we develop data wrangling solutions on these data sets, we illustrate good software engineering principles, guiding the reader through the process of developing incrementally, testing their code as they go, and error handling. We include a large number of exercises where students can sharpen their skills.

Web Resources

Accompanying this book is a website supported by the authors, containing files and supplementary content that will aid the reader:

http://datasystems.denison.edu.

This website hosts data files used in Parts I and II of the book and is used to illustrate HTTP, HTML, and web scraping in Part III. In addition, the authors have taught many iterations of a course using this book and have curated a repository of hundreds of exercises, reading questions, and hands-on activities engaging students in the material. These resources are contained within Jupyter notebooks that use the freely available nbgrader system, so that worksheets may be automatically graded. The repository also contains multiple in-depth projects guiding students to work with real-world data sets, normalize the data into the relevant data model, generate interesting questions, and answer these questions with visualizations. Via these projects, students can create a portfolio to showcase what they have learned to potential future employers and graduate programs. We host several sample projects on the book website, and we intend to add more as we continue to teach courses using this book. In this way, projects can be updated if real-world data sources change where they host data, what data they make available, or the form in which the data is provided.

To Students

This text was written with both concrete and abstract goals in mind. A large part of our motivation was to serve "data-focused" students and provide a bridge for you to take the learning from the classroom and use these skills in *concrete, real-world settings*. The focus of the book is giving you the skills you need to acquire data from a multitude of sources, mutate it into a form suitable for analysis, and access it programmatically to answer interesting questions. This includes data stored in files, on database servers, provided by Application Programming Interfaces (APIs), and data obtained via web scraping. The projects (hosted on the book web page) will guide you to applying your new skills in the real world: after learning about data models, you will be able to use that knowledge to work with real-world data sets, normalize the data into the model, and then generate interesting questions and visualizations to help answer those questions. Similarly, after learning about obtaining data over the Internet, REST APIs, and authorization for protected resources, the API projects allow you to bring it all together, and acquire data from providers such as LinkedIn, Reddit, various Google APIs, Lyft, and many others. This will entail using the material from the entire arc of the book.

We surveyed students to see how the material in this book benefited them after the course was over. The vast majority of respondents reported using material from some or all of the chapters of this book in their subsequent jobs, internships, and research projects. Furthermore, if done correctly, the projects you complete can become part of a *portfolio* that you show to potential employers, which, coupled with your knowledge of the terms and concepts covered in this book, can help you secure the kind of data-focused job you are interested in.

Equally important to these concrete learning goals are our abstract goals: to sharpen your analytic thinking, problem solving, coding, writing, and technical reading skills. For each technique in this book, we first describe the approach in general terms, then carefully work through multiple examples, and finally provide numerous exercises (building on the examples) for you to achieve mastery. In our examples, we model an *incremental approach*, where we develop a partial solution, test it, and then develop a bit more, repeating this process until we have a general solution. We encourage you to follow the same steps when you solve exercises, including the use of try/except blocks and assert statements to ensure that your code behaves as expected.

This book is written to only assume you have prior exposure to computer science principles (and Python) at the level of an introductory course. Normally, the material in this book is spread across several electives that only junior and senior computer science majors take. By condensing the material, and taking an elementary approach to it, we aim to give you the concrete and abstract skills early in your college career. However, the trade-off is that this book contains a great deal that will probably be new to you. We do not expect that you will be an expert at reading computer science books. Some parts may be difficult to understand or may require you to read multiple times. As with anything you read, if you come across a term that is new to you, we encourage you to look that term up and understand it before proceeding. It may be that the term was defined earlier in the book, in which case the index at the end of the book can tell you where the term was first defined.

Most sections begin with an abstract approach and introduce examples later. It may make sense to peek ahead at the examples if you struggle with the abstract part, or to reread it after finishing the chapter to better structure what you have learned. To help you identify the most important parts of each section, and to guide you through the types of activities that will help you to understand the material (e.g., relating the abstract concepts to experiences you may have already had in the real world), we include *reading questions* at the end of each section. These may be assigned by your instructor, to make sure you attempt the reading before class. Even if they are not assigned, we recommend working through the questions as you read each section, as they will often clarify the meaning of new terms introduced, highlight potential pitfalls, and emphasize which pieces of the reading are most essential. You can and should use Python when answering reading questions that reference code, modules, and methods.

Data systems is a rapidly evolving field, and this book is the first of its kind. Previously, students who wanted to learn this material would need to do so by reading online tutorials that often treated data science tools as a black box. By emphasizing the concepts and programming underlying the tools, we aim to give you a deeper level of understanding. This understanding, coupled with the technical reading skills you will develop, will aid you if, in future years, you find yourself needing to read tutorials and manuals in order to learn new technologies that did not exist when you were a student. For this reason, reading questions sometimes guide you to online self-learning materials to get you comfortable with this kind of exploration. This approach is especially apparent in Part III, where we learn how to acquire data from web scraping and APIs, which, necessarily, differ between different data providers. Our approach teaches you the general steps that will guide you to learn the particulars of APIs and HTML web pages, so that you can develop software to obtain data from a diverse array of sources.

As you read the text, in addition to trying to answer the reading questions, it is essential to familiarize yourself with the blocks of code provided. You should expect to practice with the code every day, as you would if you were learning a new language or musical instrument. Fundamental to the process of learning is to make connections between the new material you are seeing and your past life experience. We encourage you to be curious and playful as you explore the content: try plugging in different values for the parameters of functions, adding print statements to see how a block of code computes, and coming up with questions beyond the examples in the book.

To Instructors

This book is designed with minimal prerequisites, requiring only that students have prior exposure to Python programming at the level of a first course in computer science. This makes it suitable for any of the three student constituencies identified above, or as a supplementary textbook to go along with an introductory computer science course and add a more data-driven focus. As discussed above, we emphasize conceptual understanding, an abstract framework, concrete examples based on real-world data, and good software engineering principles. This emphasis gives readers a deeper understanding than would be obtained from a "pick it up as you go" approach, or an approach rooted in technical manuals. The structure provided in this book empowers readers to learn new technologies throughout their lives, as part of a unifying framework.

Broadly speaking, the book breaks down into six units, which we summarize in Table 1.

There is a rich set of possible orderings for covering the material in the book for a given course, and we have successfully used several different orderings. The book chapters themselves progress by covering all the data models prior to the treatment of the data sources. But, a course that wanted to interleave the two parts could do so, with arguable benefit for student engagement. For instance, after learning the tabular model, the course could cover networking and HTTP to access data sets from web servers and immediately apply the data model in client software. Similarly, after coverage of the hierarchical model, the course could access XML from web servers or non-authenticated APIs to again apply the data model.

Table 1 Textbook contents

Unit	Name	Description
00	Foundations	After an overview of data systems, we begin building a foundation in Python to be used in the remainder of the book. We review introductory basic topics typically found in intro courses. We then progress to useful and possibly new topics like list comprehensions, lambda functions, and regular expressions
01	Tabular data model	The focus is on single, rectangular table data sets. We work from native Python structures to pandas data frames. We conclude with the constraints required for rectangular data to be considered "tidy data"
02	Relational data model	This unit explores the data model associated with relational databases and the operations and organization involved in sound design of the set of tables for a database. We develop skills in querying existing databases and creating databases using the Structured Query Language (SQL) through both direct commands and Python programming with sqlalchemy
03	Hierarchical data model	The focus shifts to hierarchical data, such as that found in XML and JSON. We explore both declarative XPath and programmatic means to process the data and wrangle it into data frames. We also cover schemas for constraining data in the model, such as XML Schema and JSON Schema
04	Networking, HTTP, HTML	In this unit, our goals are to understand and program client applications over the network. We start with foundational networking concepts and the network protocol stack and progress to learning HTTP and making requests for static data in files or as HyperText Markup Language (HTML)
05	APIs	The book culminates with acquisition of data from API providers. This unit is a synthesis of the full set of data models and provider sources. This unit covers APIs and associated authentication (including OAuth) and illustrates the framework with real-world providers

We often begin with Chaps. 1–3, because this material is foundational and can identify any gaps in student background. Chapter 4 can be substantially delayed if one wishes to get more quickly to data models. While the tabular data model and relational data model can be taught in either order, we generally teach the tabular model first, as it only depends on local files instead of forming connections to database servers. Our study of hierarchical operations depends on Chap. 6, because the examples wrangle hierarchical data into tabular form, but otherwise our treatment of hierarchical models is logically independent from our treatment of the other two data models.

We note that the relational model can be taught as a stand-alone sequence consisting of Chaps. 10–14, which may be valuable to give students the basics of databases and SQL in a three-week module or attached to another course. One could similarly teach Chaps. 2, 6, 15, 16, and 17 as a stand-alone sequence

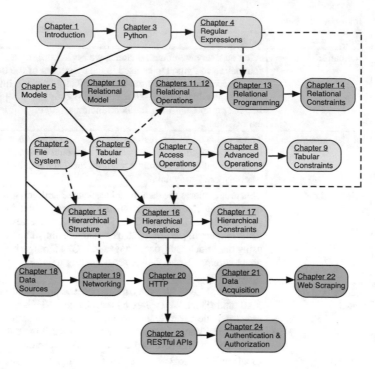

Fig. 1 Chapter dependencies

on hierarchical data or could teach Chaps. 18–20 as a stand-alone sequence on the basics of networking. If a reader is primarily interested in web scraping and APIs, the tabular and relational model could be skipped entirely (except for the introduction of `pandas` in Chap. 6), and the minimal set of chapters to cover in this situation would be Chaps. 1–6, 15–16, and 20–22 (or 20, 23, and 24 for APIs).

Figure 1 shows the dependency structure between the chapters. We have color coded the chapters to help visualize the units described above. Dotted arrows represent suggested, but optional, orderings. The dotted lines from regular expressions to hierarchical operations and to relational programming support some more advanced exercises in those downstream chapters.

For many topics, we employ a spiral approach: returning to the topic in increasingly complex ways as the book progresses. For example, regular expressions are introduced in Chap. 4, then reinforced in Chaps. 13 and 16, and then can be used heavily in Chap. 22. Similarly, the JSON format is introduced in Chap. 2, then reinforced in Chap. 13, and then used heavily in Chaps. 15, 16, and 22–23. We also introduce non-local (over the network) data sources gradually. Concepts of protocol, connection, and user identity (authentication and authorization) are introduced concurrently with the relational data model by using databases provided by a database server. The next source of data, web servers, allows us to expand the basics of networking, the network protocol stack, and then the fundamentals of the

HyperText Transfer Protocol (HTTP). Understanding the aspects of this protocol allows the facility for constructing and issuing requests and being able to retrieve and understand the constituent parts of the response. This type of data source can provide delimited flat files, HTML files, or even XML/JSON files. We combine this data source with the knowledge from the hierarchical data model to understand web scraping as an application of previously introduced ideas.

Software Assumptions

To run the code in this book, you will need to have Python 3.4 (or later) on your machine, as well as a number of packages and libraries, including `pandas, os, sys, io, json, re, sqlalchemy, sqlite3, lxml, jsonschema, requests, base64` and a custom module discussed in the Appendix and hosted on the book website. At the time of writing, this version of Python, along with all of these packages and libraries, was included with a standard installation of the freely available Anaconda distribution from Continuum Analytics.

https://store.continuum.io/cshop/anaconda

Online Corrigenda

This book was written in `bookdown`, using facilities of Python and SQL code as part of the textbook composition. In this facility, every example piece of Python and SQL is executed as part of the rendering process, ensuring that there are no errors in the code presented to the reader. Furthermore, for examples that reference Internet data sources (e.g., in the web scraping and API chapters), we attempted to select sources that seemed most likely to remain stable in the years to come. Nevertheless, we are aware of the possibility that the reader may discover typos, or that code referencing online data sources may cease to work properly. As errors are identified, they will be corrected at

https://www.datasystems.denison.edu/errata.

If you find an error not documented at the link above, please report it using the link below:

https://www.datasystems.denison.edu/feedback.

Acknowledgments

The authors are grateful to the Denison University for hosting the web resources associated with this book and providing a wonderful working environment where it was written. We also thank Dick De Veaux for facilitating the Park City Mathematics

Institute Undergraduate Faculty Program in 2016 where this book began, and Deborah Nolan and Duncan Temple-Lang for sharing an early draft of their book *Data Technologies and Computational Reasoning*. We are indebted too, to the students at Denison. We especially thank Gavin Thomas and Paul Rubenstein for their pedagogical research work related to this book (which produced a paper published in SIGCSE in 2019). Students from the Denison CS-181 and DA-210 courses have given feedback and helped tremendously. Emma Steinman was particularly helpful and dedicated in giving feedback, and Caileigh Marshall in helping with materials on SQL and OAuth2. In the Spring of 2020, a number of additional students gave significant thought and feedback, and so we sincerely thank Paul Bass, Thomas Luong, Ben Rahal, Jill Reiner, Dan Seely, Jay Dickson, Matthew Bartlett, Brandon Novak, and Dang Pham.

Granville, OH, USA   Tom Bressoud
Granville, OH, USA   David White
June 2020

Contents

Part I
Foundation

Chapter 1
Introduction

Chapter Goals

Upon completion of this chapter, you should understand the following:

- The terms *system* and *data system*.
- The roles of a Data System Provider, Data System Client, and Resource Owner.
- The importance, at a high level, of the Data System Architecture for facilitating communication between providers and clients.
- The standard Data Sources that commonly appear, and examples of each.
- What makes up a Data Model, and the meaning of *structure, operations*, and *constraints* in the context of data models.
- The most common examples of data models, including the tabular, relational, and hierarchical data models, as well as examples of each.

1.1 A Broad View of Data Systems

Definition 1.1 A *system* is defined as

- a regularly interacting or interdependent group of items forming a unified whole
- an organization forming a network for serving a common purpose [34].

© Springer Nature Switzerland AG 2020
T. Bressoud, D. White, *Introduction to Data Systems*,
https://doi.org/10.1007/978-3-030-54371-6_1

So a system incorporates *multiple interdependent pieces* that are organized and work together to serve a common purpose. Examples include ecological systems, economic systems, corporate systems, social systems, physical systems, and more.

In computer science, *systems* is the subfield focused on the principles, algorithms, and practical aspects, such as performance and organization, that govern the interacting of multiple components that *enables* applications and problem-solving algorithms to execute. So systems are the hardware and software infrastructure that enables computation, and are *not* the end computation itself.

With the general and computer science definitions of systems in mind, we can define *data systems*, the primary topic of this book:

Definition 1.2 *Data systems* are those systems whose focus is on the organization and structure of data, and the means by which such data can be obtained from the *provider* of the data by the *user* or *client* of the data.

The most commonly known form of data system is the *relational database system*, whose models, theory, and realizations have existed for 50 years [4], and are the backbone behind the reliable data needs of companies, universities, and almost all commerce, online and otherwise. It is not an overstatement to say that each of us accesses, as least indirectly, multiple relational databases every day. But relational database systems are only one possible point in a spectrum of possibilities of data systems.

In an ever-increasingly data-centric world, we must broaden our view of data systems beyond the relational database system. Data may be accessible in local data repositories, or over the Internet network (which we often refer to as the network), or reside in the cloud; it may exist in structured, semi-structured, and unstructured forms. Data can be relatively static, can be dynamic, and can be real-time, in a stream of ever-changing data. Such real-time data may be used by software executing on large server systems, by apps running on a smartphone, or by embedded applications running as part of the Internet of Things (IoT).

When we consider data systems, there are at least three separate entities, called *principals*, involved. These are illustrated in Fig. 1.1.

Data System Provider Entity, consisting of hardware, software, and possible network infrastructure, that hosts the data, provides organization, and allows access through a well defined interface.

Data System Client Local software and hardware whose goal it is to acquire data from one or more providers and to manage and transform the data to make it useful for visualization and/or analysis. We sometimes refer to the author of a piece of software as the client. For a given provider, there may be many possible clients interacting and acquiring various subsets of the data for various purposes.

Resource Owner Entity that owns the resource. We use the term resource as an abstraction. Data is one type of resource, but other resources might be involved, such as dynamic or computational resources. The resource owner might be a human user or it could be one or more organizational entities. The data may be owned by the provider, in which case, the provider and resource owner entities

Fig. 1.1 Data systems
principals

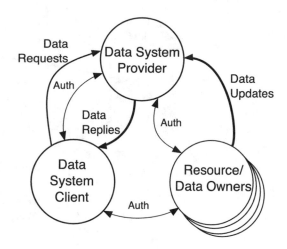

become one. Similarly, the data may be owned by the client (who is sometimes called the user in this case), and the client and resource owner entities become one.

For example, in Chap. 23 we will obtain data from The Movie Database (TMDB). If a social scientist wrote a piece of software to gather demographic information of users and their top rated movies, then the software is the Data System Client (which we will refer to as "client" in the following discussion), the users are the Resource Owners, and TMDB is the Data System Provider (which we will refer to as "provider"). As another example, in Chap. 12 we will study a database containing the data of a university. When we access the database by creating a connection to the university's server, the provider is that server, we are the client, and the owner is the university. When we access the database from a local file on our own machine, the local file is the provider, but the owner is still the university. In such a situation, it is common for the file to require a password before it can be read by the client. As one final example, when you create a spreadsheet to keep track of your schedule, that spreadsheet represents data. You are the resource owner, and if you write code to compute on your spreadsheet (e.g., to determine how many free hours you have each day), then you are also the client, while the file itself is the provider.

From the provider perspective, fundamental goals include efficient and reliable organization and structure for hosting the data, as well as design concerns in defining a usable interface by which data may be requested and returned while also providing security on behalf of the resource owner(s). From the perspective of the client, a fundamental goal is to create software to obtain the desired data, and then be able to transform and use the data for exploration, visualization, and analysis. The client must abide by the decisions of the provider given their defined interface to access and acquire the data.

The study of the provider perspective involves advanced techniques in computer science and is the subject of upper level courses and textbooks. As an *introduction* to data systems:

The focus of this book will be on topics important to the client perspective.

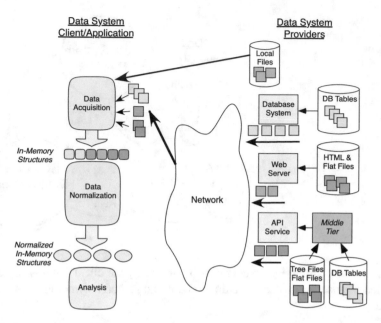

Fig. 1.2 Data systems architecture

In particular, we will focus on the two complimentary subject areas of:

1. the structure, operations, and constraints that govern how the data is organized and represented, called the *data models*, that give us our various *forms* of data, and
2. the aspects of writing software to interact with the providers of data organized as data systems, which requires an understanding of the *data sources* and how to interface with each.

We will take as given that providers already exist, and have made their data system design decisions, and we, in turn, must design clients that conform to the facilities offered by the providers.

To provide context and an understanding of the material encompassed by this book, Fig. 1.2 depicts an architecture of data systems.

On the right side of Fig. 1.2, we depict a spectrum of providers, thus giving the data sources we will explore in this book. The middle of the figure gives the method of access, where for local files, we access data through the operating system interface to the file system, but for other providers, the access must occur over the network. For example, if you as the client were working with a data source provided over the network, and your computer loses its Internet connection, then you would lose access to the data until connectivity is restored. Note that the figure is not symmetric—the left side conveys the main phases of processing needed in a *given* client, rather than showing a spectrum of clients.

In the figure, the squares represent units of data in the form and format of the provider. The different color/shading is intended to represent the variety of different data forms from the various providers. Future chapters will carefully study each data form displayed, as well as the process of acquiring the data and normalizing the data.

1.1.1 Reading Questions

1.1 How does the subfield of systems differs from the software you may have written before (say, in an introductory CS class)?

1.2 Name the three *principals* that can make up a data system.

1.3 Why do we distinguish *Resource/Data Owners* as a separate principal?

1.4 What is the oldest/classic example of a data system?

1.5 Give two examples of data systems consisting of only two principals.

1.6 Name the two complementary dimensions of data systems.

1.7 Give an example of a system, satisfying the definition in the text, but different from any other examples in the reading.

1.8 Give an example of a data system, different from the example mentioned in the reading. Use what you know about how data is stored in the real world. It may help to consult the provided list of API Service Provider Examples.

1.9 Give an example to illustrate the interplay between a data system provider, a data system client, and a resource owner. For instance, you might think about a fictional company and their data needs.

1.10 Similarly, give an example illustrating the pipeline shown in the figure on Data System Architecture. Be sure to mention each spot along the pipeline and what that might look like in your example.

1.2 The Sources of Data

At the top right of Fig. 1.2 is a representation of local files: these are files already downloaded or otherwise present in the file system of the machine on which we will develop our applications (e.g., your laptop or desktop). Local files can serve as a simple source of data for our client applications. Any individual file has a *format* that specifies the structure of the data within the file as it resides in the disk file. We will discuss file formats throughout the book as we cover the various forms of data. Common file formats that the reader may recognize include `.txt`, `.doc`, `.csv`, and `.html`.

The remaining data system providers depicted on the right side of Fig. 1.2 are accessed by a client application over the network. After the local files, we depict the aforementioned providers of relational database systems, and these systems can provide data in the form of relational tables in response to queries. Example systems in this category include servers running Oracle™, MySQL™, PostgreSQL™, and others. Next, we generically depict web servers, which can include any location on the Internet accessible from a web browser. From such servers, data system clients may, using the same protocols employed by browsers, access data contained in flat files, such as comma-separated value (CSV) or spreadsheets, or the web page files themselves, formatted in the HyperText Markup Language (HTML). From a data systems perspective, any of the data formats obtained from web servers may include one or more lists and tables of data in an organized structure.

The final provider of the figure depicts the category of API Service providers. These are providers who have designed endpoints so that clients may access their data from over the network. As depicted in the figure, because they are providing a network-based *Application Programming Interface* (API), such providers are abstracting the actual form of the data and what software may be part of the so-called *back end* implementation, depicted as access by the *Middle Tier* in the figure. Data may be stored in a relational database, in flat files or hierarchically structured files, or the architecture could be even more complex and dynamic. But when requests arrive from a client to an API service, their request is processed and the data result returned to the client, without the client needing to know the specifics of the implementation.

Such API service based providers are the fastest growing segment of data systems. Providers come from many different categories, from social media, to health and fitness, to government organizations, to storage repositories. Some providers charge for their services, while other providers are free of charge, but may have other requirements for their use. The data provided may be open or may be protected, and require authentication and authorization for access, based on the actual resource owners of the data. Table 1.1 lists a sample of some providers who publish APIs enabling client developers to access their data system(s).

Table 1.1 API Service Provider Examples

Category	Example API Service Providers
Social Media	Twitter, LinkedIn, Facebook, Reddit, Slack, Tumblr
Sports/News	NBA Stats, New York Times, NPR One
Music/Video	Spotify, YouTube, The Movie Database
Health	Fitbit, RunKeeper
Support Data	Oxford Dictionary, GeoNames geolocation services, Yelp, Wikipedia, NASA
Government	Data.gov, Census.gov, EPA, Federal Register

Even among the non-local data system providers, there are certain aspects of networking that we need to understand to intelligently interact with such providers. For instance, we may need to identify our client application, or a user or resource owner to the provider. So we must understand methods of authentication and authorization. We must understand and follow the network protocols and the provider use of those protocols to establish connections, make requests, and receive responses containing the desired data.

1.2.1   Reading Questions

1.11   Give an example of a data system provider that does *not* occur over a computer network.

1.12   What is the term for the structure/organization of data within a file or a sequential sequence of bytes?

1.13   Give three examples of file formats.

1.14   What is the most *common* type of provider, and what is the most common format, as used by that provider?

1.15   When, as a client, we acquire data from an API Service, do we know the organization of the system *behind* the programming interface?

1.16   Name three sources of data that you personally find interesting, and that, if you knew how, you would like to obtain and explore such data.

1.17   Generate a list of three real-world data sources that you have used in a previous course. Describe the data. If you were given the data, try and find the origin *source* of the data. You may need to use a search engine.

1.18   Use a search engine to find at least two additional different sources of publicly available data. If you find a site that lists other sites, go to the real *source* of the data and see what form the data comes in—e.g., downloadable file (CSV, XML, JSON), or through an API.

1.3   The Forms of Data

The left side of Fig. 1.2 depicts the main phases of the data pipeline necessary for client applications that are on the consumer side of data systems. The process starts with data acquisition. This entails the communication and working within the defined protocols of the various data sources, as we have just discussed. But the form of the data acquired can vary greatly, and we must understand these various forms. A web server or API provider could, for instance, result in data organized in rows and columns of a tabular form, and whose structure is in a delimited line-based format. But web servers or API providers could also result in data organized in a hierarchical tree structure, in formats like the eXtensible Markup Language (XML) or the JavaScript Object Notation (JSON). Similarly, local files could be organized in any of these formats. Database systems, on the other hand, always return data from queries that are organized in a relational table of rows and columns.

To provide an organized conceptual framework for understanding these various forms of data, we employ *data models* that defines the following for each form of data:

1. The *structure* used for data employing the particular model.
2. The *operations* that may be executed against data structured as given by the model.
3. The *constraints* that govern valid values for data in the model, as well as properties to be held invariant in and among the data represented in the model.

In this book, we will cover three data models, the tabular data model,[1] the relational data model, and the hierarchical data model.

We briefly summarize the three data models in broad terms. The tabular data model is structured in rows and columns, and the constraints ensure, for example, that there are no repeated rows, that columns represent individual variables, and that tables make logical sense. Tabular operations allow us to access subsets of data, to merge multiple tables together, and to mutate tables to meet the logical constraints. The relational data model consists of database tables with additional structure, facilitating the use of database queries. The structure facilitates relationships between tables, so that an organization can represent complex data using multiple tables, one for each type of data the organization wishes to keep. The constraints of the relational data model ensure sound database design, that all entries in a column have the same type, and that tables are appropriately linked to facilitate queries involving data spread across multiple tables (e.g., data on both students and courses they are registered for). Lastly, the hierarchical data model is structured into the shape of a tree, so that some data entries are stored as children of other entries. An example is the file system on your computer, where files are like children, stored in folders represented as parent nodes in the tree. Folders can be stored inside other folders, so this tree has many layers. The constraints ensure that the representation of the data really forms a tree, can constrain the shape of the tree, and can constrain the data types allowed at various places in the tree. The operations allow us to traverse the tree to locate and extract data stored at different places, or to create new trees and insert data. We will discuss all three models in more detail below.

---

[1] In data science, the term *tidy data* is used to refer to row and column tables of data conforming to a set of constraint rules. In relational database systems, a similar and related concept of *third normal form* is used. We generically use the term *tabular data model*, because it includes the necessary constraints.

We turn now to data acquisition, which means getting data into in-memory data structures (meaning, contained in the computer memory) of the client application. Often, the in-memory data structure matches the structure as given by the form (data model) of the provider. These are represented in Fig. 1.2 as rounded corner squares, but maintaining the color/shading that is based on the originating form.

The end goal of a data system client is to not just *acquire* the data, but to be able to *use* the data for some downstream objective, such as exploration, visualization, or analysis involving statistical or machine learning techniques. If acquired data is not in the proper form, and does not follow the constraints of that form, then such downstream uses of data become difficult, requiring custom programming and case-by-case software adaptation.

So the middle part of the data pipeline, as shown in Fig. 1.2, conveys the idea of taking data acquired from various sources and in various forms and encompasses the processing and transformation of the data so that it is in the correct structure and conforms properly to the constraints of the desired final form. This process is called *normalization*.[2] Performing this step is often a bridge from one form or format to another, as most software analysis packages and visualization systems work best with input data in the tabular model form.

We turn now to a more detailed discussion of the three data models.

Tabular Data Model

The tabular data model focuses on primarily single table data sets with row and column dimensions, giving a familiar structure. In this model, we want in-memory representations of tabular data with sufficient power in their operations to allow flexible access and transformation of the data. In Python, we use the `pandas` module to provide such facility, giving us the ability to take arbitrary row and column subsets of the data, filter it based on criteria, perform column-vector operations, and transform and combine data sets in various ways.

The culmination of understanding of this model is to recognize that, while data can often be represented in a tabular form, that does **not** mean that it conforms to the tabular model. It is tabular data that conforms to a particular set of *constraints* that allows the data to be effectively analyzed. These constraints are adopted by convention, and not enforced automatically, so we spend time understanding this particular normal form, recognizing when data does not conform, and using the operations of the model to transform data into its normalized form.

Relational Data Model

One of the most important forms of data system and its associated data model is the relational database system and model. Grounded in nearly 50 years of development and an important element of many larger systems, we cover this model from a primarily client/user perspective. We use the Structured Query Language (SQL) for data acquisition and then transform results into appropriate in-memory structures.

---

[2] Variously called cleaning, wrangling, and munging by the data science community.

The relational model also gives us our first exposure to clients making requests over the network to database systems servers.

By using the tabular data model as a foundation, we start with subsetting columns and selecting rows from single relations/tables, drawing parallels with the prior material, and seeing some of the differences of a declarative versus procedural style of operation. We then extend to multiple tables using join operations to see how the relational model uses constraints on keys to build one-many and many-many relationships, inherent in real data sets.

Utilizing existing well-designed databases as examples, we cover, at an intuitive level, various aspects of sound database design. We use elements of the convention-based constraints of the tabular model to arrive at an intuitive understanding of database normalized form. We see the idea of design before data population, and how a database system can enforce many of the desired constraints in this model, to avoid errors before-the-fact. Coverage of this section concludes with simple, but sound, database design, table creation, and row insertion for some existing data sets derived from the earlier section on the tabular data model.

Hierarchical Data Model

When we interact with data systems on the Internet, many providers structure and represent provided data in a flexible form based on a hierarchical (tree) data model. Dominant specific examples include XML and JSON formats, but the tree inherent in an HTML document is another significant example. Data acquisition for hierarchically formatted files can also be in local files, and we use that mechanism to more easily arrive at in-memory structure representation of the tree.

Given a tree or subtree, the operations of this model include procedural ones that allow traversal, from parent to children, and across siblings, as well as extraction of data (tags, text, and attributes) from the nodes of the tree. This model also gives us the opportunity, on the XML and HTML side, to look at declarative operations using XPath to specify what data we want, without having to write procedural code that defines how to get that data. We are also able to see how trees can represent both the notion of a single table, or a set of tables, all within a single file. Constraints in this model are either by convention, or can be explicit, using mechanisms such as Document Type Definition (DTD), or an XML Schema Definition (XSD).

1.3.1 Reading Questions

1.19 What are the three components of a *data model*?

1.20 Once data has been acquired, what is the term for the restructuring of data into a standard form amenable for analysis?

1.21 True or False, if data is in a tabular form, then it conforms to "tidy" data?

1.22 True or False, if data conforms to "tidy" data, it is in tabular form?

1.23 What is the name of the language for specifying a data request to a relational database system?

1.24 Is a typical web page, in the HTML format, an example of tabular, relational, or hierarchical data model?

1.25 Can you think of a structure for data that is *not* tabular, relational, or hierarchical?

1.26 If we acquire a CSV from a web server, are we within the hierarchical model, the tabular model, or the relational model?

1.4 Book Organization

This book presupposes programming ability in Python. Readers should be able to:

- construct and run Python programs,
- understand integer, real-valued, Boolean, and string data types, including using `format()` with strings.
- be able to write code using conditionals and iteration,
- be able to construct and use objects, including accessing object attributes and calling methods,
- be able to design, create, and invoke functions to abstract subordinate steps,
- understand variables and global versus local scope of such variables, and
- be familiar with the most common Python in-memory data structures of lists, tuples, and dictionaries.

The book itself is divided into three parts. Part I, following this introduction, covers a foundation for the other two parts. We begin the foundation in Chap. 2 with developing a better understanding of hierarchical file systems and using operating system facilities to access and read data from local files. While not providing basic instruction in lists, tuples, and dictionaries, Chap. 3 focuses on one- and two-dimensional Python structures for representing data collections, and manipulating them using Python facilities like list comprehension and working with functions in more advanced ways. We finish the foundation in Chap. 4 with *regular expressions*, a tool for dealing with unstructured data. If acquired data, in whole, or in a subordinate part, has unstructured textual data, we may need to extract information from that unstructured data to make it useful.

Depending on a reader's introduction to Python, a number of the topics and subjects of Part I may be reviewed, and could then be used as a reference for the data-centric application of these Python facilities.

In Part II, we cover the three data models listed in Sect. 1.3. Each of our three models has chapters on structures and operations, formats supporting the structure, and constraints of the model. Depending on the model, the grouping of topics into chapters varies based on what helps the presentation of the material. For the various models, we select and use the most common Python modules for supporting the needed programming.

In Part III, we cover a selection of providers as data sources and the client aspects necessary for acquiring data, as outlined in Sect. 1.2. To do so requires understanding some of the networking principles that enable a client and a provider to communicate. We cover general networking as well as the most dominant protocol of the Internet, HTTP. The client programming for acquiring the data must be integrated with parsing the various formats of the different forms of data covered in Part I. We include a chapter on web scraping, the process of extracting data from HTML web pages acquired from web servers, and on the commonalities and differences in working with Provider APIs. Because of the need for authentication and authorization when separate resource owners are present, and to fully realize the potential of API Service types of data system providers, we cover such security issues in its own chapter.

We conclude this chapter with a set of exercises on the basics of Python, that the reader familiar with Python programming should feel free to skip.

1.4.1   Exercises

Review of Basic Python

1.27 Write a function that returns the square of the number n, if $n \geq 1$, and returns 0 otherwise.

1.28 Using your `square` function, write a function that computes the sum of the squares of the numbers from 1 to n. Your function should call the `square` function—it should NOT reimplement its functionality.

1.29 Write a function `sum_of_evens` that only adds up the even numbers between 0 and the number n.

1.30 Reimplement the `sum_of_evens` function as `sum_of_evens_redux` using a different approach.

1.31 Write a function `squares(n)` that returns a list of numbers, such that $x_i = i^2$, for $1 \leq i \leq n$. Make sure it handles the case where $n < 1$ by returning None instead.

1.32 Using your `squares` function, write a function that computes the sum of the squares of the numbers from 1 to n. Your function should call the `squares` function—it should NOT reimplement its functionality.

1.33 Use three *different* ways to create three lists with the integer numbers 1 through 10, inclusive. Use the variable names `myList1`, `myList2`, and `myList3` to hold the results.

1.34 Write a function

```
decreaseList(L, amount)
```

that, for each number in list L, decreases L by `amount`. The list should be modified in place, and the function should not create a new list, nor return any value.

1.35 The *percentile* associated with a particular value in a data set is the number of values that are less than or equal to it, divided by the total number of values, times 100. Write a function

```
percentile(data, value)
```

that returns the percentile of `value` in the list named `data`.

1.36 Write a function `swap_case(s)` that swap cases in the string s. In other words, convert all lowercase letters to uppercase letters and vice versa. Do not use the `swapcase()` method. Instead, use `ord()` and `chr()`. For example, if s = "My string" you return "mY STRING".

1.37 Write a function

```
roman2digits(numer)
```

that converts a Roman numeral `numer` (encoded as a string of capital letters) into its decimal representation. For example, `roman2digits(IV)` returns 4.

Chapter 2
File Systems and File Processing

Chapter Goals

Upon completion of this chapter, you should understand the following:

- Use of a hierarchical structure for organizing resources, and the specific instance of a tree for organizing files within a file system.
- The idea of a *path* as a textual means of specifying a location of a file or directory resource within a tree. Such location may be absolute, or it may be relative to a particular position within the tree.
- How the operating system maintains information like the current position in a file that affects how file operations work.
- The idea that resources, like files and directories, exist in a single shared tree, so there must be restrictions on the set of authorized operations that a given user may perform.

Upon completion of this chapter, you should be able to do the following:

- Properly use open() and close() and with open() for a file, allowing further operations on the file.
- Use file methods to read the contents of a text file in multiple ways and using the file position to iterate over lines in a file.

(continued)

© Springer Nature Switzerland AG 2020
T. Bressoud, D. White, *Introduction to Data Systems*,
https://doi.org/10.1007/978-3-030-54371-6_2

- For a line in a file, use string methods to parse the elements of the line and then convert to the data types needed.
- Accumulate data from a file into parallel lists.

2.1 File Systems

In data systems, it is essential to know where data is stored. On our laptops, desktops, servers, and other computer systems, the users, programs, and operating systems on these computers have access to organized persistent storage known as the *file system*. This persistent storage is implemented using hard disk drives, flash drives, and other media and allows us to read and write information whose state is maintained beyond the life of our executing programs. We call such storage *persistent* because, even if a computer is turned off, rebooted, or exhausts its battery, the information is again available after the computer is next booted and for when our programs are next executed.

The *operating system* is responsible for the organization of this persistent storage, and for allowing sharing and protection and mediating access to the storage, as required by *users*, the owners of the data. While the details of such organization and access mechanisms may differ between different operating systems, such as Windows™, MacOS™, or Linux, much is similar. Logically coherent sets of data are organized into *files*. While all files are comprised of sequences of bits (0's and 1's) forming bytes, we distinguish files whose bytes represent sequences of characters as *text files*, while files whose bytes directly represent underlying data types, like integers and floats, as *binary files*.[1] The distinction between these two types of files will be very important in Chap. 21 when we acquire data over the Internet network.

We use files to store source programs, executable programs, libraries, media, data input to a program, and data output from a program, as well as many other uses. In many cases, files are used as a way of communicating between different programs.

---

[1]If a file has readable characters that are digits, it is still a text file, because digit characters are still characters. If a file, when opened in a text editor, such as vi, emacs, atom, sublime, or the editor portion of an integrated development environment, is displayed as only characters from the keyboard, it is likely a text file; but if the display does not yield a coherent sequence of readable characters, it is probably a binary file. The reader is encouraged to verify this with a text editor of your choice: sample binary and text files may be downloaded from the book web page. Some text editors (e.g., atom) have a setting to show the hexadecimal representation of data, so that you can verify that even text files consist of bytes.

One program can create a file, and another program can read and process the file in a separate step. For instance, we can create a text file using a text editor, and can then read and process the file in a Python program.

In the context of the data systems addressed by this book, files on the local file system will be our first and simplest in our spectrum of data providers. At this point, we attempt to abstract away from particular forms/formats of data that might be entailed in the contents of a file, deferring that level of detail to the respective chapters on the data models. Our goal at present is to better understand the basic operations that any client program (perhaps through a higher-level library or module) must employ to connect an executing program with a file and to use operations to read and/or write such files.

2.1.1 Hierarchical Organization

A file system, even for modest sized computers, must organize and manage millions of files. To cope with this kind of scale, and to provide flexible organizational structures, the operating system provides a *hierarchical* organization to the files in the file system. This hierarchy defines a *tree*, where the root and interior nodes of the tree provide the abstraction of a *directory* or *folder*.[2] Each folder can have contents, which are the *children* of the folder node in the tree. The child contents of a folder can consist of zero or more files as well as zero or more (sub)folders. A *leaf* is a node with no children. An *internal node* is a node that is not a leaf. A file must be a leaf of the tree, as it cannot have any subordinate folders or files. A folder whose contents are empty can also be a leaf of the tree.

Consider the top levels of the tree for a file system as represented in Fig. 2.1. Within the figure, branches to a three dot ellipse represent a subtree not pictured. This particular example is abstracted from a Linux system, but an analogous top level would be similar in Windows™ or MacOS™. The figure uses / to denote the *root* of the file system. The top level directories, direct children of the root, like bin, dev, etc and so forth, are used for operating systems resources. One of the top level directories, home[3] is used to group together the subtrees for each of the users in the system. In this case, there are three users, alice, bob, and chris represented, each of whom would have their own subtree.

Let us focus on a single user, alice, for instance, and the subtree in the file system that represents their *home directory*, which is the portion of the file system that they see when they log in, and when they start a *File Explorer*. An example subtree is depicted in Fig. 2.2.

The user alice has chosen to organize files underneath the Documents folder by class year and then by course. At the bottom of the figure, for the cs111 course, there are four leaf nodes in the tree, including two Python program files (prog1.py

---

[2]We will use the terms *directory* and *folder* interchangeably.

[3]On MacOS™, the equivalent of home is a directory named Users underneath the root.

Fig. 2.1 Top levels of a file
system hierarchy

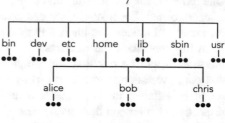

Fig. 2.2 Alice hierarchy

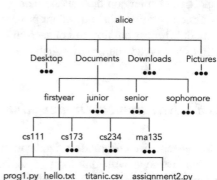

Fig. 2.3 File explorer view of alice hierarchy

and `assignment2.py`) and two data files (`hello.txt` and `titanic.csv`).
Note the use of the *extension*, the set of characters after the final period, that, by
convention, allow us to infer the type of file.

When viewed through a File Explorer of the operating system, this might appear
as shown in Fig. 2.3.

The left portion of Fig. 2.3 shows the view in a File Explorer that user `alice`
might see on navigating to the Documents folder. It depicts the folder with its four
subfolders, corresponding to the class years seen in the tree view. On the right,
we see the view of the `cs111` folder, with its four files, as would be seen after
navigating down two levels, through the `firstyear` and then to the `cs111`
folders. The process of double-clicking on a folder shifts the state of the File
Explorer program to be "in" that folder, like walking from one room in your house
into another. A sequence of such moves constitutes a "path," discussed in the next
section. The File Explorer displays all non-hidden files and folders in its current

location, and can open any files which the user has permissions to open. The reading questions will guide you in a deeper exploration of file permissions.

Other users, bob or chris in this example, have the ability and independence to organize their own home directory structure in any way they wish. Further, files and directories in subtrees of the hierarchy include attributes like the *owner* of the resource and the permissions/capability of the user and other users to access (or not access) the file or directory in the hierarchy, allowing one user to keep private their entire subtree from any other user.

It is important to realize that the File Explorer is not the only interface to the underlying file system structure. For example, files and folders can also be accessed directly using applications such as Terminal on a MacOS™ or Command Prompt on a Windows™. These applications have commands to allow the user to see a list of files and folders in a given directory, e.g., ls in Terminal and DIR in Command Prompt. There are also commands to change directory (cd in Terminal and CHDIR in Command Prompt) analogously to double-clicking a folder in File Explorer to change the location the application is currently looking at. A change of directory can move inside a single subfolder, can backtrack to a parent folder containing the current location, or can conduct a sequence (or "path") of such moves. These applications also have ways to create, delete, and move files and folders, to change permissions, and to lock files and folders. To learn about such commands, the reader is referred to [2] and [88].

2.1.2 Paths

In this section, we build on the informal discussion of paths above. A *path* is a traversal of a resource tree starting at one node, called the *source* node and ending at another node, called the *destination* node. The destination node in the path of a file system tree could represent a file, but could also be a directory node. A path whose source node is the root of the tree is called *absolute*. As in Fig. 2.1, these paths start with /. For a path that does not start at the root (a *relative* path), we must know the location of the starting node to be able to determine the *location* of the file resource within the file system.

Table 2.1 presents some example paths, using the arrow symbol ($\rightarrow$) to denote a step of a path traversal.

With a convention of how to represent the root of a resource tree, and a character to represent separation between traversal steps in a path, along with an assumption that the names of files and directories are constrained to a reasonable set of characters, we can take the concept of a path and translate it into a textual string representation. This allows us to use simple strings in our programming as a means of specifying paths.

Table 2.1 Example absolute and relative paths

Path	Description
/→home→alice	An absolute path specifying the home directory of the user alice
/→home→alice→Documents→ firstyear→cs111→hello.txt	An absolute path specifying the specific hello.txt file of alice in the cs111 class[4]
Documents → firstyear → cs111 → hello.txt	A relative path that specifies the location of the specific hello.txt file of alice in the cs111 class
hello.txt	A relative path that, relative to the cs111 directory of alice, specifies the hello.txt file

In the following list, we give the string translation for each of the example paths in Table 2.1.

- /→home→alice is string "/home/alice".
- /→home→alice→Documents→firstyear→cs111→hello.txt is string "/home/alice/Documents/firstyear/cs111/hello. txt"
- Documents → firstyear → cs111 → hello.txt is string "Documents/firstyear/cs111/hello.txt"
- hello.txt is string "hello.txt"

In this case, we use the MacOS™ and Linux convention of forward slash (/) as the path separator and / for the root. Windows™ uses <disk letter>:\ for specifying a file system tree root, and backslash (\) as the separator in paths. If the string representation of a path begins with the character(s) for the root, we know the path is absolute, otherwise it is relative. The string paths above are appropriate input to cd in Terminal (or to CHDIR in Command Prompt, with the Windows™ modifications), e.g., if a Terminal is currently located in Documents, and we execute cd firstyear/cs111, then the Terminal will shift location into the cs111 folder.

Whenever we execute our programs, the operating system maintains, on behalf of the executing program, a concept of a *working directory*. Often, the working directory is initialized as the folder in which the file representing the program or executable resides. The working directory maintained by the operating system is exactly the needed mechanism that allows unambiguous resolution of a relative path to a location in the file system, often needed when we access files programmatically.

---

[4]There could be one or many files with the same name, like hello.txt, but the path uniquely identifies this particular file.

Table 2.2 Special character sequences for paths

Special	Meaning
.	A single period means the "current directory"; so leading a string representation of a path allows us to refer to the current working directory
. .	Two periods refer to the parent directory relative to the current location in the traversal
~	A tilde refers to the home directory of the current user[5]

In addition to the concept of a working directory for resolving relative paths, operating systems, in the string representation of paths, define character sequences that have special meanings for the paths and their traversal, listed in Table 2.2.

Consider again the example file system presented in Figs. 2.1 and 2.2. Say that the current working directory is `"/home/alice/Documents"`, but the current user is `chris`. The path `"./firstyear/ma135"` defines the traversal from the current directory to `firstyear` to directory `ma135`. The path `"../Pictures/profile.jpg"` would refer, through `".."` to the home directory of `alice`, and from there, through the `Pictures` directory to the `profile.jpg` file. The parent special character sequence can be repeated, so `"../../chris"` is a path that refers to the home directory of `chris`. But since Chris is the current user, the path `"~/"` could also be used to refer to the same directory. These special character sequences are appropriate inputs to `cd` in Terminal (and analogous sequences work for Command Prompt). For example, if a Terminal's state is `Documents`, then `cd  ..` will change the state to `alice`.

2.1.3 Python File System and Path Facilities

Python, through the `os` and `os.path` modules, defines helpful functions we can use in our programs to work with the operating system's management of the file system, and with paths. In Table 2.3, we list some commonly used functions provided by the `os` module for working with directories from our programs.

In general, in this book, we will present tables and give descriptions to give basic information about functions and methods of the various modules we cover, but will, in each instance, give a URL for the official documentation. The official documentation should be used as the complete reference. You should actively explore the reference documentation to deepen your understanding and as a help in solving problems and exercises. You will often find additional examples as well as other functions that will prove useful. Documentation for the `os` module may be found at [56]. In Table 2.3, we list several useful functions of the `os` module, analogous to `pwd,  cd`, and `ls` in the Terminal application.

---

[5]This special sequence should be used with care, as some path and file operations expand this to its absolute equivalent, while others do not.

Table 2.3 Operating system directory functions

Function	Description
os.getcwd()	Return the string giving the absolute path of the current working directory (cwd)
os.chdir(path)	Change the current working directory to path
os.listdir(path)	Return a list of strings for the files and directories contained in the given path. The path may be omitted, in which case it defaults to ".", the current directory
os.getlogin()	Return the name of the user currently executing the program

Table 2.4 Operating system path functions

Function	Description
os.path.exists(path)	Return True if path refers to an existing path in the file system, and False otherwise
os.path.isfile(path)	Return True if path is an existing regular file, and not a directory
os.path.isdir(path)	Return True if path is an existing directory, and not a regular file
os.path.join(one, two, ...)	Join one or more path components correctly for the current operating/file system, using exactly one directory separator between path steps, returning a string concatenation
os.path.abspath(path)	Return a normalized absolute path version of the pathname path, converting a relative path to an absolute path if needed
os.path.expanduser(path)	Return the argument with an initial component of ~ replaced/expanded by the user's home directory

There are additional facilities in the os module, not listed here, for creating and removing directories and renaming both files and directories.

Because of the common need for working with paths, and as an aid in writing programs that work correctly across different operating systems, there is a path-specific module, os.path. Table 2.4 lists several functions in os.path that we will need, and complete documentation may be found at [57].

Of particular help is the os.path.join() function, often used to "paste together" the directory where data is stored with the name of the file. It allows us to write code that works correctly even if the underlying system uses different path separators (/ versus \), and it also takes care when a directory specification may or may not include a trailing path separator. These qualities make it vastly superior to coding a string + to create our pathnames. For example, in Alice's file system, if fy_dir is the string /home/alice/Documents/firstyear/ and hello_path is the string cs111/hello.txt, then os.path.join(fy_dir, hello_path) returns the full path from home to hello.txt. This path can then be used to open() the file, as we will see shortly. The use of many of these functions will be explored in the exercises.

2.1.4 Reading Questions

2.1 When the bytes of a file represent a sequence of characters, what do we call the *type* of the file?

2.2 When the bytes of a file encode and represent integers or floats, but *without* using the sequence of characters, what do we call the type of file?

2.3 If we want to view or edit the sequence of characters in a text file, what tool might we use? Please test the viewing capabilities of a tool/application on your own machine.

2.4 If we view a file system hierarchy as a tree, would a file like titanic.csv be an internal node of the tree, or a leaf node?

2.5 To better understand hidden files, please try to find some on your machine. For example, on a MacOS™, if you open Finder, then select Go at the top of your screen, it will show a list of folders you might want to "go" to. Now hold down the Option key and you will see a new folder appear, the Library folder. Please go there and describe how you can tell from the Finder window that this is a hidden folder. On a Windows™, if you open File Explorer, then select View -> Options -> Change folder and search options, you can select Show hidden files from the Advanced settings. Please describe how you can tell if a file was hidden.

2.6 To better understand file permissions, please attempt to find or create a locked file. On a MacOS™, right-clicking on a file in Finder, then selecting Get Info, will give you an option to lock the file (e.g., to prevent deletion). In the same place, you can change the permissions. On a Windows™, right-clicking the file and selecting Properties, then Advanced, gives you an option to "Encrypt contents to secure data." Please describe what happens when you carry out the operation described on your personal computer.

2.7 Consider the subtree rooted at your home directory on your personal computer. Draw a (sub)tree depiction of the folders and files starting at your home directory.

2.8 Assuming the current working directory is your home directory, write down at least three relative paths for files in your directory hierarchy.

2.9 For each of the three paths from the previous question, please use a Terminal and the cd command (or the analogous commands in Command Prompt) to navigate the path. Then describe what happens when you use ls (or DIR) to discover what is in each location.

2.10 The reading discusses how paths can be provided to the Terminal or Command Prompt. Paths can also be provided to a file explorer, e.g., for the MacOS™ file explorer, Finder, selecting "Go" then "Go to Folder" allows you to put in a path directly. Please carry this out using your three paths from the previous problem, and describe what happens. If you have a Windows™, spend a few minutes learning if you can put in paths to the file explorer, and report what you learned.

2.11 Use the `os` module to write code that obtains a list of strings for the files and folders in the current directory and, for each, prints to the console the absolute path for the file system item.

2.12 Like file systems, websites are also organized hierarchically, and use paths to keep track of the locations of resources. A real-world application of the previous question is the application SiteSucker, which downloads all files at a given URL, even if they require paths of length greater than one (e.g., if accessing the file involves clicking a link to load a page, then clicking a link on that page to load another page, and then finally downloading the file). Investigate the SiteSucker application and report what you learned.

2.1.5 Exercises

The following exercises allow us to practice with the `os` and `os.path` functions and to reinforce our understanding of the file system hierarchy and paths.

2.13 Write a Python expression that gives the string value of the current working directory.

2.14 Write Python code to generate two lists:

- a list of the strings giving the files and directories that are in the current directory.
- a list of the strings giving the files and directories that are in the `publicdata` child directory, relative to the current directory.

   Note that it is ok for there to be some "hidden" files or folders in these lists. These are files whose names begin with a `'.'`.

2.15 Write a function

```
printType(path)
```

that prints the string "Is a directory" (without quotes) if the given `path` is a path to a directory, and prints "Is a file" if the given `path` is a regular file, and prints "Does not exist" if the path does not yield a traversal to an actual file or directory in the file system.

   For the final question, refer to Figs. 2.1 and 2.2.

2.16 Suppose user `chris`, a sibling in the file system tree to `alice` and `bob`, has a file named `babynames.txt` in a directory named `data`, and this directory is a child of the `chris` home directory. Further suppose that user `chris` has manipulated permissions to allow user `alice` read access to the intermediate directories and ultimately to the file `babynames.txt` file. Construct file paths and assign to variables as requested in the exercise subparts below:

1. Assign to `abspath` the absolute path to the `babynames.txt` file.
2. Assuming user `chris` is executing a program in their home directory, and the home directory is the current working directory, assign to `chrisrelative` the relative path to the `babynames.txt` file.
3. Assuming user `alice` is executing a program in their home directory, and the home directory is the current working directory, assign to `alicerelative` the relative path to the `babynames.txt` file.
4. Assuming user `alice` is executing a program in their `cs111` directory, and this is the current working directory, assign to `alicerelative2` the relative path to the `babynames.txt` file.

2.2   File Level Operations

Given an understanding of the file system, its hierarchy, and the use of paths for file locations, as covered in Sect. 2.1, we can proceed to creating *connections* between our Python programs and local files in the file system hierarchy.

The Python built-in function `open()` is used to establish a connection between an executing program and a particular file in the file system. It takes, as arguments, a string specifying a path to a file, and a second string that specifies in what way the program will *use* the file, known as its *mode*. The `open()` function, even though it is provided as a function, and not a class, is an object constructor, and creates and returns a reference to a *file object*. For example, in Alice's file system, if `prog1.py` wanted to open `hello.txt`, then we could simply invoke `open("hello.txt")`, since the two files are in the same folder. However, if `prog1.py` wanted to open a file `"my_data.csv"` in the `cs173` folder, then we would need to invoke `open("../cs173/my_data.csv")`. We often create file paths using the `os.path` module.

Once a file object has been created, its methods are then used in the program to perform operations like reading a file or writing to a file. A program can potentially have many open files, as long as different file objects are maintained for each interaction. When a program is finished with a file, it calls the `close()` method to allow the operating system to clean up and release resources associated with the file connection. Closing files is essential, and forgetting to close files and connections can have major negative consequences, as we will see in Chap. 13.

Table 2.5 File open modes

Mode	Mode argument	Description
read-only	'r'	Open an existing file for reading only. An error occurs if the file does not exist, if the user does not have read permission for the file, or if a write operation is attempted
write	'w'	Open a file for writing. If the file exists, its contents are deleted so that it begins anew as an empty file. An error occurs if the user does not have permission to create a file in the current/specified directory, or permission to write to an existing file
append	'a'	Open a file for appending. Creates a new file if the path does not exist, but prepares for writes to happen at the end of an existing file. An error occurs if the user does not have read and write permission for an existing file

Files can be opened in several modes, depending on the goals of the program. For instance, in the example above, for Alice to open `"my_data.csv"` from `prog1.py` in the `cs111` folder, with the goal of *reading* data in the CSV file, then the command `open("../cs173/my_data.csv", 'r')` should be used. If Alice wanted to create a new file, `"journal.txt"`, in `cs111`, to write information into, then the command `open("journal.txt",'w')` should be used. The most prevalent modes, specifying the kind of access of a program to a file, are summarized in Table 2.5.

These access modes can be combined with a specifier for the mode for the type of file, be it *text* or *binary*. As discussed previously, text files comprise sequences of characters, including invisible characters, that can correspond to strings in our programs. Binary files contain bytes of information for non-character representation of data, like signed and unsigned integers and real-valued data. A `.txt` file can contain either text or binary data. Another common file extension used to store and transmit data is `.dat`, which we will see again in Chap. 21. A `.dat` file is generic (and often unstructured), and can hold text data, picture data, and even video data. These files can be opened with Python just like a `.txt` file.

In Python modes, we use `'t'` for text files, and `'b'` for binary files. A mode can include both the form of access as well as the file type, so a mode of `'rt'` would be used for read access to a text file, or a mode of `'wb'` would be used for write access to a binary file (e.g., `open("journal.txt",'wb')`). For the open call, the default mode is `'rt'`, and can be omitted if this is the desired mode, as we did in our first example, `open("hello.txt")`. Further, the `'t'` for a text file is the default, and may also be omitted in a mode argument, e.g., `open("hello.txt",'r')`.

Fig. 2.4 hello.txt file
contents

2.2.1   File Open and Close

Each file is a resource, and files "live" in a file system hierarchy shared with the operating system and possibly other users, who represent other resource owners. As soon as there is the potential of sharing and access by others, the file system and operating system must provide mechanisms by which resource owners can define the allowable operations on any given resource. For instance, one user may want to keep completely private some files, while other files may be shared for reading but not writing by other users, and still other files may also allow writing by certain users.

Resource owners specify *permissions* on files and on directories, using means provided by the operating system independent of a potential user program running and desiring access to the file and its constituent directories. Then, when a user running a program attempts to perform an open(), given a path location, and by specifying the mode of the desired access, the operating system can determine whether or not the open should be allowed to complete successfully.

While the examples in this book will primarily have the resource/file owner and the executing user be the same, and so permissions would then not be an issue, you need to be aware of the security/permissions when writing more involved code, where the data location may be in some other (shared) part of the file system.

Suppose, in a directory referred to by the Python variable datadir, we have a text file named hello.txt consisting of exactly 16 characters and depicted in Fig. 2.4. In the figure, each character is represented by a single box, and we use Python escape sequences for so-called invisible characters like newline ('\n'), return ('\r'), and tab ('\t'). It is important to stress that each of these invisible characters constitute a *single character* and are not actually the two character sequence beginning with a backslash. That is, len("\n") returns 1, not 2. These escape sequences allow us to manipulate, construct, and put such characters into strings and files. Blank boxes represent the space character.

So the contents of the hello.txt file consist of characters including two uppercase letters, eight lower case letters, a space, a punctuation character, and a newline. We consider a sequence of characters followed by a newline character a *line* of the text file.[6] So in this example, the file consists of exactly one line. If we

---

[6]The definition of a *line*, by default, can vary between operating systems. While lines ending in newline is the most common, and is the default in MacOS™ and Linux, Windows™, historically, has used the two character sequence of a carriage return followed by a newline (\r\n) as the line termination. In Python, with so-called universal newlines, the underlying file could use any of the line termination choices in the file itself, but will consistently use a single newline to delimit lines as presented to the program.

were to view `hello.txt` in a text editor, we would see the `Hello World!` but the text editor does not typically render anything visible for the newline.[7]

The character \ can also appear in a block of Python code to allow the use of two or more lines, through *continuation*, to form a single logical line [60]. When this occurs, the \ is not part of a string, so is unrelated to the special escape character sequences discussed above. It is also unrelated to the file system. For example, the following block of code to construct a file path uses \ to make the code easier to read.

```python
top_part = "/home/alice/"
middle_part = "Documents/firstyear/cs111/"
file_part = "hello.txt"
filepath = top_part + \
           middle_part + \
           file_part
```

The assignment of `filepath` is equivalent to `filepath = top_part + middle_part + file_part`. For this kind of readability, we will often break long lines of code using \ in this book.

Further, separate strings that immediately follow one another on a logical line form a concatenated result, so the above could be written even more simply as:

```python
filepath = "/home/alice/" \
           "Documents/firstyear/cs111/" \
           "hello.txt"
```

Returning to file operations, in the code below, we open the file by constructing the path, and by specifying a mode of `'rt'`, since we only want to read this text file. The `open()`, if successful, returns a reference to a file object that we name `inputFile`. So the code shown would then be followed by operations that use the methods of the `inputFile` object. While exception handling is not covered in detail in this book, we demonstrate a simple `try-except` block that can handle the case where the file does not exist, or the user does not have sufficient permissions for the specified access mode for the file being `open()`ed. In this code, if the open is successful, the `except` block is skipped, but if an `OSError` occurred, the block, reporting the error, is executed.

```python
filepath = os.path.join(datadir, "hello.txt")
try:
    inputFile = open(filepath, 'rt')
    print("Success opening file")
except OSError:
    print("Error opening file")
```

---

[7]Some editors include a preference setting to "Show Invisibles," which can be helpful for understanding the complete contents of a file that you intend to use for a program.

```
| Success opening file
```

More information about `try-except` blocks can be found in [53]. For instance, instead of `except OSError`, we could say `except Exception as e`, which creates an object e of type `Exception`, containing information about the error encountered. An alternative to the use of a `try-except` block is the use of an *assert statement*. Such a statement asserts that a Boolean condition must be met before proceeding, and raises an error if the Boolean condition evaluates to `False`[47]. For the case of file objects, we can use the `isfile` method to check that the path really leads to a file.

```python
filepath = os.path.join(datadir, "hello.txt")
assert os.path.isfile(filepath)
inputFile = open(filepath, 'rt')
```

However, the use of a `try-except` block is preferred, since the approach via assert statements would fail to notice if the user does not have read privileges for the file in question. Regardless of which approach you take, it is essential to develop programs one small bit at a time, and to check as you go that everything behaves as expected. Both assert statements and `try-except` blocks will appear again at several points in this book. After a successful open, a program would proceed in its invocation of methods to read or write the file. When all such operations are complete, the program would conclude the interaction with the file by invoking the `close()` method:

```python
inputFile.close()
```

For most non-trivial interaction with a file, we will employ `for` and `while` loops for the processing of the file, in the code that is executed in between the `open()` and the `close()`. We will begin covering these iteration strategies in Sect. 2.3. For now, we will cover some of the methods of file objects that we need not put in a loop, to help build our understanding of operating on a file.

One concept that comes up in reading and writing of a file is that of the current *position* in the file. This is an integer that operates much like a cursor in a text editor or word processor. It indicates *where* the next operation will occur, e.g., where a new character would appear if typed on the keyboard. In a text editor or word processor, if the cursor is prior to any of the characters in the document and we type at the keyboard, the newly entered characters will be placed at the cursor and the cursor will be advanced. Similarly, a file position starts at 0, which is the position at the very beginning of the file. When we perform any of a set of `read`-type operations, the operation reads in (consumes) some number of characters and the file position is advanced. In this way, a subsequent `read`-type operation will continue from the point where the last `read` finished.

Table 2.6 highlights some of the methods of file objects. Python documentation for more file methods may be found in [54].

In the following example, we open the `hello.txt` file, use `read()` to read all of its contents into a string referenced by `filecontents`, close the file, and then

Table 2.6 File object methods

Method	Description
read(n)	Read from the file object starting from the current file position, returning a string if the file was opened as a text file. If n is specified, the read is limited to at most n characters. If not specified, the read should read until the end of the file. If the current position is at the end of the file (beyond the last character), this operation returns an empty string (" "), i.e., a string of length zero
tell()	Return an integer giving the current file position
readline()	Read from the file object starting from the current file position, up to and including the next newline. Note that if a file ends with an unterminated line, the result of this function may not necessarily have a newline as its last character. Like the read() method, if the current position is already at the end of the file, this operation returns an empty string
readlines()	Read from the file object starting from the current file position, up through the end of the file. Accumulate the lines (including the line termination) read from the file into a list of strings. If the current position is already at the end of the file, this operation returns an empty list

print out the string. The example uses `repr(filecontents)` in the `print()` to show the contents of the string. This shows us a representation of *all* of the content of the string, including showing a \n for the newline. If we simply called `print()` with `filecontents` as an argument, the newline would result in an additional blank line in the output, and we would observe the effect of the newline, not see its representation.

```
filepath = os.path.join(datadir, "hello.txt")
inputFile = open(filepath, 'rt')

filecontents = inputFile.read()

inputFile.close()

print("Character sequence from file:", repr(filecontents))
```

```
  Character sequence from file: 'Hello World!\n'
```

Ideally, when you come across a block of code in this text, you should stop reading and run the code (if it fails, you can continue reading but should come back to this point later to make the code work). In this case, you should copy the code above into a Python file on your own machine, then create a file "hello.txt" saved in the same folder as that Python file (such a file can be found on the book web page), then run the code. The reading questions guide you through the kinds of questions you might ask yourself to fully understand this block of code.

Suppose we now have a file whose contents consist of two lines, instead of one, as depicted in Fig. 2.5.

Fig. 2.5 twolines.txt file contents

If we execute the following code block, we will create a path and open the file, perform three `readline()` method calls, close the file, and then print the resultant strings obtained from the `readline()` calls.

```
filepath = os.path.join(datadir, "twolines.txt")
inputFile = open(filepath, 'rt')

line1 = inputFile.readline()
line2 = inputFile.readline()
line3 = inputFile.readline()

inputFile.close()

print("1:", repr(line1))
print("2:", repr(line2))
print("3:", repr(line3))
```

We next illustrate how to `append()` the three lines to a list of lines, rather than simply printing them to the screen.

Now that we have the three lines stored in the list `linelist`, we can do whatever we want with the text data we have extracted from the file. We illustrate by printing the lines along with line numbers.

```
| 1: 'First line\n'
| 2: 'A second line\n'
| 3: ''
```

The above example, in addition to demonstrating multiple `readline()` operations, illustrates how the `readline()` *includes* the line-terminating newline character. If we are processing a file by lines and want to avoid introducing errors that could occur through the existence of this character, we may wish to invoke `strip()` to eliminate both leading and trailing whitespace from lines read, or `rstrip()` to eliminate trailing whitespace in a string line. For instance the code:

```
line1 = inputFile.readline()
line2 = inputFile.readline()
```

could be replaced with

```
line1 = inputFile.readline().rstrip()
line2 = inputFile.readline().rstrip()
```

and the output becomes

```
| 1: 'First line'
| 2: 'A second line'
```

The next example shows how we can `read()` the entire contents of a file and then use the knowledge that `'\n'` gives line termination to separate the obtained data into lines. The command `filedata.split('\n')` returns a list of strings, where the first item is the slice of the string `filedata` up to (but not including) the first `'\n'`, the second item starts just after the first `'\n'` and goes up to (but not including) the second `'\n'`, etc.[49]. This means none of the strings in the list `filelines` contains the newline character.

```
filedata = inputFile.read()

filelines = filedata.split('\n')

print("Total length of the file contents:", len(filedata),
      "\nNumber of splits on newline:", len(filelines))
```

```
| Total length of the file contents: 25
| Number of splits on newline: 3
```

A very common error in file processing is to fail to include the `close()` for an opened file. The Python `with` construct can help in this regard, because a `with` specifies an object as its context, and when the body of the `with` completes, the object is closed *automatically.* So the code:

```
inputFile = open(path, mode)

# Perform operations on inputFile

inputFile.close()
```

becomes

```
with open(path, mode) as inputFile:

    # Perform operation on inputFile
```

Note that the file operations and method calls are grouped together and indented as the body of the `with`. Once the file operations are complete, the code would then align outdented with where the `with` was first aligned. The file object that is the context of the `with` would no longer be valid for use after the body of the `with`, but any variables defined during the file operations would continue in scope.

As an example, we next use the `with` pattern for access to a file, `tennyson.txt`, and also illustrate the use of the `readlines()` method that, in a single invocation, can obtain a list consisting of all the separate lines in a file.

```
filepath = os.path.join(datadir, "tennyson.txt")
with open(filepath, 'rt') as file_obj:

    lines = file_obj.readlines()

for line in lines:
    print(repr(line))
```

```
 | 'Twilight and evening bell,\n'
 | '  And after that the dark!\n'
 | '\n'
 | 'And may there be no sadness of farewell, \n'
 | '  When I embark;\n'
```

2.2.2   Text File Encoding

We know that a text file is defined as a file whose contents give us a sequence of characters, which we can aggregate into strings and into lines. We also know that, ultimately, when we represent something, in memory in our programs, or in a file on disk, that the representation consists of a collection of *bytes*, each of which consists of 8 bits, each of which is either 0 or 1. We have already seen how to open bytes files for reading or writing, with the mode 'rb' and 'wb'. In this section, we assume we are working with text files (so, we would open them with 'rt' and 'wt'). It is important to note that there are multiple possible ways that the bytes making up these files can map to the characters those bytes represent. Once we specify such a mapping (from the bytes making up the text file, to the characters that those bytes represent), methods like read() and readline() are able to transform a sequence of bytes into a sequence of characters to be used as strings in our programs. Common mappings include ASCII, UTF-8, and ISO-8859-1, as we now discuss.

The *encoding* of a file specifies this mapping, and a file is said to be *encoded* with a particular (standard) mapping. When working with a text file, any read-type operation must know the encoding so that the operation may read the raw bytes of the file and perform the mapping to obtain a sequence of characters. This process of reading bytes and mapping to characters is called *decoding*. In the other direction, any write-type operation must know the desired standard encoding mapping and translate the sequence of characters (i.e., "encode") into the appropriate bytes.

To summarize, we *encode* when we go from characters in a program to bytes in a text file, and we *decode* when we go from bytes in a text file into characters in a program.

If we have an existing text file that we need to read from, the operating system must know the encoding of the file to be able to properly perform read-type operations. If we want to create a new file, the operating system must know the desired encoding for the file to be able to properly perform `write`-type operations. For both cases, the `open()` function has an optional `encoding` argument that is used to specify this.

The most dominant encoding in use for existing text files and for data obtained over the Internet is a standard known as *UTF-8*.[8] It uses a variable length of one to four raw bytes for its mapping of the entire Unicode character set. This encoding standard is the default for any file `open()` and, if specified, is indicated with the argument `encoding="utf-8"`.[9] The `-8` suffix is used as an indicator that, for most files, a single 8 bit byte is sufficient, and because it is also backwardly compatible with the ASCII encoding of the ISO-8859-1 set of characters. Alternatively, the *UTF-16* encoding standard, which also provides a mapping for the entire Unicode character set, always uses 16 bits as two 8-bit bytes to represent each character. There are many other encodings, often corresponding to specialized language character sets. The main takeaway: if you are given a text file that you need to read, and if it is encoded in something *other than* UTF-8, then you need to know its encoding in order to be able to use a proper `open()`.

Below is an example where a UTF-16 encoded version of the `twolines.txt` file is opened, and we perform the two `readline()` method calls to obtain the data.

```
filepath = os.path.join(datadir, "twolines.utf16.txt")
try:
    inputFile = open(filepath, 'rt', encoding='utf-16')
except OSError:
    print("Error opening file")

line1 = inputFile.readline()
line2 = inputFile.readline()

inputFile.close()

print("1: ", repr(line1), "; 2: ", repr(line2), sep="")
```

```
| 1: 'First line\n'; 2: 'A second line\n'
```

The *size* of the `twolines.utf16.txt` file, as reported by the operating system, is twice as many bytes as the original `twolines.txt` file, which is encoded in the default UTF-8.

---

[8]UTF stands for *U*nicode *T*ransformation *F*ormat.

[9]In Python, the case of a specified encoding used as an argument to Python functions does not matter, so the strings `"UTF-8"` and `"utf-8"` are equivalent and refer to the same encoding.

2.2.3 Reading Questions

2.17 What Python function establishes a *connection* between a Python program and a local file through the operating system? What is the data type of what is returned by the function?

2.18 Give the Python strings for the three most common *modes* used when opening files.

2.19 To get more practice with permissions for opening and closing files, please partner up with a friend, create a new Google document, and share with your friend with only *read* privileges (rather than Edit privileges, which is the same as *write* privileges). What happens when your friend tries to type new information into the document?

2.20 What access mode would be used to open a file for writing raw binary data to a file instead of a sequence of text characters?

2.21 Give two reasons why an open() call might fail. How should one write code to handle such situations?

2.22 Building on the previous question, please create a txt file, then lock the file (the procedure to do this was discussed in the reading questions of the previous section). Then use open() to open this txt file from a Python file in the same folder, with the *append* mode. Report what happens when you try to add new text to the locked file.

2.23 Say that a file consists of five lines of data, with 8 printable characters on each line (including spaces), and there are no tabs. If each character is encoded as a single byte, how many bytes should the file be? Explain.

2.24 Say that an *encoding* of a file always takes 2 bytes for each character encoded. If we look at the byte length of a 128 character file, what do we expect the byte length to be? (UTF-16 is an example of such an encoding.)

2.25 If an encoding of a file sometimes takes 1 byte for a character and sometimes takes 2 bytes for a character, and never takes more than 2 bytes per character. What are the minimum and maximum byte lengths for a 128 character file? (UTF-8 is an example of such an encoding.)

2.26 Why does the textbook advocate using a with clause for encapsulating a set of operations used to work with a file?

The next five questions are in reference to the block of code that opens hello.txt.

2.27 What would be returned for an invocation of inputFile.tell() placed *after* the inputFile.read() line?

2.28 What would be returned if another `inputFile.read()` line occurred after the first?

2.29 What would be the result returned if the `inputFile.read()` were replaced with `inputFile.readline()`?

2.30 Recall that Python string indexing allows us to access individual characters. What character do we obtain for each of the following?

- `filecontents[0]`
- `filecontents[5]`
- `filecontents[-1]`

2.31 Say that the statement for first file object method was replaced with `s1 = inputFile.read(5)`, and that was followed by `s2 = inputFile.read()`. Give the representation of the strings for `s1` and `s2`.

The next three questions are in reference to the code that reads "twolines.txt" into three string variables.

2.32 What is the length of the string variable `line1`?

2.33 Why is the value of `line3` as shown?

2.34 Suppose we inserted `pos = inputFile.tell()` after the assignment to `line1`, but before the assignment to `line2`. Give the value of `pos`.

2.35 The reading mentions that some text editors have the option to "Show Invisible" characters. Please investigate and then describe the process you need to enable this option on a text editor on your machine. It will be very useful for the remainder of the book.

The following five questions are about the block of code that introduces the `split()` function. You are encouraged to read the online documentation for this function, if it is new to you.

2.36 The `split()` operation in `filedata.split('\n')` results in a list. In this example, how would you characterize what is represented by each element of the resultant list?

2.37 Explain why the length of the result of the `split()` yields a 3, when we have a two line file? If you are unsure, check the online documentation for the `split()` function, referenced in the text.

2.38 Suppose we added together the lengths of each of the string elements in the split list, how many characters would we have? How does this compare with the total length of the file contents?

2.39 If we performed a `split()` with no argument on `filedata`, how would you characterize the elements of the resultant list? What would be the length of the resultant list?

2.40 Look at some of the functions before and after the above reference for `split()`. Identify a function that would do a superior job than `split('\n')`.

2.2.4   Exercises

2.41 Write a function

    fileCharCount(path)

that takes a single argument of a file path and, by opening, reading, and closing the file, obtains the count of characters in the file.

2.42 Enhance your solution from the previous problem so that it returns the value -1 if there is an error in opening the file.

2.43 Write a function

    fileCount(path, target)

that takes an argument of a file `path` and a `target` and, by opening, reading, and closing the file, can obtain the number of times `target` appears in the file. Like the previous problem, return the value -1 if there is an error in opening the file.

Suppose, in the data directory given by python variable `datadir`, there is a file named `tennyson.txt`. The content of the file consists of the five lines as shown here:

```
Twilight and evening bell,
   And after that the dark!

And may there be no sadness of farewell,
   When I embark;
```

So there are two lines, an empty line, and then two following lines. Each of these five lines is terminated by the newline (`'\n'`). Some lines begin immediately with a non-space character, and some lines begin with one or more spaces. Although not visible in the above, some of the lines have space characters at the end, before the terminating newline.

2.44 Write a function

    tennysonLines()

that reads the lines of the `tennyson.txt` file, and returns a list of strings, with one entry in the list for each line of the file. Use repeated invocations of the `readline()` method in your solution, and, because this is specific to the `tennyson.txt` file, your function can use the knowledge that there are five lines in the file.

2.45 Write a function

```
tennysonWordCount()
```

specific to the `tennyson.txt` file, that counts the number of words in the file, processing line by line. Words are defined as sequences of characters separated by spaces, tabs, and newlines. With this definition, you need not worry about apostrophes within a word, and a word, by this definition, would include any following punctuation, like a comma, period, semicolon, or exclamation mark.

2.46 Write a function

```
tennysonWords()
```

specific to the `tennyson.txt` file, that accumulates a list of the actual words in the file, processing line by line.

2.47 Write a function,

```
generateSquares(N)
```

which generates a list with values from 1 to *N*, inclusive, and a second list with the corresponding value from the first list squared. The two lists are returned. So for `generateSquares(3)`, the results are the lists `[1, 2, 3]` and `[1, 4, 9]`.

2.48 Write a function,

```
writeTwoLists(path, A, B, sep='\t')
```

writes the two lists, `A` and `B` and writes them to a file specified by `path`, with one line per A-B corresponding pairs of values. The `sep` argument specifies the separator between a corresponding pair of values, with a default of a tab. If the file specified by `path` exists, it should be reset to zero length and overwritten with the new set of lines.

2.49 Write a function,

```
writeTwoLists(path, A, B, sep='\t',headers)
```

that modifies your solution to the previous exercise to also write a header line with data given by the list `headers`. For example, `headers = ["Name", "Count"]` would be used for our baby names example. Use the same separating character for the headers that you use for the body of the file you are writing.

2.50 Write a function

```
make_it_lowercase(readpath, writepath)
```

that reads the text file given by `readpath` and writes a file at `writepath`, where every letter has been replaced by its lowercase version. For example, if the text of the file at `readpath` was `"LaTeX"`, then the text of the file at `writepath` should be `"latex"`.

2.51 Write a function

```
make_it_utf16(readpath, writepath)
```

that reads the text file given by `readpath` (assumed to be UTF-8 encoded) and writes a file at `writepath` with the same text, but UTF-16 encoded. You can then use your file system to check that the written file requires more bytes (e.g., with the "Get Info" option applied to both files).

2.52 Write a function

```
recordAllFolders()
```

that opens a file named `allfolders.txt` for writing. Use your knowledge of the `os` module to create a list of all of the files and folders that exist one directory up from the current working directory. Write the output of this into `allfolders.txt`, one folder name per line. Be sure to close the file. Return the list of folders.

2.3 Processing Files for Data

Building from the methods introduced in Sect. 2.2, where we focused on operations applied to an entire file, we now consider the patterns involved in *iterating* over a file in order to process it and to obtain data contained therein as one or more lists.

With the need of a greater data focus in this section, we will use, as our example data sets, files obtained from the US Social Security Administration (SSA). From records of applications by parents and guardians to obtain social security numbers for newly born babies, the SSA maintains counts of the name and registered sex for all applicants, and makes that data available to the public in the form of names and rank order by year for both male and female sexed applications [68]. We will see this data set throughout the book, often with the name `topnames` or `babynames`.

2.3.1 Single Data Item per Line

Let us start with the assumption that we have a file with one datum per line, and that the set of lines form a coherent data collection. Our goal is to iterate over the lines of the file and accumulate the data into a single list. Specifically, in this example, we have a file, `"baby_2010_female_name.txt"` with, in rank order, the most popular female names registered in 2010 with the SSA. Our objective is to create a function that performs the file processing and returns a single list with a name per entry in the list. We start with the assumption that we know and can specify the number of data lines, and thus include this as the parameter `numlines`.

Starting with the next example, we will intersperse the modeling of good practices by encapsulating a coherent set of steps into a function. Functions should accomplish a single goal, should avoid dependence on global variables and information, and should use parameters so that the functionality may be applied to multiple situations, giving a useful and general building block.[10]

```python
def readNames(path, numlines):
    namelist = []

    if not os.path.isfile(path):
        print("File not found")
        return namelist

    with open(path, 'r') as inputFile:

        for lineindex in range(numlines):

            line = inputFile.readline()
            name = line.strip()
            namelist.append(name)

    return namelist
```

An example invocation of this function, obtaining the top five female baby names in 2010:

```python
path = os.path.join(datadir, "baby_2010_female_name.txt")
lines = 5
names = readNames(path, lines)
for name in names:
    print(name)
```

```
| Isabella
| Sophia
| Emma
| Olivia
| Ava
```

Now we use the same function to read from the file of male names in 2010 to obtain the top three names :

```python
path = os.path.join(datadir, "baby_2010_male_name.txt")
lines = 3
names = readNames(path, lines)
```

[10]Because our functions, its parameters, functionality, and return values are described in the text, we will omit docstrings in the functions we present. For clarity, we may also omit error processing, like the `try-except` around a file `open()`. But well-written functions should always include docstrings to describe a function's goal and return value in terms of its parameters and should include all error processing. When we develop code, we follow the Google Python Style Guide[16].

```
for name in names:
  print(name)
```

```
| Jacob
| Ethan
| Michael
```

The core algorithm in this first example uses a *definite loop*, which is a loop where we use knowledge of the exact number of iterations needed in the construction of the loop. But we often do not have that knowledge. We would like our readNames() function to be able to work whether we have 15 names, 20 names, or 100 names. But, for efficiency reasons, we want to still process a line at a time, as opposed to a solution that employs a read() of the entirety of the file data into a single, potentially very large, string and then follows that with split() type operations.

The general solution should reflect an algorithm like the following:

1. Read in a single file line, call it *line*.
2. While *line* does not indicate no more lines in the file

   a. perform any necessary processing on *line*, and
   b. read the next file line into *line*.

This algorithm relies on a *sentinel*, which is a Boolean condition that indicates that the state has changed and that we have reached our termination goal. Fortunately, we have already seen that our read() and readline() methods provide such a sentinel. When these functions are invoked on a file whose position is already at the end of the file, they return an empty string.

Translating the algorithm into specific code for this example:

```
namelist = []

line = inputFile.readline()
while line != "":
  name = line.strip()
  namelist.append(name)

  line = inputFile.readline()

for name in namelist:
  print(name)
```

```
| Jacob
| Ethan
| Michael
| Jayden
| William
```

Note that, without introducing any facilities beyond the file methods we already know, we can now process a file consisting of an arbitrary number of lines.

This pattern, of processing a file by lines in an iterative manner, for as many lines as exist in the file, is a *very* common pattern. Python file objects accommodate this pattern by allowing the file object itself to be an iterator (a structure that we can iterate over). This allows the file object to give the sequence needed in a `for` loop, and each iteration over the file object yields exactly one line, and the iteration starts at the current file position.

So, abstractly, our algorithm simplifies to:

1. For each line in the file, call it `line`

    a. perform any necessary processing on `line`

```
for line in inputFile:
  # process line
```

We recall that `line` terminates with `'\n'`, so we use `strip()` when processing it.

```
namelist = []

for line in inputFile:
  name = line.strip() # removes \n and other whitespace
  namelist.append(name)

for name in namelist:
  print(name)
```

```
| Jacob
| Ethan
| Michael
| Jayden
| William
```

Because the use of the file object in a `for` is defined as starting from the current file position, we can combine the use of `readline()` with the use of `for` over a file object. Suppose, for instance, that a babynames file begins with two "comment" lines, whose lines start with a # character, and are then followed by the data lines.

```
# Attribution to US Social Security Administration
# Top 5 registered female names in 2010
Isabella
Sophia
Emma
Olivia
Ava
```

Our algorithm would then be to perform two `readline()` method invocations outside of any iteration, and then follow that with `for` iteration, which is used for the processing of the data itself. Applied to our example:

```
filepath = os.path.join(datadir, "baby_2010_female_top5.txt")

with open(filepath, 'r') as inputFile:

    comment1 = inputFile.readline()     # read first comment line
    comment2 = inputFile.readline()     # read second comment line

    # read and process data lines
    namelist = []
    for line in inputFile:
        name = line.strip()
        namelist.append(name)

for name in namelist:
    print(name)
```

```
| Isabella
| Sophia
| Emma
| Olivia
| Ava
```

2.3.2   Multiple Data Items per Line

Chapter 6 will give greater detail for file formats supporting data sets. Here, we introduce the idea that files often store data with each line representing a single case or observation, and then, within the line, there are multiple values of some variables for the case/observation.

Consider our babynames example. Each line represents a ranking of a particular name, with a single Name variable whose values consist of the different actual names. We could augment the babynames file format to include not just the name, but the number of babies registered with that name. Let us use Count to refer to this second variable. With this extension, our file might look like:

```
Isabella     22913
Sophia     20643
Emma       17345
Olivia     17028
Ava        15433
```

As soon as we have multiple values on each line, we must choose how we separate the values on the line. The most common choices are to use:

- spaces(s)

- a single tab

- a comma, sometimes with one or more following spaces

In the current example, values in a line are separated by exactly one tab character.[11]

The effect of this on our file processing algorithm is fairly straightforward. Conceptually, we need to accumulate, in parallel, two lists of data instead of one. So outside of any iteration, we begin by creating both empty lists, one associated with each of the collection variables. Then, at each iteration of the line-centric loop, we have to process the line by splitting a single line into its multiple parts and, for each part, accumulating into the corresponding list.

For the two-value-per-line separated by tab, the code in the main file processing loop might look like this:

```
namelist = []
countlist = []
for line in f:
  multivalue = line.strip()
  values = multivalue.split('\t')
  namelist.append(values[0])
  countlist.append(int(values[1]))
```

Say that we want to design a function that can process name and counts as in the example above. The signature of the function might look like:

```
readNamesCounts(filepath)
```

Beyond incorporating the changes discussed above, we now have to return both the list of names and the list of counts from the new function. A Python `return` can have a value that comma-separates multiple return values. This is, in actuality, creating a `tuple` object. A caller of the function can perform a multiple assignment to get all returned values, or it can assign to an aggregate tuple variable, and then use indexing of the tuple variable to access the individual return values.

Our final function for reading the names and counts from an appropriately formatted file looks as follows:

```
def readNamesCounts(filepath):
  namelist = []
  countlist = []

  with open(filepath, 'r') as inputFile:
    for line in inputFile:
```

---

[11] When using tabs, values can align if the position on the line has not exceeded the next "tab stop." This is why the first and last lines in the example align differently than the middle three. Nonetheless, there is *exactly* one tab character between a name and the corresponding count value.

```
        multivalue = line.strip()
        values = multivalue.split('\t')
        namelist.append(values[0])
        countlist.append(int(values[1]))

    return namelist, countlist                    .
```

We now provide an example invocation.

```
filename = "babynames_2010_female_namecount.txt"
filepath = os.path.join(datadir, filename)
names, counts = readNamesCounts(filepath)
print("Names:", names)
```

```
| Names: ['Isabella', 'Sophia', 'Emma', 'Olivia', 'Ava']
```

```
print("Counts:", counts)
```

```
| Counts: [22913, 20643, 17345, 17028, 15433]
```

Note that, since `readNamesCounts()` returns two lists, our invocation
assigns both return values to variables. An alternative would be to assign the return
values to a single tuple variable and then use indexes to access the individual
returned items, as we now illustrate.

```
data = readNamesCounts(filepath)
print("Names:", data[0])
```

```
| Names: ['Isabella', 'Sophia', 'Emma', 'Olivia', 'Ava']
```

```
print("Counts:", data[1])
```

```
| Counts: [22913, 20643, 17345, 17028, 15433]
```

It is common for a file to begin with several lines of text that are not data, that
we may need to skip over when extracting the data. For instance, this text may be
metadata, i.e., data about the characteristics or attributes of the data. Such metadata
commonly includes the name of the data set, an attribution to the individual or
organization that worked on the data set, the date of creation, the date of the last
update, permissions, licensing information, and information for rendering the data.
For example, .dat files are often transmitted with metadata describing the contents
of the .dat file, whether it should be renamed to another file extension (e.g.,
.mp4), and what programs can open/render the data within the file. We will discuss
metadata more in Chap. 15.

For now, if we wish to skip over lines at the top of a data file, we can introduce
an additional parameter, `skiplines`, relaying the number of lines to skip over.
In the following, we execute `readline()` the specified number of times, to "eat"
the lines we want to skip, and our iterator then begins at the first line after those
we skipped. We include this parameter with a default value. Whenever a Python
function definition has the syntax *parameter = expression* (as shown below), the

function is said to have a *default parameter value*. This means that, if the function is called without this parameter, then the default value from the function definition will be filled in. In the function below, the default number of lines to skip is zero.

```python
def readNamesCounts(filepath,skiplines = 0):
  namelist = []
  countlist = []

  with open(filepath, 'r') as inputFile:
    for skip in range(skiplines):
        inputFile.readline()

    for line in inputFile:
      multivalue = line.strip()
      values = multivalue.split('\t')
      namelist.append(values[0])
      countlist.append(int(values[1]))

  return namelist, countlist
```

Because of the default parameter, this function can be invoked with either one parameter or two. For example, we could use a function call like `readNamesCounts("babynames.txt")` (which would skip zero lines) or, for example, via `readNamesCounts("babynames.txt",5)` if we wished to skip 5 lines. Furthermore, when invoking a function in Python, parameters can be in a different order than when the function is defined, if parameter names are specified when the function is called, e.g., `readNamesCounts(skiplines = 5, filepath = "babynames.txt")`. For more details about default parameters, see [51].

The patterns in this section form the basis of all file processing that occurs in data systems, and we will see them again at many points in subsequent chapters.

2.3.3 Reading Questions

2.53 As justified in this section, what is the motivation for grouping sets of steps into a function definition?

2.54 What is the disadvantage of assuming we know the number of lines in a file and using a *definite* loop to process the lines?

2.55 What do we mean when we say that a structure or an object is an *iterator*? Give two examples of iterators we have used, not including file objects.

2.56 True or False: It is okay to mix the use of `readline()` method invocations with the use of a file object as an iterator over lines in a file. Explain.

2.57 When we have multiple items on each line of a file, how does our program know when one item ends and the next one begins?

2.58 Why is it that sometimes we use functions like `int()` or `float()` on values obtained from a file, while other times we do not?

The following four questions deal with the example of iterating through the `baby_2010_female_name.txt` and `baby_2010_male_name.txt`, which both have a single data item per line.

2.59 Suppose the files in these examples have 10 lines of data. What happens if we were to pass an argument of 15 as `numlines`? Is there an error? If so, what line causes what error? If not, what is returned by the function (be specific)?

2.60 Create a text file with a set of your favorite names, one per line, and then replicate the `readNames()` function and try it with a path to your file as an argument, along with an appropriate count of lines.

2.61 Building on the previous question, what happens if you put all the names on a single line in your input file, but keep the code the same?

2.62 Assume the format from the previous question, where the input file consists of a *single line* with all the favorite names on that single line. Rewrite the `readNames()` function so that it works with this new format. Do you still need the `numlines` parameter? Why or why not?

The following four questions deal with the babynames example that has multiple data items per line (both `Name` and `Count`).

2.63 How many elements are in the list `values`?

2.64 As part of the processing of the second value on the line (`values[1]`) we call the function `int()`. Why? What is the data type associated with `values[1]`? For a variable like `Count`, what is the appropriate data type?

2.65 What would be the effect of omitting the `strip()` as the first operation per line?

2.66 What would happen if a count field in a file did not meet expectations? For instance, what if the count field presented as a real-valued number, like `18154.5`? Alternatively, what if the count field used commas as thousands-separators, like `18,154`?

2.3.4   Exercises

2.67 Write a function

```
assignmentMapText()
```

that opens a file named `assignmentMap.csv` (available on the book web page), and returns a string consisting of all the text in the file. Make sure you close the file.

2.68 Write a function

```
lastWords(filepath)
```

that creates and returns a list of strings consisting of the last word in each of the lines of the text file whose location is given by `filepath`. The items in the list should not include any newline or whitespace as part of any of the words, and there should not be any items in the list for any lines consisting of all whitespace. A word, as required here, will include the last sequence of non-whitespace in the line.

2.69 Repeat the previous question for `lastWords()`, but this time remove any trailing punctuation from the words added to the list, and convert all words to lower case.

2.70 Write a function

```
printLower(filepath)
```

that processes a file, line by line, and prints each line of the file converted to lower case. The print should not include blank lines where they do not exist in the original, but they should include any leading or trailing spaces and/or tabs.

2.71 Write a function

```
lineLengths(filepath)
```

that processes a file, line by line, and accumulates and returns an integer list with the lengths of each line, excluding any leading or trailing whitespace (spaces, tabs, or newlines).

2.72 Write a function

```
lineLengths2(filepath)
```

that processes a file, line by line, and accumulates and returns a list of tuples, one per line. Each tuple consists of the two value of the line number and the length of the line, excluding any leading or trailing whitespace (spaces, tabs, or newlines). Line numbers in a file start at 1. So for the `hello.txt` file:

```
Hello, world!
Welcome to 'Introduction to Data Systems'.
```

the result is the list of tuples: `[(1, 13), (2, 42)]`

2.73 Sometimes, when processing a file, some number of header, comment, or documentation lines at the beginning of a file must be omitted from the processing and skipped over before processing the remaining lines in the file. Consider the addition of parameter, `skiplines` with a default value of 0 for the previous problem:

```
lineLengths(filepath, skiplines=0)
```

The function should skip the specified number of lines and then, for the remainder of the lines, compute a list of (line number, line length) tuples, one per processed lines. The line number advances, even for the skipped lines. Line lengths should exclude any leading or trailing whitespace (spaces, tabs, trailing newline). So for the `tennyson.txt` file:

```
Twilight and evening bell,
    And after that the dark!

And may there be no sadness of farewell,
    When I embark;
```

the result of `lineLengths("tennyson.txt", skiplines=2)` is the list of tuples: `[(3, 0), (4, 40), (5, 14)]` since lines 1 and 2 are skipped, and line 3 is an empty line, consisting of just a newline.

2.74 Consider a variation of the babynames and counts file as depicted below:

```
22127,      Jacob
18002,      Ethan
17350,    Michael
17179,     Jayden
17051,    William
16756,  Alexander
```

Each line of the file captures one data case/observation. The values are separated by commas, with the count occurring first and the name second, and spaces are used to align the columns of data to make it easier for a human reader.

Write a function

```
readNamesCounts(filepath)
```

that processes the `filepath` file and yields a tuple whose first element is a reference to a list of names, and whose second element is a reference to a list of integer counts.

2.4 JSON File Processing

In the sections above, we have learned about the file system, how to locate files via paths and the `os` module, and how to open, iterate through, and close files. We conclude this chapter with a discussion of a very common file format used to store and transmit data that also solidifies the previous sections. The JavaScript Object Notation (JSON) is a text format used in files for easily reading and writing data and configuration information to and from our programs. It allows us to avoid explicit encoding, conversion, and parsing steps, and for many data structures in our

Table 2.7 JSON module functions

Function	Description
json.load(file)	Read and return the JSON-formatted data structure from the file object file
json.dump(data, file)	Write the data structure data to the file object file in JSON text format
json.loads(s)	Using the JSON-formatted string given by s, interpret and construct and return the corresponding data structure
json.dumps(data)	Translate the data structure data into a JSON-formatted string and return the string

programs, allow us to directly write the data structure to a text file (or to a JSON-formatted string), and to subsequently read the data into a data structure. We will study JSON in depth in Chaps. 15, 16, and 23, so in this section we only give a brief overview.

JSON has several beneficial characteristics:

- programming language independent—libraries for working with JSON, and reading and writing JSON strings and files exist for Python, JavaScript, Java, C, C++, and many others.
- programmer familiarity—the notation/syntax is a close match for how programmers define constant data, in the form of strings, numbers, lists, and dictionaries, and so is easily understood by the programmer audience.
- readability—the format is text-based, and so files containing JSON-formatted data are readable and even editable directly in a text editor.
- lightweight but powerful—the notation focuses on simple data types of strings and numbers, and simple data structures of lists and dictionaries. But by composing and nesting these simple structures, JSON can handle complex situations.
- straightforward interface—writing a JSON file only requires a data structure and a file opened for writing, and reading a JSON file only requires a file already opened for reading and returns the contained data structure.

Because of its ability to perform its notational encoding on nested data structures, JSON can be viewed as a hierarchical structure, and we will examine it in more detail in Chap. 15. In this section, we focus on basic file operations using JSON and the json module (Table 2.7).

2.4.1   Writing Data Structures to JSON

The general steps taken when we want to write a data structure in our program out to a text file containing the JSON-encoded data structure are:

1. Create the data structure, call it D.
2. Open a file for writing, call it F.

3. Invoke the dump () json function, passing *F* and *D*.
4. Close the file *F*.[12]

Say that we have a list data structure, like names from our prior example, and we want to create a JSON file named names.json in directory datadir. The following code would accomplish this:

```
import json

# names already created

path = os.path.join(datadir, "names.json")
f = open(path, 'w')

json.dump(names, f)

f.close()
```

The result is a text file, names.json, that has the following contents:

```
["Isabella", "Sophia", "Emma", "Olivia", "Ava"]
```

In JSON, strings are always delimited by double quotes. Also note that, in a JSON-encoded file, whitespace outside of strings *does not matter*. So in this case, a file with each name on a separate line is equally valid, as long as the comma-separated list of strings is intact. JSON does not have a separate concept of a tuple, so tuples in our data structures are encoded as lists. For example, our readNamesCounts () returned a tuple with the list of names and the list of counts. If we were to perform a json.dump () with this data structure, the file would have an outer list consisting of the list of names and then the list of counts.

This same pattern seen in this example can be used for much more complex data structures, including dictionaries and two-dimensional data structures like lists of lists, lists of dictionaries, and dictionaries of lists. We will see additional examples later in the text.

2.4.2   Reading Data Structures from JSON

We often use JSON for storing both processed data and configuration data, and this allows us to easily read such data stored in text files into our programs. Suppose we have a file config.json that stores configuration information as a dictionary whose keys are the names of providers, and whose values are a (sub)dictionary with

---

[12]We could also use the with statement to handle the close automatically, and step 3 would be in the body of the with.

keys like `"login"` mapping to a login name and `"password"` mapping to a user password.

For example, `config.json` might have the contents:[13]

```
{
    "facebook": {"login": "alice", "password": "mysecret"},
    "twitter": {"login": "alice2", "password": "secret2"}
}
```

Retrieving the configuration and parsing and populating a dictionary data structure in our program is as simple as:

```
path = os.path.join(datadir, "config.json")
with open(path, 'r') as configFile:
  configD = json.load(configFile)
```

After the `load()` operation, we can use the obtained dictionary `configD` in the same way as a data structure originating in our program.

```
for key, value in configD.items():
  print('Provider:', key, 'Login:', value['login'],
        'Password:', value['password'])
```

```
| Provider: facebook Login: alice Password: mysecret
| Provider: twitter Login: alice2 Password: secret2
```

JSON-formatted data is extremely common in data systems, as will be explored below in the Reading Questions. Sometimes, JSON-formatted data is transmitted in files with a different extension than `.json`, e.g., `.dat` or `.ipynb`. Such data can still be wrangled using the contents of this section, if first read from its existing form and then written into a `.json` file. We will see JSON data again in Chaps. 15, 16, and 23.

2.4.3 Reading Questions

2.75 Are JSON files text files or binary files?

2.76 Can JSON read/write *list* data structures? How about *dictionary* data structures? How about file object/structures?

2.77 Can a data structure in our program that consists of a list containing a set of dictionaries be written to a JSON file?

---

[13]The `config.json` file might have been created previously by a client program, or might have been created in a text editor, or even some combination.

2.78 Suppose you use a text editor to create a JSON file, but make the mistake of using Python compatible strings that are delimited by a single quote instead of by the JSON syntax for a string that uses double quotes. What do you think will happen if you open and then use a `json.load()` to try and read the JSON into a data structure in your program?

2.79 When using `json.load(file)`, what file extensions are allowed for `file`?

2.80 When using `json.dump(data,file)`, what data structures are allowed for `data`?

2.81 The reading mentions carrying out a `json.dump()` with the data structure (tuple) from `readNamesCounts()`. Carry this out, mimicking the code from the reading, and then do a `json.load()` back into a Python data structure. What data structure do you get? Is it a tuple again?

2.82 Consider the data (and format) shown in `config.json` in the reading. If this data was stored in a txt file, what would you need to do to build the dictionary `configD`? Please be specific.

The next questions illustrate the ubiquity of JSON files.

2.83 If you have Firefox web browser, navigate to Bookmarks -> Show all Bookmarks -> Backup (in the dropdown menu). This produces a backup of your bookmarks as a JSON file. Is it a JSON object (like a dictionary in Python) or a JSON array (like a list in Python), and how can you tell?

2.84 Python notebook files (ipynb files), that you might open with Jupyter Notebook or Jupyter Lab, are JSON formatted. To see this, drag and drop such a file into a web browser. Is it a JSON object or a JSON array, and how can you tell?

2.85 Social media companies store your data when you post photos, text, comments, and interactions on others' posts. This data is often stored in JSON format. For example, if you have a Facebook™ account, and if you navigate to Settings -> Download my data, you can select your data to be provided in JSON format. If you do not have Facebook™, you can download JSON data from another social media company. Is it a JSON object or array?

2.4.4 Exercises

2.86 In the data directory (on the book web page) is a file named `fib.json` containing a JSON-encoded list of numbers from the Fibonacci sequence. Write a function

```
readJSONlist(path)
```

that reads and returns a list from `path`, assuming a JSON-encoded text file. If `path` does not exist, or if the result is a data structure other than a list, return the empty list. (Hint: the Python function `isinstance()` can be helpful to verify the type of a data structure.)

Then invoke the function on `fib.json` in `datadir`.

2.87 Write a function

```
writeSquares(N, filepath)
```

that creates a list of the squares of integers from 1 to N and then writes them out as a JSON-encoded text file to the file location given by `filepath`. If N is less than or equal to 0, no file should be created. This function returns no value.

2.88 In the data directory (on the book web page) is a file named `baby_2015_female_namecount.txt` with a line for each of the top female baby names of 2015. Each line contains tab-separated values for the name and the count of US Social Security applications for that name. Start by writing a function

```
readNameCount(path)
```

that reads a file formatted this way, given by `path` and returns a tuple of two lists, the first being the names and the second being the counts. The function should behave properly regardless of how many name-count lines are in the file. If the specified path does not exist, your function should return `None`. Make sure your count values are integers.

Then write code that *uses* your function to obtain the lists and writes the result to a JSON-encoded text file in the current directory named `namecount.json`.

2.89 In the data directory (on the book web page) is a file named `words.txt`, consisting of multiple lines of text, where each line has one or more words on the line. There are no punctuation characters in the file, and, within lines, words are separated by combinations of spaces and tabs. All words are in lower case.

Write a function

```
words2freq(inpath, outpath)
```

that reads `inpath`, assumed to be a file formatted as specified above, and generates a *dictionary* giving the frequencies of the words (i.e., each distinct word is a key, and each maps to an integer count of the number or times the word was found), and this dictionary is written as a JSON-formatted file at a location given by `outpath`. If `inpath` is not found, the function prints the error message `"Not found"` and returns without creating `outpath`. Otherwise, the function does its work without any printing and has no return value.

2.90 Write a function

```
babyNames()
```

that opens the file named `babynames.csv`, available on the book web page. Create a dictionary that contains the names as keys and the numbers as values. Read each line of the file, rather than reading the entire file all at once. Make sure you close the file. Populate and return the dictionary you have created.

2.91 Write a function

```
babyNamesDict()
```

that writes the dictionary created using `babynames.csv` into a JSON-formatted file without using any of the JSON utilities. Make sure you close the file. Return the name of the file you have written. Ensure the file type is `.json`, and remember that the syntax of a JSON file requires curly braces { } at the beginning and end.

2.92 The previous exercise has you writing a JSON file based on a dictionary "by hand." In practice, you would never do such a thing. Instead, you would use the built-in dump function. Write a function

```
practice_dump(myList,filename)
```

that dumps the data in the list `myList` into a file with the given `filename`, e.g., `"names.json"`.

2.93 Write a function

```
practice_load(filename)
```

that loads the data in the given `filename` (e.g., `"names.json"`) into a list, which you return.

2.94 Write a function

```
practice_load2(filename)
```

that looks inside the file with the given `filename` (e.g., `"names.json"`) to determine whether it represents a JSON array or a JSON object, and then loads the data into either a list or a dictionary (whichever is appropriate), which you return.

2.95 Write a function

```
babyNamesDictJSON()
```

that writes the dictionary created using `babynames.csv` into a JSON-formatted file named `"babynames2.json"`, using the JSON utilities. Make sure you close the file. Return the name of the file you have written.

2.96 Write a function

```
compareJSON(filename1, filename2)
```

that compares the texts of two JSON files given their filenames. Assume that the filenames contain the file extension.

Chapter 3
Python Native Data Structures

Upon completion of this chapter, you should understand the following:

- Using lists to represent vectors in a data set.
- Using dictionaries to represent a mapping from an independent variable to a dependent variable in a data set.
- The concept of an anonymous function to define temporary functional computation.
- One-dimensional processing patterns of:

  - Accumulation.
  - Application of unary operations over a vector.
  - Performing binary operations with vectors.
  - Filtering elements of a vector.
  - Reduction of a vector.

- Using nested native Python data structures to represent rows and columns of a two-dimensional data set.

Upon completion of this chapter, you should be able to do the following:

- Use list comprehensions as a mechanism to create a new list from an existing list.

(continued)

© Springer Nature Switzerland AG 2020   59
T. Bressoud, D. White, *Introduction to Data Systems*,
https://doi.org/10.1007/978-3-030-54371-6_3

- Use a lambda function to define a functional computation and to apply such function on a list to yield a new list.

  - Get practice in passing a function as an argument to another function.
  - Define a function that has a function as an argument and invoke such a passed function.

- Populate list of lists, list of dictionaries, and dictionary of list representations from simple comma-separated value files.

Fundamental to our goal of understanding data systems is our ability to work with collections of data. We begin by focusing on using native Python constructs, covering lists, and dictionaries for one-dimensional collections of values, and then lists of lists, lists of dictionaries, and dictionaries of lists for two-dimensional collections. Along the way, we cover some features of Python that can help and give additional power in our data processing goals.

3.1   List Patterns

While, in general, a Python *list* might have values of different data types and collect disparate values together under a single name, we are interested in an individual list that is used to represent a collection of values, often for a *variable* in a data set. We call such a related collection a *vector*. For example, from the previous chapter, we used a Python list called names, whose values represented a ranked collection of the most popular names registered with the US Social Security Administration, given a year and a sex Thus names is a single name for a vector defining a variable where each element is dependent on a year, a sex, and a ranking. The position in the vector gives the ranking, and the particulars of the acquisition (through a file) give the year and sex.

We can extend the use of lists defined as vectors so that we use more than one vector for a data set. Continuing the example from the previous chapter, we defined vectors names and counts that are related to one another and that together make up the data set. The counts vector is a variable, also dependent on a year, a sex, and a ranking, whose values give the number of registered instances of a name with a particular sex and year, and ranking. Clearly, there is a relationship between the elements of names and counts, as the *position* is important. For a given index i, counts[i] has the number of times names[i] was registered. We say that, by list index, elements *correspond*, and we have to take care in our programs not to disrupt this correspondence. When a collection of multiple lists have a correspondence, we often refer to them as a *parallel list* representation of a data collection.

There are a number of common patterns that occur in our algorithms as we work with vectors of data. We provide the basics of these patterns here, using Python list techniques that we should already be familiar with. In Sect. 3.3, we go further and cover features of Python that help simplify computing and using these patterns and will also be helpful techniques for the remainder of this book.

Note that these patterns are not always independent of one another. Many patterns use the first pattern, accumulation, as a part of their own pattern. Further, in practice, we may combine patterns to compose a single algorithm. For instance, the filtering pattern is often used in combination with the unary or binary operations patterns. Our treatment of operations below will form the foundation for Chap. 7. Readers familiar with R may recognize the functionality of the `dplyr` package.

3.1.1 Accumulation

Accumulation refers to starting from some base set of information, like an empty or single-element vector list, and then iterating and, at each step of the iteration, accumulating one or more additional values in the vector list.

So the accumulator pattern can be written algorithmically as:

1. Initialize the accumulator variable.
2. While there are more data to add:

   a. Compute/read/determine the next value.
   b. Append the value to the accumulator variable.

We have already employed the accumulator pattern when we used file operations to read data from a file into one or more lists. Recall the code from Chap. 2, and assume `inputFile` refers to a valid open file object:

```
namelist = []

for line in inputFile:
    name = line.strip()
    namelist.append(name)
```

In this example, there was exactly one new datum on each line of the file.

3.1.2 Unary Vector Operations

When computing and transforming vectors of data, we often want to apply some function, element by element, to all of the elements in a vector. So if we have vector

$\mathbf{x} = (x_1, x_2, \ldots, x_n)$, and function f that takes a single value as its parameter and yields a single value, we want to generate $\mathbf{y} = (y_1, y_2, \ldots, y_n)$ such that for every i:

$$y_i = f(x_i)$$

For example, if $\mathbf{x} = (1, 2, 3)$ and if $f(x) = 2 * x$, then $\mathbf{y} = (2, 4, 6)$. Algorithmically, we could write this pattern as the following steps:

1. Initialize the new vector variable y as an empty vector list.
2. For each value v in vector x:

    1. Compute the unary transform f(v) as newval.
    2. Append the value newval to the vector list y.

For example, suppose that we have list Ctemps, which contains real-valued temperatures in degrees Celsius. If we wish to create an equivalent vector in degrees Fahrenheit, we want to *apply* a function celsius2fahrenheit() to each value of Ctemps, accumulating into a new list, Ftemps. The celsius2fahrenheit() function takes a single number as a parameter and yields a single number, so is our unary function. Our code then becomes

```
Ftemps = []
for Ctemp in Ctemps:
    Ftemp = celsius2fahrenheit(Ctemp)
    Ftemps.append(Ftemp)
```

There are times, in using Python lists, which are mutable, that we want to change the vector *in-place*. In this case, we need the index position as well as the value, in order to compute the value and to update based on position. Suppose the source and destination vectors are given by temps. The Python enumerate(), as the iterable of a for, gives us both index position and value at each iteration, i.e., (0,-20.0), (1,0.0), (2,10.5), and (3,22.75), where the first element of each tuple is the index and the second element is the value of temps at that index. Using enumerate(), our mutation code is as follows:

```
for index, val in enumerate(temps):
    Ftemp = celsius2fahrenheit(val)
    temps[index] = Ftemp
```

An equivalent solution would iterate over the indices and define val = temps[index] before computing celsius2fahrenheit(val).

3.1.3 Binary Vector Operations

Another common operation, when working with vectors of data, is to apply a *binary function* or operator (a function or operator of two arguments), elementwise between two equal length vectors to obtain a new vector. So we have two vectors **x** and **y** and binary operator $\otimes$, and desire $\mathbf{z} = \mathbf{x} \otimes \mathbf{y}$, where for every index i, z_i is defined via

$$z_i = x_i \otimes y_i$$

For example, if $\otimes$ is addition, $+$, then we simply add the vectors **x** and **y** elementwise, e.g., if $\mathbf{x} = (1, 2, 3)$ and $\mathbf{y} = (-2, 0, 1)$ then $\mathbf{z} = (-1, 2, 4)$. If $\otimes$ denotes multiplication, then we would obtain $\mathbf{z} = (-2, 0, 3)$.

If we have a binary function f of two arguments yielding a single value, the equivalent functional notation would be:

$$z_i = f(x_i, y_i)$$

For our example vectors above, if $f(x_i, y_i)$ were the function $3 * x_i + y_i^2$, then we would obtain $\mathbf{z} = (7, 6, 10)$.

Python lists do not directly support the application of a binary operator or binary function to vector lists to obtain the desired elementwise computation. If, for example, we want elementwise *addition* of two vector lists, we cannot compute this using the command vec1+vec2 (which concatenates the two lists). Instead, we write the following code:

```
newvec = []
for index, val in enumerate(vec1):
    val2 = vec2[index]
    newval = val1 + val2
    newvec.append(newval)
```

Note how we use enumerate to allow us to get both an index and a value from the first vector. The index allows us to access the corresponding value from the second vector.

Now assume we have a function f() of two parameters that yields a single result, and we want to apply it elementwise to two vectors to yield a third. Python provides a built-in function, zip() that gives an iterable whose successive values compose as a tuple the individual elements of its arguments. For example, if $vec1 = [1, 2, 3]$ and $vec2 = [-2, 0, 1]$, then zip(vec1,vec2) will be the iterable: (1,-2), (2,0), and (3,1). This allows the current variation to be coded as follows:

```
newvec = []
for val1, val2 in zip(vec1, vec2):
    newval = f(val1, val2)
    newvec.append(newval)
```

As a final illustration of the binary vector pattern, suppose we want to perform a binary operation/function between a vector and a constant. Given a vector $\mathbf{x}$, a binary operator $\otimes$ (e.g., addition or multiplication), and a constant c, we want to compute $\mathbf{z} = \mathbf{x} \otimes c$, where for each i,

$$z_i = x_i \otimes c$$

In this example, suppose we want to compute a five percent increase to all elements of list vec, assigning to newvec:

```
newvec = []
for val in vec:
    newval = val * 1.05
    newvec.append(newval)
```

So this case is like performing $vec * c$, where $c = 1.05$ and is treated, instead of as a singleton constant, as a vector of the appropriate length whose values are all the same constant value, c.

3.1.4   Filter

In the filter pattern, given a vector $\mathbf{x}$, we want a vector $\mathbf{x}'$, a subvector of $\mathbf{x}$ whose elements satisfy a *predicate* P. So P can be thought of as a function that takes an element value and returns True or False. For example, we might wish to only keep the elements of $\mathbf{x}$ that are not zero, so P determines for every x_i whether or not $x_i = 0$. The code for filtering is a straightforward extension of the accumulation pattern:

```
subset = []
for val in vec:
    if P(val):
        subset.append(val)
```

Note that we need not define a function, P, but in many cases we can use a simple Python conditional. As a simple example, suppose we wish to eliminate outliers from a vector vec before a computation. In the following example we filter out elements larger than 1000.

```
subset = []
for val in vec:
    if val <= 1000:
        subset.append(val)
```

Table 3.1 Examples of common reductions

Initial value	Function	Description
0	+	Sum together the elements in the vector
1	*	Take the product of all the elements in the vector
x_0	min	Find the minimum value in the vector
x_0	max	Find the maximum value in the vector

3.1.5 Reduction

The final pattern we discuss in this section is *reduction*. This can be viewed as a form of accumulation, but instead of generating a list as a result, we "collapse" into a single value, e.g., computing the sum of all elements. This is accomplished by repeatedly performing a binary operation between a simple accumulator variable and the next element of the vector.

Given a vector, x, a binary function, f, accumulator variable a, and an initial value for the accumulation of a_0, we compute a as follows:

```
a = a_0
for val in x:
    a = f(a, val)
```

For example, if f(a,val) = a + val, then the line of code inside the loop is simply a = a + val, yielding our usual method of adding up all elements in the list. Table 3.1 lists some common reductions. For each, we need to know not only the binary function to be applied, but the initial value for the accumulator variable.

Another way of thinking about reduction is repeated composition of a function, e.g., $0 + x_1 + x_2 + \cdots + x_n$ (if $f(a, x) = a + x$), or more generally:

$$a = f(f(f(a_0, x_0), x_1), x_2)\ldots$$

In the innermost function application, we see f applied to the initial value (a_0) and the first value of the vector x. The result of this function application, along with the second element of the vector, x_1, is the arguments to the next function application, and so forth.

3.1.6 Reading Questions

3.1 What would happen if we ran the code from the text, for the elementwise addition of two vector lists, on lists vec1 and vec2 of different lengths? Explain.

3.2 Can you think of a previous time in your life when you had to be careful to make sure two lists were the same length? Was it an example of a binary vector operation?

3.3 Write code that uses the zip function to take the elementwise multiplication of two lists, `vec1` and `vec2`, of the same length. For example, if `vec1` = `[2,4,5]` and `vec2` = `[4,3,1]`, you would compute `[8,12,5]`.

3.4 What predicate would you use if you wanted to filter a list, `vec`, of words, to return a subset of words that start with a capital letter?

3.5 Suppose you have a function, `isPrime(n)`, that returns `True` if n is a prime number, and `False` otherwise. Write code to filter a list of numbers, `vec`, to return only the prime numbers in `vec`. Your code should call the function `isPrime(n)`.

3.6 Can you think of a real-world example where you would want to filter a data set and keep only a particular subset? What predicate would you use? Please choose an example different from any in this section.

3.7 The Python built-in function `len(vec)` returns the number of items in the list `vec`. This is an example of a reduction. What is the initial value `init`, and what does `f` do? Note that `f` takes two parameters. In each iteration, `f(a,val)` has access to the previous value of the accumulator `a` and the current `item` in the list.

3.8 If we wanted to count the number of 5s in a list, is that a reduction? If so, what is the initial value and what is the function `f`?

3.9 Write a function that returns the sum of the squares of the numbers in a list, `vec`. Then describe the function `f` mathematically, by filling in the formula

```
f(a,x) = ...
```

3.10 Write a function that follows the reduction pattern to return the minimum value in a non-empty list of numbers.

3.11 If you wanted to return the index of the minimum value, would that be a reduction? If so, how would you describe the function `f`?

3.12 Describe a real-world scenario where the reduction pattern is needed, different from any examples in the section.

3.13 Can you think of an algorithm that generates a new list where, after the first element (or two), subsequent elements are determined from previous element(s)? Write the code for such an algorithm. Is the algorithm/code that you wrote an example of the accumulation pattern?

3.14 Consider the unary vector operation that applies `celsius2fahrenheit()`. What other unary functions might you envision using in a pattern like this? Describe a real-world application.

3.15 Suppose you wanted to perform negation, a unary *operator* as opposed to a function like `celsius2fahrenheit()`. What would that code look like?

3.1.7 Exercises

The following exercises allow us to practice with the vector patterns:

3.1.7.1 Accumulation

3.16 When the Fibonacci sequence is defined starting at index/subscript 0, we have $F_0 = 0$ and $F_1 = 1$. All other elements in the sequence are defined as:

$$F_n = F_{n-1} + F_{n-2}$$

Write a function

```
fibList(N)
```

that generates and returns a list, starting at 0, of the Fiboanacci sequence from F_0 up to and including F_N. So `fibList(6)` returns the seven element list `[0, 1, 1, 2, 3, 5, 8]`.

3.17 Write a function

```
generateSquares(N)
```

that generates a list with values from 1 to N, inclusive, and a second list with the corresponding values from the first list squared. The two lists are returned. So for `generateSquares(3)`, the results are the lists `[1, 2, 3]` and `[1, 4, 9]`.

3.1.7.2 Unary Vector Operations

3.18 Write a function

```
vectorUpper(data)
```

whose parameter, `data`, is a list of strings, and the function generates and returns a new list whose values are the strings from `data`, converted to upper case.

3.19 Modify your `vectorUpper()` function so that it does its work *in place*. So the original list, `data`, is mutated so that its elements are the uppercase strings. This new function has no return value.

3.20 Write a function

```
vectorLengths(data)
```

that generates and returns a new list, where each element in the created list contains the integer length of the corresponding element from `data`. So, for example,

```
vectorLengths(['this', 'is', 'a', 'short', 'list'])
```

yields the list [4, 2, 1, 5, 4].

3.21 Write a function

```
vectorThreshold(data, threshold)
```

that generates a Boolean vector whose elements correspond to `data` and whose True/False values indicated whether or not the element in `data` is greater than or equal to parameter `threshold`.

3.1.7.3   Binary Vector Operations

3.22 Write a function

```
vectorSum(x, y)
```

that adds, elementwise, the corresponding elements of `x` and `y`, and returns a new vector with these sums.

3.23 Write a function

```
vectorDifference(x, y)
```

that takes the difference, elementwise, between the corresponding elements of `x` and `y`, subtracting `y` elements from `x` elements, returning a new vector with these differences.

3.24 Write a function

```
vectorGreater(x, y)
```

that generates a Boolean vector whose corresponding elements indicate whether or not an element in `x` is greater than the corresponding element in `y`. So

```
vectorGreater([4, 7, 2], [5, 6, 2])
```

would yield [False, True, False].

3.1.7.4   Filtering

3.25 Write a function

```
dropFives(data)
```

that creates, accumulates, and returns a new list that has all the elements from the list, `data`, except for the items with integer value 5.

3.26 Recall the unary vector operation pattern, where we applied `celsius2fahrenheit()` to each temperature in a list of Celsius temperatures. This is a common pattern, so it would be helpful to write a function to abstract this pattern in a callable way. Write a function

```
apply_func(data)
```

that applies `celsius2fahrenheit()` to each item in `data`. So here, we are simply creating an interface to pass the list to be applied upon. Rather than appending into and returning a new list, your function should return nothing and should update the list `data` as you iterate through it, as we did in the unary vector pattern. This is analogous to how the built-in `sort()` function does not return a list, but rather modifies the list it is called upon.

3.27 Building on the previous exercise, write a function

```
apply_func(f,data)
```

that modifies `data` by applying a function `f` to each item in `data`. To get the same functionality as the previous function, we would invoke `apply_func(celsius2fahrenheit,data)`. This new function allows us to use the unary vector operation pattern in many different circumstances.

3.28 Modify the code provided above, to write a function

```
multConstant(vec,c)
```

that multiplies a vector, `vec`, by a constant, `c`, by changing `vec` in place rather than appending to a new list each time the loop iterates.

3.29 Write a general function

```
apply_binary(vec1,vec2,f)
```

that applies a binary function `f` to two lists `vec1` and `vec2` (of the same length) and returns the resulting list.

3.30 The next several exercises use the files `namespop.csv` and `birthrate. csv`, available on the book web page. The `birthrate` file has one birthrate per five years, so we will need to fill in the birthrate for each other year, y, based on the nearest year before y. Write a function

```
near_year(y,years)
```

that returns the first year in `years` that occurs after y (including the case of equality). If there is no such year, then just return the last item in `years`. For example, if `years` has one entry per decade, and y = 2008 you would return 2010.

3.31 Building on the above, write a function

```
get_birthrate(year,path)
```

that gets the approximate birthrate for the given `year`, using `near_year` and a dictionary you build from the `birthrate` file located at `path`.

3.32 Write a function

```
br_all_years(years,path)
```

that applies your function above to all years in the list years, returning the resulting list of birthrates.

3.33 We will now make use of the namespop file. Write a function

```
num_births(year, br_path, namespop_path)
```

that uses gets the approximate number of births in the given year, by first computing the approximate birthrate for that year (as above) and then applying a binary operation with the pop column of namespop. Assume that the birthrate represents the number of births per thousand in the population.

3.34 Building on the above, write a function

```
fraction_top(year, br_path, namespop_path)
```

that uses the function above, and the count variable in namespop to compute the fraction of all births in the given year that were due to the top name in the data.

3.2 Dictionaries

Dictionaries are also an important native Python structure and can also form a one-dimensional collection of data. Here, instead of using integer indices (like we do in lists) to access elements of the collection (vector), we can employ other kinds of mapping, so the *domain*[1] of the mapping might use strings, or a sparse set of integers, or other (immutable) data types and form the independent variable. By the basic property of a dictionary, the value in the domain (the key) maps to exactly one value, which is the dependent variable. Dictionaries will also be important in mapping from a set of variable names to vectors of data and will be explored in Sect. 3.4.1. This mapping wastes no space for mappings that do not exist, and the data structure is very efficient at the lookup/mapping operations—given a key, it is efficient to determine if the dictionary contains the key, and, if found, it is efficient to obtain the value associated with the key.

Lists, by contrast, for an N element collection, use integer indices from 0 to $N-1$ to "map" (index) to associated values. To find a particular value in a list, we may have to search the entire list. This can be problematic as our data collections grow in size.

When our goal is to maintain a collection comprising a data set, a dictionary can be a powerful tool. As we will explore in greater detail in Part II, a well constructed data set is a collection that consists of one (or more) *independent variables* and the value(s) of the independent variable determines the value(s) of one or more *dependent variables*. This is precisely what a dictionary provides. The set of keys are

---

[1] In a mapping, the set of values that we map *from* is called the *domain*. The set of values we map *to* is called the *range*.

Table 3.2 Example dictionary operations

Operation	Example
Create an empty dictionary	`gpaD = {}`
Add an entry to a dictionary; replaces the mapping if it exists	`gpaD['alice'] = 3.85`
Access an existing mapping; exception if mapping is not found	`gpaD['alice']`
Boolean check to see if mapping is present	`'chris' in gpaD`
Iteration over key-value pairs	`for key, value in gpaD.items():`
Delete an element based on key	`del gpaD['bob']`
Retrieve the list of keys in a dictionary	`gpaD.keys()`
Retrieve the list of values in a dictionary	`gpaD.values()`

the independent variable, and the mapped values are the dependent variable. Given a key (a value of the independent variable), we obtain the value of the dependent variable. Dictionary keys give us flexibility in the data type of the independent variable. They can be strings, or integers from a sparse space, or starting from a non-zero position, or even real-valued numbers, or tuples.

Consider the following illustrative example. Suppose a data set consists of student grade point averages (GPAs). Using lists, we could realize this data set as parallel lists, with one list for the student name, and another list for the corresponding GPAs. Here, the independent variable is the student name, and given a particular student, the dependent variable of the GPA may be found. But in a list implementation, given a particular independent variable value of a name, we have to perform a linear search through the name list to find the index of that student name, and then use that index to then access the correct GPA. By contrast, a dictionary that maps keys of student names to values of GPAs can much more efficiently find the value given the key.

By way of review, Table 3.2 briefly describes and gives examples for many of the Python operations and idioms, using this name to GPA example. Documentation for Python dictionaries may be found at [48] and [52].

Like `enumerate()`, the dictionary `items()` function is most useful when we need the key(s) where some condition on the values is satisfied, e.g., to find the student with the highest GPA, we would use a loop of the form `for student, gpa in gpaD.items()`.

Let us consider another example, derived from the US Social Security Administration `babynames` data introduced in Chap. 2. Suppose we are interested in a data set representing the top female baby name for each year from 1880 to 2018, which are the range of years available in the records provided. We want to map from an integer year in the range 1880 to 2018 inclusive to a string name. Suppose further that the data is currently provided as a list of strings `names`, where the index 0 element corresponds to the year 1880, and entries increase by year, and we can create a dictionary for the collection as follows:

```
topnames = {}
year = 1880
for name in names:
    topnames[year] = name
    year += 1
```

We note that the line `year += 1` is shorthand for `year = year + 1`. With the `topnames` dictionary in hand, accessing elements is straightforward. For instance, we can print the top name from years 2007 through 2012:

```
for year in range(2007,2013):
    print('year', year, topnames[year])
```

```
| year 2007 Emily
| year 2008 Emma
| year 2009 Isabella
| year 2010 Isabella
| year 2011 Sophia
| year 2012 Sophia
```

If we had two (or more) dependent variables for a data set, we have a couple of possible choices using dictionaries as our representation. First, we could employ a solution analogous to our parallel lists and have parallel dictionaries. In this example, we could have a dictionary for `topnames` and a parallel dictionary for the `count` of applications. Both would use year as a key. Alternatively, we could have a single dictionary, still with `year` as a key, but the *value* for each dictionary entry could be a tuple with two constituent parts—the name and the count.

3.2.1   Reading Questions

3.35 Does D = `{'a':'ant', 'c':'cat', 'a':'alligator'}` represent a legitimate dictionary? Explain.

3.36 The text describes the benefits of a dictionary over a list for student GPA data. Suppose our data set has a large number, n, of students. What is the worst case number of steps it would take to locate the index of a particular student if a list representation were used? Why is it more efficient to use a dictionary?

3.37 The text describes two ways that dictionaries could be used for data with more than one dependent variable, e.g. the data

```
id year   name      count
r0 2010   Isabella  22913
r1 2010   Sophia    20643
r2 2015   Emma      20455
r3 2015   Olivia    19691
```

Demonstrate how to store this data in a list of four lists (e.g., if you were typing it into Python), ignoring the header.

3.38 With reference to the question above, demonstrate how to store the data using three parallel dictionaries, where each has `id` as key.

3.39 With reference to the question above, demonstrate how to store the data using the alternative approach where the keys are `id`s and the values are lists representing the rest of each row.

3.40 Which of the two ways of storing the data using dictionaries, from the previous two problems, do you think is better and why?

3.41 With reference to the question above, how would you add a new row

```
r4 2018 Claire 18101
```

to the dictionary where values are lists? Remember that dictionaries do not have any `append()` method. Give your answer as a single line of Python code that references the dictionary name you assigned in the previous question.

3.2.2 Exercises

Warm-Up Exercises
3.42 Write a function

```
printDict(D)
```

that makes a list V of the values of D, and then, for each value, prints the value and a key that is associated to that value in D (one key per line). For example, if D is {1 : 'cat', 2:'dog', 3: 'dog', 4: 'fish'}, your function might print

```
'cat': 1
'dog': 2
'fish': 4
```

Note: because dictionaries are unordered, your test might come out a little different.

3.43 Write a function

```
makeDict(L)
```

that creates a dictionary whose keys are the items of L and whose values are the corresponding *indices* of L. You may assume L is a list without any duplicate items.

3.44 Write a function

```
frequencies(L)
```

that takes as a parameter a list of words and returns a dictionary with words in the list as keys and the number of times the word appears in the list as its value.

3.45 The `in` keyword tests if a key appears among the keys in a dictionary. Write a function

```
isValue(D,v)
```

that returns `True` if v is a value of some key:value pair in D, and `False` otherwise.

3.46 Write a function

```
filmography(L)
```

whose parameter L is a list of tuples of the form `(actor, movie)`, and creates and returns a dictionary where the keys are actor names and the values are lists of movies that actor has been in. For example, given

```
L = [('DiCaprio','Revenant'),
     ('Cumberbatch','Hobbit'),
     ('DiCaprio','Gatsby'),
     ('DiCaprio','Django')]
```

then you should return

```
{'DiCaprio':['Revenant','Gatsby','Django'],
 'Cumberbatch':['Hobbit']}
```

3.47 Write a function

```
intersection(A, B)
```

that takes as parameters two dictionaries A and B and returns a new dictionary D, where K is a key of D precisely when it is a key in both A and B. The value associated to K should be a *list* containing the value associated to K in A and the value associated to K in B. For example, passing in the parameters `{'A': 10, 'B': 20, 'C': 3}` and `{'C': 100, 'B': 40}` should return `{'C': [3, 100], 'B': [20, 40]}`

3.48 Write a function

```
organize(L)
```

that takes as parameter a list L of words and creates a dictionary D to store the words. The keys of D are the first letters of words in L, and the value associated to a key is the complete list of words in L that start with that letter. For example, `organize(["aardvark", "cat", "alligator", "dog"])` should return the dictionary `{"a": ["aardvark", "alligator"], "c": ["cat"], "d":["dog"]}`.

3.49 Write a function

```
distinct(L)
```

that takes as a parameter a list and returns a list with only the distinct elements in it. Do this using a dictionary. For example, if the list is `["c", "a", "a", "b", "c"]`, your function should return the list `["c", "a", "b"]`.

3.50 Write a function

```
negativeValues(D)
```

that uses a dictionary filter pattern to return a dictionary consisting of all key:value pairs in D where the value is a negative number. For example, if D = `{'a':5, 'b':-3, 'c': -2, 'd':0 }` you return `{'b':-3,'c':-2}`. You may assume all values appearing in D are integers.

3.51 Write a function

```
sumValues(D)
```

that uses a dictionary reduction pattern to return the sum of all values in D. For example, if

$$D = \{'a':5, \ 'b':3, \ 'c': \ 2, \ 'd':1\}$$

you return 11. You may assume all values in D are integers.

3.52 Write a function

```
sumValues2(D1,D2)
```

that uses a dictionary binary vector pattern to return a new dictionary D with the following properties:

```
- every key of `D1` or `D2` appears as a key in `D`,
- if a key appears in both `D1` and `D2` then its
    value in `D` is the sum of its values in `D1` and `D2`,
- otherwise the value associated to the key is its
    value in whichever of `D1` or `D2` it appears in.
```

For example, if D1 and D2 have values as follows:

$$D1 = \{'a':5, \ 'b':3, \ 'c': \ 2, \ 'd':1\}$$
$$D2 = \{'a':1, \ 'B':4, \ 'c': \ 6, \ 'e':1\}$$

then you return `{'a':6,'b':3,'B':4,'c':8,'d':1,'e':1}`.

Dictionaries and Data

3.53 Write a function

```
scanString(s)
```

whose parameter s is a string, and whose result is a dictionary that counts the occurrences of each letter in the string. Be sure your dictionary includes "zeros" for all the letters which do not appear in the string. Convert any uppercase letters in your string to lowercase so that your dictionary only counts letters as lowercase instances. Ignore any characters in the string which are not 'a..z'.

3.54 Write a function

```
readNames(path)
```

that reads from the file at location `path` and returns a dictionary mapping from years to names. The format of the file is a *CSV file*. The first line is a comma-separated set of string names for the columns contained in the file, in this case `year,name,count`. Subsequent lines have, on each line, comma-separated values for a data mapping, where the first value gives the independent variable value of the year, and the second and third values on the line give the string name and the integer count.

Your function will ignore the count value on each line and will accumulate a dictionary from years to names. Make sure your year keys are integers.

On the book web page, there are files `topfemale.csv` and `topmale.csv` that are formatted in this way. Using your `readNames()` function, assign to variable `female` the dictionary obtained from `topfemale.csv` and assign to variable `male` the dictionary obtained from `topmale.csv`.

3.55 Assuming the same file format as in the last question, enhance your answer from the last question, writing function:

```
readNamesCounts(path)
```

that reads from the file at location `path` and returns a dictionary mapping from years to 2-tuples consisting of a name and an integer count. Make sure your year keys are integers. Assign to variable `female` the dictionary obtained from `topfemale.csv` and assign to variable `male` the dictionary obtained from `topmale.csv`. Both files may be found on the book web page.

3.3   Python Features

Python is an object-oriented language. It emphasizes usability and ease of translating algorithmic solutions into code. As such, Python has a number of features helpful in working with one-dimensional representations of data as covered in this chapter.

3.3.1   Functions as Objects

In an object-oriented language, we can create objects, assign variables to the created object references, and can then invoke methods and pass objects as arguments to functions and other methods. These object references allow us to put variables in the "name space," global or local, of our running Python program, while the values that make up the object instance reside in the "value space." We use the names as proxy for the object itself.

In Python, the functions that we define are simply another type of object. When we define a function, the name after the `def` is placed in the current name space and the instructions that make up the function are packaged together as an object in the value space. Understanding these two points:

1. functions are objects that capture the set of steps, parameters, and return semantics of the function;
2. the name of a function is like the name of a list, string, or any other referenced object in Python, which can be passed as arguments, assigned to another name, and so forth,

conveys a great deal of power in how we use functions in our programs. In particular, we can refer to a function by its name (without calling it) and can pass it as an argument to a function, or even have a function return another function as its result.

In Sect. 3.1.2, we developed an example where we applied a unary function, `celsius2fahrenheit()`, to a list of Celsius temperatures. This operation is commonplace, and we might want to write a function to enable applying a general "to-be-specified" unary function to a "to-be-specified" list. Thus, we want both the *function* and the *list* to be the parameters of this desired `apply_func` function.

In Python, with functions as objects, this is straightforward, as shown in this example:

```python
def apply_func(func, data):
    newlist = []
    for datum in data:
        newdata = func(datum)
        newlist.append(newdata)
    return newlist
```

and invoking for this particular example:

```python
Ftemps = apply_func(celsius2fahrenheit, Ctemps)
```

Observe that Python needs to distinguish, when we use the name of a function, when we want to *invoke* the function versus when we are interested in using a *reference* to a function. Python uses the existence (or lack thereof) of the (and) to make this distinction. So in the example invocation above, `celsius2fahrenheit`, without the parentheses, is the name of the function, and no invocation occurs. Instead, the reference of the actual parameter (`celsius2fahrenheit`) is associated with the formal parameter from the `apply_func` function definition, `func`. So when `func(datum)` occurs in the body of `apply_func`, Python uses the parentheses to determine that it should invoke the function, `func`, because of the association of actual parameters (arguments) with the formal parameter, both referring to the same actual function object (`celsius2fahrenheit`).

The beauty of this is that we can now use the same `apply_func`, without modification, to apply other unary functions on a list. For instance,

```
abs_temps = apply_func(abs, Ctemps)
```

would apply the built-in unary function `abs()` to each element of `Ctemps`. If the function passed to `apply_func` was *not* a function with a single parameter or did not return a valid result to be accumulated, the invocation of `func`, when called, would reflect such an error. Python has another built-in function, `map()`, which has functionality close to our `apply_func`, and which we will explore more in the exercises. We will need `apply_func()` in Chap. 7.

3.3.2 Lambda Functions

A *lambda function*, in computer science in general, and in Python in particular, refers to an *expression* that defines a function. Lambdas are used to define functions that are "lightweight" and simple—they consist only of an expression and do not have multiple statements, and (generally) avoid side effects. For example, the function $f(x) = 2 * x$ is a simple function that, given x, multiplies it by 2. As we will soon see, this function can be defined via `f = lambda x: 2*x`. The name "lambda" (a Greek letter) comes from a subfield of computer science known as functional programming.

By having an expression that can define a function, we can use this expression wherever a callable function would be used in a program. This gives us an alternative to defining functions with the `def` mechanism and can make code simpler and easier to follow. Lambdas can also be combined with the "functions as first class objects" from Sect. 3.3.1, and the ideas on *comprehensions* are to be covered in Sect. 3.3.3.

For a comparison with the `def` mechanism, let us define the `celsius2 fahrenheit` function in our previous example in the conventional way:

```
def celsius2fahrenheit(Ctemp):
    Ftemp = (Ctemp * 9 / 5) + 32
    return Ftemp
```

Note that, in this simple function, there is no need to use the local variable `Ftemp`, and the function could be written more simply with a body that consisted of the single statement `return (Ctemp * 9 / 5) + 32`.

As an expression, a lambda takes the syntax:

$$\texttt{lambda}\ \textit{parameters}\ :\ \textit{expression}$$

So the keyword `lambda` followed by one or more (comma separated) parameter names, then the `:` (colon) character, then a valid Python expression uses the parameters and yields a result. A lambda *does not* include a `return` as part of its expression—whatever value obtained from the expression is the value obtained from invoking the function. So a lambda expression capturing the Celsius to Fahrenheit function would be:

```
lambda Ctemp: (Ctemp * 9 / 5) + 32
```

Since this expression yields a function object, we can use it anywhere a function object would be used in our program, so we could use the lambda in an invocation of our `apply_func()` as follows:

```
Ftemps = apply_func(lambda Ctemp: (Ctemp * 9 / 5) + 32, Ctemps)
```

Further, if we want to use the same lambda multiple times, we can give it a name by simply assigning it to a variable to use as its reference, and then use the name as we would any other function object:

```
c2f = lambda Ctemp: (Ctemp * 9 / 5) + 32
Ftemps = apply_func(c2f, Ctemps)
```

One of the advantages to using a lambda is that, for relatively simple expression-style functions, we can see and understand the meaning of the function. If we used a `def` for defining the function, the definition would be further from where we are using it, and could even be defined in a different file altogether. In cases like this, the `lambda` makes for more readable code. We will explore lambda functions with multiple parameters in the exercises.

A situation that arises because of the expression requirement of a lambda is that sometimes we need a lambda to "branch" and, based on a condition, return one value in one case and another value in another case. For example, to compute the absolute value of a number, x, we want to return x if $x \geq 0$ and to return $-x$ if $x < 0$. In traditional statement-based functions, using the `def` keyword, we accomplish this with an `if-else` statement. For example:

```
def abs_value(x):
    if x >= 0:
        return x
    else:
        return -x
```

The nature of the `if-else` is that it is a statement, and not a function. Fortunately, Python provides an *expression* that resembles an `if-else` but is an expression that yields a value upon evaluation. The Python syntax for the `if-else` *expression*:

$$\textit{true-expr} \text{ if } \textit{condition} \text{ else } \textit{false-expr}$$

So we begin the overall expression with *true-expr*, an expression that should be the result if the condition is true, we follow with keyword `if` and then the Boolean condition itself, then the keyword `else` and *false-expr*, the expression that should be the result if the condition is false. This is called a *ternary*-if. It is ternary because there are three operands. In our absolute value example, the `if-else` could be rewritten as `x if x >= 0 else -x`. The function becomes:

```
abs_value = lambda x: x if x >= 0 else -x
```

This matches how we might describe the conditional in a sentence. As a more complicated example, we take the result of the ternary-if and assign the resultant value to a variable:

```
s = "Found" if os.path.isfile(path) else "Not found"
```

First, the condition is evaluated to see if a file system location specified by `path`, in fact, exists and is a file. Based on the outcome, *true-expr* or the *false-expr* is evaluated and the value of the expression is either `"Found"` or `"Not found"`.

Say in our temperature example, we wanted a conversion where all negative temperatures mapped to 32.0, but all positive temperatures were converted in the normal way. The lambda, assigned to `c2f`, then becomes

```
c2f = lambda Ctemp: 32.0 if Ctemp < 0 else (Ctemp * 9 / 5) + 32
```

We will use lambda functions heavily when wrangling data in Chap. 9.

3.3.3   List Comprehensions

Because of the prevalence of the list processing patterns of accumulation, application of (unary and binary) functions on elements, and filtering, Python defines a "shortcut" syntax that gives a readable and concise way of writing these patterns that does not use a `for` loop. This shortcut is called a *list comprehension* and is more efficient for a number of reasons. The core idea is that, for all these patterns, we want to generate a *new* list that is somehow based on an existing *source* list or sequence. To make the facility general, the source can be an existing list or an iterable like a `range()` or the result of a `map()` or `zip()`, or even a file object. The general syntax for a list comprehension is

[*expression* for *element* in *iterable* if *conditional*]

where

- *iterable* is the source, be it an existing list, or some other iterable;
- *element* allows the writer to give one (or more[2]) variable name to be associated, on each iteration, with the values from the *iterable*;
- *expression* yields a value, in terms of *element*, that will be a single element of the newly generated list; the expression can invoke functions, perform arithmetic, and so forth;

---

[2]If the values in the iterable yield tuples, we can give separate multiple variable names to the tuple elements, just like we might in a `for` loop.

- *conditional* is a Boolean expression, in terms of *element* that allows a filtering decision on whether or not *expression* should be included in the resultant list.

Consider the following example. We have an existing source list of numbers, nums, and want to generate a new list that doubles the value of elements in nums, but only if the original number was non-negative—all negative values are filtered out of the list.

For this case,

- nums is the *iterable*,
- we can choose val as a variable name defining *element*,
- 2 * val is an *expression*, defining our objective for an individual element of the new list,
- val >= 0 is the *condition* to be used for inclusion in the resulting list.

Bringing this all together in a list comprehension, and assigning to doubles, we have

```
doubles = [ 2 * val for val in nums if val >= 0 ]
```

Demonstrating the example in context:

```
nums = [5, 3, -1, 2, 4, -3, 6]
doubles = [ 2 * val for val in nums if val >= 0 ]
print(doubles)

| [10, 6, 4, 8, 12]
```

To construct a list comprehension yourself, you should begin with the keywords (for, in, if) and work outward from there. Next, fill in the iterable, then the element, then the condition, and lastly the expression. Think of the process as asking yourself a sequence of questions, like "what am I iterating through?" and "which values do I want?" For the example above, this sequence of steps would look as follows.

```
[             for     in      if              ]
[             for     in nums if              ]
[             for val in nums if              ]
[             for val in nums if val >= 0 ]
[ 2 * val for val in nums if val >= 0 ]
```

As we have said, list comprehensions provide a concise shortcut for many common patterns that recur in our programming. We can always accomplish an equivalent result by using a traditional for loop with appropriate list accumulation. This is one example of using code that more closely resembles vector-style operations, and, as we explore tabular data, such vector type operations are natural when working with column vectors from a tabular data source. We will see many more examples in Chaps. 7 and 8.

If is important, when reading code that contains list comprehensions, to realize that, while we are using keywords like for and in that have meaning in other

syntactic constructs, this is NOT a `for` loop. It requires an expression before the
`for` keyword and is a construct that is an expression, yielding a value. Neither of
these points is true for a `for` loop.

We finish this section with the examples used in our various list patterns from
Sect. 3.1, but solved this time with list comprehensions.

Accumulationl

```
namelist = [ line.strip() for line in inputFile ]
```

Unary Operation

```
Ftemps = [celsius2fahrenheit(Ctemp) for Ctemp in Ctemps]
```

but since we allowed an expression yielding a value, we could avoid the function
call and write the expression explicitly:

```
Ftemps = [(Ctemp * 9 / 5) + 32 for Ctemp in Ctemps]
```

Binary Operation

```
sumvec = [num1 + num2 for num1, num2 in zip(vec1, vec2)]
```

Filtering

```
subset = [val for val in vec if val <= 1000]
```

3.3.4 Reading Questions

3.56 The text discusses how Python is an "object-oriented language." Give an
example of a Python object that is not a function. Hint: think about times when
you imported a module and then created an instance of an abstract data type taken
from that module.

3.57 Consider the code for `apply_func`. How could you modify this to create
a version for dictionaries, `apply_func(f,D)`, that modifies a dictionary in
place, to replace every value `val` by `f(val)`? For example, if `D =` `{'a':5,`
`'b':-3, 'c': -2, 'd':0}` and `f` is the function that squares its parameter,
then `apply_func(f,D)` modifies D to be `D =` `{'a':25, 'b':9, 'c':`
`4, 'd':0}`.

3.58 What happens in Python when you try to name a variable `def` What do you
think would happen if you tried to name a variable `lambda`?

3.59 Use a lambda expression to create a function `x3` that has one parameter, `x`,
and returns x^3.

3.60 Create a list `cubes` obtained via applying your function `x3` from above to
the list `[1,2,3,4]`. Then write a version where the lambda statement defining the

function is inside `apply_func`. Your answer should be two lines of code. You can assume you already have `apply_func` defined elsewhere in your program.

3.61 Write 1–2 sentences arguing that the use of lambda expressions can make code easier to read. Hint: think about large programs with thousands of lines of code, and where functions may be located within that program.

3.62 Convert the following block of code into a single line of code, following the model in the reading. Assume `x` is a variable that holds an integer value.

```
if x == 0:
    s = "Do not divide"
else:
    s = "Ok to divide"
```

3.63 When creating a list using an accumulator pattern, Python has to execute the `append()` command each time the loop iterates. As the list accumulator grows, it becomes more and more likely that the space allotted in computer memory is too small for the accumulator to fit. Thus, each time `append()` is executed, there is a small chance the computer will need to copy the list to a new place in memory. Contrast this situation with list comprehensions. When executing a block of code featuring a list comprehension, does Python already know how much space will be required?

3.64 If

```
Ctemps = [22,13,35],
vec = [673, 1132, 1390, -2002],
vec1 = [3, 4], and
vec2 = [-2,5]
```

what is computed by each of the blocks of code below? Figure this out by hand, rather than typing into Python.

```
Ftemps = [(Ctemp * 9 / 5) + 32 for Ctemp in Ctemps]
sumvec = [num1 + num2 for num1, num2 in zip(vec1, vec2)]
subset = [val for val in vec if val <= 1000]
```

3.65 What is the value of L at the end of the following? Work this out by hand rather than by using Python.

```
L = [2 * i for i in range(15) if (2 * i) % 6 != 0]
```

3.66 Rewrite the following with a list comprehension.

```
K = []
for i in range(2, 8):
    for j in range(i*2, 50, i):
        K.append(j)
print(K)
```

3.3.5   Exercises

3.3.5.1   Functions as Objects

The following set of questions involve the `sorted()` built-in Python function and the list of tuples:

```
students = [('bob', 'B', 19), ('chris', 'B', 18),
            ('chris', 'A', 18), ('alice', 'A', 19)]
```

Documentation for `sorted()` may be found at

- https://docs.python.org/3/library/functions.html#sorted

3.67 What is returned when `sorted()` is invoked on the `students` list? In this case, we are sorting a list whose elements are tuples. Describe how the algorithm used by `sorted()` works in the case that two elements are equal in their first element.

3.68 Look at the documentation for `sorted()`. If we wanted to sort the `students` list based on the second element of the tuple (the grade), we want a function that is used to extract the "key" from a list element (tuple) to retrieve that second element. Write a function

```
getGrade(record)
```

that, given a student record that consists of a 3-tuple, accesses and returns the middle element of the record/tuple.

3.69 Now use the `key` parameter to `sorted()` with your `getGrade` function to sort the `students` list. What do you get? When there are ties, do you get the elements involved in the tie in the same order as in the original list, or in different orders?

3.70 Write a function

```
whichIsBigger(f1,f2,n)
```

which computes `f1(n)` and `f2(n)` and returns whichever function (`f1` or `f2`) results in the larger value when an integer `n` is passed as a parameter. In case of a tie, return `f1`. Test with `f1(x)` defined as x^3 and `f2(x)` defined as $5x^2 + 4$ and using $n = 2$. Test again with $n = 100$.

3.3.5.2   Mapping Functions

Python has a built-in function `map()`

- https://docs.python.org/3/library/functions.html#map

whose parameters are a function object followed by one or more iterables. If there are only two arguments, the provided first argument function (which should take a single argument) is invoked over the elements of the second argument. This gives us a built-in way to apply a unary function elementwise over a list or other sequence.

If we pass, as arguments to map(), a function followed by *two* iterables, then the function is expected to have *two* parameters, and the function is called for each pair of elements coming from the first and second iterables. This gives us a built-in way to apply a binary function elementwise over two lists/vectors/sequences.

Note that, for efficiency reasons, map() actually returns an *iterator* that, like range(), can be used as the sequence in a for statement, and waits to compute elements until needed. If we want a *list* instead of an iterator, we must pass the iterator as the argument to the list() function.

3.71 Write a function

```
incrementAge(record)
```

that, given a 3-tuple, record, of a student, composed of a name, a grade, and an age, increments the age component and returns a tuple with the same name and age and the incremented age.

3.72 Use map() to apply the incrementAge() function to the students list.

3.73 Write a function

```
elementwiseMult(vec1,vec2)
```

that uses the zip function to compute and return the elementwise multiplication of two lists, vec1 and vec2, of the same length. For example, if vec1 = [2,4,5] and vec2 = [4,3,1], you would return [8,12,5].

3.74 Write a function

```
vector_mult(L1, L2)
```

that performs elementwise multiplication between the elements in L1 and L2. You may assume that L1 and L2 are the same length. You *must* use map() as the one line body of the vector_mult function, although you may write a helper function.

3.3.5.3 Lambda Functions

3.75 Write a lambda expression and assign it to the variable squared such that it squares the value of its single argument.

Afterward, for example, expression squared(5) should yield 25.

3.76 Write a lambda expression and assign it to the variable evensquared such that it squares the value of its single argument if the argument is even, and *cubes* the argument if it is odd.

Afterward, for example, expression `evensquared(5)` should yield `125`, while `evensquared(4)` should yield `16`.

Hint: We need an *expression* to handle an `if` type condition

3.77 Write a lambda function that takes two parameters and multiplies them together.

3.78 Write a lambda function that takes three parameters and multiplies them together.

3.79 Define a lambda function `sum_sqr` that takes two parameters, squares them, and adds the result.

3.80 Write a single line of code to define a function `inverse(x)` that returns $1/x$ as long as x is nonzero, and returns 0 if x is zero. You may assume x is an integer.

3.81 Write a lambda function named `num_digits(n)` that will return the number of digits in an integer n. For example, `num_digits(1230)` is 4. Hint: Using `str` and `len` might help.

3.82 Define a function that will compute the area of a circle of a given radius. Use a lambda expression and `math.pi`.

3.83 The formula for the volume of a barrel is as follows:

$$V = \pi h (2D^2 + d^2)/12$$

where *h*=height, *D*=middle radius, and *d*=top or bottom radius.

Write a lambda expression assigned to `vol` so that the resulting function, `vol(h,D,d)`, computes the volume of a barrel. Use `math.pi` in your computations.

3.84 Many people keep time using a 24 h clock (11 is 11am and 23 is 11pm, 0 is midnight). If it is currently 13 and you set your alarm to go off in 50 h, it will be 15 (3pm). Write a lambda function to solve the general version of the above problem using variables and functions. *Hint:* You will want to use modular arithmetic.

3.3.5.4   List Comprehensions

3.85 Write a function

```
square(data)
```

that takes a list of numbers named `data` and returns a new list containing the squares of each number in data. Use a list comprehension. For example, if the list `[4, 2, 5]` is assigned to a variable named `numbers`, then `square(numbers)` should return the list `[16, 4, 25]`.

3.86 Use a list comprehension to write a function

```
mult_by_2(data)
```

that multiplies each item in data by 2, returning the resulting list. For example, mult_by_2([1,2,3]) returns [2,4,6].

3.87 Use a list comprehension and a lambda function to define a function add_3 that starts with a list, data, and then adds 3 to every item in the list, returning the result. Do not use def in your solution.

3.88 Write a function

```
squares(n)
```

that returns a list containing the squares of the integers 1 through n. Use a list comprehension.

3.89 Define a lambda function sum_of_squares that starts with a list data and computes the sum of the squares of the numbers in data. Use a helper function in your solution. For example, sum_of_squares([2, 3, 4]) = 4 + 9 + 16.

3.90 Write a function

```
add2odd(data)
```

that modifies a list data by adding 2 to every number that is odd, and leaving even numbers unchanged. Use a list comprehension with a Boolean condition.

3.91 Write a function

```
get_tuples(data)
```

that uses a list comprehension to extract a list of tuples from data, where the first item of each tuple is the index, and the second item is the corresponding item in data. Hint: there is a very useful Python built-in function for this task, discussed in the reading. For example, if data = ["a", "b", "c"] you return [(0, 'a'), (1, 'b'), (2, 'c')].

3.92 Write a function

```
cube_root_odds(data)
```

that modifies a list data by taking the cube root of every odd number in data, and leaving even numbers unchanged. Use a list comprehension with a Boolean condition.

3.93 Write a function

```
remove(data, value)
```

that returns a new list that contains the same elements as the list data except for those that equal value. Use a list comprehension. Your function should remove all items equal to value. For example, remove([3, 1, 5, 3, 9], 3) should return the list [1, 5, 9].

3.94 Write a function

```
delete(data, index)
```

that returns a new list that contains the same elements as the list data except for the one at the given index. If the value of index is negative or exceeds the length of data, return a copy of the original list. For example, delete([3, 1, 5, 9], 2) should return the list [3, 1, 9]. Use a list comprehension. Hint: Your source list must contain all the needed information for the list comprehension to do its job. Explore the zip function and/or the enumerate function.

3.95 Write a function

```
word_lengths(s)
```

that, for each whitespace separated word in string s, determines the length of the word and returns a list of such lengths, *omitting occurences of the string: the*. For example, word_lengths("Computer Science is the best of the best") returns the list [8, 7, 2, 4, 2, 4], with lengths for 'Computer', 'Science', 'is', 'best', 'of' and 'best', and omitting the occurrence of 'the'. Use a list comprehension.

3.96 Write a function

```
sum_even(data)
```

that sums up all the even numbers in a list. Include helper functions in your solution. For example, sum_even([1,2,3,4]) is 6. Hint: You need a filter here.

3.97 Define a lambda function scalar_mult_row that uses a list comprehension to multiply a given list, row, by a given number, scalar. For example, if row = [1,2,3] and scalar = 5, then you return [5, 10, 15].

3.98 Define a lambda function scalar_mult that uses a list comprehension to multiply a given list of lists, mat, by a given number, scalar. For example, if mat = [[1,2,3],[4,5,6]] and scalar = 5, then you return [[5, 10, 15], [20, 25, 30]]. Hint: use the solution to the previous problem.

3.99 Use list comprehensions and lambda expressions to create a sequence of functions that combine to average two matrices. A complete solution will provide functions for each level of abstraction. Package your results together in one function using a def statement and include a test function.

3.4 Representing General Data Sets

Up to this point, parallel lists have formed our representation of data sets that extend beyond a single variable. Such a representation is, however, limited:

- As we add additional variables, we must continue to add additional lists, each with their own name.
- When we developed the idea of parallel lists, we noted the requirement of maintaining the *correspondence* between elements. If we add or delete items in *one* list, we must update *all* parallel lists, so that the "by-index" correspondence among the lists is maintained.
- We desire a *single name* to refer to all the data in the data set, both for the convenience of our own code, but also needed as we begin to use other packages involved in writing data systems clients.

In this section, we will develop three particular representations of general data sets using *nesting* of native Python data structures. These representations will continue to be needed even once we have learned more powerful structures, such as those provided in other library modules, because the tasks of data acquisition and normalization often require moving between packages and representations. We will often need to wrangle data into one of these in-memory data structures, especially in Chaps. 15, 16, and 21.

A two-dimensional data set is closely related to the tabular data model that will be covered in Part II of this book, but we identify here some top-level characteristics that help us understand our decisions in defining data set representations.

- Fundamentally, we want to represent data as rows and columns.
- A data set consists of a set of variables, with the independent variables determining the dependent variables.
- By *strong* convention, each variable *forms its own column* in the two-dimensional representation, and so *rows must form a data record/observation*, giving the values of the independent variables and the determined values of the dependent variables.
- Corollary to the last point, any individual column is "coherent," representing the same kind of thing, and so the data type of values in the column is generally the same—all integers, or all strings, etc.;[3]
- Each column has a *name*, representing the name of the variable represented in the column.
- The values in different columns can have different data types.

For the rest of this section, we will consistently use the same data set, adapted from the US Social Security Administration top baby names data. This adaptation is only interested in the *top name* (i.e., not a ranking), but unifies both female and male names, and includes information over the available data years of 1880 through 2018.

Table 3.3 details the variables, their types, and notes them as independent or dependent:

---

[3]Column values can also have some indication that a value in "missing" or not observed, but further discussion of this will defer to the tabular model chapters.

Table 3.3 Top baby names data set

Variable	Data type	Independent/dependent	Description
year	Integer	Independent	The year of a data record between 1880 and 2018
sex	String	Independent	Female versus Male, represented as the string "Female" or "Male"
name	String	Dependent	Top name applied for in the given year for the given sex
count	Integer	Dependent	Count of applicants for the top name in the given year for the given sex

Table 3.4 Top baby names data set (last three years)

Year	Sex	Name	Count
2018	Male	Liam	19837
2018	Female	Emma	18688
2017	Male	Liam	18798
2017	Female	Emma	19800
2016	Male	Noah	19117
2016	Female	Emma	19496

A variable whose values come from a (small) discrete set of possibilities is known as a *categorical* variable, and sex is such an example. There are two independent variables—year and sex, and so the *combination* of values of the year and sex are needed to uniquely identify a data record and, from which, the dependent variables of name and count are determined. In English, we would say "Given a year and a sex, the values of name and count are determined." With 138 years represented, and, for each year, values of sex of both female and male, there are a total of 276 data records in the data set.

Table 3.4 presents a subset of six rows of the data set giving the top name and count values for Male and Female applications for the years 2018 to 2016 in descending order.

Our goal is to understand general techniques to represent row- and column-based data sets like the above in our Python programs using the native constructs of lists and dictionaries, and using nesting of data structures to accomplish this. The next two subsections give two possible representations, but note that many variations are possible. These two are relatively intuitive and can be used to create tabular data sets in more powerful packages later.

For purposes of exposition below, we will use Python constant definitions of lists and dictionaries to show these representations. However, with 276 rows of data, such constant definition is not feasible for any real data set of even moderate size, and the data, ultimately, must be acquired from outside our program and the representations populated. One such external source might be a textual comma-separated value (CSV) file, where each line represents a row of data, and commas are used to separate columns. The first few lines are illustrated below:

```
year,sex,name,count
```

```
2018,Male,Liam,19837
2018,Female,Emma,18688
2017,Male,Liam,18798
2017,Female,Emma,19800
2016,Male,Noah,19117
2016,Female,Emma,19496
         . . .
```

We see that the data set has four columns and that the first line in the file represents the column names, rather than data values. We will discuss CSV files in much more depth in Chap. 6.

3.4.1   Dictionary of Lists

A logical first step in representing a data set as a single Python variable would be to extend our notion of parallel lists. As such, each data set variable is still represented by a single list, and we can gather these together into a single outer level structure. For the outer level structure, we can choose a dictionary, as we can use a string version of the variable name as the key, and map from that key to the list of data values for that variable/column. The result is a *dictionary of lists* (DoL).

The following constant definition of topnames illustrates this representation:

```
topnames = {'year': [2018, 2018, 2017, 2017, 2016, 2016],
            'sex': ['Male', 'Female', 'Male',
                    'Female', 'Male', 'Female'],
            'name': ['Liam', 'Emma', 'Liam', 'Emma',
                     'Noah', 'Emma'],
            'count': [19837, 18688, 18798, 19800, 19117, 19496]}
```

There are four dictionary entries, one for each column. The keys are strings 'year', 'sex', 'name', and 'count', appropriate for our four variable names. The values mapped to are lists containing the data for each column.

With this representation, it is easy to get data for an entire column:

```
print(topnames['year'])
```

```
| [2018, 2018, 2017, 2017, 2016, 2016]
```

But getting data for a row is more involved. Consider the row with index 3:

```
row_index = 3
print(topnames['year'][row_index], topnames['sex'][row_index],
    topnames['name'][row_index], topnames['count'][row_index])
```

```
| 2017 Female Emma 19800
```

Note how we accessed the individual row elements. We first write the access for the column, like topnames['count'], and then compose that with the square bracket list access, using the integer index to access a particular element within the list, `[row_index]`, for a complete element access of `topnames['count'][row_index]`.

So in this representation, we can efficiently access entire columns, as well as an element (or slice of elements) within a column, but accessing rows requires multiple independent expressions.

In building a representation from, say, a CSV file, we can simply extend our parallel list accumulation pattern. We start with a dictionary that maps from our column names to empty lists and then, as we iterate over the lines in the file, we

- parse the string `line` into (string) value fields,
- perform conversion for any non-string data types,
- append to each of the column lists.

This pattern is illustrated in the code below, omitting the details of opening the file and skipping over the line containing the column names in a header line:

```
topnames = {'year': [], 'sex': [], 'name': [], 'count': []}

for line in csvFile:
    fields = line.strip().split(',')               # four field list
    topnames['year'].append(int(fields[0]))        # convert to int
    topnames['sex'].append(fields[1])              # leave as string
    topnames['name'].append(fields[2])             # leave as string
    topnames['count'].append(int(fields[3]))       # convert to int
```

This representation has both advantages and disadvantages. Column operations are straightforward, and referring to a column by dictionary key name is intuitive. We can employ list comprehensions and our list patterns to easily perform unary and binary operations on columns and even create and add new columns to our data set. On the other hand, anything that needs to deal with a *row* as a logical unit is more difficult and requires additional expressions and complexity. Because of the use of columns as the "natural" form of the representation, we call the dictionary of lists representation *column-centric*. We will use this representation when we have easy access to the columns, especially when using XPath and web scraping, in Chaps. 16 and 22.

3.4.2 List of Lists

Often our data is presented to us as a collection of *rows*. Consider the CSV file prefix for `topnames` given above. Each line, after the header of column names, adds exactly one additional row to the data set. So collecting the data together into a full data set, we may choose to use the accumulation pattern, defining an empty list for the aggregate data set, and then adding a *list representing a row* as we accumulate the data row by row. So in this representation, our outer level structure is a list, and this list consists of a set of lists, one per row. We refer to this representation as a *list of lists* (LoL). The following constant definition of `topnames` illustrates this alternative representation:

```
topnames = [[2018, 'Male', 'Liam', 19837],
            [2018, 'Female', 'Emma', 18688],
            [2017, 'Male', 'Liam', 18798],
            [2017, 'Female', 'Emma', 19800],
            [2016, 'Male', 'Noah', 19117],
            [2016, 'Female', 'Emma', 19496]]
columns = ['year', 'sex', 'name', 'count']
```

So `topnames` is a list consisting of six row lists. Note that, in the example, in addition to the *data* of the data set, we also set up a variable, `columns`, to contain the names of the columns. This was not needed in the dictionary of lists representation, since the keys of the dictionary maintained this information.

In this representation, we can easily access and refer to a particular row, as long as we know its index:

```
row_index = 5
print(topnames[row_index])
```

```
| [2016, 'Female', 'Emma', 19496]
```

To access a column, we need to know its index. In this example, we may need to know that the count values are at index 3 within the list in each row, and to access all the values in a particular column, we must refer to all of the individual rows:

```
col_index = 3
print(topnames[0][col_index], topnames[1][col_index],
      topnames[2][col_index], topnames[3][col_index],
      topnames[4][col_index], topnames[5][col_index])
```

```
| 19837 18688 18798 19800 19117 19496
```

We could use a `for` loop or a list comprehension to gather column data together to use a unit.

Note the similarity with the dictionary of lists representation in writing expressions to access a single datum in the data set. We first access the *outer* structure, in this case the integer index for the row, like in `topnames[1]`. We then compose with the notation to access a list element for the column within the row, like `[col_index]`. The full accessor is `topnames[1][col_index]`. Because this representation is a list of lists, both the values inside the accessor brackets are integers.

If we use this representation and want to populate from a CSV source, we might use code like that illustrated below, where we create the list of column names and then accumulate the rows into an outer structure that is a list:

```
columns = csvFile.readline().strip().split(',')

topnames = []

for line in csvFile:
    fields = line.strip().split(',')    # four field list
    fields[0] = int(fields[0])          # convert year to int
    fields[3] = int(fields[3])          # convert count to int
    topnames.append(fields)             # single append of row
```

Some of the advantages of this representation include:

- when printed or displayed with each row list on a line by itself, the display of the data set more closely resembles a table or matrix organized in row and column form,
- when we want to process rows in turn, we merely have to iterate over the top level list,
- if the data set changes by adding additional observations, we can simply append such new rows to the existing list of row lists.

On the other hand, if a processing algorithm needs to work by columns, it is more difficult to process, since we can only access a single column field with each access to a row list. Further, the names of the columns are separate from structure holding the list of lists itself.

3.4.3   List of Dictionaries

If our data is presented to us as a collection of *rows*, another two-dimensional data structure to store it is a *list of dictionaries* (LoD). The outer level is a list, just like our last representation, and each element within this outer list again represents a single row. The difference is that the inner structure is a dictionary for the row of data. In each dictionary, each key:value pair represents one cell of data, mapping the column name to the value in the associated row. This avoids the need for a separate header list and also makes missing data easy to cope with (we simply omit the key:value pair for any cell where the data is missing). Handling missing data in our other representations is more difficult—we need to explicitly enter a value such as None in the relevant spot, or else the row–column structure would be lost.

For our topnames example, a LoD representation is

```
topnames = [
{'year':2018, 'sex':'Male', 'name':'Liam', 'count':19837},
{'year':2018, 'sex':'Female', 'name':'Emma', 'count':18688},
{'year':2017, 'sex':'Male', 'name':'Liam', 'count':18798},
{'year':2017, 'sex':'Female', 'name':'Emma', 'count':19800},
{'year':2016, 'sex':'Male', 'name':'Noah', 'count':19117},
{'year':2016, 'sex':'Female', 'name':'Emma', 'count':19496}
]
```

As with the LoL representation, we can easily access a row if we know its index. The data values for each row are the values of the associated dictionary, `topnames[row_index]`:

```
row_index = 5
print(topnames[row_index].values())
```

> | dict_values([2016, 'Female', 'Emma', 19496])

To access a column, we need to know the column name, e.g., count. We then simply print `rowD['count']` for each row dictionary:

```
for rowD in topnames:
    print(rowD['count'])
```

> | 19837
> | 18688
> | 18798
> | 19800
> | 19117
> | 19496

Lastly, to populate a list of dictionaries from a CSV source, we follow a procedure very similar to how we populated a list of lists, but now for each line of the file we define the associated row dictionary and append to the list of dictionaries. In the following, we assume we already know the column names, so that we do not need the first line in the CSV file (which contains the headers). However, we still need to read this line so that our loop begins with the first row of data.

```
csvFile.readline()                          # consume header line

topnames = []
for line in csvFile:
    rowD = {}
    fields = line.strip().split(',')    # four field list
    rowD['year'] = int(fields[0])       # convert year to int
    rowD['sex'] = fields[1]
    rowD['name'] = fields[2]
    rowD['count'] = int(fields[3])      # convert count to int
    topnames.append(rowD)
```

In the solution above, we assumed that we knew the four column names. However, this is not necessary in general. We could have instead used the header line of the CSV file, and an inner loop iterating over the columns, extracting the name of each, and creating the relevant dictionary entry. However, it would take more work (with if-statements) to convert years and counts to integers.

```
columns = csvFile.readline().strip().split(',')

topnames = []

for line in csvFile:
    rowD = {}
    for c in range(len(columns)):
        column_name = columns[c]
        rowD[column_name] = fields[c]
    topnames.append(rowD)
```

We will use both the LoL and LoD representations frequently, e.g., in Chaps. 6, 16, and 22.

Having covered lists of lists, lists of dictionaries, and dictionaries of lists, the reader may be wondering about *Dictionaries of Dictionaries* (DoDs). Anything that can be stored as a list can be stored as a dictionary, whose keys are the indices of the list and whose values are the associated items. So, the LoD representation can be turned into a DoD representation by mapping indices (row numbers) to the associated dictionaries (rows):

```
topnames = {
0:{'year':2018, 'sex':'Male', 'name':'Liam', 'count':19837},
1:{'year':2018, 'sex':'Female', 'name':'Emma', 'count':18688},
2:{'year':2017, 'sex':'Male', 'name':'Liam', 'count':18798},
...
}
```

However, there is no significant value in this representation relative to the LoD representation. We will see dictionaries of dictionaries (and even deeper nestings) in Chap. 15, where they are used to represent hierarchical data, but we will not use them to represent two-dimensional data.

3.4.4   Reading Questions

3.100 The reading discussed the importance of maintaining the correspondence between parallel lists. Type the data below into Microsoft Excel and ask Excel to sort the data by total_births.

year	name	total_births
2010	Isabella	22913
2010	Ava	15433
2015	Emma	20455
2015	Olivia	19691

Excel will ask if you want to "expand the selection"—try it with and without this option and keep track of what happens. How is this related to the correspondence between parallel lists?

Note: if you do not have Excel, you can do the same test using Google Spreadsheets.

3.101 Show how to represent the data above as a dictionary of lists, as in the reading.

3.102 Suppose D is a dictionary of lists representing a data set as in the reading. How can you determine the total number of columns? How can you determine the number of rows?

3.103 Suppose we realized there was an error in the dictionary of lists provided in the text, and in fact the top male name in 2016 was "Noah" with a count of 19251. Write a single line of Python code that updates the dictionary topnames to correct the wrong number it contains. Hint: since you have the data, you can just count to find which index needs to be changed.

3.104 With reference to the dictionary of lists topnames from the reading, show how to determine the row with the largest value for the variable count.

3.105 The reading introduced the term "categorical variable" and gave one example. Provide a different example here.

3.106 In the list of lists approach, what could go wrong if, instead of having a separate 1D list columns, we simply made the list columns the first list in the list of lists? Hint: Refer back to the "corollary" in the first part of the section.

3.107 Suppose we realized there was an error in the list of lists topnames provided in the reading, and in fact the top male name in 2016 was "Noah" with a count of 19251. Write a single line of Python code that corrects the list of lists topnames by modifying the relevant cell. Hint: You have the data so you can simply count to figure out which row to change.

3.108 With regard to whether it is efficient to access a row (vs. efficient to access a column), give an example of a data set where you might choose to represent it as a dictionary of lists. Give an example of a data set where you might choose to represent it as a list of lists. Justify your choices.

3.4.5 *Exercises*

Simple CSV Format
When, in the following questions, we say that a file is in a *simple csv format* that
means that:

- there is exactly one header line, with a comma-separated set of column-variable
  names,
- no column-variable name is the empty string,
- there are no embedded spaces in the column-variable names,
- all subsequent lines carry exactly one row of data,
- the fields in all rows are comma separated,
- all rows, including the header row, have exactly the same number of fields,
- fields do not include any leading nor trailing spaces, so all spaces, for instance,
  in a string, should be considered part of the field, and
- there are no quotes nor escape sequences, and so no string can have an
  "embedded" comma.

This means that there are no "comment" lines, nor blank lines, and so processing
can be cleanly performed by a `strip()` followed by a `split()` on commas.

3.109 Write a function

```
readBabynames2DoL(path)
```

that reads from the file at location `path` and returns a dictionary mapping from
column names to lists containing the data in those columns. The format of the file is
a *CSV file*. The first line is a comma-separated set of string names for the columns
contained in the file, in this case, `year,name,count`. Subsequent lines have, on
each line, comma-separated values for a data mapping, where the first value gives
the value of the year, and the second and third values on the line give the string name
and the integer count.

Your function will accumulate a dictionary mapping from `year` to the list of
years, from `name` to the list of names (in the same order), and from `count` to the list
of counts (in the same order). Make sure you convert `year` and `count` to integers.
On the book web page, there are files `topfemale.csv` and `topmale.csv` that
are formatted in this way. Note that, unlike the example in the reading, `sex` is not a
variable.

3.110 Write a function

```
readNamesCount2LoL(path)
```

that reads from the file at location `path` and returns a list of lists (where inner lists
are rows of the data set). The format of the file is a *CSV file*. The first line is a
comma-separated set of string names for the columns contained in the file, in this
case `year,name,count`. Subsequent lines have, on each line, comma-separated
values for a data mapping, where the first value gives the independent variable value

of the year, and the second and third values on the line give the string name and the integer count. Be sure to convert `year` and `count` to integers. Your function should return both a list of the column names, and the list of lists, in that order. On the book web page, there are files `topfemale.csv` and `topmale.csv` that are formatted in this way.

3.111 Exercises in the Patterns section introduced data sets `namespop.csv` and `birthrate.csv`. On the book web page, you will find `namesbirths.csv`, which is very like our `topnames` data set but with an extra column telling how many births there were in total in each year. Write a function

```
read_namesbirthsLoL(path)
```

that reads this file into a LoL.

3.112 Just like the previous exercise, write a function

```
read_namesbirthsDoL(path)
```

that reads `namesbirths.csv` into a DoL.

3.113 Building on the above, write code to create a new column named `percent` in the DoL data set, whose value is defined by:

$$count/births * 100$$

3.114 Building on the above, write code to create a new column named `percent` in the LoL data set, whose value is defined by:

$$count/births * 100$$

3.115 Building on the above, write a function

```
filterSexLoL(LoL, sex)
```

that filters a list of list representations so that only the given sex *passes* the filter. This function should create and return a brand new list that includes the subset rows. Then see if you can write an in-place version, where nothing is returned and work is done through mutation of the passed reference LoL.

3.116 Building on the above, write a function

```
filterSexDoL(DoL, sex)
```

that filters a dictionary of list representation so that only the given sex *passes* the filter. This function should create and return a brand new dictionary that includes the subset of the data. Then see if you can write an in-place version, where nothing is returned and work is done through mutation of the passed reference DoL.

3.117 Write a function

```
indexing(grid)
```

that converts a given list of lists, `grid` into a dictionary with `r` many entries, where `r` is the number of rows. The key for any given mapping should be the first entry in the corresponding row. The value should be a list containing the other entries. For example,

```
[[5,'cat','dog'],[6,'toy','the'],[8,'he','she']]
```

Should be converted to

```
{5:['cat','dog'],6:['toy','the'],8:['he','she']}
```

3.118 Write a function

```
filterLoL(data, threshold)
```

that creates and populates a list of lists for the `topnames` data set in `data` (as an LoL), such that rows with `count` (the index 3 element) are greater than or equal to `threshold`. Because columns have not changed, we need not pass in nor return the list of column names.

3.119 Write a function

```
filterDoL(data, threshold)
```

that creates and populates a dictionary of lists representation for the `topnames` data set in `data` (as a DoL) such that rows with `count` greater than or equal to `threshold` are included in the result.

3.120 Write a function

```
filterLoD(data, threshold)
```

that creates and populates a list of dictionaries for the `topnames` data set in `data` (as a LoD) such that rows with `count` (the index 3 element) are greater than or equal to `threshold`.

3.121 Write a function

```
convertDoL2LoL(D)
```

that converts from a dictionary of lists representation in D to the equivalent list of lists representation, returning both the list of column names, as given by the keys of D, as well as a list of lists storing the data in the values of D. Note that, because we do not know the order of mappings in a dictionary, the order of fields in the column names, and the order of fields in the rows of the data set may not be apparent. But as long as all rows as well as the list of column names are *consistent*, the conversion is valid. Hint: a list comprehension might come in handy.

3.122 Write a function

```
convertDoL2LoD(D)
```

that converts from a dictionary of lists representation in D to the equivalent list of dictionaries representation.

3.123 Write a function

```
convertLoL2DoL(columns, data)
```

that converts from a list of lists to a dictionary of lists D, which you return. Here, columns is a list of column names (which will become the keys in D), and data is a list of row lists.

3.124 Write a function

```
read_simplecsv_LoL(path)
```

that can read a file at path in simple csv format and construct a list or row lists representation, but without knowing the columns a priori. The function should return both the list of column names and the list of row lists data from the data set.

3.125 Write a function

```
read_simplecsv_DoL(path)
```

that can read a file at path in simple csv format and construct a dictionary of row lists representation, but without knowing the columns a priori. The function should return the resultant dictionary.

3.126 Write a function

```
read_simplecsv_LoD(path)
```

that can read a file at path in simple csv format and construct a LoD representation, but without knowing the columns a priori. The function should return the resultant LoD.

3.127 Modify your function above to a function

```
read_write_LoD(readpath, writepath)
```

that reads the CSV file at readpath (e.g., "topnames.csv"), creates an in-memory LoD representation, and then writes that to a JSON file at writepath (e.g., "topnamesLoD.json"). Optionally, do the same for the DoL and LoL representations.

3.128 Imagine an accounting routine used in a book shop. It works on a list with sublists, which look like this:

So using Python to construct this list of lists, we have

```
orders = [
["34587", "Learning Python, Mark Lutz", 4, 40.95],
["98762", "Programming Python, Mark Lutz", 5, 56.80],
["77226", "Head First Python, Paul Barry", 3,32.95],
["88112", "Processing in Python3, Bernd Klein", 3, 24.99]]
```

Order Number	Book Title and Author	Quantity	Price per Item
34587	Learning Python, Mark Lutz	4	40.95
98762	Programming Python, Mark Lutz	5	56.80
77226	Head First Python, Paul Barry	3	32.95
88112	Processing in Python3, Bernd Klein	3	24.99

Write a Python function,

```
costlist(orders)
```

that returns a list of 2-tuples. Each tuple consists of the order number and the product of the price per items and the quantity. You *must* use both lambda expressions and list comprehensions to do your work. The solution should only need two lines in the body of the function.

Chapter 4
Regular Expressions

Chapter Goals

Upon completion of this chapter, you should understand the following:

- The idea of finding matches in a target string by defining a regular expression composed of a combination of literal characters and metacharacters.
- The meaning of the most common regular expression metacharacters.
- The desired pattern to be matched by a given regular expression match string.
- How to create a regular expression match string given a structure pattern to be matched.

Upon completion of this chapter, you should be able to do the following:

- Use the Python data type of "raw" strings to define regular expression match strings.
- Use the Python `re` module to complete various types of matches/searches on target strings, including being able to iterate over a set of matches.
- Work with returned match objects to usefully obtain data from matches.

© Springer Nature Switzerland AG 2020
T. Bressoud, D. White, *Introduction to Data Systems*,
https://doi.org/10.1007/978-3-030-54371-6_4

4.1   Motivation

When we are dealing with data systems, the information we acquire may be in many forms (see Chap. 1). Depending on the data source and the providers' system design and their intent, the level of *structure* for the data may vary widely. In Part II of this textbook, we will learn about some of these varying degrees of structure as we explore three particular data models.

But there are times when we must deal with data and text that have less structure. We may encounter a single text file or a collection of text files that are "plain text," and not in a structured format. Even when we have a format with some higher-level structure, we may find embedded data that are simply strings of text. Needing to work with unstructured data can also occur when, in the context of data science and analytics, we *repurpose*, taking data originally generated and presented for one purpose, and use it to obtain data for analysis in a new purpose.

We wish to use and process these collections of unstructured text because they have useful data embedded within them. For instance, we might be interested in log files from web servers, network service logs, and operating system error reporting logs. Email messages are another example where we have text consisting of various lines of header information followed by text for the body of the message. We might also be interested in collections of user-generated text, in the form of tweets, blog posts, or messages in a social network feed. In all of these examples, the text is not in the structured format of some data model but contains useful data within. Our goal, then, is to be able to process the text by searching through and finding and then extracting the useful data.

This goal is often helped when the text has a regular pattern that we can exploit to help us in the processing. Consider the following Python text string named `target`:

```
target = """Alice says the price of oranges is $2.50
Bob says the price of apples is $1.75"""
```

Our goal might be to extract the data from `target` to get a two-row table with contents (Table 4.1) which we can represent as a Python list named `dataset`, with two tuples, each containing values for `source`, `fruit`, and `cost`.

```
[('Alice', 'oranges', 2.5), ('Bob', 'apples', 1.75)]
```

Table 4.1 Extracted data on fruit prices

Source	Fruit	Cost
Alice	Oranges	2.50
Bob	Apples	1.75

Examining `target`, we see a pattern. The text consists of two lines, the first one terminated with a newline.[1] In each, the line begins with the *source*, followed by the string ' `says the price of` ', then the name of the *fruit*, then the string ' `is $`', followed by the real-valued number giving the *cost*.

Given this particular pattern, we could solve this problem using Python strings and string methods like `find()` to search for the needed indices in the text. The code below demonstrates such a solution. In general, for each data item we wish to extract, we determine the beginning and ending indices and then use slicing to get the substring. In the case of cost, we also have to convert the obtained substring into a float.

```python
dataset = []

for line in target.split('\n'):
    begin = 0                        # source begin
    end = line.find("says") - 1      # source end
    source = line[begin:end]
    begin = line.find("of") + 3      # fruit begin
    end = line.find(" is")           # fruit ends
    fruit = line[begin:end]
    begin = line.find('$') + 1       # cost begin
    end = len(line)                  # cost end
    cost = eval(line[begin:end])
    item = (source, fruit, cost)     # compose tuple
    dataset.append(item)             # accumulate

print(dataset)
```

```
| [('Alice', 'oranges', 2.5), ('Bob', 'apples', 1.75)]
```

Regular Expression is the term given to a general and standard way of expressing the pattern matching and data extraction from text exemplified in the above example. Regular expressions define a *language agnostic*[2] way of specifying patterns to be matched in text, and libraries and modules exist to enable programming with regular expressions in almost any programming language. Regular expressions can also be used in tools of the operating system to provide flexible searching of files and resources at a command-line as well.

For the above `target` example, we could describe the desired pattern as:

- start with one or more alphanumeric characters defining the *source*,

---

[1] Recall that, in Python, a triple quoted string includes a newline in the string constant for each time the source advances a line.

[2] This means that the same specification for patterns and mechanisms for search/extraction are available in Python, Java, Perl, .NET, C#, JavaScript, and many other programming languages, as well as in command-line facilities like `grep` for searching files in the file system [30].

- continue with the spaces and intervening characters up through "of" and any following whitespace,
- the next set of alphanumeric characters up until the next whitespace defines the *fruit*,
- continue with the spaces and any intervening characters up through the "$" character, and
- the next set of digits, followed by a ".", then two more digits defines the string for *cost*.

The regular expression below:

```
(\w+).*of\s*(\w+).*\$(\d+\.\d{2})
```

is one way to specify this pattern. Do not worry about the apparent complexity of the expression; the material below will take you through the elements step by step.

Integrating the regular expression into Python code and using the regular expression to iterate over any/all matches in a target string are shown below:

```
import re

dataset = []
pattern = r'(\w+).*of\s*(\w+).*\$(\d+\.\d{2})'
for match in re.finditer(pattern, target):
    item = (match.group(1), match.group(2),
            float(match.group(3)))
    dataset.append(item)
```

After execution of this code, the dataset variable holds the list of tuples with the desired data, exactly as in the string-based solution. In addition to being more compact and efficient, this regular expression solution does not suffer from the lack of robustness present in our original solution.

In Sects. 4.2 and 4.3, we will learn about regular expressions in their general form. Then, in Sect. 4.4, we will cover the use of regular expressions through Python programming. The general form is essential, as the same type of logic for regular expressions will reappear in Chaps. 11 (in the guise of SQL queries), 16 (in the guise of XPath), and 22 (in the guise of web scraping). Regular expressions also appear in the projects, e.g., extracting data from a corpus of emails.

4.1.1   Reading Questions

4.1 Give an example of a time when you have extracted data from text, e.g., reading in a txt file and producing lists of data from it.

4.2 Give a real-world application or question where you would want to automate the process of extracting data from tweets, social media, blog posts, web servers, or emails.

4.3 What slice of

```
"Alice says the price of oranges is $2.50\nBob says the
price of apples is $1.75"
```

is "2.50"?

4.4 Go and find the online documentation for the module `re`, then write down three built-in functions and examples of how to call them. You should see several that are familiar from an introductory course in computer science.

4.5 What type of object does `re.finditer` return, and how is this related to the `for` loop in the code provided in the text?

4.6 Consider the code we used to extract names, fruit, and prices from a given sentence `target`. How robust is our solution in getting the correct column vectors if a `target`, following the same general pattern, had the following variations:

- One or more of the spaces in the target had multiple spaces, or a tab separating words.
- There was a period or other characters following the cost.
- The second line was terminated by a newline.
- There were additional lines consisting of just a newline (i.e., blank lines).

4.7 What changes would be required in the code to make it work correctly under each of these variations?

4.2   Terminology

The language of regular expressions uses a sequence of characters to define a *pattern*. When programming, this sequence of characters is typically assigned to a variable whose data type is a string. In Sect. 4.3, we will denote such patterns in code typeface but will refrain from showing delimiters, as these are specific to the programming language and how it defines string constants. In common usage, the term *regular expression* (or *regex* or *re*) is also used to refer to the pattern itself, but these uses overload the meaning of regular expression as a language and, in Python, the modules that are named `regex` and `re`, so we will use the term *pattern*.

We will call the text that we wish to search, and from which we wish to extract data, the *target*. A target might be a simple sequence of characters with no notion of line. It might be a single line, terminated by a newline, or it could be multiple lines. A target could be as small as an empty string or one character string, or it could be as large as the contents of a large file, or a string representing the text of a full novel. Sometimes, we evaluate a pattern against a *set* of targets.

The application of a pattern against a target is called a *search*. We also say that we are looking for one or more *match*es of the pattern in the target. When we do a search, the general expectation should be for *multiple* matches to result. It is also the general expectation that the target can contain many lines. When we want to restrict further, for instance, to only find a match at the beginning of a target, or to treat lines or line termination specially, we will use a combination of restrictions in our patters, the specification of flags that alter how the search algorithm operates, or specialized facilities of the host programming language.

To be useful, a pattern must allow both the exact matching of specific characters, called *literal characters*, or *literals*, and matching where a sequence of characters can be generalized and satisfy a portion of a match. We, in general, refer to the ability to match multiple sequences as a *wildcard* and characters in the pattern that specify wildcards are called *metacharacters*.

4.2.1   Reading Questions

4.8 A pattern can be as simple as a single character, and a target can be as simple as a string. Give an example of a target (at least 20 characters long) and a pattern with three matches in the target. Choose an example different from the book.

4.9 Open Google Chrome, navigate to any web page (e.g., www.google.com), then in the top menu click View -> Developer -> View Source. If you do not have Google Chrome, you can do this in many other web browsers (and you can use Google to learn how). It may simply work to navigate to

- view-source:https://www.google.com/.

Look at the text it shows you and try to spot data inside that text. Write 2–3 sentences explaining in high-level terms how regular expressions could be used to extract that data.

4.10 Give an example of a real-world scenario where you would be searching a target for some pattern of literals, and be clear in describing the target you would be searching, and what a *match* would entail.

4.11 Give a real-world scenario where you would be searching a target for a pattern made of both literals and metacharacters. Choose an example different from any in the book.

4.3   The Regular Expression Language

Consider the following target:

    There is gold in them there hills.

A regular expression pattern, e, would match all five instances of the literal character e in the target. The pattern em would match a single instance, and the pattern et would match zero instances. The regular expression search starts at the beginning of the target, proceeding until it matches a first character between the pattern and the target. The search then attempts to match the next character in the pattern with the next character in the target. This continues until either a full match is found (representing a *slice* of the target) or until the target can no longer match the pattern. In the latter case, the search, in effect, retreats to the last point of the match and may try alternative matching. If a match is found, it is accumulated into the collection of matches, and the search resumes from the *end* of the match, starting anew with the pattern.[3]

Much of regular expression patterns are rooted in matching *single characters*. All of the sections below cover details of matching a single character. Then, we define the regular expression concepts of repetition of sub-patterns and grouping sub-patterns in the sections that follow.

4.3.1   Literal Characters

As described above, we use specific literal characters in a pattern where we want to force an exact match in the target. This, at a minimum, gives us the ability to do a substring search. For the significant majority of characters, we just use the character itself in the pattern, like we did with e, m, and t in the example above. However, some characters in a pattern are "special" and are used for wildcards and for affecting the search algorithm. Examples include [,], ^, ., and a number of others described below. To be able to get these as literal characters we "escape" them by preceding them with a backslash. So, to match the final period in the target above, our pattern would be \..

"Special" characters that require a \ escape to obtain the literal character in a regular expression pattern:

   [\ ^ $. | ? * + ( ) { }

In a manner similar to strings in most programming languages, we can also use escape sequences to be able to specify literal characters for whitespace, line \n for newline, \r for carriage return, and \t for tab. We have already seen \$ used in our fruit example, to match the literal dollar sign.

If you are given a pattern and want to understand what it matches, or if you need to design a pattern, it is often helpful to start with the literal characters in the pattern, since they must find a match in the target exactly. Understanding a given pattern or designing a pattern then proceeds "outward," to the left and right, of the literal sequences. This is how we designed the regular expression for the fruit example:

---

[3]The description of the regular expression searching algorithm, as given here, is a simplification. There is much more detail pertaining the backtracking used in the algorithm and to the *greediness* of trying to match sub-pattern repetitions.

by first recognizing that we would need to match of, \$, and \ . exactly, and then building in the other match patterns \w, \s, \d, which we will discuss shortly.

4.3.2 Single Character Wildcard Matching

The real power of regular expressions comes from our ability to define wildcards that are used to match more than one thing. In the following subsections, we define and give brief examples of wildcards that *match exactly one character*, and so consume one character in the pattern and one character in the target, for each successful match.

4.3.2.1 Dot

The period (.) or "dot" metacharacter matches *any* character except for the line termination character (newline on most systems). The dot wildcard is often used in conjunction with literal characters to amplify its effect. In the target,

    There is gold in them there hills.

the pattern . would match the first character, the second character, the third character, and so forth, yielding a match for every character in the target. The pattern . h would match the initial Th, the th in them and in there, and the h matching the space and initial h of hills. If we had a pattern of

    There is gold in them there hills.
    hooray!

then the pattern . h, since it does not match a line termination character, would *not* match the newline at the end of the first line, followed by the h of hooray.

4.3.2.2 Predefined Single Character Sets

The extremes for single character matching are a literal character, which only matches exactly one thing, and the . metacharacter, which matches anything. We often want to match, in a wildcard fashion, one character from a specific *set* of characters. For instance, we might want to match a particular digit character with the set of any digit characters. Or we may want to match a character with either a space or a tab or a newline, the set of whitespace characters. Or we may be interested in the so-called word characters, those that come from the set of upper- and lowercase letters, plus the set of digits, plus underscore. These are often called the alphanumerics.

Table 4.2 Single character set patterns

Pattern	Set description	Pattern for negation
\d	Any decimal digit from 0 to 9	\D
\w	Alphanumerics: upper- and lowercase letters, digits, and underscore	\W
\s	Any whitespace character, including space, tab, newline, or carriage return	\S

The regular expression language has predefined many of these useful sets, plus their logical negation (match any character *not* in the set). The most common of these are given in Table 4.2.

For example, in the target from the beginning of this chapter:

```
Alice says the price of oranges is $2.50
Bob says the price of apples is $1.75
```

the pattern \d\d would match a two digit sequence, and it would match 50 and 75 but would not match the 2 or the 1. The pattern \s\w\w\w\s would match any sequence of three alphanumerics surrounded with whitespace on either side, so it would match the two instances of the and would also match the newline followed by Bob followed by a space.

4.3.2.3   User-Defined Single Character Sets

Sometimes, we want the limited wildcard ability of matching a single character from a set, but none of the predefined sets give us exactly what we want. Regular expressions allow you to define your own set, simply by delimiting the set of individual characters within square brackets ([and]). So the meaning of the pattern [abc] means to match exactly one character if it is a or b or c. Note that this is different than the pattern abc, which is matching three character in succession.

For example, the pattern [AB]\w\w matches Ali and Bob in the apples/oranges example, consuming one character for the A or B and then matching two word characters. Note that the pattern [AB]\w\w\w would only match Alic, since, on the sequence of characters starting with Bob, there would be a non-word character that could not be matched for the final \w in the pattern.

These single character sets could be used to define (again) a predefined set. For instance, the pattern [0123456789] matches exactly one digit, just like \d. Regular expressions also allow us to specify a contiguous set of characters in a user-defined set by using a - in between the first and last of the contiguous characters. So, [0-9] means the same as the earlier example. We can combine ranges and individual characters in a set, so the pattern [A-F2-7p] matches a single character from the set of the capital letters A through F, the digits 2 through 7, and the lowercase letter p.

If a ^ character immediately follows the opening bracket in a user-defined set, the meaning is the *negation* of the set. So, the pattern [^0-9] has the same meaning as \D.

4.3.3   Repetition

The literals, the predefined sets, and the user-defined sets give us great power in matching any single character. To that, we want to add the power to specify *repetitions* of a single character (literal or wildcard), and eventually, repetition of sequences of characters. In regular expressions, when we want to specify *how many* of something we want to match, we append a repetition suffix. The repetition suffix always applies to *whatever immediately precedes it*. Table 4.3 gives the various forms of regular expression repetition suffixes.

The type of repetition is known as a *quantifier* for the preceding element.

Suppose we want to write a single regular expression pattern that matches the entirety of all three of the following targets:

- abc
- abbbc
- ac

We see, in common, that each of these targets starts with an a and ends with a c, but there may be zero, one, or three b characters in between. The pattern ab*c matches all three targets in their entirety. By contrast, the pattern ab+c matches the first and second targets in their entirety and does not match the third target at all, because the + repetition suffix requires at least one instance of the preceding b.

The {n} pattern lets us specify an exact number of repetitions of an element. For instance, we could use the pattern \d{2} to match instances of two digits together. Consider again our original example target:

```
Alice says the price of oranges is $2.50
Bob says the price of apples is $1.75
```

The \d{2} pattern would match the 50 and the 75. Now, consider the following target:

Table 4.3 Repetition patterns

Suffix	Repetition interpretation
*	Match zero or more repetitions of the preceding element
+	Match one or more repetitions of the preceding element
?	Match zero or one repetition of the preceding element—i.e., an element is *optional*
{n}	Match exactly *n* repetitions of the preceding element
{n,m}	Match at least *n* and no more than *m* of the preceding element

```
$7.75
$12.50
$101.20
```

The same pattern, in this case, would match the three expected, 75, 50, and 20. It would also match the 12 in the second line and the 10 starting the number in the third line.

4.3.4   Disjunction

Disjunction (also known as alternation) allows a pattern to match either one sequence *or* another sequence. The single character set matching discussed earlier already gives us an *or*-based match between single characters, so this metacharacter is an or between *sequences*. Regular expressions use the vertical bar (|) to denote disjunction.

Suppose our target is There is gold in them there hills. The pattern [Tt]he | is would match one instance of The, two instances of the, and one instance of is. Note how the sequences on each side of the disjunction can, themselves, include other types of character wildcards. The [and] have higher precedence than the | .

4.3.5   Boundaries/Anchors

Sometimes, we want patterns that go beyond matching a character and instead are interested in matching relative to a particular *position* in the target. Positions at the beginning and end of the target are *anchors*; we use ^ in a pattern to anchor at the position beginning the target and we use $ in a pattern to anchor at the position of the end of the target. This is why we cannot use $ in a pattern to match the literal dollar sign but must use \$ instead. Note that these metacharacters, while setting a position from which the rest of the pattern is matched, do not consume any characters in the target.

Say, using an example from email header lines, that we have targets of

```
Subject: Party Tonight
```

and

```
Re: Subject: Party Tonight is cancelled
```

If we wanted to find a match in the first target, but not the second, we can use anchors for both the beginning and the end of the line, so that the pattern

`^Subject: Party Tonight$` ensures that we do not get a false positive match, like the pattern `Subject: Party Tonight` would.

The normal behavior of `^` and `$` in patterns is to anchor at the beginning and end of the entire target. So, if the target were multi-line, like

```
To: alice@school.edu
Subject: Party Tonight
Date: 14 June 2019
```

there would be no match for the pattern `^Subject: Party Tonight$`, because `Subject` does not follow the start of the target and `Tonight` does not precede the end of the target. The use of *flags*, discussed below, allows us to affect this behavior.

In a similar position-centric way, we can make a pattern align to the *start* or *end* of a word. We use `\b` in a pattern to match either start or end word boundaries.

Say that you wanted to find words that begin with h but want to exclude punctuation in the case that the word is at the end of a sentence with a following `.` or `?`, or the word was at the end of the target. Finding word boundaries allows a solution like the pattern `\bh\w*\b`, which would find the match `hills` in the target `There is gold in them there hills`. The pattern finds a word boundary, which could occur after whitespace but could also occur at the beginning of the target, then the literal character h followed by zero or more word characters followed by another boundary.

4.3.6   Grouping

When writing arithmetic expressions in a programming language, we use parenthesis to group terms together and to enforce precedence. In regular expressions, we have additional reasons to want to group things together. One of the goals of regular expressions is to allow us to *extract* data when we find a match. But our pattern, and the character sequence in the target corresponding to the match, most often encompasses *more* than the part we want to extract. The syntax of regular expressions allows us to specify both of these types of grouping:

1. Grouping to define a subsequence of characters to allow extraction of data from a target.
2. Grouping to enforce association and precedence. Once we have the ability to group to enforce association, we can apply our repetition suffixes against sequences of characters instead of individual characters. For example, the pattern `abc*` is wrong if we want to match `abcabc`.

We call a grouping of type 1 a *capture group*, and we use `(`, and `)` in our patterns to delimit a capture group. For example, the pattern `(abc)*` matches `abcabc` (and the substring `abc`) and also captures matches so that we can use them elsewhere in our program. Note that the pattern [abc]* matches any string made of a, b, and/or c, e.g., bab, so square brackets should not be used for grouping.

Recall the pattern used in our fruit example, and note where we place the beginning and ending parenthesis in the first part, replicated below:

```
(\w+).*of\s*(\w+)
```

Recall our example target:

```
Alice says the price of oranges is $2.50
Bob says the price of apples is $1.75
```

We want to *capture* the initial name (Alice or Bob) to be able to extract it for the source field, and the first (\w+) matches and captures the name, ending with the first whitespace. We next use .* to allow any characters between the name and the literal of, and then allow as many spaces as needed after of, before moving to our second capture group (\w+). This captures the sequence of word characters that follow after the literal of and the intervening whitespace, to be able to extract for the fruit field. So, for both source and fruit, we simply enclose the desired subsequence in the pattern in (and).

To *reference* a particular capture group, we use integers and number them. Group number 0 refers to the *entire* match of the pattern against the target, and subsequent groups are numbered in the order they appear in the pattern, based on their opening parenthesis. So, in this example, group 1 is the source field, and group 2 is the fruit field. In the full solution for the fruit example, group 3 is the cost field, captured by (\d+.\d{2}), coming directly after the literal dollar sign.

If we want to group together a subsequence or wish to enforce precedence, but the goal is not extraction of the subsequence, we should not use capture groups. Otherwise the result becomes confusing, and additional care must be taken with capture group numbers, particularly when there is a mixture of sequences to be captured and sequences that are grouped only for association or precedence.

The regular expression notation for a *non-capturing group* begins with (?: in the pattern and ends with). So, if we wanted to associate (but not capture) abc as a subsequence that is repeated one or more times, we write the pattern (?:abc)+. The repetition of + applies to the immediately preceding element, which is a non-capture group. This pattern matches abcabc, for example.

4.3.7 Flags

We have alluded to the possibility that sometimes we wish to *alter* the default behavior of the regular expression search algorithm. In the programming or command environment in which we are using regular expressions, a mechanism for specifying "flags" is typically introduced that allows us to control certain behaviors. We only describe three flags here. The particular method of the setting of flags, like specification of patterns and targets, is specific to the programming language or environment where regular expressions are being used. We will see the Python interface/mechanism in Sect. 4.4.

4.3.7.1 Case Insensitive

In normal behavior, case matters and the upper- and lowercase versions of a letter are not considered the same. If we set the case insensitive flag (often denoted with lower- or uppercase i), we can match targets regardless of case. For example, with this flag set, the pattern `latex` would match both `LaTeX` and `LATEX`.

4.3.7.2 Multi-line

When the target consists of multiple lines, with associated line termination following each line, we often wish to anchor a pattern with the beginning and ending of a line. If we set the multi-line flag (often denoted with lower- or uppercase m), the interpretation of ˆ and $ changes to anchor at beginning and ending of each line and not at the beginning and ending of the entire target. For example, suppose our data consisted of a collection of usernames and passwords, one pair per line, numbered and formatted as follows:

```
1. datasystems_user mysecret,
2. joe.<3.amy passw0rd,
3. k1nd.guy 8.cant.wait,
4. ...
```

To capture the line number, username, and password (assuming no spaces in the username/password), we could use the pattern ˆ`(\d*)\.\s(.*)\s(.*),`$, which captures the first set of digits (up to the first period), then captures the username (up to the first space), then the password (up to the comma). The multi-line flag allows the anchors ˆ and $ to be used on each line. Note that anchoring is important, to avoid accidentally capturing `3.` from `joe.<3.amy` or the `8.` from `8.cant.wait`.

4.3.7.3 Single Line

A target may or may not include line termination characters like newline, but we sometimes want to treat it as a single string and, in particular, want the dot (`.`) to match newline characters as well as any other character. In this case, we set the single line flag (often denoted s or S) to get this "dot matches newline" behavior. For example, if we are extracting text from a corpus of emails, we want to be able to match critical information regardless of whether different individuals use more line-breaks than others.

4.3.8 Reading Questions

4.12 In the target `Hahahahahaha`, how many matches are there for the pattern `hah` and why?

4.13 In the target `aaaaaaaa`, how many matches are there for the pattern `aaa` and why?

4.14 What pattern would you use to find all occurrences of the string "+" in a string like "3+3 = 2+4 = 1+5 = 0+6"? Explain.

4.15 True or False: the pattern `[David]` matches exactly the occurrences of `David` as a substring of the target. Justify your answer.

4.16 True or False: the pattern `\D\d\D\d` matches `R2D2`. Justify your answer.

4.17 True or False: the pattern `[A-Z2-3^0-90-9]` matches `R2D2`. Justify your answer.

4.18 True or False: the pattern `[0-90]` matches `57`. Justify your answer.

4.19 True or False: the pattern `*ing` matches `running`. Justify your answer.

4.20 True or False: the pattern `R|D2` matches `R2D2` twice. Justify your answer.

4.21 Why does `\bh\w*\b` match `hills`? Give two other words it might match.

4.22 In finding `hills`, we used the pattern `\b\b\w*\b`. What happens if you used a pattern that looked for whitespace followed by h and word characters, followed by whitespace? (i.e., if you used a pattern like `\sh\w*\s`). Without `\b`, how would you try and solve the problem?

4.23 Make an analogy between capturing multiple groups when making a regular expression and functions that return multiple values.

4.24 Explain why the pattern `\d{2}` against the target `$101.20` matches `10` and `20`.

4.25 Why precisely does

```
(\w+).*of\s*(\w+).*\$(\d+\.\d{2})
```

capture `Alice`, `oranges`, and `2.50` in the textbook's example?

4.26 Give an example where you might want to use a non-capturing group.

4.27 Write a regular expression that captures the counts in sentences like `The most popular baby name in 2018 was Claire with 18101 births`.

4.28 What does the pattern `s.` match in the target `There is gold in them there hills.`?

4.29 What about the pattern `. . e`? Why?

4.30 What does the pattern `\d\D` match in the oranges/apples example target? (Do not forget the newline character.)

4.3.9   *Exercises*

4.31 Write a regular expression pattern that matches one or more *literals plus some wild card*.

4.32 Outside of the USA, it is common to write dates in the form `year.month.day`, e.g., `2020.01.05` for `January 05, 2020`. Write a regular expression that matches a date written in this form.

4.33 An IP address (using the standard IPv4) looks like `x.y.z.w` (where each of `x,y,z,w` is numbers in the range of 0-255). For example, the smallest is `0.0.0.0`, the largest is `255.255.255.255`, and a generic one is `174.17.6.152`. If you Google "What is my IP address?" you can see yours. Write a regular expression that matches an IPv4 address, e.g., `"Joe's IP address is 174.17.6.152, let's keep data on all his searches!"`.
Your regular expression should capture each of the four numbers separately. You can assume only numbers `0-255` are used, e.g., no one will try to sneak in an invalid IP address like `825.603.971.243`. However, notice that an IP address does not end with a period.

4.34 Write a regular expression to capture an IP address within a larger document, e.g.,   `"Joe's IP address is 174.17.6.152, let's keep data on all his searches!"`. Your regular expression should capture each of the four numbers separately.

4.35 Write a regular expression to capture only correct IPv4 addresses, e.g., it should match `174.17.6.152` but not `825.603.971.243`. Capture each of the four numbers separately.

4.36 Hexadecimal numbers are numbers that have 16 symbols, and the symbols use the digits 0 through 9 and then the first six alphabetic letters of A through F. The letters may be in either upper- or lowercase, or some mix. Numbers are written the same way as integers—with one or more of the 16 symbols in a sequence, with no intervening spaces or punctuation.
   Write a regular expression pattern that, if applied to a target, completely matches all hexadecimal numbers within that target.

4.37 It is considered likely that the world will run out of IPv4 addresses, because there are so many devices using the internet. In order to have more IP addresses, a new protocol, IPv6, represents addresses as eight groups of four hexadecimal digits, separated by colons. For example, the smallest

is   0000:0000:0000:0000:0000:0000:0000:0000,   the   largest   is
ffff:ffff:ffff:ffff:ffff:ffff:ffff:ffff, and a generic one
is 2003:0de8:81a3:0000:9f78:6a5c:1376:7b34. Write a regular
expression to match IPv6 addresses.

4.38 Write a regular expression pattern that will capture all triple double quote
comments in a `target` comprised of Python source code. Your pattern should
handle cases where the comment is "empty" (i.e., there are no characters in between
the beginning and ending triple double quotes), as well as handle cases where the
triple double quotes reside on a single line and on multiple lines. You are capturing
the comments, not the quotes themselves.

4.39 Write a regular expression pattern that matches all `import` statements in
a target comprised of Python source code. An import will be contained within a
single line, although there could be leading whitespace (if indented in a block) and
there could also be trailing whitespace. Your pattern should handle both `import`
`module` and `import  module  as  alias` variations of import statements.

4.40 Write a regular expression pattern that matches any `def` lines, including the
sequence of characters making up the parameters, even if they go over multiple
lines.

4.41 Write a regular expression pattern that matches function calls and argument
list, capturing both the function name and the contents (argument list) *within* the
invocation parenthesis.

4.42 Write a regular expression pattern that matches 9 digit telephone numbers.
The numbers will be formatted as `dddsdddsdddd` or `(ddd)sdddsdddd`, where
`d` is a digit and `s` is a separator, e.g., `555 555.5555` or `(555)  555-5555`.
The separators are space, period, or dash. Capture all three portions of the phone
number—the area code, the three digit exchange prefix, and the four digit line
number.

4.43 In blogs and posts, users are identified by the @ symbol immediately followed
by the user identifier or handle name. Typically, a user identifier must start with a
letter and then be followed by a sequence of letters, digits, plus the underscore (_)
character.

   Write a regular expression pattern that, if applied to a target, completely matches
all identifier users. The pattern should include a single capture group that captures
the sequence of characters of the identifier but does not include the @ sign.

4.44 Find instances of the word `class` or `klass`, but either could occur within a
larger word (like `classless` or `inclass`), and these should *not* be matched.

4.45 Suppose the target you are processing is a JSON file or string containing a
representation of a dictionary. The goal is to match and capture the *keys* in the
dictionary, assuming the keys are all of the `string` data type. In JSON, strings are
always delimited by double quotes, with everything in between the double quotes
part of the string. You may assume that the key string does not include any backslash

escapes but can include all other keyboard characters (letters, digits, punctuation, and spaces). The key always appears before a : and inside the dictionary enclosing characters of { and }.

4.4   Python Programming with Regular Expressions

4.4.1   Specifying Patterns

Within most programming languages, we will specify our regular expression patterns as strings. We must take care, however, because strings themselves have their own notions of escape sequences. For instance, if we want the regex pattern \t to match a tab character, we might write `pattern = "\t"`. But Python will translate that string into the single character string with the actual tab character, instead of the desired backslash character followed by the t character. To get the correct regular expression using Python strings, we would have to write `pattern = "\\t"`. It would get immensely confusing and error prone if we had to, by hand, translate every backslash in a regular expression pattern into two backslashes. It would also make patterns even more difficult to read and to relate to the language-agnostic specification covered above.

Fortunately, Python allows the definition of "raw" strings, which are strings that do *not* perform the standard interpretation of escape sequences. It is easy to get raw strings in Python string constants by simply prefixing the start delimiter (single quote, double quote, and triple quote) with an `r`. So, to get the pattern for the h-word example assigned to a Python string variable, we can code

```
pattern = r"\bh\w*\b"
```

For consistency and clarity, in the examples that follow, we will use the Python variable `pattern` for our regular expression patterns, although you could use any variable name you would like.

4.4.2   The re Module Interface

We assume our target is in the form of a Python string. It may have come from a string constant, from a file by performing a `read()` method invocation, or from a network connection as we will see in future chapters. But regardless of how the data was acquired, we have a string whose contents form our target. In most of our examples, we will use the Python variable `target` as representing this string.

Given a pattern and a target, and possibly flags changing the search behavior, we want to search the target for the pattern.

The Python module named `re` provides regular expression processing sufficient for our needs, so we will `import re` to use the module. When we perform a regular expression search of a pattern against a target, we may be interested in

- the full string matched,
- one or more of the strings for the capture groups extracted during the match, and
- the indices giving the location of the match within the target.

If the search results in more than one match, we may only be interested in the first match, but, more likely, we may want to iterate over all of the found matches. This is particularly important when we are using regular expressions to search, find, and extract data for a data set embedded in a target.

The `re` module defines a *Match Object* that, for any successful match, gives an object-oriented way of retrieving the full match string, any/all of the strings for the capture groups to allow data extraction, as well as indices giving positions of the match and capture groups within the target.

Table 4.4 summarizes the most common functions in the `re` module for performing regular expression search. All of these functions take the three arguments of a pattern, a target, and, optionally, flags. As in any summary, there are many additional

Table 4.4 Common functions in the `re` module

Function in `re`	Type of Return	Description
`search()`	Single Match Object	Apply pattern against target and return *first* successful match as a Match Object, or None if no match was found
`match()`	Single Match Object	Apply pattern against target, but only at the *beginning* of the target. A potential match that starts after the beginning of the target string will not give a successful match
`fullmatch()`	Single Match Object	Apply pattern against target, but a match is successful only if the matched pattern spans the *entire* target string
`findall()`	List of string matches	Apply pattern against target and, for each match, include an element in the list for the strings involved in the match. If there are no capture groups, this is a list of the full matched strings. If there is a single capture group, this is a list of the strings for that capture group. If there are multiple capture groups, this is a list of tuples, where each tuple consists of the captured strings
`finditer()`	Iterator for set of Match Objects	Apply pattern against target and build an iterator that allows a subsequent `for` loop to be able to access a Match Object for each successful match. This is the most general and powerful (and efficient) of the search functions

details to be found to use a library or module, and the Python documentation is available for the `re` module [58]. There is also a "HOW TO" that gives useful advice and examples [59].

We now provide several examples, starting by defining `pattern` and `text` and then carrying out a simple search to find the *first* match.

```
import re
pattern = r'is'
text = "Isabel, is this right?"
print(re.search(pattern,text))
```

    | <re.Match object; span=(8, 10), match='is'>

The function `findall()` finds all *string* matches but not Match objects.

```
print(re.findall(pattern,text))
```

    | ['is', 'is']

To find Match objects, which include both the match and the location where it occurs, we use `finditer()`.

```
for m in re.finditer(pattern,text):
    print(m)
```

    | <re.Match object; span=(8, 10), match='is'>
    | <re.Match object; span=(13, 15), match='is'>

Finally, we illustrate an example that uses flags. The following conducts a search that will match regardless of case:

```
print(re.findall(pattern,text,flags=re.IGNORECASE))
```

    | ['Is', 'is', 'is']

We could extract Match objects using `finditer(pattern,text,flags)`.

The Match Object provides a set of methods that, given a successful match, allow a Python program to obtain the various possible pieces of information in the successful match. We will see this functionality again in Chap. 13, where we obtain an object similar to a Match Object containing data matching a specified query to a database.

We use the notation `Match.` to denote that what follows is a method of a Match Object. In a program, you would have variable with a reference to the object, and that variable name would replace the `Match` shown in Table 4.5.

To illustrate, we recall the Python code given for our first example, of extracting fruit prices.

```
target = """Alice says the price of oranges is $2.50
Bob says the price of apples is $1.75"""

pattern = r'(\w+).*of\s*(\w+).*\$(\d+\.\d{2})'
```

Table 4.5 Methods of Match Objects

Method	Description
Match.group(i)	Return the string for the ith capture group. For example, if the match object is m, then m.group(2) gives the string of capture group 2. Match objects define an accessor shorthand, so m[2] does the same thing. Group 0 is the full match
Match.start(i)	Return the index in target of the start of the string of the ith capture group
Match.end(i)	Return the index in target of the end of the string of the ith capture group

```
for match in re.finditer(pattern, target):
    item = (match.group(1), match.group(2),
            float(match.group(3)))
    print(item)

  | ('Alice', 'oranges', 2.5)
  | ('Bob', 'apples', 1.75)
```

We could also extract *where* the matches occur. The following prints the end index of the first group (name) and the start index of the second group (fruit), as well as the number of characters between them. This could be useful to determine irregularities in the text data.

```
for m in re.finditer(pattern, target):
    print("Gap indices:",m.end(1),m.start(2))
    print("Difference:",m.start(2)-m.end(1))

  | Gap indices: 5 24
  | Difference: 19
  | Gap indices: 44 63
  | Difference: 19
```

As a final example, we combine the methods of this section to write a general matching function that find as many matches as possible in the given text, passing flags as appropriate to modify the means of matching. The function returns a list with an entry for each found match. For a given match, the entry is a list of tuples, which we call a "match list." The first tuple is the string and starting offset (index within text) for the complete match (i.e., group 0). The remaining tuples are the string and index for each of the capture groups found within the match. Hence, the function returns a list of lists of tuples.

This function follows an accumulation pattern, finds and iterates over an iterable of Match objects, and uses a list comprehension to streamline the matching. The range of this list comprehension is justified by the observation that the first capture group is in match.group(1) rather than match.group(0). We use the Python insert() function to insert the whole match into the beginning of our list.

```
def getMatches(pattern, text, flags=0):
    matches = []

    for match in re.finditer(pattern,text,flags):

        # build a list for all the capture groups
        matchlist = [(match.group(i), match.start(i))
                  for i in range(1,len(match.groups())+1)]

        # add the whole match to the beginning of the list
        matchlist.insert(0,(match.group(0), match.start(0)))
        matches.append(matchlist)

    return matches
```

We illustrate with a sample invocation, using LoLoT as shorthand for "list of lists of tuples."

```
target = """Alice says the price of oranges is $2.50
Bob says the price of apples is $1.75"""
pattern = r'(\w+).*of\s*(\w+).*\$(\d+\.\d{2})'
LoLoT = getMatches(pattern,target)
for matchlist in LoLoT:
    for tup in matchlist:
        print(tup)
```

```
| ('Alice says the price of oranges is $2.50', 0)
| ('Alice', 0)
| ('oranges', 24)
| ('2.50', 36)
| ('Bob says the price of apples is $1.75', 41)
| ('Bob', 41)
| ('apples', 63)
| ('1.75', 74)
```

4.4.3   Reading Questions

4.46 By way of analogy with what the book discussed regarding the tab character, how do you think you would match newline characters?

4.47 Run the following code, and then explain why only the second invocation of match results in a match.

```
import re
print(re.match(r'From\s+', 'Fromage is delicious'))
print(re.match(r'From\s+', 'From Russia with Love'))
```

4.48 Run the following code, and then explain why the first print results in None.

```
import re
print(re.match('age', 'Fromage is delicious'))
print(re.search('age', 'Fromage is delicious'))
```

4.49 Run the following code, and then explain the difference between what findall and finditer return.

```
import re
print(re.findall('\d{2}', 'An amazing thing happened
    Thursday May 13 at 17:11:10 1997'))
print(re.finditer('\d{2}', 'An amazing thing happened
    Thursday May 13 at 17:11:10 1997'))
```

Hint: it might help to run the following code:

```
iter = re.finditer('\d{2}', 'An amazing thing
    happened Thursday May 13 at 17:11:10 1997')
for match in iter:
    print(match)
```

4.50 With the following variable declaration:

```
text = "Say, is this for real?"
```

Give an example of a pattern that would have a match using re.match(), and give an example of a pattern where re.match() would return None, but re.search() would find a match.

4.51 Come up with an example that calls a function in the module re but uses a flag.

4.52 Explain (or give code) how to use finditer, Match, and Match.start to find all indices where a given char appears in a string s

4.4.4   Exercises

4.53 Write a function

```
findG(text)
```

that finds all instances of either 'grey' or 'gray' in text by setting the variable pattern to the appropriate regular expression pattern and using the re module.

4.54 Write a function

```
findCapWords(s)
```

that, using regular expressions and the `re` module, finds all the instances of capitalized words in `s`. The function should return a list of *strings* of the found words, and an empty list if no matches were found.

4.55 Write a function

```
myFind(literal, text)
```

that uses regular expressions to perform the same functionality as the python string `find` function. Namely, the function should search `text` for the string specified by `literal` and, if found, return the *index* at which the first match occurred. If not found, the function should return −1.

4.56 Suppose valid 9 digit phone numbers are formatted as dddsdddsdddd or (ddd)sdddsdddd, where d is a digit and s is a separator, where the separators are space, period, or dash.

Write a function

```
findPhoneNumbers(s)
```

that, using regular expressions and the `re` module, finds all occurrences of valid phone numbers in a string but omits other number-containing subsequences. The return should be a *list*, and each element of the list should be a *list* where the elements are the full matched string, the area code, the exchange prefix, and the line. The last three should be integers.

Match and No Match

Sometimes, regular expression patterns are expected to match certain patterns and, equally important, to *not* match other patterns. In the next couple exercises, the goal is to find a single pattern that matches all targets in the first list and matches none of the targets in a second list. In this setup, each possible target is a string without any line termination.

4.57 Suppose we want to distinguish between the two columns below:

```
Match      | No Match
-------------------
pit        | pt
spot       | Pot
spate      | peat
slap two   | part
respite    | top it
```

Write and assign to a Python variable `pattern` a regular expression that completely matches all the targets in the first column and does *not* match any of the targets in the second column. When you test your function for errors, be sure to consider potential errors both with your pattern and with your `findMatches()` function.

4.58 Suppose we want to distinguish between the two columns below:

```
Match        | No Match
-------------------
1. xyz       | 2.xyz
2.  xyz      | a. xyz
3.    xyz    | 3. x yz
```

Write and assign to a Python variable `pattern` a regular expression that completely matches all the targets in the first column and does **not** match any of the targets in the second column.

4.59 Write a function

```
findMatches(pattern, targetlist)
```

that, for each of the targets in `targetlist`, performs a regular expression search to determine if the `pattern` has a full match with the target. The function should return a list of targets in `targetlist` that the `pattern` matched fully. For example, if `targetlist = ['ab', 'abc', 'abcd', 'abcdef']` and `pattern = 'abc?'` then the function would return `['abc', 'abcd']`.

4.60 On the book web page, there is a file named `rime.txt`, whose contents are a small prefix of the *Rime of the Ancyent Marinere*, which we read into a string variable, `data`.

Write a function

```
findFirstAncient(s)
```

that, using regular expressions and the `re` module, finds the first instance of "ancient" in the given string, `s`. The sought "ancient" can be spelled either with an `'i'` or a `'y'` (i.e., either `'ancient'` or `'ancyent'`), and the search should be insensitive to case. The function should return a tuple with the start and end indices (inclusive) of the match, or None if not found.

4.61 Extend your solution from the previous problem by writing a function

```
findAncients(s)
```

that, using regular expressions and the `re` module, finds **all** instances of "ancient" in the passed string s, spelled either with an `'i'` or a `'y'` (i.e., either `'ancient'` or `'ancyent'`), and insensitive to case. The function should return a list of tuples with the start and end indices (inclusive) of the match, or an empty list if no instances were found.

4.62 Write a function using regular expressions

```
uniqueWord(literal, text)
```

that checks if `literal` occurs *only once* in a text and returns `False` if it does not occur or if it is not the only instance of the word in the text. If `literal` does only occur once, the function returns the start and end indices of the word as a tuple. Your function should be insensitive to case.

Part II
Data Systems: The Data Models

Chapter 5
Data Systems Models

Chapter Goals

Upon completion of this chapter, you should understand the following:

- The purpose of data models for describing the forms of data in data systems.
- The three elements of data models—their structure, operations, and constraints, and what they contribute to the model.
- The difference between procedural and declarative operations.
- For each of the three models, tabular, relational, and hierarchical, the basics of their structure, operations, and constraints.
- The idea of constraints being automatically checked and enforced versus adopted by convention and not enforced.

5.1 Data Model Framework

Part II of this book uses data models as a framework for discussing the most common *forms* of data we may encounter as we acquire data and build the client side of our data systems. The current chapter provides an overview and describes the objectives of using a data model framework. The chapter also provides an overview

© Springer Nature Switzerland AG 2020
T. Bressoud, D. White, *Introduction to Data Systems*,
https://doi.org/10.1007/978-3-030-54371-6_5

of the three models described in this book and, at that level, gives us the opportunity to compare and contrast them from one another.

The term data model is somewhat overloaded, being used in two distinct but very related ways. In software engineering, a data model may be used to formalize a logical abstraction for the objects and relationships in a *particular application domain*. So, part of the process for solving a software engineering problem would be to define a data model for the objects, their specific relationships, and the flow of data in that application domain. In data systems, while still being a logical abstraction, it is about the forms of the data themselves, not about instantiations in business or organizational processes.

Definition 5.1 A *data model* is a notation for describing data or information, generally consisting of three parts—the *structure* of the data, the *operations* to obtain and update data within the model, and the *constraints* that, within the model, limit the data in various ways [67].

It is important to understand that a model deals with the *logical* aspects of the data form, allowing us to abstract away from the details of a particular package, module, or interface. Understanding the forms of data from a model perspective allows us, in fact, to understand the same concept even as we move *between* languages (like Python or R) or packages within the language.

5.1.1 Structure

In our programs, such as those written in Python, C, or R, we use *data structures*, explicit or through objects, to define the memory storage for the data processed by our program. In a comparable way, a database management system (DBMS) might define how blocks of disk and file storage are used to store and organize units of data and auxiliary structures of information to allow efficient access. These lower-level organizational structures and choices are known as the *physical data model*. There can often be multiple possible implementation choices in physical data model to represent the same logical structure.

Consider Sect. 3.4 of this textbook. Our goal was to use Python to represent a data set, logically structured as a two-dimensional set of column variables and rows mapping from instances of dependent variables to values of independent variables, i.e., a table. We explored two different representations as our data structures, a dictionary of column lists and a list of row lists, giving us alternatives for the same logical structure of a table.

The structure part of a data model is higher-level than these data structures and physical data models. It entails the logical structure and is sometimes referred to as a *conceptual model*. In the three models covered in this book, we will see two major types of conceptual model structures:

- row–column-based tables and
- general tree structures.

These structures involve logical organization of the data and may impose limitations on the data and its mapping to the structure of the data model. In this way, the structure itself imposes constraints on the data organization.

5.1.2 Operations

We understand the *operations* of a data model once we are contextualized within the structure of the model. For instance, if our structure is a tree, the operations must somehow deal with traversals of the tree; if our structure is a table, the operations must somehow allow us to deal with rows and columns. The operations are what allow us to obtain specific parts of the data as needed by our algorithms. When we are using operations to *access* data, we call that a *query* of the data. When we are using operations to modify data, be it to perform an update of one or more existing data elements, or to add information into a data set, or to delete information from a data set, we call that a *mutation* of the data.

In general, operations can be categorized as *procedural* or *declarative*. Procedural operations are those that define methods, functions, and/or operators composed of a sequence of steps determining what data to acquire and specifying exactly how to acquire it. The function readNames of Sect. 2.3.1 is an example. By their nature, procedural operations are closely tied to a particular programming language and package, although we will try and convey general concepts before building examples that are concrete in code and package. By contrast, declarative operations specify the *what* but not the *how*. They are a logical language of their own, where the language "declares" *what* data is desired. But the user does not specify the order, the traversal strategy, or the iterative steps for the *how*. The underlying system is tasked with, given a declaration of what data is needed, figuring out how to break that into an efficient set of steps to actually obtain the data. The distinction between procedural and declarative operations is essential in this book and will arise again in later chapters.

When operations are procedural, it requires programming to use the operations to achieve some objective. On the other hand, when operations are declarative, a user must understand the declarative language for requesting data but is relieved from the task of writing a program to be able to get the job done. Thus, data models that support declarative operations can be used by a broader set of users.

In the relational model, the Structured Query Language (SQL) is a declarative language giving users a way to specify what data is desired. In the hierarchical model, and when the format of the data is XML, then XPath is a declarative language giving users a way to specify what data is desired. In many ways, both are similar to the language of regular expressions from Sects. 4.2 and 4.3—they form their own language for specifying what data is desired and are ultimately programming language agnostic. We can create the same regular expression, SQL statement, or XPath query statement in many *different* programming languages, and the declaration has the same meaning and obtains the same data.

By contrast, the operations used in the tabular model, while we can generalize them by talking about, for instance, creating a subset of columns, or filtering to a subset of rows based on some condition, are procedural. We compose them into steps in functions and programs and illustrate them with their use in a particular programming language and package, Python and `pandas` in this case. Working with JSON in the hierarchical model is similar, and the procedural steps are unadorned Python loops and list and dictionary access. In XML, while we can use XPath to obtain data, we often integrate that with procedural steps, and so XML operations are often a hybrid of declarative and procedural.

5.1.3   Constraints

The third leg of a data model is its constraints, which specify limitations on valid values for data. Constraints are an important tool for avoiding errors and ensuring data is both consistent and remains in a form amenable to analysis. It can help in understanding constraints to segment them into structural constraints, element constraints, and relationship constraints.

Structural constraints might involve:

- In a tabular model, ensure that a column represents a single thing and is not some composite of multiple things.
- In a tabular model, ensure that a row uniquely represents a relational mapping.
- In a hierarchical model, ensure that some child node does not reference back to some ancestor, essentially violating the tree nature of the model.

Some example constraints on individual elements might involve:

- Limit values in a column of data to be one of a limited number of specific values, as would be desired for categorical variables.
- Limit values in a column to be a specific data type, like only real-valued numbers, or only strings,
- Specify where data elements are allowed to be missing.
- Require that certain values not be missing.
- Require that certain values be unique relative to some set.

Some example constraints on relationships between structure components might involve:

- Require that all values in a column must also appear as a value or key in the column of another table.
- In a tree, require that if the type or tag of a node in the tree is X, then its children *must* be of type/tag Y.

Some of the most significant differences between data models involve the mechanisms (or lack thereof) for specifying, checking, and enforcing constraints. Some models, like the relational model, strongly check and enforce constraints. Some models are more moderate, allowing checking but not enforcing, while others have no mechanisms for checking nor enforcement at all, and rely on providers and clients "playing nicely" and following structure and constraints by convention. But the flip side to the benefit of checking and enforcement is that, when constraints are strict, flexibility may be lost. For instance, the relational model does not adapt well to changing requirements after a database has been designed, while for the tabular model, and often for the hierarchical model, changes to a data set and its structure can be made with (dangerous) ease.

5.1.4 Reading Questions

5.1 Why is it a good thing that data models are defined using a logical abstraction?

5.2 Give an example of a time you have *accessed* data in Python and of a time you have *mutated* data. You can either describe in words or can give a snippet of code.

5.3 Give an example of a *procedural mutation* of a data set. Hint: think of a time you learned or programmed an algorithm to change the data.

5.4 Give an example of a *declarative mutation* of a data set. Hint: think of times you changed the data without writing your own function. Your answer does not have to be Python-specific.

5.5 List the three types of constraints mentioned in the reading and an example of each (different from the examples in the reading).

5.2 Tabular Model Overview

In data science and data analytics, the term *tidy data* is used to refer to data that have a recognized structure and data organization, and whose data satisfy accepted constraints in addition to these structural/organizational ones. The *tabular model* described in this book is a model-generalization of these same principles. So, the tabular model is not a new invention but rather accepts the principles of tidy data and describes the result in a way that fits with the model framework employed in this book. So, if data is truly "tidy," then it is in a standard (*normal*) form and conforms to the tabular data model. While we will use both terms in this book, we will use "tabular model" when we wish to emphasize the data model nature and use the term

Table 5.1 Top baby names
data set (last three years)

Year	Sex	Name	Count
2018	Male	Liam	19837
2018	Female	Emma	18688
2017	Male	Liam	18798
2017	Female	Emma	19800
2016	Male	Noah	19117
2016	Female	Emma	19496

"tidy data" when we are focusing on recognizing and transforming data that is not in the standard form.[1]

As used in this book, the tabular data model will focus on primarily single table data sets with row and column dimensions. In this model, we are interested in in-memory representations of tabular data with sufficient power in their operations to allow extracting and updating subsets of rows and columns, working with columns as vectors, and transforming from one tabular representation to another. In Python, we use the `pandas` module to provide such facility.

5.2.1   Structure

The basis of the tabular data model is the structure of a *table*, where a table is *much more* than just data in a two-dimensional form. The *same* data can be represented in many *different* two-dimensional forms. To conform to the tabular/tidy data model, a table must have *columns* that define the *variables* of the data set. The variables of a table in a data set can be independent variables or they can be dependent variables. A *row* of a table, then, gives exactly one set of values for a *mapping* (or *relational mapping*), where we are mapping from the value combination of independent variables to their dependent variable values.

As a simple example, recall our top baby names example from Sect. 3.4. We show six rows of the data set in Table 5.1.

The data set variables are `year`, and `sex` as independent variables, along with `name` and `count` as dependent variables. These form the columns of the table. Each row has a unique value combination for `year` and `sex`, and the values in `name` and `count` give the mapping to these dependent variables.

---

[1]One of the problems with the term "tidy data" is that it conveys a semantic notion of there being a *spectrum* of how tidy data might be—we can make data "somewhat" tidy and then do some more cleaning and make it more tidy. But if data really conforms to a model, then the data is in *normal* form, and such normal form is *not* a spectrum. The ideas and terminology of data models, conformance, and normalization long predate the field of data science. Nonetheless, the tidy data term has wide acceptance, and so we will use it, where appropriate, in this book as well.

5.2.2 Operations

The operations in the tabular model are procedural and work on the in-memory structure to accomplish some goal. In Chaps. 7 and 8, we will describe operations in terms of the interface provided by the `pandas` package and its `DataFrame` object class. Here, our objective is to enumerate the *logical* categories of operations of the tabular model.

- **Access Operations**
  Those operations that *read* data values out of a table. There are many types of operations needed to access data values, from obtaining a single element to one or more columns of the data set to some subset of the rows, or a combination of the above.
- **Computational Operations**
  Given the access operations allow us to subset the data in various ways, computational operations allow us to compute with these various subsets of the data. These can include column-vector operations, aggregation operations, and Boolean operations.
- **Mutation Operations**
  The mutation operations provide a means to update a data set. These may include adding a column, updating a column, deleting a column, adding a row, deleting a row, and changing the order of rows or columns in the collection.
- **Advanced Operations**
  Because the operations of the tabular model are procedural, many of the above operations can be combined in interesting and helpful ways. More advanced operations might include partitioning and grouping information, performing more generalized aggregation operations, possibly in combination with grouping, and operations that perform transformations on the row–column structure itself.

5.2.3 Constraints

In the tabular model, there is neither checking nor enforcement to ensure any limitations on the data. In Sect. 5.1.3, we discussed model constraints as those limitations on the data that constrain individual element values and those that constrain relationships between structural components.

Since any creator of data can choose how to take data and put it into a row and column tabular form, the biggest concern is whether or not the data conform to the structural limitations of the model. In other words, we must always be asking the question of whether or not the data violates any of the four tenets of tabular/tidy data, namely:

- Each column must define *exactly one* variable.

- Each row must give *exactly one* mapping/observation from values of independent variables to the values of the dependent variables.
- *Exactly one* table defines all the information for a single coherent subject.
- Multiple separable subjects are each in their own table.

Examples and a full treatment of these conditions are the subject of Chap. 9.

5.2.4 Reading Questions

5.6 Are tabular data and tidy data the exact same thing? Explain.

5.7 Give an example demonstrating that the same data can be represented in multiple different tabular forms.

5.8 List the tabular operations from the reading, and describe examples of each, referring back to earlier chapters if necessary. Go beyond the words in the text, e.g., instead of "aggregation operations," you could discuss a time when you needed to compute the sum of all entries in a column of data.

5.9 List the four tenets of tabular data, and describe examples of what it would look like to violate each one.

5.3 Relational Model Overview

The relational model is the basis for data stored in relational databases, which are the predominant form of data system in existence. These databases provide consistent, reliable, persistent storage of data for businesses, non-profits, universities, and organizations, both large and small. From the early days of computing and storage, the relational model and its databases have developed into an efficient and effective class of system with solid design methodology.

Chapters 10 through 14 will provide details on this model.

5.3.1 Structure

Like the tabular model, the structure of the relational model is based on tables, composed of rows and columns. We may rely on that structural similarity to be able to understand much of the relational model. Sometimes, the terminology used in relational databases differs somewhat from what is used in the tabular model, and we will detail many of these in Chap. 10.

At a broad level, the structural aspects that distinguish the relational model from the tabular model involve database design, an emphasis on multiple tables, and an underlying foundation based on mathematics.

- **Database Design**
  In the relational model, a database is carefully designed and specified before any data is populated. This design entails describing the set of tables, constraints on columns, and relationships between tables. The result is a formal structure called the *schema* for the database and its tables. This specification, in fact, becomes part of the database itself.
- **Multiple Tables**
  Through the process of database design, the structure and set of tables are, in part, determined by analysis of what data is dependent on specific data attributes. This process eliminates redundant data that could violate constraints or introduce consistency errors in the data set. One significant effect of this process is a set of "minimal" tables and the existence in the design of many tables. This many-table aspect of the model is evident in the schema and also impacts the operational aspect of the model—we need strong operations to efficiently combine these multiple tables.
- **Mathematical Foundation**
  From its first conception [4], the relational model has been based on the mathematical field of *set theory*. Every table defines a set, and rows in the table are the set elements. Operations that define subsets of data and that operate between tables always yield new tables (sets) with well-defined properties. The use of set theory allows formal definitions of the operations of the relational model and the "sanity" of composing operations in near-endless variation.

Figure 5.1 gives a simple example schema for a school database, illustrating some of the structure concepts. There are three tables defined: students, classes, and schedule. Inside the representation of each table, the column fields are listed,

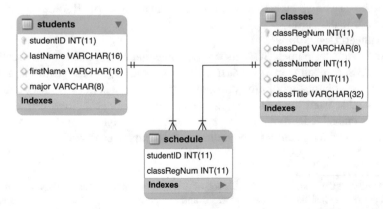

Fig. 5.1 School database schema

along with a data type for each field. The lines between tables give the relationships between the tables. Here, we are showing that the `schedule` table, in effect, *links* students with the classes they are taking through entries in the schedule table.

5.3.2   Operations

In the relational model, operations are declarative—we specify *what* data is desired and leave it to the system to work out *how* to get the data. Over the years, the declarative language of SQL has become the standard for specifying data queries of a relational database.[2]

These declarative SQL operations have much in common with their tabular model counterparts. Generally speaking, the operations can be categorized as access operations, computational operations, and advanced operations. In a separate category, SQL defines means for mutations on the tables of a database and even for changing the database schema itself.

- **Access Operations**
  The SQL to obtain columns of a table (called *projection*) and to obtain subsets of rows (called *selection* or *filtering*) based on some Boolean condition is some of the most common kinds of queries. For example, we might want to select all classes in the computer science department.
- **Computational Operations**
  SQL provides, in its language, the ability to perform a restricted set of unary and binary computations on the columns of data obtained in a query. For example, we might want to concatenate the class department and class number, for display purposes in a catalog.
- **Table Combination Operations**
  SQL provides powerful operations for combining tables, known collectively as *join* operations. For example, we might want to combine the `schedule` table with the `classes` table to learn the student ID numbers of all students registered for a particular class.
- **Advanced Operations**
  Like the tabular model, SQL allows grouping and aggregation for the relational model. Specifically, in a `GROUP BY` operation, SQL can partition a table, aggregate columns within the partition (e.g., finding a sum of values), and yield a new table from the result. SQL, because of its composability, can also have queries embedded within queries. For example, we might want to group students by major and then compute the number of students in each major.

---

[2]Associated with the mathematical foundation of the relational model, there is an *algebra* that also defines the operations that may be performed and composed to obtain data from the relational tables. This is called the *relational algebra*, and further coverage is beyond the scope of this textbook.

In the example above, we could, for instance, build an SQL query that returned results for the upper-level classes being taken by students whose major is MATH. Chapters 11 and 12 will develop these SQL skills.

5.3.3 Constraints

Constraints for the relational model are specified through the creation of the database schema. When we create tables using SQL, we specify constraints, including type constraints (e.g., that the number of students in a class must be an integer), value limitations on a per-column basis (e.g., that the number of students in a class can't be negative and can't be larger than 100), and uniqueness aspects of the table (e.g., insisting that two different students must have different student ID numbers). Additional schema elements are defined that formalize the relationships between tables.

Given a schema, like that shown in Fig. 5.1, we can construct the SQL for creating the tables, as well as the SQL for the relationships given by the lines in the figure.

Because the schema is part of the database itself, and because these constraints are known ahead of time, every modification (insertion or deletion) to the database can be checked and the operation only permitted if no constraints are violated. In this way, errors are avoided.

5.3.4 Reading Questions

5.10 Describe an example of a database schema to store information about people, their homes, and their car registration information. For each table, describe at least three attributes, and describe the links between the tables.

5.11 Give an example of a database constraint. Hint: think about the example in the text and "uniqueness."

5.12 Give an example where an insertion or deletion should be prevented, because it would violate a constraint.

5.13 List two similarities and two differences between the relational data model and the tabular data model.

5.4 Hierarchical Model Overview

The basis for the hierarchical model is the tree structure. The hierarchical model has become popular because of its flexibility and, in the case of tree-based markup languages, its self-describing nature. The hierarchical formats of JSON and XML

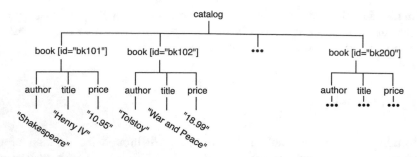

Fig. 5.2 Book Catalog Tree

have become a dominant form of data interchange. Because of the flexibility and power, there are many other examples. For instance, the language of web pages, when conforming to a strict tree, is XHTML. Languages for specifying structure of written documents can also form a tree, and the language of LaTeX is an example from that domain.

5.4.1 Structure

Because of the common underlying structure of the tree, the terminology introduced in Chap. 2 applies to the structure of the hierarchical model as well:

- Vertices are called *nodes* or *elements*.
- A tree has a single root node/element.
- Each node:
  - can have zero or more *children*
  - , may have a symbolic name, or *tag* associated with it, and
  - may have *attributes* associated with it.
- A node with no children is called a *leaf* node.
- A *path* gives traversal(s) within the tree, which start at some source element and proceed through children down the tree to some destination element.

Consider the example in Fig. 5.2 representing the data set of a book catalog. The tree comprises the entire structure. The root of the tree has symbolic name/tag of catalog. This root node has 100 children, each of which has a tag of book. In this depiction, we include the name of an attribute id and the value of the attribute in brackets following the tag name. So, we have a single attribute for each book, whose value gives a string identifier for the book. Each book has exactly three children, author, title, and price, whose text children give string values to be associated with the corresponding parent node.

5.4.2   Operations

The operations for the hierarchical model can have both procedural and declarative aspects. Many operations are defined relative to a given element. The procedural elements can be broadly partitioned into access operations and mutation operations. Typically, we start with a hierarchical structure defined in some *format*, like XML or JSON, and, before procedural operations are applied, we first parse the data, based on its format, into an in-memory representation of its tree.

- **Access Operations**
  Those operations that *read* data, which could be values associated with attributes, text values associated with an element, or could be structural information, like finding the children of an element or their parent.
- **Mutation Operations**
  Operations that provide a means to update a tree structure. These could update attributes or text of a node but could also affect the structure and add children elements for a given element.

In the hierarchical model, declarative operations may be employed to specify data to be queried without having to write a procedural operation-based traversal. In XML, the predominant query language is called XPath. Here, we specify a traversal (or collection of traversals) using a notation that is similar to the paths in a file system, as covered in Sect. 2.1.2, but combined with ideas similar to metacharacters and wildcards like we saw with regular expressions in Chap. 4. The result of query, then, is all the data that "matches" the specified path, augmented with these new wildcards.

5.4.3   Constraints

In the hierarchical model, the foremost constraint is that the data conform to a tree and the syntax of how that tree is specified must be followed. This constraint is always checked as a part of the process of taking data in a particular hierarchical format (XML, JSON, XHTML, and LaTeX) and parsing the data from a file or stream of characters into an in-memory structure of a tree. This constraint prevents paradoxes, e.g., an element being its own grandparent in the tree. If the format of the data violates the syntax or tree structure, an error occurs before any further processing on the tree can happen.

Other constraints limiting the data might be desired that go beyond the syntax and tree structure. Consider our catalog example. We might want to ensure that every book node has an id attribute and always has at least the three children of title, author, and price. We might want a catalog to only have book children, or it might be okay to also have magazine children. We could want the text child of the price node to be constrained to only be a string that represents a number.

We can specify these kinds of additional constraints in JSON using a *JSON Schema*. Furthermore, XML has two commonly used means of specifying constraints: *Data Type Definition (DTD)* and *XMLSchema*. These are typically specified in an auxiliary documents and then, when available, in a step following the parsing, a tree can be checked for conformance to this constraint document. These will be discussed in Chap. 17.

5.4.4   Reading Questions

5.14 Describe part of the tree structure of a web page you are familiar with. Instead of drawing a picture as in the text, you could type your response using indentation, as in this example:

```
catalog
    book [id = "bk101"]
        author
            Shakespeare
        title
            Henry IV
        price
            10.95
    book [id = "bk102"]
        author
        title
        price
    ...
    book [id = "bk200"]
        author
        title
        price
```

5.15 Based on the Book Catalog Tree in the text, give an example of a node, its children and its attributes.

5.16 Give an example of a mutation operation that changes the tree structure in the Book Catalog Tree.

5.17 List two similarities and two differences between the hierarchical data model and the tabular data model.

5.18 List two similarities and two differences between the hierarchical data model and the relational data model.

Chapter 6
Tabular Model: Structure and Formats

Chapter Goals

Upon completion of this chapter, you should understand the following:

- The three principles of tidy data.
- What is meant by the variables and values of a data set.
- Independent and dependent variables and the associated concept of a functional dependency or logical mapping.
- The key considerations in file formats to support tabular data.

Upon completion of this chapter, you should be able to do the following:

- Write functions to read data from delimited row-oriented formatted files.
- Use `pandas` functions to create new tabular structure data frames:

    - from list of row lists representation of tabular data,
    - from dictionary of column lists representation of tabular data, and
    - from a CSV file.

Section 5.2 introduced the structure of the tabular model as a two-dimensional table of rows and columns. The current chapter begins by delving more deeply into the structural requirements of tidy data. The chapter then discusses the formats by which we can encode tabular data into a file. The chapter finishes with creation of

© Springer Nature Switzerland AG 2020

T. Bressoud, D. White, *Introduction to Data Systems*,
https://doi.org/10.1007/978-3-030-54371-6_6

tabular structures known as *data frames* using the `pandas` module, the package of choice for working with tabular data in Python. In subsequent chapters (Chaps. 7, 8, and 9), we will explore the operations of the model as realized by `pandas`, and we will then finish coverage of the model by extending material from the Tidy Data section (Sect. 6.1) so that we can recognize when data we obtain does not conform to the requisite structure and can then use our operations to normalize a data set into its proper form.

6.1   Tidy Data

Recall our top baby names data set and its tabular representation, a subset of which is reproduced in Table 6.1. This representation has columns `year`, `sex`, `name`, and `count` and, in each row, we give the name and application count for that name in the row's year and the given sex.

To motivate the structural requirements of tidy data, let us consider just a few of the alternative ways that the same top baby name data set might be represented in a row–column tabular form. These tabular representation variations might occur because of how the data is generated by the provider, or could be the result of a client choosing a representation for presenting in a document or visualization.

First, we might partition the data into the top names for females and the top names for males. The result would be *two* tables, each with columns `year`, `name`, and `count`, as shown in Tables 6.2 and 6.3.

As a second representation, consider Tables 6.4 and 6.5. In this representation, we use the various `year` values as column headers, and the values of `sex` (`Female` and `Male`), for the rows. Here, we again have two tables, one for the `name` and the other for the `count`.

Tables 6.4 and 6.5 are also examples of a common presentational choice: when we have values of a dependent variable that rely on *two* independent variables, we often use the *values* of the independent variables (the values of `year` and `sex` in

Table 6.1 Top baby names

Year	Sex	Name	Count
2018	Male	Liam	19837
2018	Female	Emma	18688
2017	Male	Liam	18798
2017	Female	Emma	19800
2016	Male	Noah	19117
2016	Female	Emma	19496
2015	Male	Noah	19635
2015	Female	Emma	20455
2014	Male	Noah	19305
2014	Female	Emma	20936

Table 6.2 Top female baby names

Year	Name	Count
2018	Emma	18688
2017	Emma	19800
2016	Emma	19496
2015	Emma	20455
2014	Emma	20936

Table 6.3 Top male baby names

Year	Name	Count
2018	Liam	19837
2017	Liam	18798
2016	Noah	19117
2015	Noah	19635
2014	Noah	19305

Table 6.4 Top baby names by year

	2014	2015	2016	2017	2018
Female	Emma	Emma	Emma	Emma	Emma
Male	Noah	Noah	Noah	Liam	Liam

Table 6.5 Top baby name application counts by year

	2014	2015	2016	2017	2018
Female	20936	20455	19496	19800	18688
Male	19305	19635	19117	18798	19837

Table 6.6 Top baby names by year

	2014	2015	2016	2017	2018
Female	Emma, 20936	Emma, 20455	Emma, 19496	Emma, 19800	Emma, 18688
Male	Noah, 19305	Noah, 19635	Noah, 19117	Liam, 18798	Liam, 19837

this example) as column and row names, and the dependent variable value is at the intersection of row and column.

In a third representation, we could merge these together, and make the table entries a mashup[1] of both the name and the corresponding application count, as shown in Table 6.6.

As a fourth alternative, we could have a single row for each year, and encode the Female/Male values of the sex attribute as part of the column. So in this representation, the columns would be year, FemaleName, FemaleCount, MaleName, MaleCount, and the five year subset of the data would be represented as shown in Table 6.7.

Clearly, even with a relatively simple data set, there are many ways of representing the *same* data values in two-dimensional row–column structures. Depending on

[1]We use the term *mashup* when two or more distinct values or variables/variable names are combined into a single value. A mashup might appear in a column name, in a row name, or in a table entry value.

Table 6.7 Top baby names with female/male columns

	FemaleName	FemaleCount	MaleName	MaleCount
2018	Emma	18688	Liam	19837
2017	Emma	19800	Liam	18798
2016	Emma	19496	Noah	19117
2015	Emma	20455	Noah	19635
2014	Emma	20936	Noah	19305

which aspects of the data that we might want to highlight in a presentation of the data, one could argue for one or another representation.

However, from a processing-not-presentational perspective, it should be clear that, if we wish to write software that processes data sets in various ways, it would help greatly to have a common standard for representing various data sets. Otherwise, software would have to be changed and customized for each possible representation. Instead, with a standard representation of the data model, packages and tools may be developed that, using the assumptions of the standard, can then perform similar work on *different* data sets without having to be rewritten.

Toward this goal, it has been found that keeping exactly one variable in each column allows us to effectively use column-vector operations (see Sect. 3.1) on such a representation, and we can then "slice and dice" the rows, partitioning based on values of other variables. This allows software to be written in a manner agnostic to the domain, including statistical software, machine learning software, and even graphics/visualization packages. This is the basis of the tidy data structure employed by the tabular data model.

In [82], the principle requirements of tidy data are developed and discussed, and are rephrased here.

Definition 6.1 A data set in the tabular data model is said to be *tidy data* if it conforms to the following.

1. Each column represents *exactly one variable* of the data set. (*TidyData1*)
2. Each row represents *exactly one unique (relational) mapping*, that maps from a set of *givens* (the values of the independent variables) to the values of the dependent variables.[2] (*TidyData2*)
3. *Exactly one table* is used for each set of mappings involving the same independent variables. (*TidyData3*)

In a data set, a *variable* is a symbolic name with an associated *value* that can be a number, an integer, a real-valued number, a string, or it can be *categorical* which means that the value comes from a discrete set of valid values. Categorical variables often yield a partitioning of the data. For instance, `Male/Female` partitions the

---

[2]In the data science literature, it is stipulated that each row is exactly one *observation*. This is a more experimental view of data sets, and so the tabular model characterizes this more generally as a mapping from independent variables to dependent variables.

applications by sex. In a data set associated with, for instance, a survey, an age categorical might have values "20-29", "30-39", etc., that partition the respondents by age.

In order to assess possible representations, like the ones in Tables 6.1, 6.2, 6.3, 6.4, 6.5, 6.6, 6.7, for their adherence to the tidy data requirements (TidyData1 through TidyData3), we have to think carefully about the data set, and understand what make up the *variables* and the *values* of variables in the data set. In the current example, it should be fairly clear that year is a variable, and its values are integers that range from 1890 through 2018. We argue that, for this data set, the other variables are sex, name, and count. Central to this argument is the consideration of what Female and Male are with respect to this data set. The answer is that they are two different *values*, and the values must belong to a variable, that we name sex. The variable name has possible values from the set of first names by applicants for a US social security card, and the variable count has integer values based on number of applications for a particular name.

A logical mapping maps from values of independent variables to the values of dependent variables. So we observe that, in order to interpret a name value or a count value, we need to know, not only the year, but also the sex. So we are mapping from values of year and sex to name and count. We use the following notation to describe this logical mapping:

$$\text{year, sex} \rightarrow \text{name, count}$$

where the independent variables are on the left-hand side of the arrow and the dependent variables are on the right-hand side of the arrow. If we have understood our variables and values correctly to arrive at this logical mapping (aka functional dependency[3]), then a value combination of the independent variables should always be *unique*—there should not be another, different value for name or count for the same year and sex.

This logical mapping/functional dependency can be written equivalently as

$$\text{year, sex} \rightarrow \text{name}$$

$$\text{year, sex} \rightarrow \text{count}$$

This notation is asserting that *all* the variables on the left-hand side are needed to determine the variable(s) on the right-hand side.

If, in an attempt to understand our data set, we mistakenly arrive at the dependency:

---

[3] The term *functional dependency* term comes to us from relational database theory and is part of the foundation for understanding the data and normalizing a database design. We will use functional dependency and logical mapping interchangeably; a row provides an *instance* of a mapping.

$$year \rightarrow name, count$$

we would find that we have *two* values of name and two values of count for a given year, meaning that we have either made an error in determining the variables, or we have missed one or more independent variables to define the logical mapping.

Reviewing the alternative representations against the principles of tidy data, we see the following:

1. In Tables 6.2 and 6.3, we have *two* tables, and the tables are themselves defined by the value of a variable (Female and Male), and so this is a violation of TidyData3.
2. In Tables 6.4 and 6.5, we see columns headed up by year *values*, so that we are violating TidyData1—the columns here are *not* a variable. We also violate TidyData3 by separating our data set into two different tables based on the variables of name and count.
3. Table 6.6 uses a conjoined value and addresses the TidyData3 violation bringing the data into one table, but the meaning of the values populating the entries of the table is a mashup, which is violating the *Exactly one* stipulation of TidyData1, that a column (and thus the values in that column) represents exactly one variable.
4. In the representation given by Table 6.7, we would write the functional dependency/logical mapping as:

$$year \rightarrow FemaleName, FemaleCount, MaleName, MaleCount$$

This is a valid dependency given the data in that, given a year, the four columns are uniquely determined. The problem here is that columns like FemaleName no longer represent exactly one variable. The name of the column itself shows two parts, one of which is a *value* of a categorical variable, and the other part is a variable that is then repeated in MaleName. This is a violation of TidyData1.

Only the original table, Table 6.1, satisfies all three requirements of tidy data.

Chapter 9 illustrates more of these concepts with examples and advice and enumerates some "red flags" to look for in recognizing violations. Once we detect a data set that is not in the standard form, we should resolve these violations and achieve the desired standard form. The coverage of the operations of the tabular model, in Chaps. 7 and 8, will provide the tools we need for such efforts to normalize our data sets.

Other Model Structure Considerations

In addition to the data itself, a structure supporting the tabular model may maintain information for the *names* of each of the columns. These can be a list of strings, but may have additional complexity. Similarly, since rows define unique mappings, a structure may maintain information for some type of symbolic or logical identifier for each *row* of the data set. Often, this symbolic index involves the value(s) of the independent variables that are unique to the row.

Within a table, the *order* of the rows may also be important. For instance, in the version of the baby names data set where we had the rank order of, say, female names in a given year, the position in the table gave (implicitly) the rank of the mapping. The first row of the table was the top name, the second row of the table was the rank 2 name, etc. While it is not precluded by the structure of the model, such an implicit assumption can be dangerous, and it would be better to have some type of "order" variable to make it explicit.

In the remainder of our coverage of the tabular model, we will primarily cover instances in which data is in a *single* table, although this is not a restriction of the model. Multiple table data sets will be covered more fully in the relational model, and we will draw correspondences between the operations developed there and their tabular model counterparts.

6.1.1 Reading Questions

6.1 The text describes many ways to store the baby name data in tables. Give another way, different from any in the text, and then describe whether it fails any of the assumptions TidyData1 through TidyData3 (and if so, which one it fails).

6.2 The reading mentions that in a survey one might use a categorical variable for age, like "20-29". Why would you ever do this?

6.3 If you have a categorical variable for age, with values like "20-29", why not just replace all values by a numerical value like 25?

6.4 Think back to the student table example from the previous chapter, and give a logical mapping describing the relationship between the independent and dependent variables.

6.5 Give an example of a data set and a column in it with unique identifiers for each row.

6.1.2 Exercises

6.6 Consider a medical data set assessing efficacy of two different treatment possibilities, called Delta and Gamma, on a set of persons. For each person, when they receive a treatment, the efficacy is measured (e.g., the person's blood pressure). A person can independently receive both treatments, and a different efficacy is measured for each treatment.

Define the variables in this data set, and write the functional dependency/ies.

6.7 Reconsider the medical data set from the last question. But now assume that, given a particular person, the person is *only* given either the Delta treatment or the Gamma treatment. No person is ever given both treatments.

Define the variables in this data set, and write the functional dependency/ies.

6.8 Suppose you have two data sets. Table 6.1 consists of the number of deaths due to opioid overdose every month in Ohio, from 2014–2018.

```
Month              | Deaths
January, 2014      | 382
Feburary, 2014     | 391
...                | ...
```

Table 6.2 consists of the number of searches/seizures by police of drugs, every day in Ohio, from 2014–2018.

```
Day                | Seizures
January 1, 2014    | 65
January 2, 2014    | 92
...                | ...
```

With reference to the tidy data assumptions, discuss the pros and cons of the following two strategies for getting the data into a single table.

Strategy 1:   create a new column, deaths, in Table 6.2, and for every row (i.e., day) set deaths equal to the number of deaths in that month from Table 6.1. For example, deaths would be set to 382 for each day in January of 2014.

Strategy 2:   create a new column, seizures, in Table 6.1, whose value in a given row (i.e., month), is the total number of seizures in that month. For example, seizures in January, 2014, might be 2746.

If you would like to use knowledge from statistics in answering this question, please do so.

6.9 Consider the student data set. Explain the pros and cons of each of the following two functional dependency structures:

(1) Student ID -> first name, last name, class year
(2) last name, first name -> Student ID, class year

6.10 Imagine a data set, featuring the five countries Brazil, Russia, India, China, and South Africa (one per row). For each, the data set contains the country code, name, capital, area, and population. Decide which variables are independent and which are dependent, and write out the functional dependency.

6.11 Imagine a data set consisting of every country in the world, every year 1980–2010, the population, the GDP, and the number of cars. Think carefully about the dependencies that would be present in this data, and then record the functional dependency structure.

6.12 Imagine a data set of users who have subscribed to a web service. For each user, we have a username, a name, a phone number, an address, the date they joined, and an identifier that tells us whether they are an active user or not. Think carefully about the dependencies that would be present in this data, and then record the functional dependency structure.

6.13 Imagine a data set about automobiles. Using what you know about cars (or Google if you must), write down at least 5 columns that might be present in your data, and the functional dependency structure.

6.2 Tabular Data Format

6.2.1 Format Background

Definition 6.2 A *format* (aka *file format*, aka *document format*) is a specification for the encoding of information for a data structure into a sequence of characters or bytes, in a file or across a stream, to be communicated between a creator/writer/provider and a corresponding consumer/reader/client.

There are many file formats already familiar to computer and smartphone users. For instance, there are a number of audio file formats, such as mp3 or wav, that allow structured digital audio to be encoded into a file or sent over the network. Similarly, pictures and video can be encoded in a picture/video format like jpeg, png, mov, and wmv. When a creator and a consumer both understand and use/parse the same format (including possibly a *version* of the format), successful communication is achieved.

In data models, we need the same kind of agreement and standard format. We need to have a format that gives the encoding (the sequence of bytes) of the structure, so that a creator/provider can take data and output it in the format, and a corresponding consumer/client can take data in a file or from a stream (like bytes over a network) and can interpret and reassemble the data into its structure. The more precise the rules for the encoding, the greater the probability that both provider and client can successfully communicate a data set.

6.2.2 Format for Tabular Data

The most common format for tabular data is the comma-separated value (CSV) format [20, 84] and its many variations. Its strength is that, in its basic form, it is simple to describe and understand, and can be created easily by text editor, by program, and even, through export feature, by spreadsheet programs. Its greatest weakness is that, in part because it is so easy to create, there are many ways for a user or program to change a file in the CSV format and make it difficult for subsequent decoding and parsing by a program. Also, the apparent simplicity can become much more complex when dealing with variations and exceptions to the basic format.

We have already encountered a simplified version of the CSV format when we discussed file processing in Sect. 2.3 and were processing multiple data items per line in a text file, and in the exercises of Sect. 3.4 where we processed files in the simplified CSV format to construct native two-dimensional Python data structures. The format from earlier simplified our programmatic decoding/parsing of files, but is insufficient to the general task—when providers create CSV files that, for instance, might include integrated documentation, or embedded spaces or the separation character itself in the data. To handle this, we must understand CSV more deeply.

The purpose of the remainder of this section is to heighten awareness of some of the decisions that might be made in the format for tabular data, so that, whether by writing functions to parse a file or stream into tabular data, or the determination of arguments for software that provides this capability, we can adapt appropriately to take CSV files of data created by others and recreate tabular structures in our client programs.

6.2.2.1   Tabular Format Design

The CSV format is based on encoding to a sequence of characters, i.e., a *text* file. Thus, there are really two levels of encoding happening as a file or stream is created:

1. A tabular structure is encoded into a sequence of characters, governed by the CSV format, and
2. A sequence of characters is encoded into bytes, governed by the file or stream encoding (e.g., UTF-8).

The decoding on the client side is comprised of the two steps in reverse. The characters to bytes decoding might already be handed by a file interface, but if the data is coming over the network, this decoding might be visible to the client.

The CSV format also assumes exactly *one* tabular structure per separate file. This is more constrained than a spreadsheet model, in which a single file might contain multiple "worksheets" that could each hold its own tabular structure.

The following paragraphs describe some of the particulars and design decisions that may be involved in the CSV format, depending on the source provider.

Rows

With few exceptions, one of the fundamental assumptions regarding this tabular data format is that, in the data portion of the file, one row of data is encoded in one *line* of the file. Even the definition of a line, though, can depend on platform and result in variations for the client to consider. In the file itself, lines may be separated by '\n', '\r', or '\r\n'. Also, it may or may not be enforced that the *last* line in the file ends with the line termination sentinel. If enabled, and if line separation can unambiguously be determined when a file is opened, Python 3 provides "universal newlines," where the Python file interface yields ' \n ' for all the above cases.

Column Fields

The designer/producer of a tabular file format must determine how column fields in a row-line may be distinguished from one another. Often, a single character is designated as a *separator*, to be placed between any two column field values. Alternatively, column fields could be of a known fixed length. When using a sentinel separator, the most common of these is the comma (' , '), but tab (' \t '), semicolon (' ; '), and one or more spaces could also be used. Sometimes, optional spaces before or after a separator may be allowed. The value of a field itself may need to incorporate the separator, in which case the design must provide for "escape" of such inclusion, or some type of delimiter for field values so that they are not misinterpreted as a separator. If spaces may be part of a field, then string fields including spaces must be delimited.

Column Headers

Tabular file format may include a single line or even multiple lines that give the names of each of the columns. These may or may not be allowed to be an empty string. Typically, these column header lines are separated by the same column field separator as the data lines. They may include embedded spaces as well.

Preamble

For purposes of self-documentation, a creator using a tabular file format may wish to include lines describing the provenance of the data, or some abbreviated type of data dictionary describing the fields or other assumptions. These documentation lines may precede the header line or the data itself.

Comment Lines

Whether for a preamble, or for additional documentation within the data lines of the file, a tabular file format may include lines that are *not* to be interpreted as data. These may use a designated character to start the line as a sentinel that what follows is a comment. Sometimes a ' # ' character is used for comment lines.

Empty Lines

For clarity or presentation, a file may include lines consisting of entirely whitespace that, if present, should be allowed and skipped over.

Missing Data

In many data sets, particularly in experimental and observational data, some rows of the data set might have (dependent) variables where values are "missing." This can be a normal case, but a tabular file format may need some way to designate such missing data. Sometimes this occurs by simply having an "empty" field, following immediately with the next field separator or line separator, but it might occur by using a special string like "NA" or other designated value.

6.2.3 Tabular Format File Processing

For data systems clients, data acquisition gives us tabular-formatted files (or streams), often CSV, and we must perform the decode steps, constructing either a native Python two-dimensional structure, or a data frame supported by a package. There are three choices for accomplishing this:

- **Choice 1**: Write our own code to parse the file/stream and build native structures, as in Chap. 3. These native structures can then be used to construct a data frame. In an effort to understand some of the variations listed above, custom parsing will be explored through some of the exercises.
- **Choice 2**: There is a Python module, csv [46], that provides a few variations of reader objects that can be constructed from a file and that provides an interface for acquiring rows of data and fields within the row, and handles many of the lower-level parsing duties. This would also be used to create native structures, which could then be used to construct a data frame. This can often be a better choice than Choice 1 as full parsing can be quite involved.
- **Choice 3**: When the target is a data frame, a data analysis module (like pandas) often provides a constructor (read_csv) that handles the parsing and the result is the desired data frame form. In order to handle many of the variations described above, arguments to the read_csv specify the expected format particulars for the CSV at hand. This is explored in Sect. 6.3 and in the exercises.

6.2.3.1 CSV Parsing [Optional]

In this section, we give a high level overview of CSV Parsing (Choice 1 above). This topic will be explored in more detail in the exercises. We assume we are given a file path or stream of characters representing CSV format. Our goal is to construct a native Python two-dimensional data structure as in Chap. 3.

Top Level
The basic phases of an algorithm to parse and process a CSV comprise three stages:

1. In the first stage, we perform all one-time setup, including determining column names from a header and creating the initial data structure.
2. In the second stage, we iterate through the data lines of the file, creating rows of filed values that get accumulated into our data structure.
3. In the third stage, we perform all one-time processing to complete the process. Sometimes column based data type conversions may occur in this stage, depending on whether conversion was done during Stage 2.

We now provide additional detail for each stage.

Stage 1
- The file or stream is opened or otherwise prepared for processing.

- Any preamble lines are skipped over, advancing the current position in the file or stream.
- The header line is read in and split into column name fields. The line itself might need removal of a trailing newline, and column name fields may need leading and/or trailing spaces removed prior to constructing a list of column names. If these names are to be used, as part of the return value or in constructing the data structure (like in a dictionary of lists), then any empty names must have alternatives generated.
- An empty version of the native data structure is generated, possibly using the column names obtained from the prior step.

Stage 2

The basis of this stage is a loop that continues until there is no more data in the file/stream. At each iteration of the loop, we obtain a string variable `line` that must be processed for one *row* of data. The processing for `line` involves the following steps.

- If the line is empty or begins with a comment character, the line is skipped so that we continue with the next iteration of the loop.
- The newline in `line` must be removed, e.g., using `strip()`.
- The line is split into string field values based on the delimiter (comma, most commonly). This is easy if there are no instances of the delimiter as part of an escaped or delimited field value, but much more complex otherwise (e.g., if commas appear in an Address field, and different rows have different numbers of commas).
- For each of the fields in the row, we must:
  - Remove leading/trailing whitespace.
  - Depending on the data type of the column, possibly convert the string into a different data type, e.g., using `int()` or `float()`.
  - If our native structure has lists per column (e.g., DoL), append the field value to the column list.
- For the row as a whole, if the native structure is a list of rows (e.g., LoL or LoD), append the current row to the aggregate structure.

Stage 3

- The file or stream should be closed.
- If columns, after Stage 2, are comprised of strings, but should be some other type, these can be converted at this time.
- The resultant native structure is returned. If a list of row lists, then the column names list should also be returned.
  Examples of this procedure were provided in Chap. 3.

6.2.4 Reading Questions

6.14 If you create a CSV file that is "double-spaced" (meaning, a blank line between each line of data), does this represent tidy data? Explain.

6.15 It is common to write an address with commas, and different addresses sometimes require different numbers of commas, e.g., `Department of Mathematics and Computer Science, 100 W. College Street, Granville, OH 43023` vs. `1600 Pennsylvania Ave NW, Washington, DC 20500`. Give a detailed explanation of how to include the address field in a CSV file in such a way that different rows do not end up with different numbers of columns.

6.16 When writing your own code to parse a CSV file, does the number of lines of code you write grow with the number of columns in the data? Explain.

6.2.5 Exercises

6.17 Write a function

```
read_earthquake_lat_long_mag(filename)
```

that opens and reads this file, accumulating three lists, the latitudes, longitudes, and magnitudes, for all the earthquakes included in the file. The function should *return* all three lists in the order of latitudes, longitudes, magnitudes. The lists should *only* contain numbers, and so should not include the header string.

6.18 Repeat the problem above, but now instead of returning the three lists, put them into a dictionary of lists (DoL) with keys `"latitude"`, `"longitude"`, `"magnitude"` and return the DoL. Name your function `read_earthquake_DoL(filename)`.

6.19 Using the same data set, write a function

```
read_earthquake_LoL(filename)
```

that extracts a list of lists, one per row, and also a header column with the three column names as strings. Return the pair `LoL, headers`.

6.20 Using the same data set, write a function

```
read_earthquake_LoD(filename)
```

that extracts a list of dictionaries, one per row, where each dictionary contains three mappings (from each of the three header strings to the value in that row). Return the LoD.

6.21 Write a function

```
read_csv(filename)
```

that reads a CSV file whose header is used for column names, and returns a dictionary of column lists. Hint: you must first determine how many columns there are. Do not assume you can carefully study the CSV file as you did in the previous problems.

6.22 Write a function

```
read_csv2(filename,skiplines)
```

that improves your function above the function by incorporating the parameter `skiplines`, which says how many lines at the top of the file should be skipped before the header line.

6.23 Write a function

```
read_csv3(filename,skiplines,sep)
```

that improves your function above the function by incorporating the parameter `sep`, which is the separating character. In all our examples above, it was a comma `","`, but now you should allow it to be set by the parameter (e.g., a tab or a space).

6.24 Write a function

```
read_csv4(filename,skiplines,sep,commentmark)
```

that improves your function above the function by incorporating the parameter `commentmark`, which is a character that, when present at the start of a line, indicates that the line in question is a comment. Your function should ignore comment lines. Comment lines can occur anywhere in the file, not just in the `skiplines` part.

6.25 Write a function

```
read_csv5(filename,skiplines,sep,commentmark,missing)
```

that improves your function above the function by incorporating the parameter `missing`, which is a string that indicates missing data at a particular cell. For example, this string might be `"NA"` or `"N/A"` or `"None"`. When your function finds this string, append `None` to the relevant column list. In this way, Python will not be confused into doing something like trying to cast `"NA"` into an integer. The Python keyword `None` is immune to such errors.

6.26 Repeat the previous exercise, but now return a LoL representation of the CSV file (including a list, `headers`). You are strongly encouraged to build your function incrementally as we did above, adding one new parameter at a time.

6.27 Repeat the previous exercise, but now return a LoD representation of the CSV file. You are strongly encouraged to build your function incrementally as we did above, adding one new parameter at a time.

6.3   Tabular Structure as `pandas DataFrame`

In Sect. 3.4, we explored three possible alternatives for representing tabular row/-column data as native Python data structures. However, this low-level approach has some distinct disadvantages:

1. The DoL choice, a column-centric representation, is effective for accessing and manipulating entire columns, but working with rows or subsets of one or more columns requires additional code and effort.
2. The LoL choice, a row-centric representation has other concerns:

   - While effective for dealing by rows, there is more work for manipulating the columns of the data set.
   - Rows can only be referred by integer index, not symbolically.
   - We must explicitly maintain the additional list of column names.

3. The LoD choice improves on the LoL choice in some ways, but can be more difficult to construct due to the need to create a list of column names, and then constantly refer to it when creating each row dictionary. We saw in Chap. 3 that this requires nested loops.
4. For all three representations, there are no operations beyond list indexing and dictionary access. This means we need to write custom code for almost all the operations described in Sect. 5.2.
5. Algorithms on these data structures become complex to read and maintain. They almost always have nested loops and are fragile to changes, like the addition or deletion of columns.

These disadvantages distill down to a lack of abstraction. We desire an in-memory structure where we can define algorithms that work on abstract rows and columns equally efficiently and that reflect the tabular structure itself, as opposed to the underlying realization and organization.

Building abstract data types and defining logical higher-level operations while hiding implementation is one of the strengths of an object-oriented programming language like Python. The need for abstract data type(s) for working with tabular data has lead to the development of the `pandas` module [39], the de-facto standard for tabular data in Python. The `pandas` module defines two primary object classes: the `DataFrame` class represents two-dimensional tabular data, and the `Series` class represents one-dimensional data for the columns within the `DataFrame`.

In the remainder of this section, we will present some alternative ways that `pandas DataFrame` objects may be created, and then present some basic methods for working, as a whole, with the created data frames. This will provide us with sufficient means for looking at the operations for tabular data that are covered in Chaps. 7 and 8.

As with any of the packages used in this book, the material presented here is not intended to be comprehensive nor to teach all the ins and outs of the package. There are excellent trade books and online resources for goals such as these. Rather, we show enough building blocks to enable basic use of the package for the purposes of understanding and using the tabular model. Beyond that, the reader is encouraged to explore further using online tutorials, online references, and trade books detailing the API/package.

6.3.1 *DataFrame Creation*

The manner in which we create a DataFrame depends on the form of the data serving as the source row and column values. We may, from a file or from other data acquisition, have the data in one of our three in-memory native representations as described in Sect. 3.4. Alternatively, we may have the data in a CSV format in a file or as a sequence of bytes from a network. We give examples of each of these, in turn.

Dictionary of Column Lists Source
In the following snippet of code, we define topnamesDoL as a Python dictionary whose column name keys map to a column list of values for a three year subset of the top babynames data set.

By convention, the pandas module is imported with an alias of pd. In the snippet, the DataFrame constructor, as a function in the pd (pandas) module, is invoked with the dictionary data, and the resultant data frame assigned to topnames and the result displayed.

```
import pandas as pd

topnamesDoL = {'year': [2018, 2018, 2017, 2017, 2016, 2016],
               'sex': ['Male', 'Female', 'Male',
                       'Female', 'Male', 'Female'],
               'name': ['Liam', 'Emma', 'Liam', 'Emma',
                        'Noah', 'Emma'],
               'count': [19837, 18688, 18798, 19800,
                         19117, 19496]}

topnames = pd.DataFrame(topnamesDoL)
topnames
```

```
|     year     sex   name   count
| 0  2018    Male   Liam   19837
| 1  2018  Female   Emma   18688
| 2  2017    Male   Liam   18798
| 3  2017  Female   Emma   19800
| 4  2016    Male   Noah   19117
| 5  2016  Female   Emma   19496
```

Note that the display of the data frame results in an aligned set of columns, with column labels and the rows of data beneath. While pandas allows rows to have labels, just like they do columns, we have not specified the row labels, and so pandas has provided a set of integers for the row labels. In pandas, the row labels are referred to as the *index* of the data set, and pandas defines an Index class that maintains the sequence of values along with other information, like a name for the Index. As we will see in Chap. 7, having an Index makes it much easier to access data within a DataFrame.

The next example is similar to the first, using a data set of country-based indicators like pop for population (in millions), gdp for gross domestic product, life for life expectancy, and cell for number of cell phone subscribers in a country.[4] The dictionary indicatorDoL is defined, and then an index list of three-letter country codes is also defined. In the constructor, these codes are passed as the argument for the named parameter index, which defines unique symbolic names for each of the rows in the data set.

```
indicatorDoL = {
 'country': ['Canada', 'China', 'India',
             'Russia', 'United States', 'Vietnam'],
 'pop': [36.26, 1378.66, 1324.17, 144.34, 323.13, 94.59],
 'gdp': [1535.77, 11199.15, 2263.79, 1283.16, 18624.47,
                                                  205.28],
 'life': [82.30, 76.25, 68.56, 71.59, 78.69, 76.25],
 'cell': [30.75, 1364.93, 1127.81, 229.13, 395.88, 120.60]}

codes = pd.Index(['CAN', 'CHN', 'IND', 'RUS', 'USA', 'VNM'],
                 name='code')

indicators = pd.DataFrame(indicatorDoL, index=codes)
indicators
```

```
|               country       pop        gdp   life     cell
| code
| CAN            Canada     36.26    1535.77  82.30    30.75
| CHN             China   1378.66   11199.15  76.25  1364.93
| IND             India   1324.17    2263.79  68.56  1127.81
| RUS            Russia    144.34    1283.16  71.59   229.13
| USA     United States    323.13   18624.47  78.69   395.88
| VNM           Vietnam     94.59     205.28  76.25   120.60
```

Now, when we display the data set, the provided row names (the Index) prefix the rest of the row instead of integer indices. This will become useful when we later consider row-centric operations on the structure. We will often use this approach, of creating an Index object and then passing it to the pd.DataFrame() constructor using the index= parameter. Later, we will also learn how to remove or reset an index.

---

[4]The country indicators data set, and a number of variations, will be used to demonstrate many of the operations covered in Chap. 7.

Defining our data directly in a dictionary form, like we have for these examples, is inconvenient, particularly as the size of a data set grows. If the data is in a CSV file, one option would be to write a function that reads the data and populates and returns a dictionary of column lists representation. This dictionary could then be passed to a constructor call as we have done above.

For instance, if the contents of a simple CSV file with the data, named `topnames.csv` looks as follows:

```
year,sex,name,count
1880,Female,Mary,7065
1880,Male,John,9655
1881,Female,Mary,6919
1881,Male,John,8769
   ...
2017,Female,Emma,19800
2017,Male,Liam,18798
2018,Female,Emma,18688
2018,Male,Liam,19837
```

and if we have a (user-defined, not shown) function `readcsv_DoL(path)` to parse a CSV file located at `path` and return a dictionary result, we could create our `topnames` data set:

```
path = os.path.join(datadir, "topnames.csv")
topnamesDoL = readcsv_DoL(path)
topnames = pd.DataFrame(topnamesDoL)
```

The CSV contains all 278 rows of the data set, and so now the `topnames` `DataFrame` object contains our full data set.

List of Row Lists Source

If, instead of in a dictionary, our data is organized as a list of row lists and if we have a list of the column labels, then we can use the `DataFrame` class constructor, where the first argument is the list of lists data, and we can use a named argument of `columns=` to specify the column labels, and the result is a `DataFrame` object. If our data is acquired from a file, or from a server across the Internet, we can, using the techniques of Sects. 3.4.2 and 6.2.3, process the rows from the data source and build the list of row lists representation, then we can also use the technique shown to create a `DataFrame` object.

```
topnamesLoL = [[2018, 'Male', 'Liam', 19837],
               [2018, 'Female', 'Emma', 18688],
               [2017, 'Male', 'Liam', 18798],
               [2017, 'Female', 'Emma', 19800],
               [2016, 'Male', 'Noah', 19117],
               [2016, 'Female', 'Emma', 19496]]
dataset_columns = ['year', 'sex', 'name', 'count']
```

```
topnames = pd.DataFrame(topnamesLoL, columns=dataset_columns)
topnames
```

```
|    year      sex   name   count
| 0  2018     Male   Liam   19837
| 1  2018   Female   Emma   18688
| 2  2017     Male   Liam   18798
| 3  2017   Female   Emma   19800
| 4  2016     Male   Noah   19117
| 5  2016   Female   Emma   19496
```

If we had not specified `columns=dataset_columns`, it would have provided integers for the column labels.

If our data is in a CSV file and we write a function, `readcsv_LoL`, to parse
and return the column names and the list of lists, we can then create a `DataFrame`
object from the CSV data in a couple of steps.

```
path = os.path.join(datadir, "topnames.csv")
dataset_columns, topnamesLoL = readcsv_LoL(path)
topnames_df = pd.DataFrame(topnamesLoL,
                            columns =dataset_columns)
```

The situation for a list of row dictionaries is analogous, and the `pd.Data
Frame()` constructor can accept a list of dictionaries like `topnamesLoD`, and
can infer the column names from the keys. Hence, to construct a `DataFrame`, the
invocation `topnames_df = pd.DataFrame(topnamesLoD)` suffices.

CSV File Source

If we wish to avoid the complexities of writing functions to parse CSV and the
two-step process exemplified above, we can go more directly from a CSV file to a
pandas `DataFrame` using the `read_csv` function of the module.

As a prelude and preparation for the needs of Sect. 6.3.2, we give two examples
here. In the first, we create `DataFrame` object `topnames0`, in which we do
not define an `Index` for the rows. Often, in starting with data that needs to be
manipulated and transformed for normalization, this is the appropriate starting point.
In this case, *all* columns of the CSV result in the columns of the data frame.

The first parameter of `read_csv` always provides the channel to the formatted
data. In this first example, we provide a `str` path giving the location of a file. The
pandas module also allows an open file object, or a stream of bytes made to look
like an open file object, for this first parameter, increasing its generality. To show
the result, we invoke the `head()` method on the `DataFrame` to show a prefix of
the data obtained. This method is extremely helpful for exploratory data analysis,
and to check that our programs perform as expected, and so we shall use it often.

```
import pandas as pd

path = os.path.join(datadir, "topnames.csv")
```

```
topnames0 = pd.read_csv(path)
topnames0.head()
```

```
|     year      sex    name    count
|  0  1880   Female    Mary    7065
|  1  1880     Male    John    9655
|  2  1881   Female    Mary    6919
|  3  1881     Male    John    8769
|  4  1882   Female    Mary    8148
```

The read_csv function has *many* more parameters that allow flexible parsing of a CSV source. Some of these include prelude lines to skip, using a separator other than a comma, specifying comment lines, and many others. There are also parameters that can control data conversion, missing value processing, and interpretation of header lines, and the reader is referred to the full documentation[40].

In the second example, we show the specification of an Index using the independent variables that uniquely identify rows in the data set, by using the index_col named parameter. These form the columns that make up the index for the rows, allowing year and sex, in this example, to be treated specially in the operations involving rows of the data set. Notice the difference in how the data and the column labels are displayed in this case. Here, from the CSV, two of the columns form the index of the data frame, and the other two columns form the columns of the data frame.

```
path = os.path.join(datadir, "topnames.csv")
topnames = pd.read_csv(path, index_col=['year','sex'])
topnames.head()
```

```
|                 name    count
| year sex
| 1880 Female    Mary    7065
|      Male      John    9655
| 1881 Female    Mary    6919
|      Male      John    8769
| 1882 Female    Mary    8148
```

If a data set has a single column that uniquely identifies rows, then just that column name can be specified as the index_col instead of a list of column names, like we show next with the indicators data set (which will be introduced formally in Chap. 7).

```
path = os.path.join(datadir, "indicators2016.csv")
indicators = pd.read_csv(path, index_col='code')
indicators
```

```
|              country      pop      gdp    life     cell
| code
| CAN           Canada    36.26  1535.77   82.30    30.75
```

```
| CHN                China  1378.66  11199.15  76.25  1364.93
| IND                India  1324.17   2263.79  68.56  1127.81
| RUS               Russia   144.34   1283.16  71.59   229.13
| USA        United States   323.13  18624.47  78.69   395.88
| VNM              Vietnam    94.57    205.28  76.25   120.60
```

For these data set, we prefer to have an index, to facilitate easier access operations in Chap. 7.

6.3.2   Operations Involving Whole Data Frames

Before we explore the various operations of the tabular model working on rows and columns of a data frame, it will be helpful to talk through just a few operations that operate on a `DataFrame` as a whole. These will allow us to better summarize results and understand the effects of the operations to come.

In the following, we will use the `topnames0`, the `topnames`, and the `indicators` data frames constructed above to help illustrate the operations.

Prefix and Suffix of Data

The first two operations allow us to examine a small prefix or suffix of a `DataFrame`. The `head(n)` method returns the first n rows (and all columns) of a `DataFrame`, where the number of rows is five if not specified. We already used `head()` to show a portion of the results of a `read_csv` above.

```
topnames0.head()
```

```
|    year       sex   name    count
| 0  1880    Female   Mary     7065
| 1  1880      Male   John     9655
| 2  1881    Female   Mary     6919
| 3  1881      Male   John     8769
| 4  1882    Female   Mary     8148
```

The `tail(n)` method returns the last n rows, again with a default of five, if the n is omitted.

```
topnames.tail(6)
```

```
|                 name   count
| year sex
| 2016 Female    Emma   19496
|      Male      Noah   19117
| 2017 Female    Emma   19800
|      Male      Liam   18798
| 2018 Female    Emma   18688
|      Male      Liam   19837
```

These are often used as code is developed to test out and see if the latest `DataFrame` meets expectation. They are also used to avoid a document showing the entirety of a data frame in a presentation of an exploration or analysis.

Metadata

Some operations give data *about* the data, sometimes called *metadata*. To start, we can use the Python built-in `len()` to find the number of rows in a `DataFrame`:

```
len(topnames)
```

```
| 278
```

If we want to know both the row dimension *and* the column dimension, we use the `shape` *attribute* of the `DataFrame`:

```
topnames0.shape
```

```
| (278, 4)
```

In the design of an object class, the designers can use methods (like `head()`) which we invoke as a function on an object instance. Designers can also use attributes, like `shape`, where there are no parentheses, and accessing the attribute yields a value, a tuple in this case. Such use can simplify programming and can help keep an algorithm clear.

Note the tuple we get for the `shape` attribute of `topnames`:

```
topnames.shape
```

```
| (278, 2)
```

This `DataFrame` has the same number of rows, but, because `year` and `sex` are part of the index, there are only *two* columns in the data frame, compared with four in `topnames0`.

General information about a `DataFrame` can be obtained with the `info()` method. This gives information about the row `Index`, as well as information about each of the columns. For `topnames0`:

```
topnames0.info()
```

```
| <class 'pandas.core.frame.DataFrame'>
| RangeIndex: 278 entries, 0 to 277
| Data columns (total 4 columns):
|  #   Column  Non-Null Count  Dtype
| ---  ------  --------------  -----
|  0   year    278 non-null    int64
|  1   sex     278 non-null    object
|  2   name    278 non-null    object
|  3   count   278 non-null    int64
| dtypes: int64(2), object(2)
| memory usage: 8.8+ KB
```

We see the index is an integer (`RangeIndex`), and that there are four columns, with the given number of non-null (i.e., not missing) entries. The `year` and `count` columns are integers (`int64`, where the 64 gives the bit size of the integer), and `sex` and `name` are "`object`", which generally is an indicator of a string object. The method also shows the program memory used by the `DataFrame`.

Contrast the `info()` for `topnames0` with that for `topnames`, that has the multi-index of `year` and `sex` defined as the row index:

```
topnames.info()
```

```
| <class 'pandas.core.frame.DataFrame'>
| MultiIndex: 278 entries, (1880, 'Female') to
|                           (2018, 'Male')
| Data columns (total 2 columns):
|  #    Column  Non-Null Count  Dtype
| ---   ------  --------------  -----
|  0    name    278 non-null    object
|  1    count   278 non-null    int64
| dtypes: int64(1), object(1)
| memory usage: 6.4+ KB
```

In this case, the index is `MultiIndex` that has tuple combinations of year and sex, and there are only two columns in the `DataFrame`.

We can also query the column labels for a `DataFrame` by using its `column` attribute, which we `print` for both `topnames0` and `topnames`:

```
print(topnames0.columns)
```

```
| Index(['year', 'sex', 'name', 'count'],
|                         dtype='object')
```

```
print(topnames.columns)
```

```
| Index(['name', 'count'], dtype='object')
```

Similarly, we can examine the `index` attribute of a `DataFrame`:

```
topnames0.index
```

```
| RangeIndex(start=0, stop=278, step=1)
```

If we were to do the same for the `topnames` `DataFrame`, which has a more complex index, we would see an enumeration of the levels that make up this index. So the example is omitted here.

Index and Order

If we have a `DataFrame` object that, for instance, has an integer index for the rows, we may want to convert it into a `DataFrame` that has an index based on the independent variables of the data set. The `set_index` method allows this conversion. By default, this creates a new memory copy of the `DataFrame`, so both the original and the new DataFrame still exist, and we have to remember to assign the result to a variable:

```
topnames1 = topnames0.set_index(['year', 'sex'])
topnames1.head(6)
```

```
|               name   count
| year sex
| 1880 Female   Mary    7065
|      Male     John    9655
| 1881 Female   Mary    6919
|      Male     John    8769
| 1882 Female   Mary    8148
|      Male     John    9557
```

If we want to avoid consuming memory for two versions, and know that, after conversion, we will not need the original, we can use the inplace named parameter, set to True, to use the mutability of data frames to change the original itself:

```
topnames0.set_index(['year', 'sex'], inplace=True)
topnames0.head(6)
```

```
|               name   count
| year sex
| 1880 Female   Mary    7065
|      Male     John    9655
| 1881 Female   Mary    6919
|      Male     John    8769
| 1882 Female   Mary    8148
|      Male     John    9557
```

Note that, when we are mutating in place, we do not assign the result to a variable.

The *order* of rows in a DataFrame is significant, and the order is based, at creation, on the index. For a given data set, we can organize and change the order based on the index. Additionally, we can choose a sort order that is *not* the default of an ascending order. When we sort based on the index, and the index consists of more than one variable, the sort first happens on the first variable, and then, for the "ties" with the same value of the first variable, the second variable is used as a secondary sort. So in our topnames data set, the original order was from earliest year to latest year. If, instead, we want the order to reflect the most recent year first, we can sort_index and specify the named parameter ascending to False to get a descending sort:

```
topnames0.sort_index(ascending=False, inplace=True)
topnames0.head(8)
```

```
|               name   count
```

```
| year sex
| 2018 Male     Liam   19837
|      Female   Emma   18688
| 2017 Male     Liam   18798
|      Female   Emma   19800
| 2016 Male     Noah   19117
|      Female   Emma   19496
| 2015 Male     Noah   19635
|      Female   Emma   20455
```

When we have defined a non-integer index for the rows of a data set, the data making up the index is no longer manipulable as column data. We can "undo" the effect of a defined index by using the `reset_index` method of a `DataFrame`. This keeps the present order, and re-creates columns for each of the variables that used to make up index, and then provides a new default integer index:

```
topnames0.reset_index(inplace=True)
topnames0.head(8)
```

```
|     year     sex   name   count
| 0   2018    Male   Liam   19837
| 1   2018  Female   Emma   18688
| 2   2017    Male   Liam   18798
| 3   2017  Female   Emma   19800
| 4   2016    Male   Noah   19117
| 5   2016  Female   Emma   19496
| 6   2015    Male   Noah   19635
| 7   2015  Female   Emma   20455
```

Data Types

One last brief example of a whole `DataFrame` manipulation involves the data type associated with one or more columns of the data frame. Depending on source and processing when we create a `DataFrame`, the data types associated with some of the columns may not be what we wish. The `read_csv()` will often be able to interpret correctly and automatically convert string-based numbers in a column to integers or floats. But sometimes, the function cannot reliably make the determination that such a conversion is appropriate. Similarly, if we are parsing CSV ourselves, the result might be strings in a column that should be integers or some other data type.

As an example, we call the locally defined `readcsv_DoL` function to create a dictionary of columns from a CSV, but does *not* perform conversions for the `year` and `count` columns. We invoke the `info()` method to see that these columns are indeed of type `object` (i.e., string), instead of the desired integer (`int64`):

```
path = os.path.join(datadir, "topnames.csv")
topnamesDoL = readcsv_DoL(path)
```

```
topnames = pd.DataFrame(topnamesDoL)
topnames.info()
```

```
| <class 'pandas.core.frame.DataFrame'>
| RangeIndex: 278 entries, 0 to 277
| Data columns (total 4 columns):
|  #   Column  Non-Null Count  Dtype
| ---  ------  --------------  -----
|  0   year    278 non-null    object
|  1   sex     278 non-null    object
|  2   name    278 non-null    object
|  3   count   278 non-null    object
| dtypes: object(4)
| memory usage: 8.8+ KB
```

Given a DataFrame, the astype() method can take a dictionary that maps from column names to the desired data type of the column. The method returns a new DataFrame with the values in the specified columns converted, and leaves the rest of the values in the data frame unaltered.

```
topnames2 = topnames.astype({'year': int, 'count': int})
topnames2.info()
```

```
| <class 'pandas.core.frame.DataFrame'>
| RangeIndex: 278 entries, 0 to 277
| Data columns (total 4 columns):
|  #   Column  Non-Null Count  Dtype
| ---  ------  --------------  -----
|  0   year    278 non-null    int64
|  1   sex     278 non-null    object
|  2   name    278 non-null    object
|  3   count   278 non-null    int64
| dtypes: int64(2), object(2)
| memory usage: 8.8+ KB
```

6.3.3   Reading Questions

6.28 Pick a representation (e.g., DoL or LoL) and an algorithm you are familiar with on that representation. Explain precisely in what way this algorithm is fragile to changes like the addition or deletion of columns.

6.29 What happens if you pass an Index object to pd.DataFrame but the Index object is created from a list of size smaller than the number of rows? Experiment and report your findings!

6.30 What is the value of working with row labels rather than simply using the default index in Pandas?

6.31 Can you imagine ever using both the `index=` and the `column=` options inside of `pd.DataFrame`? Explain.

6.32 What is the `head` function good for? Explain with a specific example.

6.33 When using the function `read_csv` in `pandas`, what is the connection between what you pass to `index_col` and with functional dependencies in the data?

6.34 Give a specific example where it is important to know the metadata associated to a data set.

6.35 With regards to the `info` function, why is it important to know the number of bits used when encoding integers?

6.36 What is `indicators1` if you get confused about the `inplace` parameter and write a line of code like:

```
indicators1 = indicators.set_index(['code'],
    inplace = True)
```

Where have you seen this kind of logical error before?

6.37 Experiment with the last code provided in the section, to bring in `'count'` as a `float` or as some other type. Are you able to change the `memory usage` reported from `info`? Does that change make sense?

6.38 To get practice with the `pandas DataFrame` constructor, please type a header list and a corresponding list of lists, with at least two rows. Then use the `DataFrame` constructor to create a data frame from this LoL.

6.3.4   Exercises

6.39 On the book web page you will find a file `worldpopulation.csv`, based on data hosted by www.census.gov. The data set has a column for the `year` and another for `population`. Use `pandas` to read this data into a data frame named `dfpop`, and be sure your data frame has column headers matching the names just given.

6.40 On the book web page you will find a file `meteorites.csv`, based on data hosted by `Open Street Maps`. The first row of the data set consists of column names. Use `pandas` to read this data into a data frame named `dfmet`, and be sure your data frame has column headers matching the names just given.

6.41 On the book web page you will find a file `sat.csv`, based on data from www.onlinestatbook.com. This data contains the high school GPA, SAT math score,

SAT verbal score, and college GPAs, for a number of students. Read the data into a pandas data frame dfsat with column headers.

6.42 On the book web page you will find a file education.csv, based on data hosted by www.census.gov. Each row is a US metropolitan area, with data on population (row[3]), number unemployed without a high school degree (row[15]), number unemployed with a terminal high school degree (row[29]), number unemployed with some college (row[43]), and number unemployed with a college degree (row[57]). Read the data into a pandas data frame dfedu with column headers.

Later, we will learn how to aggregate each column to its total, and report the fraction of unemployed people from each category.

6.43 On the book web page you will find a file floods.csv, based on data hosted by www.nwis.waterdata.usgs.gov. This data includes measurements related to water heights of a river in Wyoming. The US Geological Survey keeps this data to forecast floods, and build appropriate containment infrastructure. The first row represents the column names. Read the data into a pandas data frame dffloods with column headers. Note that the csv file has many comment rows starting with #.

6.44 On the book web page you will find a file earthquakes.csv, based on data hosted by the US Geological Survey. This data includes latitudes, longitudes, and magnitudes of earthquakes from all around the world, as well as many other variables of interest. The first row represents the column names. Read the data into a pandas data frame dfquake with column headers.

6.45 Using the list below:

```
days = ['Mon', 'Tue', 'Wed', 'Thur', 'Fri']
```

Create a pandas Series with days as the index.

6.46 It turns out that Index objects are immutable. Prove this to yourself by trying to directly change Mon to Monday. Then, when that fails, create a new Index with full weekday names and reset the index of your Series above.

Chapter 7
Tabular Model: Access Operations and `pandas`

Chapter Goals

Upon completion of this chapter, you should understand the following:

- The logical access operations to query data.
- The logical operations to manipulate vectors of data.

Upon completion of this chapter, you should be able to do the following:

- Query data using attributes and methods of `DataFrame` and `Series` objects.
- Perform computations on single and multiple `Series`.

We learned in Sects. 5.2 and 6.1 that our fundamental structure of the tabular data model is a two-dimensional structure made up of rows and columns, and that logical operations on that structure include access of one or more columns, selecting rows through filtering, performing column-vector operations between columns and between columns and scalars, and updating and adding columns, among others. We have also been cautioned that we may acquire data that is two-dimensional in rows and columns, but the data may not conform to the constraints of the tabular model, and we may need operations that can manipulate and transform the data to normalize it.

© Springer Nature Switzerland AG 2020
T. Bressoud, D. White, *Introduction to Data Systems*,
https://doi.org/10.1007/978-3-030-54371-6_7

7.1 Tabular Operations Overview

We start with a conceptual summary of the types of operations we want and need. This will allow us to better understand the pandas package in context. Subsequent sections will then show pandas realizations of these operations. Many of the summaries will give high-level examples using the top baby names data set, which has columns year, sex, name, and count.

7.1.1 Access Operations

Access operations are those that *read* or *query* data values out of a table. There are many types of operations needed to access data values, and we list the most common here. We should note that there are many variations, even among this list. For instance, it may be possible to identify columns and/or rows both logically or by integer position. Each of these combinations gives operational variations to consider.

- **Single-element access** At one end of the spectrum, we may wish to access a single data value, which is the intersection of a column and a row. For instance, we may wish to read the value from the name column and the row associated with the year 2018 and the sex of "Female" (recall that we need values of both independent variables to uniquely refer to a row).
- **Column access** When we seek a subset of a data frame based on one or more columns, we call that a *projection* of the desired columns.[1] For instance, we may wish to project the vector associated with the count column, including the values for all the rows. We also might want a subtable obtained by projecting the year, sex, and count columns, with all rows included. A projection might also be used to obtain a new table with the columns re-ordered.
- **Single row** We may wish to *select*[2] a single row, obtaining the values of all columns in that row. So we may want the year, sex, name, and count column values obtained by selecting, say, the first row in the data set.
- **Multiple rows** In this operation, we want a subset of the data consisting of all the columns, but a limited set of the rows. This, in general, is also called *selection*, but is often called *filtering*, when we want the chosen rows to satisfy some condition. For instance, we may want all the rows where the year is between 1920 and 1950.
- **Subset rows and columns** The most general form of access operation would allow us to both project a subset of the columns *and* filter for a particular subset of the rows.

---

[1]The term *projection* for access of one or more columns has been employed in the domain of *relational algebra* for relational databases from the beginnings of that theory [4].

[2]The *select* term for access of one or more rows also has its origins in relational algebra.

For access operations in the tabular model, we need to pay attention to the data type of the result. For example, when we use an operation to obtain a single element, it will make a difference if we expect a *scalar* (i.e., a non-vector unit of data), like an integer or a string, or expect a *data frame* consisting of exactly one row and one column. It is important to distinguish between these cases, as downstream functions will break if they expect a string and we pass them a data frame, or vice versa. The `pandas` operation examples will illustrate some of these data type result differences.

7.1.2 Computational Operations

Given that the access operations allow us to subset the data in various ways, computational operations allow us to compute with these various subsets of the data:

- **Column-vector operations** Given one/two columns, we may want to perform unary/binary operations between the columns to obtain a new column of data. For instance, if we had the `count` column and an equal sized column for the total number of applicants in a given year, we could compute the percentage of applicants for the top name for each sex for each year as an equal sized vector or column. Analogously, in our `indicators` data, we could compute GDP per capita in this way.
- **Aggregation operations** Given a column, and perhaps a subset of the rows, we may want to compute some aggregate value for the set of values. We may want `max`, `min`, `sum`, or `mean`, of the `count` column, for instance.
- **Boolean operations** We may want to compute a Boolean predicate elementwise on a column vector. For instance, we may want to compute a Boolean vector whose values are `True` if the `count` is above 20000, and `False` otherwise. This is really just a particular type of column-vector operation, but is listed separately as such a Boolean vector could then be used in an access operation where we want to filter rows based on a condition.

7.1.3 Mutation Operations

The mutation operations provide a means to update a data set. Allied with mutation, we may also wish to create a separate in-memory data structure that is a copy of some subset of the rows and columns, but that allows a mutation to be independent of the original data set. We list common mutation operations.

- **Add a column** We want to be able to create and add a new column to a data set. For instance, after computing a column of GDP per capita, we may wish to add this new column to the `indicators` data set.

- **Update a column** In this operation, an update to the values of an existing column would be made in place, overwriting original values in the column. A variation of this operation would update only a subset of the rows, based on some predicate. We commonly need this operation if our data source releases updated data.
- **Delete column(s)** In this operation, we wish to delete one or more existing columns from the data set. If an analysis were only interested in the count of applications, the name column could be deleted. The deletion of a column is referred to as a *drop* of the column.
- **Add rows** Adding to a data set involves augmenting the data set with one or more additional rows. Care needs to be taken on the order in which new rows are placed in the data frame.
- **Delete rows** We sometimes need an operation to delete one or more rows. The operation must allow us to specify exactly which rows are to be deleted, such as by index, or by satisfying a Boolean condition.
- **Change order** The most common example of an operation to change the order of the rows in the table would be to sort the rows based on values in one or more columns. For instance, we might want to sort the rows in our baby name data set in decreasing order of the year value, to create a time series.

7.1.4   Advanced Operations

Because the operations of the tabular model are procedural, many of the above operations may be combined as a set of steps in interesting and helpful ways, building more powerful aggregate operations. A few of these more advanced operations are listed here.

- **Iteration** Built on lower-level access operations, we can define operations (iterators) that allow us to iterate over rows, over fields within rows, and over other subsets of the data in a data frame.
- **Generalized Aggregation** This operation allows us to aggregate multiple columns at once, possibly using a different aggregation function on different columns, e.g., finding the maximum population and minimum GDP.
- **Group By** This operation refers to a means of partitioning a data set into multiple disjoint parts and then performing some aggregate operation that summarizes each partition, and then combining these summaries into a new table. For instance, in the baby names data set, we could "GroupBy" year, so each partition would include the two rows for any given year, and then we could sum the count field in the partition. This would give us a table whose independent variable was year and whose dependent variable is the total applications for the top names in the year. In Chap. 8 we will see an example where we group countries by their region, and so obtain a new data frame with summary information on Asia, Europe, etc.

- **Table Combination** These operations allow us to take data that are in more than one table and combine them into a single table result. For example, if we wanted to combine our `indicators` data set with another data set containing health-related information on countries, we could *join* the two data sets together and so obtain more information per country than either data set had by itself.
- **Transformations** These can be complex operations that allow us to transform columns into rows or rows into columns, or perform a transpose on a table. These operations are used when we have data in a two-dimensional structure, but it does not satisfy the needed structure of the tabular data model (i.e., is not in tidy data form). Or, we might have data in a tidy data form, but want to transform it to create a view of the data that is easier to communicate information, or in preparation for a visualization. Because of their applicability in normalization, these operations will be covered in Chap. 9.

7.1.5 Reading Questions

7.1 Give an example of filtering in the context of the data set of top names.

7.2 Give an example where you might wish to update a column in place.

7.3 Give an example of a real-world situation that would require you to add rows to a data set.

7.4 Give an example of a real-world situation requiring you to delete rows.

7.5 Give an example when you might want to iterate over the rows.

7.6 What does it mean to "aggregate multiple columns at once?" Refer to the earlier discussion of aggregation and give an example.

7.7 We illustrated GroupBy with the example of grouping by the `year`. Give another example of GroupBy with a different variable in the same data set, and describe the resulting data set.

7.8 Give an example of a time when you have needed to transform a data set.

7.9 In previous data analysis work, have you ever needed to combine two different tables? If so, describe the situation. If not, think up an example where this would be important. If you cannot, refer to the exercises in Chap. 3, where such an example is provided.

7.2 Preliminaries and Example Data Sets

For use in illustrating operations in `pandas`, we will employ two different data sets (along with some variations thereof). One is the top baby names data set from earlier. The data set consists of 278 rows and 4 columns (year, sex, name, and count) of data, and has functional dependency:

$$\text{year}, \text{sex} \rightarrow \text{name}, \text{count}$$

We will use Python variable topnames0 to refer to the DataFrame that has only the default integer index for rows, and topnames to refer to the DataFrame that has the two-level index of year and sex, the independent variables of the data set.

The other data set consists of *indicator* data for a set of countries. The data set is organized as a single row per country with columns:

- code: the unique international three character code for the country.
- country: the English name of the country, including spaces and capitalization.
- pop: the population of the country, in units of millions of persons.
- gdp: the gross domestic product of the country, measured in units of billions of US dollars.
- life: the life expectancy of the country, in years.
- cell: the number of cell phone subscribers in the country, in millions of subscribers.

For this data set, the functional dependency is:

$$\text{code} \rightarrow \text{country}, \text{pop}, \text{gdp}, \text{life}, \text{cell}$$

The pandas DataFrame with default integer index is referred to by the Python variable indicators0, and indicators is used for the DataFrame that uses code as the row label index. The data is obtained from the World Bank Open Data, and is available through the Creative Commons License. As in Chap. 6, we create the indicators DataFrame so that the index is the code column, giving us symbolic names for the rows.

To make it more manageable for the purposes of illustration, the indicators data frame is limited to just five countries (rows), and the data is only from the year 2017. The current entirety of the data set is presented in Table 7.1.

Table 7.1 World indicators 2017

Code	Country	Pop	Gdp	Life	Cell
CHN	China	1386.40	12143.49	76.41	1469.88
IND	India	1338.66	2652.55	68.80	1168.90
RUS	Russian Federation	144.50	1578.62	72.12	227.30
USA	United States	325.15	19485.39	78.54	391.60
VNM	Vietnam	94.60	223.78	76.45	120.02

In variations we will see later, we extend the data set to include information on the indicators for years from 1960 through 2017, and for approximately 120 countries. Because the indicators change over time, each of the indicator values is dependent on both country and year, so the functional dependency of the expanded data set becomes:

$$\texttt{code}, \texttt{year} \to \textit{indicators}$$

Pandas Terminology
We identify some of the terminology, attributes, and class types specific to `pandas`:

- `Index` is the class name for a collection of symbols/values used in referencing the *set of rows or* the *set of columns* of a data frame.
- `index` is an attribute that refers to the `Index` for the rows of a data frame. It can be as simple as a range of integers supplied by default at data frame creation, or can incorporate one or more of the original columns to uniquely identify rows through a hierarchical index (like `year` and `sex` in the top baby names example). These are also called the *row labels*.
- `columns` is an attribute that refers to the `Index` for the columns of a data frame. It, like `index`, can be as simple as a one-dimensional sequence of string column names, or can have additional complexity. These are also called *column labels*, and *column names*.
- `Series` is the class name for a *one-dimensional* sequence of data. It has a sequence of values, like the values in a column, or the values in a row, and also has an `Index`. The `Index` comes from the row labels when the `Series` is a column, and comes from the column labels when the `Series` is a row.
- `DataFrame` is the class name for a *two-dimensional* collection of data. It has a collection of columns, each a `Series`, supports an `Index` each for column labels and for row labels, and supports the tabular model operations.

When a package, like `pandas`, and its object classes are being designed in an object-oriented language, the designers have multiple tools enabled by the language at their disposal to provide functionality for their classes. These tools and features are used to provide the operations supported by the class. So in the `pandas` realization of data frames following, we will see the use of:

- *methods*—functions that operate on a data frame and are invoked by naming an object, naming the method, and passing arguments that govern the behavior of the operation.
- *attributes*—these appear as variable-like names composed of the name of an object and give access to data associated with an object.
- *access operators*—drawing from their use in lists and dictionaries, the access operator ([ ]) is used to give access to subsets of the data maintained by the object.

- *operators*—an object class can overload the meaning of familiar operators like +, −, *, and so forth, to define binary and unary operations that extend to the type of object.

7.2.1 Reading Questions

7.10 Using the `topnames DataFrame`, constructed in the previous chapter, give an example (listing all data) of a row `Series`, including the `index`.

7.11 Give an example (listing all data) of a column `Series` in `indicators`, including the `index`.

7.12 Give an example of a `pandas` method. We note that several were used in the previous chapter.

7.13 Give an example of a `pandas` attribute, in the context of the `indicators` `DataFrame`.

7.3 Access and Computation Operations

Some of the most common and fundamental operations are those that access and query data and allow computation on that data.

7.3.1 Single Column Projection and Vector Operations

This subsection looks at some `pandas` paradigms for projecting single columns and performing column-vector operations with those columns.

We can use access operators (square brackets) on a `DataFrame` object with a string column label as the *indexer*, just like we would a Python dictionary, yielding the column associated with the column name. When we project a single column in this way, we always obtain a `Series` as the projected column.

We illustrate with an example. From `topnames0` we project and assign to `names` the `name` column of the data frame, and use `head()`[3] to get a prefix of the result:

```
names = topnames0['name']
names.head()
```

---

[3]In Sect. 6.3.2, we saw the use of `head()` to get a prefix of a `DataFrame`. Here, we see the same method is also defined for a `Series`, and has similar semantics.

```
| 0      Mary
| 1      John
| 2      Mary
| 3      John
| 4      Mary
| Name: name, dtype: object
```

Note how pandas displays a Series, with the index/value pairs first, followed by the metadata of the "Name" of the Series (the column name name, in this case), and the type of the data values in the Series (namely, object, which is the type pandas uses to denote a string). When we project the 'name' column from topnames, as opposed to topnames0, we get:

```
names = topnames['name']
names.head()
```

```
| year   sex
| 1880   Female      Mary
|        Male        John
| 1881   Female      Mary
|        Male        John
| 1882   Female      Mary
| Name: name, dtype: object
```

This result is still a Series, but here the index of the Series, coming from the index of the DataFrame, are comprised of year and sex values.

If a column is already defined (i.e., we are not in the process of creating it), and if the name of the column satisfies the requirements of a variable name,[4] and is not the name of an existing attribute or method, then pandas gives us a shortcut for projecting a single column, by making the *column name* an *attribute* of the DataFrame object. So, from the indicators data frame, we can easily project the gdp column:

```
indicators.gdp
```

```
| code
| CHN       12143.49
| IND        2652.55
| RUS        1578.62
| USA       19485.39
| VNM         223.78
| Name: gdp, dtype: float64
```

---

[4]Python variable names must begin with a letter or underscore, be composed of letters, numbers, and underscores, and have no embedded spaces or special characters.

In this example, the row label (code) is part of the Series object. If we had not created the data frame with code as its index, the data frame, and consequently, this Series, would have integers indexing the Series, just like our previous topnames0 example.

If we want to see or verify the type of an entity in Python, we can call the type() function with the object as an argument.

```
type(indicators.gdp)
```

```
| <class 'pandas.core.series.Series'>
```

So we see that this single-column projection indeed yields a Series.

If we have a Series and wish to reference a single element within it, we can again use the access operator. The row label is used to specify the desired element. In this next example, we project the pop column, and then access and print the 'RUS' and 'USA' elements of the Series:

```
pop_column = indicators['pop']
print(pop_column['RUS'], pop_column['USA'])
```

```
| 144.5 325.15
```

Similar to other two-dimensional structures, we can compose access operator use in a single expression. So, for example, accessing the 'USA' row of the 'pop' column could be accomplished by the expression indicators['pop']['USA'], where the indicators['pop'] part of the expression yields the column Series, and the ['USA'] operates on the Series to obtain the 'USA' element.

If the row label/index of a Series are integers, we would use integer indexing to access individual elements:

```
names_column = topnames['name']
print(names_column[0], names_column[1],
      names_column[276], names_column[277])
```

```
| Mary John Emma Liam
```

Series are powerful objects in their own right. We can work with Series as vectors, and perform operations in a vector-oriented way. Say for instance, we wanted to create a Series with 25% greater population based on the population column.

we want the vector

$$population * 1.25$$

each of whose entries is 25% larger than the corresponding entry in the original vector *population*. In pandas, this is exactly what we do:

```
new_pop = indicators['pop'] * 1.25
new_pop
```

```
| code
| CHN     1733.0000
| IND     1673.3250
| RUS      180.6250
| USA      406.4375
| VNM      118.2500
| Name: pop, dtype: float64
```

Similarly, we can perform binary operations on two columns in vector fashion. If we wanted to create a `Series` vector calculating the number of cell phone subscriptions per person in a country, and given that the units of both are millions, we want:

$$cell/population$$

and in `pandas`, we write:

```
cell_per_person = indicators['cell'] / indicators['pop']
cell_per_person
```

```
| code
| CHN     1.060214
| IND     0.873187
| RUS     1.573010
| USA     1.204367
| VNM     1.268710
| dtype: float64
```

Contrast the ease of notation and readability of the operations shown above with what would be required of vectors represented as Python lists or dictionaries. These would require a `for` loop or a list comprehension to achieve the same result.

Vector-style operations on `Series` vectors are not limited to arithmetic operations. Consider the following example, where, on the right-hand side of the assignment, we specify a one-column projection followed by a *relational* operator and a scalar:

```
short_life = indicators['life'] < 75
short_life
```

```
| code
| CHN     False
| IND      True
| RUS      True
| USA     False
| VNM     False
| Name: life, dtype: bool
```

The operation, elementwise, computes whether an element in the `Series` is less than the scalar value 75, and builds a new `Series` with the Boolean results of each of these condition results. So, in this example, the values at index `IND` and `RUS` met the threshold of being less than 75, and the other four did not. Such Boolean `Series` results are important, and form the basis of row-filtering.

We can also create more complex Boolean `Series` vectors using Boolean operators. In particular, we can use:

- `~`, a unary operator to logically negate the elements of its `Series` operand,
- `&`, a binary operator to perform a binary `and` operation elementwise over the values of its two operands, and
- `|`, a binary operator to perform a binary `or` operation elementwise over the values of its two operands.

For instance, to build a Boolean `Series` for rows where life expectancy is *not* less than 75, *and* the GDP is at least 10 trillion, we can compute:

```
long_life = ~(indicators['life'] < 75)
big_gdp = indicators['gdp'] >= 10000.0
condition = long_life & big_gdp
condition
```

```
| code
| CHN        True
| IND       False
| RUS       False
| USA        True
| VNM       False
| dtype: bool
```

Instead of using intermediate variables like `long_life` and `big_gdp`, we can build a single more complex expression, using parentheses to achieve the needed functionality:

```
condition = ~(indicators['life'] < 75) &
            (indicators['gdp'] >= 10000.0)
```

Another common-vector operation is the application of a unary function on each of the elements of a source vector to obtain a vector of resultant values, as discussed in Chap. 3. If the source vector is a `Series`, we can use the `apply` method on the source `Series`. The argument to `apply` is a function, and that function should take an element as a parameter and produce a value result. For instance, the `len()` function takes an object, like a string, and computes and returns an integer result. So `len` would be a function that satisfies the requirements for the type of a function that is needed by `apply()`.

In the following code, we apply the `len()` function to the string names of each of the countries to yield a vector of integers, where each element has the length of that country's name:

```
country_len = indicators['country'].apply(len)
country_len
```

```
| code
| CHN      5
| IND      5
| RUS     18
| USA     13
| VNM      7
| Name: country, dtype: int64
```

This is the `Series` vector equivalent of what we did with Python lists in Sect. 3.1.2. This technique can be used in general to obtain a unary operation on a `Series`. The function applied can be a built-in, like demonstrated here, or could be user-defined, or even be an anonymous function defined by a `lambda` expression.

7.3.2   Multicolumn Projection of a `DataFrame`

When we want to project one or more columns and want to yield a `DataFrame` instead of a `Series`, we employ a slightly different operation on a tabular data structure. In `pandas`, we still use the square bracket ([ ]) accessor with an existing `DataFrame` object, but the expression *inside* the square brackets is different. Instead of a single value of a column label, we can use a *sequence*, most often a list, that specifies the *collection* of columns to project.

To get the columns `gdp` and `country`, we create a list `['gdp', 'country']` and use it as the sequence argument of the square bracket accessors:

```
GDP_country = indicators[['gdp', 'country']]
GDP_country
```

```
|               gdp                 country
| code
| CHN     12143.49                    China
| IND      2652.55                    India
| RUS      1578.62     Russian Federation
| USA     19485.39          United States
| VNM       223.78                  Vietnam
```

So the outer brackets are again being used to specify a query of the data frame, but the inner brackets are giving an explicit *list* of the desired columns by label. When using a list to project one or more columns, the elements in the list *must* match column labels of the `DataFrame`.

In this example, we requested the columns in an order different from their order in the `DataFrame`, and this requested order was respected in the result, so projection can also be used to retrieve columns in a new order.

We can also use a list of length *one* to obtain a `DataFrame` consisting of a single column. Consider the following two variations for getting the `'gdp'` column from our `indicators` `DataFrame`:

```
gdp1 = indicators['gdp']
gdp2 = indicators[['gdp']]
```

The `gdp1` assignment line obtains a single column projection in the manner discussed in Sect. 7.3.1, which yields a `Series` object. The `gdp2` assignment line, on the other hand, obtains a single column projection as a one-column data frame in its own right, which means the type of `gdp2` is a `DataFrame`.

```
type(gdp2)
```

> | <class 'pandas.core.frame.DataFrame'>

7.3.3   Row Selection by Slice

Another common form of query access to a data frame is when we want a subset of one or more rows. When the *order* of rows within the data frame allows us to define desired subsets based on that order, `pandas` provides a syntax for easily expressing the desired rows using a *slicing* notation. In Python, we use slicing, incorporating one or two `:` characters along with specification of where to start, where to go to, and how to "step," to define subset elements of a list. `pandas` uses this familiar notation to similarly specify a subset of elements in a `DataFrame`.

Whether or not the `index` of row labels for a data frame have been defined to be other than their default integers, there always exist integer-based *positions* for each of the rows and each of the columns in any `DataFrame`. These start at 0 and go up to the number of rows or number of columns minus one.

The syntax for a slice is:

$$start : end \; [: stride]$$

The math-style square brackets here are distinct from Python bracket syntax; here indicating an *optional* element in the syntax. In the syntax, *start* specifies the position or label for the beginning of the set of items to be included, *end* specifies the position or label determining the end of the set of the items to be included, and *stride* specifies a step of how to proceed from one item to the next. If the `: stride` is omitted, the default stride is 1. If *start* or *end* is omitted, it indicates the beginning or end of the collection, respectively.

7.3.3.1   Position Slicing for Selecting Rows

In pandas, we can specify an integer-based slice inside the accessor operator brackets applied to a DataFrame, and this defines an operation that selects the specified *rows* (and all columns) from the DataFrame. This operation always yields a DataFrame result. Slicing using integers operates in the same way as slicing for a Python list, in that the slices defines the position "up to, but not including" the *end* item. For instance, in the indicators frame, we select the rows starting at position 3 and (using the default stride of 1) up to, but not including, position 5 (i.e., positions 3 and 4):

indicators[3:5]

```
|                 country      pop      gdp   life     cell
| code
| USA    United States   325.15  19485.39  78.54   391.60
| VNM           Vietnam    94.60    223.78  76.45   120.02
```

We select rows starting at the beginning and proceeding up to, but not including position 2:

indicators[:2]

```
|          country      pop       gdp   life      cell
| code
| CHN        China  1386.40  12143.49  76.41   1469.88
| IND        India  1338.66   2652.55  68.80   1168.90
```

7.3.3.2   Index Slicing for Selecting Rows

One of the advantages of pandas DataFrames over native structures is the facility for the row *index*, the logical names identifying rows based on a value or combination of values that uniquely identify a row. This uses the Index class of pandas. For instance, in our indicators data set, the code column gave three letter strings uniquely associated with each row. In our topnames data set, when we set the index to ['year', 'sex'], the combination of values of the integer year and string 'Female' or 'Male' identifies the row.

Using the row labels/index attribute, we can define a slice and specify a *start* and *end* where the values of these come from the row labels. For instance, we select the rows in indicators from CHN to RUS as follows:

indicators['CHN':'RUS']

```
|                     country      pop      gdp   life     cell
| code
| CHN                   China  1386.40  12143.49  76.41   1469.88
| IND                   India  1338.66   2652.55  68.80   1168.90
| RUS    Russian Federation   144.50   1578.62  72.12    227.30
```

A notable difference from position-based slices for selecting rows: the *end* is *inclusive*, not "up to, but not including." So in the above example, the 'RUS' row is included in the result.

The *stride* can be specified, and works as one might expect, so we can select rows in a descending position sequence by using a *stride* of −1.

```
indicators['RUS':'CHN':-1]
```

```
|                        country      pop       gdp life     cell
| code
| RUS   Russian Federation  144.50  1578.62 72.12   227.30
| IND                India 1338.66  2652.55 68.80  1168.90
| CHN                China 1386.40 12143.49 76.41  1469.88
```

When we have a row Index that consists of multiple levels like in topnames with year and sex making up the two levels of the index, a value of a row label is a tuple with specific values at each of the levels. Then we can select rows using a slice where the *start* and *end* values are tuples:

```
topnames[(1950, 'Male') : (1952, 'Female')]
```

```
|               name   count
| year sex
| 1950 Male    James   86224
| 1951 Female  Linda   73978
|      Male    James   87261
| 1952 Female  Linda   67082
```

All of these types of row selection using slices depend on the order of the rows in the data set. The operation of setting an index does not change the order of the rows, and so one could imagine a data set where the row ordering did not follow a regular pattern and so did not enable effective slice-based row selection. One solution to this problem would be to perform a sort_index as a means of changing the order, and thus allowing the techniques in these last two subsections.

7.3.4 Row Selection by Condition

The ability to select rows by an order-based slice may not be sufficient when we are trying to use operations to query the data to obtain a subset that helps answer a question. We often want to perform row selection based on finding the rows that match a *Boolean condition*, a condition defined by values of the row label index and by values of certain column fields within the row. This is more commonly known as *filtering*, and was previously discussed in Chap. 3.

We know from Sect. 7.3.1 that we can define a column vector (Series) whose values are Boolean True and False based on column values:

```
large_country = indicators['pop'] > 1000
large_country
```

```
| code
| CHN      True
| IND      True
| RUS      False
| USA      False
| VNM      False
| Name: pop, dtype: bool
```

Given such a Boolean vector, we can use square bracket accessor operators with a Boolean vector inside the brackets, and this serves to select exactly those rows for which the Boolean vector is True.

```
indicators[large_country]
```

```
|          country      pop         gdp    life      cell
| code
| CHN      China    1386.40    12143.49   76.41   1469.88
| IND      India    1338.66     2652.55   68.80   1168.90
```

Often, if only used once, the Boolean vector is not assigned to an intermediate variable:

```
indicators[indicators['pop'] > 1000]
```

```
|          country      pop         gdp    life      cell
| code
| CHN      China    1386.40    12143.49   76.41   1469.88
| IND      India    1338.66     2652.55   68.80   1168.90
```

When we need more complex Boolean conditions, say to perform a logical OR operation, like finding the rows where population is greater than 1000 million people or where life expectancy is greater than 77, we need to perform an elementwise OR between two Boolean vectors. If we use intermediate variables for each of the conditions:

```
many_people = indicators['pop'] > 1000
long_life = indicators.life > 77
```

We then use the | as an *or* operator (see Sect. 7.3.1):

```
combined = many_people | long_life
combined
```

```
| code
| CHN      True
| IND      True
```

```
| RUS      False
| USA       True
| VNM      False
| dtype: bool
```

Then, `indicators[combined]` would select the desired rows returning the resulting `DataFrame`. Alternatively, we can specify the conditions directly inside the accessor operator:

```
indicators[(indicators['pop'] > 1000) |
           (indicators.life > 77)]
```

```
|                 country      pop      gdp  life     cell
| code
| CHN              China  1386.40 12143.49 76.41  1469.88
| IND              India  1338.66  2652.55 68.80  1168.90
| USA     United States   325.15 19485.39 78.54   391.60
```

Sometimes we need to add parentheses to force the precedence; to act as a Boolean logical operator on `Series`, the vertical bar must have operands that are Boolean `Series` vectors. In `many_people | large_area`, this is satisfied. If we were to write `indicators['pop'] > 1000 | indicators.life > 77`, the `|` would, by Python precedence, try to operate between the integer 1000 and the `indicators.life Series`, and would result in an error.

A common use of Boolean filtering is to filter out missing data. Whenever a data entry (characterized by its (row, column) pair) is missing, `pandas` represents that data entry by the special NaN. In `pandas`, the built-in `pd.notna()` (read "not N.A.") takes as argument a `Series` and returns `True` for every row of the `Series` that does not represent missing data. The built-in `pd.isna()` has the logically opposite functionality. For example, if we were concerned that some rows of `indicators` might be missing the life expectancy data, the filter `indicators[pd.notna(indicators.life)]` would yield only the rows where life expectancy data is present. We will return to missing data in Chap. 8.

In Sects. 7.3.1 through 7.3.4, we have seen the *same* operator syntax, the access operator (`[]`), be used to project a single column, project multiple columns, select rows by slicing, and select rows by condition filtering. To effectively use this operator, it helps to understand that `pandas` is using the *type* of the operand between the square brackets to determine the actual operation performed. Table 7.2 summarizes these operations supported by the access operator, when operating directly on a `DataFrame`.

The `pandas DataFrame` provides two additional methods that give a specialized form of row selection based on a Boolean condition, and these do not use the access operator. Often we wish to select a set of rows whose values for a particular column are the largest (or smallest) among all of the rows in the data set. For these methods, we provide n, the number of largest (or smallest) rows to select, and the column label to be used in evaluating the value of the row.

Table 7.2 Operations supported by the `DataFrame` access operator

Access operator Arg type	Operation	Result type
String	Project single column by label	`Series`
List of strings	Project one or more columns by label	`DataFrame`
Integer slice	Select one or more rows by position	`DataFrame`
Non-integer slice	Select one or more rows by ordered row label/`Index` value	`DataFrame`
Boolean `Series`	Select zero or more rows by Boolean condition as filter	`DataFrame`

In the following example, we select the five rows with the largest `count` column value in the `topnames DataFrame`:

```
topnames.nlargest(5, 'count')
```

```
|                 name   count
| year sex
| 1947 Female    Linda   99689
| 1948 Female    Linda   96211
| 1947 Male      James   94757
| 1957 Male      Michael 92704
| 1949 Female    Linda   91016
```

In similar fashion, the following expression selects the three rows in the `indicators DataFrame` with the smallest value in the `life` column:

```
indicators.nsmallest(3, 'life')
```

```
|                 country      pop      gdp  life    cell
| code
| IND                India  1338.66  2652.55 68.80 1168.90
| RUS  Russian Federation   144.50  1578.62 72.12  227.30
| CHN                China  1386.40 12143.49 76.41 1469.88
```

7.3.5 Combinations of Projection and Selection

For any `pandas DataFrame`, `.loc[]` and `.iloc[]` are Python attributes of the `DataFrame` object that use square bracket accessors to flexibly specify rows and/or columns, enabling the combination of projection and selection in a single operation. The difference between `loc` and `iloc` is that `loc` is `Index` based, so its specification of rows and of columns is relative to the row label and column labels defined for the `DataFrame`. On the other hand, `iloc` is integer position based, and so for both rows and columns, the specification of rows and columns is based on the integer position of the rows/columns involved.

We always use square bracket accessors with either the `.loc` or the `.iloc` attributes. Inside the square brackets, we specify a first component that determines the desired rows, then a comma, and then a second component that determines the desired columns. This gives a row-comma-column notation familiar from algebra when subscripting two-dimensional variables. Syntactically, we have

dataframe`.loc`[*rowspec, colspec*]

dataframe`.iloc`[*rowspec, colspec*]

This syntax specification can be translated as:

1. Name the dataframe object.
2. Append the `loc` or `iloc` attribute with a period.
3. Follow the attribute with the access operator:

   • inside the access operator, compose a row specifier (*rowspec*) followed by a comma, followed by a column specifier (*colspec*).

The *rowspec* and *colspec* can range from identifying a single row and/or column to identifying all rows or all columns. For each of *rowspec* and *colspec*, the syntax variations define specification of either:

1. *Exactly one* row or column of the data frame:

   • these are specified with a single value of an `Index` or a single positional integer.

2. *Potentially more than one* row or column of the data frame:

   • these may be specified with a list, a slice, or a Boolean `Series`.

The next few subsections will cover some of the possibilities among these variations.

7.3.5.1   Access a Single Element

When accessing a *single element* in a data frame, we use a *rowspec* and a *colspec* that both specify *exactly one* row or column. This means, for `loc` that we use a value of the `Index` in each dimension, and for `iloc` we use a single integer position. In the `indicators` set, our `Index` are strings in both dimensions, so accessing the `'gdp'` for `'VNM'`:

```
indicators.loc['VNM', 'gdp']
```

```
| 223.78
```

The result is a scalar with the data type of the entry. (So *not*, for instance, a `DataFrame` of one row and one column.)

In the `topnames` dataframe, the `Index` for the row has two levels, `year` and `sex`, and a value for a row is then a tuple, so we can access the top `'Male'` name from 1961:

```
topnames.loc[(1961, 'Male'), 'name']
```

> 'Michael'

Using `iloc` on the `indicators` data set, if we know that the `'VNM'` row is at position 3, and the `gdp` column is at position 2:

```
indicators.iloc[4, 2]
```

> 223.78

7.3.5.2 Querying a Single Column or Single Row

We can use slicing to specify potentially more than one row or to specify potentially more than one column, when the desired items are based on the *order* in the data frame. In the limit, a slice with no *start* nor *end* (i.e., : by itself) specifies *all* the positions.

So if we desire a single column, like `cell`, from the `indicators` data set, and all rows, we can use a *rowspec* of : for all rows, and the column label `'cell'` to specify exactly one column:

```
indicators.loc[ : , 'cell']
```

> code
> CHN   1469.88
> IND   1168.90
> RUS   227.30
> USA   391.60
> VNM   120.02
> Name: cell, dtype: float64

The result is a `Series`, since the row dimension has (potentially) many elements and the column dimension specifies exactly one.

Similarly, if we want one row, like `(1961, 'Female')` and all columns from `topnames`:

```
topnames.loc[ (1961, 'Female') , : ]
```

> name   Mary
> count   47680
> Name: (1961, Female), dtype: object

We again get a `Series`, but since the data types of different columns may be different from one another, the type of the `Series`, dtype, is a generic `object`.

In both examples, one of the *rowspec* or *colspec* had *exactly one* in one dimension and *potentially more than one* in the other dimension, so it makes sense that the result is a one-dimensional `Series`.

The `.iloc` can be used in similar fashion, so to project the single column at position 1 and all rows from `indicators`:

```
indicators.iloc[ :, 1 ]
```

```
| code
| CHN    1386.40
| IND    1338.66
| RUS     144.50
| USA     325.15
| VNM      94.60
| Name: pop, dtype: float64
```

obtaining the `Series` for the population column.

Once we have a `Series` result, we can obtain elements from *within* the `Series` using another access operator, and referring to elements by the appropriate label. So, if in the example above, we obtain a particular row, the top `'Female'` baby name in 1961, and assign to a variable, we can then access the individual column elements `'name'` and `'count'`:

```
row = topnames.loc[ (1961, 'Female') , : ]
print(row['name'], row['count'])
```

```
| Mary 47680
```

7.3.5.3   Querying a Subset of a Single Column or Single Row

As opposed to using an unbounded slice (`:`) to specify all the elements in one dimension, we may wish to query a *subset* of the rows for a single column, or a subset of the columns for a single row. We have three options for specifying such a subset:

1. Use a slice, but bound the *start* and/or *end* of the slice.
2. Use an explicit list and, by `Index` value or by position, identify the exact items for the subset.
3. Use a Boolean `Series` (like we did in Sect. 7.3.4), where the items corresponding to `True` values are to be included, and the items corresponding to `False` values are not.

Regardless of which of these three methods we use to specify the subset, if we combine, in the other dimension, with an *exactly one* specification for row or column, we will again obtain a `Series` result.

There are many permutations of specifying the *rowspec* and *colspec* when using `.loc` and `.iloc`, so in the following we just show a limited sample of examples.

The one restriction is that, for `loc`, the values used in slices and lists *must* by values of the appropriate `Index`, and for `iloc`, the values used in the slices and lists must be integer positions.

In general, for both readability and for avoiding errors and fragile code that could arise when changes occur in the data frame, we prefer `loc` over `iloc`.

Suppose we want the `'count'` column for the ordering of rows from `(1990, 'Female')` to `(1991, 'Male')` in `topnames`. We use a slice to specify the subset of rows, and a single value to specify the column.

```
topnames.loc[(1990, 'Female'):(1991, 'Male'), 'count']
```

```
| year   sex
| 1990   Female      46475
|        Male        65290
| 1991   Female      43478
|        Male        60785
| Name: count, dtype: int64
```

Or we might want the `'cell'` column for those countries whose `'gdp'` is less than 2000 million US dollars. We use a Boolean `Series` to specify/filter the rows, and use a single value to specify the column.

```
indicators.loc[ indicators.gdp < 2000, 'cell']
```

```
| code
| RUS     227.30
| VNM     120.02
| Name: cell, dtype: float64
```

If we want an explicit subset of the columns, say `'country'`, `'life'`, and `'pop'` for the `'USA'` row, we can do the following:

```
indicators.loc[ 'USA', ['country', 'life', 'pop'] ]
```

```
| country      United States
| life                 78.54
| pop                 325.15
| Name: USA, dtype: object
```

Lastly, we illustrate how to use `.iloc` to obtain the same result, by knowing that the `'USA'` row is at position 3, and the desired columns are at positions 0, 3, and 1.

```
indicators.iloc[ 3, [0, 3, 1] ]
```

```
| country      United States
| life                 78.54
| pop                 325.15
| Name: USA, dtype: object
```

7.3.5.4   Generalized Projection and Selection

In the general case, we want, in a single operation, to project a subset of the columns as well as a subset of the rows. We already have the machinery needed to do this, and simply need to use both a *rowspec* and a *colspec* that specify potentially more than one item in the particular dimension. Our choices for these specifications are a combination, one per dimension, of

- a slice to specify the subset based on *order*,
- a list to explicitly specify the items in the subset, and
- a Boolean Series to allow a condition, obtained in a subordinate operation, to determine the subset.

Again, many permutations are possible, so we present an illustrative set.

Suppose, in the topnames0 data set, whose index is the default integer indices, we want to select rows where the count of applicants is more than 92000, and to select the explicit columns of 'count', 'sex', and 'year':

```
topnames0.loc[topnames0['count'] > 92000,
                    ['count', 'sex', 'year']]
```

```
|       count     sex   year
| 134   99689  Female   1947
| 135   94757    Male   1947
| 136   96211  Female   1948
| 155   92704    Male   1957
```

We might want to select the explicit rows of 'RUS' and 'VNM', and project all the numeric columns, which are ordered from 'pop' to the end of the columns:

```
indicators.loc[['RUS', 'VNM'], 'pop':]
```

```
|            pop       gdp    life     cell
| code
| RUS      144.5   1578.62   72.12   227.30
| VNM       94.6    223.78   76.45   120.02
```

Many more examples are explored in the Reading Questions, and the reader is encouraged to approach them with a "problem-solving" mindset rather than a "guess and check" mindset. Think first about whether your DataFrame has an index, then which rows and columns you want, and then whether you want the result as a DataFrame, Series, or scalar. Once you are ready, carefully write down the access operator and test it.

7.3.6 *Iteration over Rows and Columns*

There are times when we need to process a data frame by operating row-by-row, or column-by-column over the entire frame. This, in fact, allows us to compose and build more advanced operations, like those for transforming and combining data frames. When operations are procedural, this requires the use of a `for` or `while` loop and, at each iteration of the loop, we must be able to obtain the information for the unit of iteration. For a column-by-column iterations, the information would be the label for the column, along with the column `Series`. For row-by-row iterations, the information would be the row label, along with a row `Series`.

While the `pandas` `DataFrame` class has a number of variations and methods that enable iteration, the most efficient of these provides an *iterator* that can be used in a `for` loop, and does so very efficiently, even for large data sets.

In our first example, we iterate over the rows of the `indicators` `DataFrame`. The `iterrows()` method gives the iterator to use in the `for` loop. At each iteration of the loop, the iterator yields the label of the row and the `Series` for the row, which in our example are associated with variables `rowlabel` and `rowseries`, respectively. In the body of the loop, at each iteration, we simply print the label and access the `'gdp'` and `'pop'` columns of the row.

```
for rowlabel, rowseries in indicators.iterrows():
    print(rowlabel + ':', rowseries['gdp'], rowseries['pop'])
```

```
| CHN: 12143.49 1386.4
| IND: 2652.55 1338.66
| RUS: 1578.62 144.5
| USA: 19485.39 325.15
| VNM: 223.78 94.6
```

The second value returned in each iteration, `rowseries`, is a `pandas` `Series` object. As we learned previously about a `Series` object, there is a `pandas` `Index` object along with the data itself, giving each element in the `Series` a logical name. When the `Series` is a row of the data set, the names of the individual elements are the *column labels*. This is why we can use the column name as the accessor value inside the square brackets (e.g., `rowseries['pop']`) to get, in dictionary-like fashion, the value associated with that element of the `Series`.

Column-by-column iteration is very similar. The `iteritems()` method provides the iterator for the `for` loop, and at each iteration, provides a tuple of a column label and a column `Series`. This is analogous to the dictionary `items()` method from Chap. 3. In the provided example, we iterate over the columns of `topnames` `DataFrame`. We use the `Series` to get the column label, and we use the `Index` to access the data desired. For a two-level index like in `topnames`, access involves a tuple, as shown below. We retrieve values for two specific rows, and for all columns.

```
for collabel, colseries in topnames.iteritems():
   print(collabel + ':', colseries[(1961, 'Female')],
                        colseries[(1971, 'Female')])
```

```
| name: Mary Jennifer
| count: 47680 56783
```

If we employ the operation of iteration row-by-row, and, inside that loop, include an inner loop that iterates over all the elements of the row `Series`, we can individually access all table entries of a frame. We could accomplish the same thing with column-by-column iteration. But with the `Series` type of operations supported by `pandas`, there is little need for such algorithms.

7.3.7 Reading Questions

7.14 In order to project onto a column using its name, like `indicators.gdp`, it must be the case that the column name is not the name of an existing attribute or method. Give an example where this could fail.

7.15 When accessing an element in a data frame via double subsetting, what happens if you swap the rows and columns, e.g., `indicators['USA']['pop']`? Explain.

7.16 When accessing an element in a data frame via double subsetting, what happens if you combine the row and column identifiers, e.g., `indicators['pop','USA']`?

7.17 We can multiply a `Series` by a number (e.g., *population* ∗ 1.25) and can divide one `Series` by another (*cell*/*population*). Do you know another data structure with the same functionality?

7.18 What happens if you apply ~ to a `Series` of strings (instead of Booleans)?

7.19 Give a different example application of `apply`, using either a built-in function (but not `len`) or defining your own function.

7.20 Give an example of an application of `apply` using an anonymous `lambda` expression.

7.21 To better understand the difference between the `Series` gdp1 and the `DataFrame` gdp2, try to use the various `Series` built-ins on gdp2, including multiplication by a number, filtering by a Boolean condition, and `apply`. Which of these work?

7.22 Where have you previously seen something like `stride`?

7.23 It was a design decision to make slicing include its *end* row when `index` is used rather than position. Explain on a human level why this is a good design decision.

7.24 In the generic syntax for a slice, it says *start:end[:stride]*, but the example in the text says `indicators['RUS','CHN',-1]`. Explain.

7.25 How does Python decide which countries should be included in a slice like `indicators['CHN':'RUS']`? Is it always alphabetical? What if someone wanted a slice by GDP?

7.26 The examples show how to get a subset of rows where a Boolean condition is True, e.g., with `indicators[large_country]`. How could you get exactly the rows where `large_country` is `False`?

7.27 Would it ever make sense to introduce a *row* of Booleans instead of a *column* of Booleans?

7.28 How do you think Python extracts the relevant rows when you call `nlargest`? If you were designing `pandas` would you have it sort every time? Justify your answer.

7.29 When using `loc`, what happens if you swap the rows and columns, like `indicators.loc['gdp','VNM']`?

7.30 Write a use of `loc` that specifies all columns, for the bottom three rows of `indicators`.

7.31 When using `iloc` and slicing, does it include both endpoints?

7.32 Write down three more example uses of `loc`, including 1. The GDPs and populations for China and the USA 2. All years where the count for male names in `topnames` was bigger than 60,000. 3. An example using slicing of both rows and columns.

7.33 What type of object do you get from `indicators.loc[ ['VNM'], : ]`? Explain.

7.34 The ability to iterate over rows and columns is a major upside of `pandas` data frames. Discuss how this functionality compares to list of lists and dictionary of lists representations of two-dimensional data.

7.35 Write three lines of code that iterate over the rows of `topnames` and print counts of male births in years like 1950, 1960, etc. Use an if-statement.

7.36 Write three lines of code that iterate over the columns of `indicators` and print every country name where $life > 75$ and $gdp > 10,000$.

7.3.8   *Exercises*

7.37 Convert the following into a data frame and name it `df`:

```
[[1,  2,  3],[4,  5,  6],[7,  8,  9],
 [10,  11,  12],[13,  14,  15],[16,  17,  18]]
```

Then display the data frame.

7.38 With reference to the above, assign the columns the following labels: "A," "B," and "C."

Assign the rows the following labels: "U," "V," "W," "X," "Y," and "Z"

Then, display the data frame.

7.39 In a single line of code, slice off everything besides the first 3 rows and 2 columns of the data frame. Call this new data frame `df1`. Then display the new data frame.

7.40 Convert the following dictionary to a data frame called `df` and compute the mean test score for `Test1`:

```
{'Name'  :  ['Owen',  'Rachel',  'Kyle'],
           'Test1'  :  [82,  85,  95],
           'Test2'  :  [89,  85,  90],
           'Test3'  :  [93,  85,  87]}
```

7.41 Print the following:

1. The test scores of Test1
2. Rachel's test scores
3. Kyle's third test score
4. Using indexing, the scores of Owen and Rachel

7.42   Add a column, "Final," that equals the average of the three test scores, rounded to 2 decimal places. (You may need to explore the built-in function `round()` in addition to adding a column to a data frame based on a computation.)

7.43 Using the `drop` method, delete the column you just made.

7.44 Read the following into a data frame:

```
data = {'animal': ['cat','cat','snake','dog',
     'dog','cat','snake','cat','dog','dog'],
'age': [2.5, 3, 0.5, 7, 5, 2, 4.5, 4, 7, 3],
'visits': [1, 3, 2, 3, 2, 3, 1, 1, 2, 1],
'priority': ['yes','yes','no','yes','no',
             'no','no','yes','no', 'no']}
```

Using as index:

```
labels = ['a', 'b', 'c', 'd', 'e',
          'f', 'g', 'h', 'i', 'j']
```

Then, carry out a selection of rows d through g, inclusive.

7.45 Select the data in rows 2, 3, and 4 and project the columns "animal" and "age."

7.46 Create and add a column `age_visit` which, for each row, is the age of the animal divided by the number of visits for the animal.

7.47 Select only the rows where the `age` is greater than 3.

7.48 Compute the sum of the number of visits of animals in the data set.

7.49 Project the animal and visits columns from those rows where priority is yes.

7.50 On the book web page you will find a data file `us_rent_income.csv`. Read it into a data frame, then:

1. Extract the `shape` of the data into a variable called `sh`.
2. Extract the *list* of `estimate`, into a variable called `L` (hint: be very careful about type).
3. Define a list of Booleans `K` where `K[i]` is `True` exactly when `variable[i]` is `income`.
4. Define a Boolean `N` which is `True` for any rows with missing data. Do this using a `pandas` built-in function, rather than by using the `math` module. The documentation will help.
5. Use a projection or built-in function to define a new dataframe `df2` consisting of only the rows of `df` without missing data.
6. Create a new data frame `df3`, via slicing and `iloc`, corresponding to the rows that are about `Ohio` and the columns for `variable` and `estimate`. Your data frame should have two rows and two columns.
7. Use a projection to define a new dataframe `df4` consisting of only columns without missing data. Use `loc`.
8. Create a new data frame `df5` with the same rows as before, but with a new column `CI` defined as a string interval `[estimate-moe, estimate+moe]`. For example, in the first row, it would be `[24476-136,24476+136]`, i.e., `[24340,24612]`.
9. Iterate over the rows to find the state with the largest income. Beware that this will involve checking the `variable` column to see if it says "income" or "rent."

7.51 On the book web page you will find a file `education.csv`, based on data hosted by www.census.gov. Read it into a data frame, then:

1. Extract the `shape` of the data into a variable called `sh`.
2. Extract the `Series` of `Geography`.
3. Write lambda functions to split a given entry (of the form shown in `Geography`) into city and state.
4. Apply your functions to the `Geography Series`, obtaining new `Series` for city and state.

5. Create a new data frame, via slicing and `loc`, corresponding to the rows that are about `Ohio` and the `Estimate` columns (but not the `Margin of Error` columns).
6. Select 2–3 estimates you are interested in, then use a projection to reduce down to a smaller data frame with just the columns you are interested in.
7. Get rid of the descriptive first row.
8. Select a geography you are most interested in, then iterate over the columns to find the one with the largest `Estimate`.

7.52 On the book web page you will find a data file `world_bank_pop.csv`. Read it into a data frame, then:

1. Extract the `shape` of the data into a variable called `sh`.
2. Extract the list of years the data set concerns.
3. Extract the list of countries the data set concerns. Be sure not to include duplicates.
4. Extract a list of indicators present in the data. Do not include duplicates.
5. Use row and column projection to reduce the data to just a few countries, a few years, and one indicator. Decide which based on your interests.
6. Create a `Series` whose values are the log (base 10) of each of the total population values in the data set. Create such a `Series`, utilizing the `numpy` function `log10()`, which can operate on an argument that is either a one-column `DataFrame`, a `Series`, or a `numpy` array. Add this `Series` as a column.
7. Iterate over the rows or columns of your resulting data frame to answer an interesting question.

Chapter 8
Tabular Model: Advanced Operations and `pandas`

Chapter Goals

Upon completion of this chapter, you should understand the following:

- Mutation operations to update, add, and delete data.
- Operations to group and aggregate data.
- The need for operations specific to handling missing data.

Upon completion of this chapter, you should be able to do the following:

- Perform aggregations on single and multiple `Series`.
- Aggregate and summarize information on group-based partitions of data frames.
- Combine and update data frames.
- Detect and mutate data frames to handle missing data.
- Carry out these operations in `pandas`.

© Springer Nature Switzerland AG 2020   205
T. Bressoud, D. White, *Introduction to Data Systems*,
https://doi.org/10.1007/978-3-030-54371-6_8

8.1 Aggregating and Grouping Data

In many analyses as well as in data exploration, we need to summarize data from a data frame in various ways. We may, for a single column, want to get some individual values that help characterize the data in that column, called *aggregating*, by finding its mean, median, max, min, and so forth. Sometimes we want a single aggregate for a column, but sometimes we want to compute multiple aggregates for a single column. We can extend this, and if we have a data frame with multiple columns of numeric data, we may want to characterize those multiple columns in a single operation, perhaps finding the mean for one column and the max and min for another column.

When the variables of a data frame allow us to define different *row subsets* of the data, we often want to compute aggregates for one or more of these subsets. This can allow us to focus on and characterize a single subset, or it can allow us to draw comparisons between the different subsets. An example is grouping countries by their regions of the world and computing aggregates over different regions. This kind of grouping often relies on the independent variables of the data frame, or values of other columns in the data frame, like columns containing categorical variables.

The collection of operations that support such aggregation and grouping are the subject of the current subsection.

8.1.1 Aggregating Single `Series`

The simplest form of aggregation occurs when we have a single `Series`, often a single column projection from a `DataFrame`, and want to compute one or more functions over the values in the `Series`. The pandas `Series` class defines *methods* that operate on a `Series`, and these methods include many/most that one might need. You will notice in the description that aggregation functions often omit missing values (values with a `NaN` entry in the `DataFrame`) from their computation.

Table 8.1 summarizes many of the most common aggregation functions/methods:

Table 8.1 Aggregation function/methods

Method	Description
mean()	Arithmetic mean, not including missing values
median()	Value occurring halfway through the population, omitting missing values
sum()	Arithmetic sum of the non-missing values
min()	Smallest value in the set
max()	Largest value in the set
nunique()	Number of unique values in the set
size()	Size of the set
count()	Number of non-missing values in the set

Table 8.2 Aggregation to `Series` methods

Method	Description
`unique()`	Construct a subset `Series` consisting of the unique values from the source `Series`
`value_count()`	Construct a `Series` of integers capturing the number of times each unique value is found in a source `Series`; the `Index` of the new `Series` consists of the unique values from the source `Series`

Many of the methods are overloaded—they have the same name as built-in functions, so some may be invoked by passing the `Series` as the argument to the function, instead of invoking a method (with no argument) on a `Series` object. For example, if `s` is a `Series` of integers, then `max(s)` and `s.max()` compute the maximum value in `s`.

Table 8.2 displays two other `Series` methods that come in handy in data exploration and characterizing data. Both return a collection, rather than a single value.

Consider the `topnames0` DataFrame, which has the default integer index, and so `year` and `sex` are the regular columns. We can determine the `max` and `median` counts of the top social security applications, for a year and sex, by a single column projection of `'count'`, followed by a method invocation for the two aggregates. In the following, we project the `'name'` column and, in the same statement, invoke the method for determining the number of unique names in the collection:

```
count_series = topnames0['count']          # project
count_max = count_series.max()             # agg method
count_median = count_series.median()       # agg method
num_unique = topnames0['name'].nunique()
                                           # project and method
print("Mean:", count_max,
      "Median:", count_median,
      "Number of Unique Names:", num_unique)
```

```
Mean:99689 Median:52556.0 Number of Unique Names:18
```

We see that the number of unique names, over the 238 rows of the data frame, is a mere 18. So if we wanted to see what these top 18 names are, we could invoke the `unique()` method and get the vector of names; if we wanted to see *how many* of each of these unique names, we would invoke the `value_count()` method. We do so in the following code and then use the `head()` method on the latter `Series` to output just a prefix of the result:

```
name_counts = topnames0['name'].value_counts()
name_counts.head()
```

```
| Mary         76
| Michael      44
| John         44
| Robert       17
| Jennifer     15
| Name: name, dtype: int64
```

8.1.2 Aggregating a Data Frame

In this and the following two subsections, we build from aggregation of a `Series` to some of the most powerful and useful operations of the tabular model—grouping rows into discrete partitions, aggregating each partition into a row of computed values, and combining the result. We help the discussion using a sequence of figures that themselves build upon one another, starting in Fig. 8.1.

The left side of the figure depicts an original data frame with four columns, A, B, C, and D. Assuming that column A is a level of an index, or a column with a categorical variable, and, by color and shading, we see three different "categories" of values for A. Let us assume that columns B and C are numerical and are columns that we want to employ aggregation to characterize the data in the frame. Column D might be a string value and, by assumption, is not of interest in our aggregation. For example, suppose the source frame is our `indicators` data set, column A is the index (`code`), column B is the life expectancy column, and column C is the cell subscriber column.

The goal then is, starting with a source data frame, to perform an operation on the frame and get a result that is composed of (multiple) aggregations, e.g., the minimum life expectancy and median number of cell subscribers.

On the right side of the figure, column A is gone because we have aggregated over it. We no longer have one row per country, but now instead have one row per aggregation function. Figure 8.1 shows the general case, where, for any given column, we might apply multiple different aggregations. We show the result as a

Fig. 8.1 Aggregating a data frame

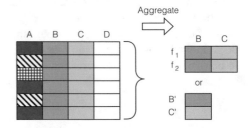

table in its own right, with columns corresponding to the columns from the original (B and C), and rows corresponding to different aggregation functions (f_1 and f_2, in this case). When, for any given column, we only have a single aggregation function, an alternative view of the result is as a vector, where the elements contain the particular aggregation for each of the requested columns. We label these as B' and C' to emphasize that they are different, by their aggregation, from the original column.

If $f1$ were the min aggregation and $f2$ were the median aggregation, the entries in the result would allow us to lookup, by the original column and aggregation function, each of the individual results. Suppose further that we wanted the min and median for the life column, but only the median for cell column. The figure would still be a valid representation of the result, but at column cell and row min, there would be a *missing value* (NaN), because we did not request to compute the minimum number of cell phone subscribers.

In pandas, we use the agg() method of a DataFrame to perform this general form of the aggregation operation. The argument to agg() must convey, for each column of the original DataFrame, what aggregation function or functions to perform. One way of doing is this is with a dictionary, where we map from a column name to a function or string function name, or to a list of specified functions.

We demonstrate the indicators example below, performing two aggregations on the 'life' column ('min' and 'median'), and just the 'median' aggregation on the 'cell' column. The result is a DataFrame, and the row labels on the result are the names of the aggregation functions, matching the top right of Fig. 8.1.

```
indicators.agg({'life':['min', 'median'], 'cell':
                                          'median'})
```

```
|              life    cell
| median     76.41   391.6
| min        68.80     NaN
```

If a data frame aggregation only has a *single* aggregation for any given column, the result can be simpler—it can be a one-dimensional object with an entry for each of the columns being aggregated. This is still true, even if the same aggregation function is repeated for different columns. This is shown pictorially in the bottom right of Fig. 8.1. In the following example, we compute the min aggregation for the life and cell columns, and the sum aggregation for the gdp column.

```
indicators.agg({'life': 'min', 'cell': 'min', 'gdp':
                                          'sum'})
```

```
| life          68.80
| cell         120.02
| gdp        36083.83
| dtype: float64
```

The `Index` of the `Series` reflects the names of the columns being aggregated, but the information on what aggregation is performed is not part of the result.

Many other shortcuts are provided by the `pandas` module. For instance, if we want to perform the *same* aggregation for *all* columns of a `DataFrame`, we can simply invoke the aggregation method directly on the `DataFrame` itself. Here, we invoke `max()` aggregation for all columns of the `indicator` frame:

`indicators.max()`

```
| country      Vietnam
| pop           1386.4
| gdp         19485.4
| life           78.54
| cell         1469.88
| dtype: object
```

This is like the prior example: since there is only a single aggregation function per column, the result is a `Series` containing the results. Also notice that `pandas` will, whenever possible, perform the aggregation over *all* columns. Here, the `max()` aggregation, when applied to the string column of `country` names, results in the alphabetically last country name of the collection.

8.1.3 Aggregating Selected Rows

Often, we wish to focus on a subset of a data frame and then characterize, through aggregation, the data within that subset. At one level, this is a composition of operations we have already seen:

1. Create a subset by selecting just the rows in the data frame that form the subset, yielding a new data frame.
2. Perform aggregation on the new subset data frame.

This "subset and aggregation" is also a critical piece in understanding the more general group partitioning and aggregation that is to follow, and so we will look at this limited example first.

The situation is depicted in Fig. 8.2. We see the filter step, where the value from column *A* is used as the criterion for the subset—all rows that have a "downward hash" value in column *A* are selected for a new data frame, and then we perform an aggregation to obtain the desired result.

Say that the original data frame is `topnames`, defined with the two-level index of `year` and `sex`. We may be interested in all the rows where `'Mary'` was the top applicant. We can select rows with the access operator and using a Boolean `Series` compute which rows have a value in the `'name'` column of `'Mary'`. We display a portion of the result and see that we have the index values and the two columns of `name` and `count`.

Fig. 8.2 Filtering and aggregating

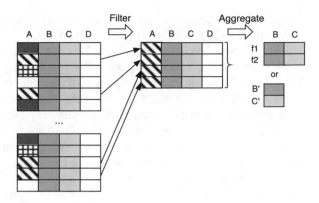

```
mary_rows = topnames[topnames['name'] == 'Mary']
mary_rows.tail()
```

```
|               name    count
| year sex
| 1957 Female  Mary    61094
| 1958 Female  Mary    55857
| 1959 Female  Mary    54474
| 1960 Female  Mary    51475
| 1961 Female  Mary    47680
```

This new `DataFrame`, `mary_rows`, is the middle data frame in Fig. 8.2. We show two variations where we aggregate to obtain the min, max, and median of the count column of the selected rows. For our first example, we aggregate against the subset `DataFrame` and use a dictionary to specify the mapping from a column to a list of desired aggregations:

```
mary_rows.agg({'count': ['min', 'max', 'median']})
```

```
|             count
| min        6919.0
| max       73985.0
| median    54423.0
```

The result is a `DataFrame` with a single column, and the labels of the rows are the three aggregation functions.

For our second example, we first project the `'count'` column and then use `agg` as a method of the `Series` and specify the list of aggregations.

```
mary_rows['count'].agg(['min', 'max', 'median'])
```

```
| min        6919.0
| max       73985.0
| median    54423.0
| Name: count, dtype: float64
```

The result is a `Series`, with the row labels denoting the requested aggregations.

Sometimes when we wish to define a row selection, it is based on part of the row `Index`, not based on a regular column. This can occur when we have a multi-level index. Consider our `topnames DataFrame`. We may want a selection of rows that consists of just one of the `sex` values, like just the `Female` rows, which would be a subset that includes half the rows of the original `DataFrame`. Or, we might be interested in a row selection of a single year, say 1961, in which case the subset would consist of exactly two rows.

When the condition for the row selection is based on a value of an index level, rather than a value of a column, we cannot use a Boolean `Series` to obtain our subset. While we could `reset_index` to make each of the levels regular columns (and then use the Boolean `Series` method of selecting rows), `pandas` offers an alternative. In `pandas`, a subset of the `DataFrame` that is based on one or more index values is called a *cross section*, abbreviated (`xs`). So `pandas` provides a cross-sectional method, `xs()`, that allows us to specify a value and an index level and can very efficiently create the subset. We do so here, to obtain the selection of rows for `'Female'` entries and then aggregate the `count` column by min, max, and median.

```
female_subset = topnames.xs(key='Female', level='sex')
female_subset['count'].agg(['min', 'max', 'median'])
```

```
| min        6919.0
| max       99689.0
| median    48347.0
| Name: count, dtype: float64
```

This allows us, in this case, to compare the application counts for when `'Mary'` was the top baby name against the population of `'Female'` entries overall.

8.1.4 General Partitioning and GroupBy

Generalizing the pattern above, an analysis may want to go beyond a single selection of rows to one where the filtering is repeated to obtain a *set* of row selections, with a goal of comparing the various sets. This is depicted in Fig. 8.3, where we see three separate row selections based on the three different values present in the *A* column. For example, the rows might be individuals and *A* could be gender, taking values "Male," "Female," and "Non-Binary."[1]

We call this *partitioning* to draw on the mathematical meaning: each and every row from the original is selected into *exactly one* of the sets. Following the figure,

---

[1]Gender includes a person's concept of their own gender identity, whereas sex is biological and anatomical. This is why sex, rather than gender, is included in the US Social Security Administration data on baby names.

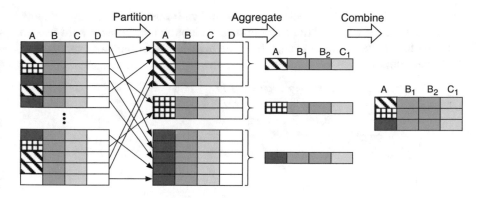

Fig. 8.3 Partition/aggregate/combine

from left to right, we start on the left with a *single* data frame. The partitioning operation occurs, called a *groupby operation* (pronounced "group by"), yielding, logically, a *set of data frames*, each called a partition. Each of these partitions then has an aggregation performed upon it, and so, from each partition, we obtain a *single row* of values. Finally, the set of individual rows are combined to build a single data frame result. For our gender example, we would end up with three data frames after the groupby operation, one for each of "Male," "Female," and "Non-Binary." The aggregation could then compute, for example, the number in each group, so that the resulting data frame would have one column ("number") and three rows ("Male," "Female," and "Non-Binary").

In general, for the aggregation, we need to yield a result that is a single row per partition. Each row contains a column corresponding to the column (or columns) that we use for the partitioning, known as the *groupby column*. By the definition of how the partitioning works, each discrete value in the groupby column yields exactly one partition and then exactly one row after the aggregation. So the values in the groupby column are unique, and in pandas, they form a row label index for the final combined result. In Fig. 8.3, A is the groupby column and so the unique values in A form the row labels in the final combined data frame.

For the rest of the columns in Fig. 8.3 that result from the aggregation, we use subscripts to denote that these are some aggregation function over the elements of some particular column and are not the same as the column itself. In the figure, we depict two different aggregations that are based on the B column, and one aggregation that is based on the C column. This would be like if, in our gender example, we originally had columns B and C for the salary and number of years of education and then in the aggregation computed the median salary and average number of years of education, for each of the three gender groups.

In the topnames DataFrame, we could group by name, or by sex, or by year. The pandas method for the partitioning/groupby operation is called groupby(), and its first parameter is the name of the column (or columns) to use for the partitioning. The result is a GroupBy object, which is different from a

DataFrame because it, in essence, is a *set* of data frames. In the example below, we create the three aforementioned groupby partitions and then use the built-in `len` function to see the length of each, where the built-in length function, applied to a GroupBy object, gives the number of partitions.

```
groupby_name = topnames.groupby('name')
groupby_sex = topnames.groupby('sex')
groupby_year = topnames.groupby('year')
print(len(groupby_name), len(groupby_sex), len
                                        (groupby_year))
```

```
| 18 2 139
```

We see that partitioning by `name` yields 18 partition groups, which is the same as the number of unique names from the data frame. When we group by `sex`, we, as expected, get 2 partition groups, and when we group by `year`, we get 139 partition groups, one for each year in the data set.

To perform the aggregation step, we use the `agg()` method, defined for GroupBy objects in a fashion similar to the `Series` and `DataFrame agg()` methods. This lets us specify which columns and which aggregation methods to apply to each partition group. The result is the data frame obtained after both the aggregation step (to get a single row per partition group) and the combine step that concatenates these aggregation rows together.

For the `groupby_name` partition group, our example below shows aggregation of `count`, computing median and sum, which, in this case, helps us to understand, through the sum, the total number of applications for the top applications. We subset the result using `head()` after sorting the result by descending `count sum` values:

```
namegroup = groupby_name.agg({'count': ['median',
                                        'sum']})
namegroup.sort_values(by=('count', 'sum'),
                      inplace=True)
namegroup.head()
```

```
|              count
|              median        sum
| name
| Mary        54423.0   3098428
| Michael     68117.5   3084824
| James       86224.0   1056228
| Robert      60699.0   1041984
| John         9032.5    861403
```

The result shows how row and column `Index` objects are constructed when performing this kind of partition/aggregate/combine operation. The row labels come from the column specified in the `groupby()`, with a unique row per discrete value from the column. The column labels yield a two-level index, with the outer level determined by the column being aggregated, and the inner level determined by the set of aggregation functions applied to the column. This allows fully general aggregation over multiple columns and using multiple, different, aggregation functions.

8.1.5   Indicators Grouping Example

Because the groupby and aggregation operations are somewhat complex, we conclude this subsection by reinforcing our understanding with an additional example, using a derivative of the indicators data set. In this derivative, we drop the columns of life and cell but, based on a categorization from the World Bank Open Data project[89], include two additional columns, region, that provide a categorical variable for countries, grouping them into regions of the world, and income, that categorizes each country into "High Income," "Upper Middle Income," "Lower Middle Income," and "Low Income." We reference the data set with the indicators2 variable. Further, the new data set has many more countries—all the countries with at least $US 100 million in GDP. In this new version of the indicators2 data set, there are a total of 63 countries within a total of 7 regions of the world represented. Table 8.3 presents a five row prefix of the data set.

Before we perform a groupby() partitioning, we explore the data by first projecting the region column and use value_counts() to obtain the Series with the unique regions and the number of non-missing values in each:

```
indicators2['region'].value_counts()
```

```
| Europe & Central Asia          24
| East Asia & Pacific            12
| Middle East & North Africa     10
```

Table 8.3 World indicators

Code	Country	Region	Income	Pop	Gdp
AGO	Angola	Sub-Saharan Africa	Lower middle income	29.82	122.12
ARE	United Arab Emirates	Middle East & North Africa	High income	9.49	382.58
ARG	Argentina	Latin America & Caribbean	Upper middle income	44.04	642.70
AUS	Australia	East Asia & Pacific	High income	24.60	1330.80
AUT	Austria	Europe & Central Asia	High income	8.80	416.84

```
| Latin America & Caribbean          8
| Sub-Saharan Africa                 4
| South Asia                         3
| North America                      2
| Name: region, dtype: int64
```

The regions range from just two countries in the "North America" region to 24 countries in the "Europe & Central Asia" region.

We could further identify the countries in a particular region by selecting the rows for that region. For instance, we select the rows where region is 'Sub-Saharan Africa' and project all columns except the 'region' column:

```
indicators2.loc[indicators2.region ==
                            'Sub-Saharan Africa',
              ['country', 'income', 'pop', 'gdp']]
```

```
|             country               income     pop      gdp
| code
| AGO         Angola  Lower middle income    29.82   122.12
| NGA        Nigeria  Lower middle income   190.87   375.75
| SDN          Sudan  Lower middle income    40.81   123.05
| ZAF   South Africa  Upper middle income    57.00   348.87
```

In a similar fashion, we could explore the distribution of the countries in the different income categories. To start, we determine how many countries are in each category:

```
indicators2['income'].value_counts()
```

```
| High income              34
| Upper middle income      17
| Lower middle income      12
| Name: income, dtype: int64
```

Say that our goal was to compare the population, GDP, and number of countries based on the two categorizations of region and income. We first group by income and then apply our desired aggregates, seeking the mean of pop and gdp, and an aggregation of count for the number of countries in each partition:

```
aggs = {'pop': 'mean', 'gdp': 'mean', 'country': 'count'}
income_groups = indicators2.groupby('income')
income_groups.agg(aggs)
```

```
|                            pop          gdp   country
| income
| High income          33.574118  1458.293824        34
| Lower middle income  217.416667   486.519167        12
| Upper middle income  143.549412  1256.629412        17
```

Similarly, we can group by region and aggregate. This time, instead of counting the number of countries in the partition, we find the number of unique income categories in each region.

```
aggs = {'pop': 'mean', 'gdp': 'mean', 'income':
            'nunique'}
region_groups = indicators2.groupby('region')
region_groups.agg(aggs)
```

	pop	gdp	income
region			
East Asia & Pacific	180.982500	1922.537500	3
Europe & Central Asia	32.806667	861.221667	3
Latin America & Caribbean	61.947500	607.893750	2
Middle East & North Africa	34.971000	287.073000	3
North America	180.845000	10566.130000	1
South Asia	568.743333	1069.073333	1
Sub-Saharan Africa	79.625000	242.447500	2

8.1.6 Reading Questions

8.1 Thinking back to what you know about data analysis, give an example when you might want multiple aggregates of a single column.

8.2 Give an example of an aggregate you might use on a column of string data.

8.3 At a high level, how do you think the method unique() is implemented?

8.4 At a high level, how do you think the method value_count() is implemented?

8.5 In the example that calls max, median, and nunique, what would happen if you called nunique on topnames0 instead of topnames0['name']?

8.6 Give an example of a data set of the sort described in the figures, i.e., with columns A, B, C, and D, where A has three levels (and multiple rows of each level). Then, give an example of two possible aggregation functions f_1 and f_2.

8.7 Give an example of a dictionary mapping from column names to lists of functions (rather than function names). Use more than just built-in functions. Hint: you do not need parentheses because you are not calling the functions.

8.8 The command indicators.max() applies the function max to every column. How would you change this command to perform the max aggregation for all the columns except the country column? Use the agg method.

8.9 Why is it that, "when the condition for the row selection is based on a value of an index level, rather than a value of a column, we cannot use a Boolean Series to obtain our subset"?

8.10 Write a sentence to explain why len(groupby_name) is 18, and why len(groupby_year) is 139, in the topnames example.

8.1.7 Exercises

8.11 Use `agg` to return a data set with the max and min counts per year in `topnames` for males born in the 1960s. Is your data set a `DataFrame` or a `Series`, and how do you know?

8.12 Write code to add a new column `popSize` to `indicators`, which takes value "high" if `pop > 1000`, "low" if `pop < 100`, and "medium" otherwise.

8.13 Building on the question above, use `groupby` to partition `indicators` by this new column `popSize`.

8.14 Building on the question above, use `agg` to report the average population in each of the three groups.

8.15 Use one line of code to sort `indicators` by `pop` and `gdp`. Warning: pay attention to your `inplace` settings.

8.16 Modify the `indicators2` example to find the maximum and minimum population and GDP in each of the three income categories.

8.17 Modify the `indicators2` example to project onto the columns `country`, `pop`, `gdp`, `region` for all lower middle income countries. Then find the number of countries in each region of the resulting data frame.

8.18 This is a pencil and paper exercise. Suppose we have the following data set named `df`, and our goal is, for column C, to add together values for various row subsets of the data, and for column D, to average values from that column for various row subsets of the data.

A	B	C	D
foo	one	0.4	−1
foo	two	−1.5	0
bar	three	−1.1	1
bar	two	−0.1	−1
foo	three	−1.0	1

Answer the following:

   Show the partitioning (i.e., the sets of rows) that would result if we did a *GroupBy* the A column.

8.19 Give the result of

- dropping the A and C columns,
- grouping by the B column, and
- specifying aggregation by taking the `mean()` of the D column

   df[['B', 'D']].groupby('B').mean()

8.20 On the book web page, you will find a file sat.csv, based on data from www.onlinestatbook.com. This data contains the high school GPA, SAT math score, SAT verbal score, and college GPAs, for a number of students. Read the data into a pandas dataframe dfsat with column headers. Use agg to compute the average of each numeric column in the data set.

8.21 On the book web page, you will find a file education.csv, based on data hosted by www.census.gov. Each row is a US metropolitan area, with data on population (row[3]), number unemployed without a high school degree (row[15]), number unemployed with a terminal high school degree (row[29]), number unemployed with some college (row[43]), and number unemployed with a college degree (row[57]). Read the data into a pandas data frame dfedu with column headers. Then use agg to compute the total number of people in each unemployment category and to report the fraction of unemployed people from each category.

8.22 The reading showed three ways to group the topnames data but only carried out agg using one of them. Use grouping and aggregation to produce a DataFrame containing the average number of male applications and the average number of female applications.

8.23 Similarly to the previous exercise, produce a DataFrame containing the maximum number of applications (regardless of gender) each year.

8.24 Similarly to the previous exercise, find the year that had the maximum number of applications (regardless of gender).

8.25 To practice with a two-level index, use read_csv to read

```
earthquake_all_month.csv
```

into a data frame. Group the data frame by magType, and give statistics regarding the magnitudes, including min, max, median, and mean. Sort the resulting data frame by the mean, ascending. Store your result in the variable newDf.

8.2   Mutation Operations for a Data Frame

Up to this point, we have learned operations to project columns and select rows from a data frame, as well as partitioning and aggregating data frames, but none of the operations have *modified* an existing data frame. To achieve our needs for specific domain and analysis goals, and for normalizing data frames, like we do in Chap. 9, we also need to perform such update operations, as summarized previously in Sect. 7.1.

Because pandas is implemented to take advantage of many of the object-oriented facilities in Python, much of the syntax learned in Sect. 7.3 will be used in similar fashion for mutations, often with the specification of data frame rows, columns, and subsets on the *left-hand side* of an assignment.

In what follows, we will primarily use the indicators data set for each of the examples. Each of the examples will start from the same initial state, with the indicators data frame as shown initially in Table 7.1, with information on five countries from 2017, an index of code, and columns of country, pop, gdp, life, and cell. The first step in each example will be to copy the original data frame, since many mutations will change the data frame directly, and we want to maintain the original.

8.2.1 Operations to Delete Columns and Rows

8.2.1.1 Single Column Deletion

When we wish to delete a single column from a DataFrame, the easiest method is to use the Python del statement, using an argument specifying the single column. This is accomplished with the access operator using a column label.

In the following example, we remove the 'cell' column from the ind2 DataFrame using del and display the result:

```
ind2 = indicators.copy()
del ind2['cell']
ind2
```

```
|                    country       pop        gdp    life
| code
| CHN                  China   1386.40   12143.49   76.41
| IND                  India   1338.66    2652.55   68.80
| RUS    Russian Federation    144.50    1578.62   72.12
| USA         United States    325.15   19485.39   78.54
| VNM               Vietnam     94.60     223.78   76.45
```

Alternatively, we may use the pop() to delete a single column. This method should be familiar from Python list methods; it deletes and modifies the data structure in place and returns the element that was deleted. In a similar fashion, we can delete a column and assign the deleted column (a Series) to a variable. In the following example, we copy the indicators DataFrame to ind2 and reset the index, so the result is a data frame where code is just another column. In the example, we then invoke the pop() method, specifying the column we want to delete (and return) and assign to code_series. We then show the resultant DataFrame:

```
ind2 = indicators.copy()
ind2.reset_index(inplace=True)
```

```
code_series = ind2.pop('code')
ind2
```

```
|                     country       pop        gdp   life      cell
| 0                     China   1386.40   12143.49  76.41   1469.88
| 1                     India   1338.66    2652.55  68.80   1168.90
| 2   Russian Federation    144.50    1578.62  72.12    227.30
| 3         United States    325.15   19485.39  78.54    391.60
| 4               Vietnam     94.60     223.78  76.45    120.02
```

So we see that the column has been dropped, but we still have the variable code_series, which we can use, if we wish, in later code. We extend the example showing how we can assign code_series to the index attribute of the DataFrame, thereby establishing a set of row labels:

```
ind2.index = code_series
ind2
```

```
|                     country       pop        gdp   life      cell
| code
| CHN                   China   1386.40   12143.49  76.41   1469.88
| IND                   India   1338.66    2652.55  68.80   1168.90
| RUS     Russian Federation    144.50    1578.62  72.12    227.30
| USA         United States    325.15   19485.39  78.54    391.60
| VNM               Vietnam     94.60     223.78  76.45    120.02
```

8.2.1.2   Multiple Column Deletion

Dropping multiple columns can be accomplished with the DataFrame drop() method, where the first argument can be a single column label, or can be a list of column labels. For example, to delete both the 'cell' and 'life' columns from the indicators data frame, we execute:

```
ind2 = indicators.copy()
ind2.drop(['cell', 'life'], axis=1, inplace=True)
ind2
```

```
|                     country       pop        gdp
| code
| CHN                   China   1386.40   12143.49
| IND                   India   1338.66    2652.55
| RUS     Russian Federation    144.50    1578.62
| USA         United States    325.15   19485.39
| VNM               Vietnam     94.60     223.78
```

The axis=1 argument indicates the "dimension" for what is to be dropped, in this case, 1 indicates the column dimension (and an axis of 0 would indicate the row dimension).

For many methods that support changes to a `DataFrame`, pandas includes an `inplace` named parameter. If `True`, then the change *does not* create a new structure and does its work in place. This would be preferred when dealing with large data frames to save memory, and if the original non-modified frame is not needed in future steps. When `inplace=False` is specified, the change creates a new structure in memory and returns the result. In this case, the method invocation would appear on the right-hand side of an assignment, so that we have a variable referencing the new, changed, structure. In our example, since we are using `ind2` as a new copy of `indicators`, we could specify `inplace=False` on the invocation of `drop()` on `indicators`, which would create a new copy of the data frame with the specified columns dropped. We assign this new copy to `ind2`. Thus, the following is equivalent to the last example:

```
ind2 = indicators.drop(['cell', 'life'], axis=1, inplace=False)
ind2
```

```
|                     country       pop       gdp
| code
| CHN                   China   1386.40   12143.49
| IND                   India   1338.66    2652.55
| RUS     Russian Federation    144.50    1578.62
| USA          United States    325.15   19485.39
| VNM                 Vietnam     94.60     223.78
```

8.2.1.3 Row Deletion

Because of the `axis` named parameter, the `drop()` method can also be used to drop a list of specified rows, and the first argument would contain a list of row labels. The following example uses `axis=0` to specify that rows should be dropped, and drops the rows corresponding to the USA and Russia:

```
ind2 = indicators.copy()
ind2.drop(['USA', 'RUS'], axis=0, inplace=True)
ind2
```

```
|         country       pop       gdp    life      cell
| code
| CHN       China   1386.40   12143.49   76.41   1469.88
| IND       India   1338.66    2652.55   68.80   1168.90
| VNM     Vietnam     94.60     223.78   76.45    120.02
```

When we have a multi-level index for the rows, like we do in the `topnames` data set, we can specify a `drop` that is explicit as to *which* level is used to interpret the first argument. Here, the first argument to `drop()` specifies a range list of the integer years to drop, from 1880 up to but not including 1900. The `level='year'` argument is how we specify that the first argument should apply to the `'year'` level of the two-level index.

```
top_subset = topnames.drop(range(1880,1900), level='year', axis=0)
top_subset.head()
```

```
|                name    count
| year sex
| 1900 Female    Mary    16706
|      Male      John     9829
| 1901 Female    Mary    13136
|      Male      John     6900
| 1902 Female    Mary    14486
```

So in this example, we dropped all rows in the range of years from 1880 up to 1900 and did not have to specify the sex level.

8.2.2   Operation to Add a Column

One of the most common mutation operations is the addition of a new column to an existing data frame. The values of a new column is normally a Series (or other list-like set of values of the proper length) that is specified on the right-hand side of an assignment. We can also specify a scalar assignment to a new column, which would create the column as a Series with the scalar value repeated for the length of the vector, e.g., adding a column with value "Male" for every row of a data set of male individuals, perhaps in preparation for combining the data set with a different data set of females.

The left-hand side of these assignments uses the same notation we use for a single column projection we learned previously: the name of the data frame followed by the access operator with a column label between the brackets.

So the right-hand side could use the single column projection of one or more existing columns and perform column-vector computations to realize a new Series. This is then assigned, using the column label on the left-hand side to give a name to the new column. In the following example, we compute a Series based on a scalar multiple of an existing Series to get a 15% growth in GDP. We also carry out a vector division of two Series to compute the number of cell phone subscriptions per person in a country.

In the example below, we create a new ind2 data frame by initially dropping the country and life columns to help the readability of the result. We then compute and assign two new columns, based on column-vector computation, and finally, we display the result. Of course, it would be wise to include a check, as in Chap. 3, that we are not accidentally dividing by zero, but we have omitted this step.

```
ind2 = indicators.drop(['country','life'], axis=1)
ind2['growth'] = ind2.gdp * 1.15
ind2['cellper'] = ind2['cell'] / ind2['pop']
ind2
```

	pop	gdp	cell	growth	cellper
code					
CHN	1386.40	12143.49	1469.88	13965.0135	1.060214
IND	1338.66	2652.55	1168.90	3050.4325	0.873187
RUS	144.50	1578.62	227.30	1815.4130	1.573010
USA	325.15	19485.39	391.60	22408.1985	1.204367
VNM	94.60	223.78	120.02	257.3470	1.268710

Other techniques using column-vector operations, like using `apply` to perform a unary operation on the elements of a vector, work equally well. For instance, we can apply a `lambda` function that converts a string to uppercase to define a new column with country names in upper case. We start with the original `indicators` frame and create a new frame while dropping (for readability) several columns. We then use a lambda function applied to the `country` column and show the result:

```
ind2 = indicators.drop(['life', 'cell','pop','gdp'],
                        axis=1, inplace=False)
ind2['countryCaps'] = ind2['country'].apply(lambda
                                      s: s.upper())
ind2
```

	country	countryCaps
code		
CHN	China	CHINA
IND	India	INDIA
RUS	Russian Federation	RUSSIAN FEDERATION
USA	United States	UNITED STATES
VNM	Vietnam	VIETNAM

When `pandas` uses a list of column names for projection, the result is actually a *view* or *reference* to the columns as they reside in their original data frame. For read-only access to the data, this is much more efficient, particularly for large data sets. However, if, after projecting the columns, the result is then *updated*, perhaps by adding a new column, then `pandas` does not know if the programmer's intent is to add a column to the subset data frame, or to add a column to the original data frame. The reader may be familiar with this phenomenon from a previous study of lists in Python, as an analogous issue motivates the use of the `copy()` function.

For example, when we set `country_cell = indicators[['country', 'cell']]`, the new data frame `country_cell` is bound to `indicators`, so that changes in `indicators` will cause changes in `country_cell`. This means, if we later add a column to `country_cell`, `pandas` will be confused as to whether or not to add the same column to `indicators` and will produce an error stating `"A value is trying to be set on a copy of a slice from a DataFrame."`

Just like the situation of lists in Python, a good solution is to specify that the new data frame be a `copy()` of the projected columns from the original data frame. Then `pandas` knows that the assignment should be treated as a one-time operator, rather than as binding these two data frames together forever. We illustrate the correct code for our example, which avoids the error discussed above.

```
country_cell = indicators[['country', 'cell']].copy()
country_cell['cellper'] = indicators.cell / indicators['pop']
country_cell
```

```
|                       country      cell    cellper
| code
| CHN                     China   1469.88   1.060214
| IND                     India   1168.90   0.873187
| RUS       Russian Federation    227.30   1.573010
| USA            United States    391.60   1.204367
| VNM                   Vietnam    120.02   1.268710
```

8.2.3 Updating Columns

Instead of creating an entirely new column in a data frame, we often want to change values in an existing column.

8.2.3.1 Update Entire Column

When the goal is to update all the values in an existing column, we specify the existing column on the left-hand side of an assignment and a valid column-vector operation on the right-hand side. So to update the `life` column, whose values are:

```
ind2 = indicators.copy()
ind2['life']
```

```
| code
| CHN     76.41
| IND     68.80
| RUS     72.12
| USA     78.54
| VNM     76.45
| Name: life, dtype: float64
```

we simply use an assignment with `ind2['life']` on left-hand side, and the column computation on the right-hand side:

```
ind2['life'] = ind2['life'] + 5
ind2
```

```
|                  country      pop      gdp    life      cell
| code
```

```
| CHN                China  1386.40  12143.49  81.41  1469.88
| IND                 India  1338.66   2652.55  73.80  1168.90
| RUS   Russian Federation   144.50   1578.62  77.12   227.30
| USA       United States   325.15  19485.39  83.54   391.60
| VNM              Vietnam    94.60    223.78  81.45   120.02
```

Like creating new columns, the right-hand side of the assignment for a column update operation can utilize all the column projection and column-vector computation operations we have learned previously.

8.2.3.2   Selective Column Assignment

When we wish to change a *subset* of the values of an existing column, we need a mechanism to specify *which* row positions should be updated. In pandas, we use the same mechanism that we use to select row positions in a filter—the Boolean vector, where values in the vector of True mean that the column's row position should be updated, and values in the vector of False mean that the column's row position should be left at its current value.

Since we want to specify both *rows* (through a filter) and exactly one column (by column label), we use the .loc attribute so that we can provide both *rowspec* and *colspec*. We start by using the appropriate .loc access to query the values we want to update, namely the countries with GDP greater than 1000:

```
ind2 = indicators.copy()
ind2.loc[ind2.gdp > 10000, 'life']
```

```
| code
| CHN      76.41
| USA      78.54
| Name: life, dtype: float64
```

So, in this example, we want to update just the 'CHN' and 'USA' rows of the 'life' column. The update places this on the left-hand side of the assignment, with the "increment by 5" column-vector computation on the right-hand side:

```
ind2.loc[ind2.gdp > 10000, 'life'] = ind2.life + 5
ind2
```

```
|                    country     pop       gdp   life     cell
| code
| CHN                China  1386.40  12143.49  81.41  1469.88
| IND                 India  1338.66   2652.55  68.80  1168.90
| RUS   Russian Federation   144.50   1578.62  72.12   227.30
| USA       United States   325.15  19485.39  83.54   391.60
| VNM              Vietnam    94.60    223.78  76.45   120.02
```

We get the desired result where just 'CHN' and 'USA' rows of the column are updated. From our understanding of .loc and .iloc in Sect. 7.3.5, we can generalize our selective update and use .loc and .iloc and specify rows by explicit list and by slice as well as by Boolean Series.

8.2.4   Reading Questions

8.26 Why is it important to maintain a copy of the original data frame? Give an example of a previous time you needed to retain a copy of the original data.

8.27 Does the drop() method have the same functionality as iterated calls to del?

8.28 Give a real-world example where you would use drop() with axis = 0.

8.29 After popping 'code' from ind2, into a variable code_series, how could we create a new column in the data frame with the same value as code_series? In other words, how could we put 'code' back into the data frame?

8.30 Check out the online documentation for the insert() command in pandas, and then write down a line of code that would insert a column named 'code', with values code_series, in position index 2, into ind2.

8.31 Can you use apply on a whole data frame instead of just a single column? Explain.

8.32 The reading highlighted the importance of the copy() method, when working with a subset of the columns, as in country_cell = indicators[['country', 'cell']].copy(). Give another example of a time in your life that failing to work with a copy could confuse Python about where you meant to change something.

8.33 Run the following code, and then explain precisely why Python gets confused.

```
country_cell = indicators[['country', 'cell']]
country_cell['cellper'] = indicators.cell /
 indicators['pop']
```

8.34 The reading shows how to update the 'CHN' and 'USA' rows of ind2 to add 5 to the values in the life expectancy column. Show how to do the same thing using iloc.

8.35 Read the online documentation about sort_values() in pandas and then show how to mutate the ind2 data frame to sort it by the values in the life expectancy column.

8.2.5   Exercises

8.36 On the book web page, you will find members.csv, with (fake) information on a number of individuals in Ohio. We will use this for the next several exercises.

Read this data set into a pandas DataFrame using read_csv. Name it members0, and do not include an index.

8.37 Repeat the above, but now do include an index, by specifying index_col in the constructor. Name your DataFrame members.

8.38 Write a projection that will isolate just the 'Phone' column into a variable phone_series. What type are the values in this column?

8.39 Write a selection that will isolate just those rows where the phone number starts with a 614 area code. Selection involves picking which rows will be shown.
    In order to complete this:

1. Write a lambda function that will, given a string, isolate the first three characters and compare these to '614'.
2. Apply the lambda function to the column you isolated in the last question.
3. Use the result as an index to the pandas DataFrame.

8.40 Split up the column Name into two different columns, FName and LName. We follow a similar process for this example as we did for selecting rows.

1. Write a lambda function that will, given a string, split on space and select only the first element in the resultant list.
2. Apply the lambda function to the Name column, and save the result.
3. Access the data frame text using a new column name (FName), and assign the result from step 2.
4. Acquire the last name from the Name column and create a new column in the data frame called LName.
5. Acquire the city from the Address column and create a new column in the data frame called City.
6. Acquire the state from the Address column and create a new column in the data frame called State.
7. Drop the original "Name" and "Address" columns.

8.41 On the book web page, you will find a file sat.csv, based on data from www.onlinestatbook.com. This data contains the high school GPA, SAT math score, SAT verbal score, and college GPAs, for a number of students.

1. Read the data into a pandas data frame dfsat with column headers.
2. Add a new column 'total_SAT' containing the total SAT score for each student, calling the resulting data frame dfsat2.
3. Add a new column 'GPA_change' with the change in GPA from high school to college, calling the resulting data frame dfsat3.
4. Update the column 'total_SAT' to scale it so that the max score is 2400 instead of 1600 (i.e., multiply each entry by 1.5).
5. Delete the column 'total_SAT', calling the resulting data frame dfsat5.

8.42 Building on the exercise above, create a new column that partitions students by their high school GPA, into groups from 2.0 to 2.5, 2.5–3.0, 3.0–3.5, and 3.5–

4.0, such that each group includes its upper bound but not its lower bound. Then use grouping and aggregation to compute the average of each numeric column in the data set.

8.43 On the book web page, you will find a file education.csv, based on data hosted by www.census.gov. Each row is a U.S. metropolitan area, with data on population (row[3]), number unemployed without a high school degree (row[15]), number unemployed with a terminal high school degree (row[29]), number unemployed with some college (row[43]), and number unemployed with a college degree (row[57]). Read the data into a pandas data frame dfedu with column headers. Create a new column 'total_unemployed' with the total number of unemployed people in each metro area (i.e., the sum of the columns just listed).

8.44 Building on the exercise above, use a regular expression to extract the state (e.g., 'TX') from the geography column, and add state as a new column in your data frame. Then use grouping and aggregation to compute the total number of people in each unemployment category, in each state.

8.45 Building on the exercises above, enrich your data frame with a column containing the population of each state and then use grouping and aggregation to compute the per capita total unemployment in each state.

8.3 Combining Tables

The operations associated with combining tables will, in pandas, use two source DataFrame objects with a goal of a single resulting DataFrame that has some union or intersection of the rows, columns, and values from the source data frames.

The need for combining tables comes up most often when:

- We have one data set with a particular set of columns and need to append by adding additional rows, but with the *same set of columns*. This could be indicative of a "growing" data set, perhaps along a time dimension, where the variables remain the same, but additional rows mapping the independent variables to the dependent variables are to be appended.
- We have two data sets with, logically, the same set of rows, but with different columns, and want to create a data set with those same rows and the columns from both the source data sets. Here, we are adding new variables with their corresponding values to the data set, e.g., to enrich an old data set with more information on each individual.
- We have two tables that are related to each other through the *values of a particular column* or *index level* that already exist in both tables. Here we want to combine the tables so that the values of columns of the source tables are combined, but constrained to the matching values. We saw an example, in Chap. 6, of combining tables on classes and students to get information about which students are in which class sections.

We will deal with each of these cases in turn, showing a representative example, and limiting ourselves to two-table combinations. There are many variations possible when combining tables that are `DataFrame` objects. One source of variation comes about depending on whether or not we have a "meaningful" index for row labels or column labels. By this we mean, for instance, that for a row index, the index and its levels capture what makes rows unique vis-a-vis the independent variables of the data set.

In Chap. 12, we will examine more deeply some of the same table combination concepts in the context of the relational model.

8.3.1 Concatenating Data Frames Along the Row Dimension

For the examples and use cases we illustrate here, we make the assumption that, for the tables being combined, the source data frames represent *different* information. When combining along the row dimension, that implies that there is no intersection of value combination of the independent variables (i.e., the same mapping is not present in both data frames). Because `pandas` is designed to be general purpose for more than data frames that conform to the tidy data standard, the operations to combine tables have many additional arguments that can perform, for instance, union or intersection of columns when this assumption is not met. But these variations will not be covered here.

When we want to concatenate along the row dimension, we are referring to the situation where we have two source data frames, each with the same set of columns, but, at least logically, rows that are non-intersecting between the two source data frames.

8.3.1.1 Meaningful Row Index

In normalized tidy data, a row index is *meaningful* if the index is composed of the independent variable(s) of the data set. So if a data set has one independent variable, the value of that variable is unique per row, and the remaining columns give the values of the dependent variables. For instance, the `code` variable is independent for the `indicators` data set, and the values of this variable form the row label index. In the `topnames` data set, the `year` and `sex` variables are independent, and the combinations of year and sex give the row labels for the data frame.

Suppose, from the `indicators` data set, we define a subset, `ind1`, with columns for `country`, `pop`, `gdp`, and `life` for the rows `'CHN'`, `'IND'`, and `'USA'`:

```
ind1 = indicators.loc[['CHN', 'IND', 'USA'], :'life' ]
ind1
```

```
|                   country       pop         gdp    life
| code
| CHN                 China   1386.40    12143.49   76.41
| IND                 India   1338.66     2652.55   68.80
| USA        United States    325.15    19485.39   78.54
```

Now suppose that you construct, from a list of row lists, a new DataFrame with values for the *same* columns, but for the countries 'DEU' and 'GBR':

```
data2 = [[ 'Germany',            82.66, 3693.20, 80.99 ],
         [ 'United Kingdom', 66.06, 2637.87, 81.16]]
ind2 = pd.DataFrame(data2,
                    columns=['country', 'pop', 'gdp',
                                                  'life'],
                    index=['DEU', 'GBR'])
ind2
```

```
|                   country      pop        gdp    life
| DEU                Germany    82.66    3693.20   80.99
| GBR        United Kingdom    66.06    2637.87   81.16
```

How the two source data frames might be obtained or constructed does not matter, but that they have the same columns and different rows, and a meaningful index, where the row labels give the independent variable code does.

The combination of the two frames is obtained with the concat() function of the pandas module, where the first argument is a list of the frames to be combined, and the axis= named parameter, with a value of 0, specifies to combine in the row dimension:

```
combined = pd.concat([ind1, ind2], axis=0)
combined
```

```
|                   country       pop         gdp    life
| CHN                 China   1386.40    12143.49   76.41
| IND                 India   1338.66     2652.55   68.80
| USA        United States    325.15    19485.39   78.54
| DEU               Germany     82.66     3693.20   80.99
| GBR        United Kingdom     66.06     2637.87   81.16
```

The order of the result is based simply on the order of the data frames in the list, and no sorting occurs.

8.3.1.2 Meaningful Index with Levels

Suppose we have two source data frames that we want to combine along the row dimension, and these have a meaningful index, the code giving the country designation in this case, but the frames might have the same code, and the source frames are separate due to another independent variable. So, for example, suppose with have indicator data from 2015 for 'CHN', 'IND', and 'USA', referenced by indicators2015:

indicators2015

```
|                country       pop        gdp   life
| code
| CHN            China    1371.22   11015.54  76.09
| IND            India    1310.15    2103.59  68.30
| USA    United States     320.74   18219.30  78.69
```

Now, we might acquire data for the same countries, but in a different year, say 2017, and this is a DataFrame referenced by indicators2017:

indicators2017

```
|                country       pop        gdp   life
| code
| CHN            China    1386.40   12143.49  76.41
| IND            India    1338.66    2652.55  68.80
| USA    United States     325.15   19485.39  78.54
```

If we want to combine these into a single, *tidy* data frame, we are, in essence, adding an independent variable of year, so that each row in the result is uniquely identified by the combination of a year and a code.

If we combine the same way as we did in the first example, the result is a valid data frame, but one where the row labels actually have *duplicates*. While such non-uniqueness is allowed by pandas, we now no longer have a tidy data set.

```
combined = pd.concat([indicators2015, indicators2017], axis=0)
combined
```

```
|                country       pop        gdp   life
| code
| CHN            China    1371.22   11015.54  76.09
| IND            India    1310.15    2103.59  68.30
| USA    United States     320.74   18219.30  78.69
| CHN            China    1386.40   12143.49  76.41
| IND            India    1338.66    2652.55  68.80
| USA    United States     325.15   19485.39  78.54
```

The concat() function supports a keys= named parameter that allows us to specify a new, additional level for the row labels. The value is a list where each element specifies an index level value, so keys=[2015, 2017] will use an outer index of 2015 for the first data frame's rows and 2017 for the second data frame's rows.

```
combined = pd.concat([indicators2015, indicators2017], axis=0,
                     keys=[2015, 2017])
combined
```

```
|                      country       pop       gdp   life
|         code
| 2015  CHN              China   1371.22  11015.54  76.09
|       IND              India   1310.15   2103.59  68.30
|       USA      United States    320.74  18219.30  78.69
| 2017  CHN              China   1386.40  12143.49  76.41
|       IND              India   1338.66   2652.55  68.80
|       USA      United States    325.15  19485.39  78.54
```

This gives us the tidy two-level index with year and code as the levels. If we want the outer level to have a symbolic name, we can construct an Index object and specify both the list of outer level values as well as the name of the outer level index.

```
combined = pd.concat([indicators2015, indicators2017], axis=0,
                     keys=pd.Index([2015, 2017],name='year'))
combined
```

```
|                      country       pop       gdp   life
| year code
| 2015  CHN              China   1371.22  11015.54  76.09
|       IND              India   1310.15   2103.59  68.30
|       USA      United States    320.74  18219.30  78.69
| 2017  CHN              China   1386.40  12143.49  76.41
|       IND              India   1338.66   2652.55  68.80
|       USA      United States    325.15  19485.39  78.54
```

8.3.1.3   No Meaningful Index

As we manipulate data frames, even ones that satisfy the structural constraints of tidy data, we often process through cycles where the index of a data frame has been reset, or, on first construction, has only the default integer index for the row labels. When we combine such tables through concat(), we need to understand what happens.

For the example source data frames to illustrate this case, let us suppose a data frame topnames1, a subset of the data from the topnames data set with the data from years 2015 and 2016, with both Female and Male rows, and topnames2, with the data from 2017 and 2018. We assume the only index for both source data frames is the default integer index.

```
topnames1
```

```
|     year      sex   name   count
| 0   2015   Female   Emma   20455
| 1   2015     Male   Noah   19635
| 2   2016   Female   Emma   19496
| 3   2016     Male   Noah   19117
```

```
topnames2
```

```
|      year      sex   name    count
| 0   2017   Female   Emma   19800
| 1   2017     Male   Liam   18798
| 2   2018   Female   Emma   18688
| 3   2018     Male   Liam   19837
```

If we perform concat(), we see, not surprisingly, the index values of 0 through 3 repeated.

```
pd.concat([topnames1, topnames2], axis=0)
```

```
|      year      sex   name    count
| 0   2015   Female   Emma   20455
| 1   2015     Male   Noah   19635
| 2   2016   Female   Emma   19496
| 3   2016     Male   Noah   19117
| 0   2017   Female   Emma   19800
| 1   2017     Male   Liam   18798
| 2   2018   Female   Emma   18688
| 3   2018     Male   Liam   19837
```

To avoid errors on subsequent operations, like performing a .loc that uses the index, we should specify that the operation ignores the incoming row label index of the two source frames and generates a new one:

```
pd.concat([topnames1, topnames2], axis=0,
          ignore_index=True)
```

```
|      year      sex   name    count
| 0   2015   Female   Emma   20455
| 1   2015     Male   Noah   19635
| 2   2016   Female   Emma   19496
| 3   2016     Male   Noah   19117
| 4   2017   Female   Emma   19800
| 5   2017     Male   Liam   18798
| 6   2018   Female   Emma   18688
| 7   2018     Male   Liam   19837
```

8.3.2 Concatenating Data Frames Along the Column Dimension

For combining data frames along the column dimension, we make assumptions similar to those made in Sect. 8.3.1. When we are using tidy data and we want to combine tables, we assume that the two source data frames have the *same* rows and an entirely *different* set of columns. In tidy data, the stipulation of the same rows means that the values of the independent variables are the same, and here we assume that they are incorporated into a meaningful index, and the stipulation of different columns means the data frames have different dependent variables. Since we use the names of variables as the column labels, we rarely have columns whose Index is generated by pandas.

8.3.2.1 Single Level Row Index and New Columns

Using subsets of the indicator data set, let cols1 refer to a data frame with rows for 'IND', 'CHN', and 'USA', and columns of country and gdp:

cols1

```
|                country        gdp
| code
| IND            India    2103.59
| CHN            China   11015.54
| USA    United States   18219.30
```

Let cols2 refer to a data frame with rows for the same codes of 'IND', 'CHN', and 'USA', and columns of imports and exports:

cols2

```
|       imports   exports
| code
| USA   2241.66   1504.57
| IND    392.23    266.16
| CHN   1601.76   2280.54
```

Both cols1 and cols2 have an index defined by code. For the purpose of this example, we have intentionally used a *different* order for the rows in the two data frames. The concat() of these two frames, specifying combination along the column dimension by the argument axis=1, follows:

pd.concat([cols1, cols2], axis=1, sort=False)

```
|                country        gdp   imports   exports
| IND            India    2103.59    392.23    266.16
| CHN            China   11015.54   1601.76   2280.54
| USA    United States   18219.30   2241.66   1504.57
```

This example also illustrates the use of the sort= named parameter. When False, the order of the result is based on the order of the first data frame listed in the concat and is an efficient operation. We can also use pass sort=True to get a sorted result, but this could be slower.

```
pd.concat([cols1, cols2], axis=1, sort=True)
```

	country	gdp	imports	exports
CHN	China	11015.54	1601.76	2280.54
IND	India	2103.59	392.23	266.16
USA	United States	18219.30	2241.66	1504.57

8.3.2.2   Introducing a Column Level

Having discussed data frames with a two-level row index, we now briefly explain the concept of a two-level column index, and how to create one during a concatenation, again using the keys= named parameter. We will also use this as an opportunity to introduce an unfortunate common form of untidy data, where columns represent years. We will see in Chap. 9 how to wrangle such data into tidy form, with a column called year and years represented in the rows where they belong.

We suppose we are given the topnames data in a different form, where there are two data frames, one for the female top applications and another for the male top applications. We further suppose that, in both, the columns are years, as shown below. We assume the female data has data for years 2013 through 2015 and is referred to by the Python variable females:

```
females
```

year	2013	2014	2015
name	Sophia	Emma	Emma
count	21223	20936	20455

For the male data frame, suppose we have data for a slightly different set of years, from 2012 through 2015:

```
males
```

year	2012	2013	2014	2015
name	Jacob	Noah	Noah	Noah
count	19074	18257	19305	19635

While the tables females and males are not tidy, they do satisfy the initial assumptions of this section that the two tables have the same rows (name and count) and the columns have all logically *different* data, even if some of the column names are the same.

To combine these two frames along the column dimension, and because such combination would have two groupings of columns, one for the Female columns and one for the Male columns, we would want to add a new level to the column Index. We do so in exactly the same way as we did in the previous section, constructing an Index object for the new level and passing it as the argument to the keys= named parameter.

```
pd.concat([females, males], axis=1,
          keys=pd.Index(['Female', 'Male'],name='sex'))
```

sex	Female			Male			
year	2013	2014	2015	2012	2013	2014	2015
name	Sophia	Emma	Emma	Jacob	Noah	Noah	Noah
count	21223	20936	20455	19074	18257	19305	19635

The use of the argument `axis=1` tells pandas that both the concatenation and the index setting are occurring in the column direction. The name for the new column level is `sex`. In Chap. 9, we will see how to extract data out of a column index and reshape the data into tidy form.

8.3.3   Joining/Merging Data Frames

When combining data from two frames becomes more complicated than concatenation along the row or column dimension, as with the example of combining data on students with data on classes (Fig. 5.1), we need a more powerful tool. The `pandas` package provides two variations, depending on the frames we are combining, and whether or not we match up rows of the two tables based on their row label index, or on values from a regular column.

One variation, the `join()`, is a `DataFrame` method and uses the row label index for matching rows between the source frames. The other variation, the `merge()`, is a function of the package and can use *any* column (or index level) for matching rows between the source frames. The combined frame has columns from both the original source frames with values populated based on the matching rows. If the two frames have columns with the same column name, the join/merge will include both, with the column names modified to distinguish the source frame of the overlapping column. If we had three `pandas DataFrames` representing the three tables of Fig. 5.1, we could join/merge `schedule` and `classes` along the column `classRegNum` to get a single table with the six distinct columns contained in these two tables.

8.3.3.1   Using Index Level

Suppose we have a one-level index, the `code`, in the `indicators` data set, and we have one data frame, `join1`, that has columns for `imports` and `exports`, and another data frame, `join2`, that has columns for the country name, `country`, and the land area of the country, `land`. These two data frames might, in fact, come from different sources and might have `code` row labels that are not identical.

For the first data frame, `join1`, we might have:

join1

```
|          imports    exports
| code
| CAN      457.46     418.86
| GBR      643.52     441.11
| USA     2342.67    1545.61
```

And the functional dependency for this data set is:

$$code \rightarrow imports, exports$$

For the second data frame, join2, obtained from a different source, we might have:

join2

```
|                 country         land
| code
| BEL             Belgium      30280.0
| GBR      United Kingdom     241930.0
| USA       United States    9147420.0
| VNM             Vietnam      310070.0
```

The functional dependency of join2 is

$$code \rightarrow country, land$$

Both have the same independent variable, code, and so we could want to combine the tables so that matching codes yield rows with the union of the dependent variable columns, e.g., GBR would now have data on imports, exports, the country name, and the land mass. It is essential to note that the code values between the tables may not all match up. In this case, join1 has a row for CAN that is not in join2, and join2 has rows for BEL and VNM that are not in join1. There are several choices for what to do with the partial data present for these three rows, and the choice is made based on the analysis one desires to carry out.

All choices of how to combine these two tables are predicated on the code index level providing matches for rows. Suppose the analysis emphasizes the import and export values and, if available, will use land area, but can handle the case where land area is a missing value. In this case, we want *all* rows of join1 to be represented in the combined result that includes country and land, but if a match is not found in join2, we just fill with missing values for country and land. If join1 is the first (or left-hand) frame written as we combine, this type of combination is called a *left join*. We invoke the join() method on the first data frame and specify a second argument of the "right" data frame. The named parameter how= is passed a string "left" to specify a left join, as illustrated below:

```
join1.join(join2, how="left")
```

```
|         imports  exports          country        land
| code
| CAN      457.46   418.86              NaN         NaN
| GBR      643.52   441.11   United Kingdom    241930.0
| USA     2342.67  1545.61    United States   9147420.0
```

So we see the appropriate values for the matched rows, and missing values where the right-hand frame did not have a match to correspond to the left frame. In this situation, we think of the left frame as being dominant, and taking whatever values it can from the right frame, to enrich rows present in the left frame.

If the desired analysis requires that the combined result only has rows where a row label index matches from *both* frames, this is called an *inner join*. It is a form of intersection, where the intersection is defined by matching *just* the value of the common row label index, and disregarding any/all other column values. This inner join simply changes the how= named parameter argument to "inner":

```
join1.join(join2, how="inner")
```

```
|         imports  exports          country        land
| code
| GBR      643.52   441.11   United Kingdom    241930.0
| USA     2342.67  1545.61    United States   9147420.0
```

The result includes just the two rows where code is in common between the two frames, namely GBR and USA. In this case, we think of the new table as sitting between the two old tables, and drawing from both sides, whenever the same code is present in both.

If we were to join the two tables in a different order, making join2 be the left-hand frame and join1 be the right-hand frame, then, in a left join, the result would have all the rows from join2 with the join2 columns of country and land and would fill in with missing values for the columns in join1 (imports and exports) where there was no corresponding row in join1. In this case, join2 is the dominant table.

```
join2.join(join1, how="left")
```

```
|                country         land  imports   exports
| code
| BEL            Belgium      30280.0      NaN       NaN
| GBR     United Kingdom     241930.0   643.52    441.11
| USA      United States    9147420.0  2342.67   1545.61
| VNM            Vietnam     310070.0      NaN       NaN
```

Rows BEL and VNM are present in join2 but not in join1, so these are the ones that are augmented with missing values for the join1 columns.

There also exist join types of a *right join* where the right-hand frame has all its rows included, and we fill in with missing values for unmatched rows from the left-hand frame, and an *outer join* that includes all rows from *both* source frames.

In the previous example, the matching of the row label index was one to one. For any given match in one frame, there was at most a single match in the other frame. However, this need not be the case in general. We are now expanding our `join1` example to encompass two years of import and export indicator data:

```
join1
```

```
|             imports   exports
| year code
| 2016 CAN    426.94    389.66
|      GBR    636.64    409.04
|      USA   2189.18   1453.70
| 2017 CAN    457.46    418.86
|      GBR    643.52    441.11
|      USA   2342.67   1545.61
```

Now the functional dependency of `join1` is

$$\text{year}, \text{code} \rightarrow \text{imports}, \text{exports}$$

and `join1` uses a two-level index for the two independent variables. We may still want to combine this new `join1` with `join2` that maps from `code` to `country` and `land`, to essentially "fill in" with these values for all the matches of `code` between the two data sets. This often occurs when we want to do groupby and aggregation operations.

The `join()` method continues to work as expected in the "many-to-one" relationship between `join1` and `join2`. So a left join yields:

```
join1.join(join2, how="left")
```

```
|             imports   exports        country        land
| year code
| 2016 CAN    426.94    389.66            NaN         NaN
|      GBR    636.64    409.04  United Kingdom    241930.0
|      USA   2189.18   1453.70   United States   9147420.0
| 2017 CAN    457.46    418.86            NaN         NaN
|      GBR    643.52    441.11  United Kingdom    241930.0
|      USA   2342.67   1545.61   United States   9147420.0
```

and an inner join yields:

```
join1.join(join2, how="inner")
```

```
|             imports   exports        country        land
| year code
| 2016 GBR    636.64    409.04  United Kingdom    241930.0
|      USA   2189.18   1453.70   United States   9147420.0
| 2017 GBR    643.52    441.11  United Kingdom    241930.0
|      USA   2342.67   1545.61   United States   9147420.0
```

Note how, in the result, we have redundant data in the `country` and `land` columns. Care needs to be taken so that redundant data does not cause inconsistency problems if data frames are updated, but the possibility exists that such updates might not be applied to *all* instances of redundant data. In Chap. 9, we discuss how such redundant data is a sign that *two different* tables have been combined but, by TidyData3, should be separate.

8.3.3.2 Using Specific Columns

Sometimes, we wish to combine two tables based on common values between the two tables, but the values are in a regular column and not part of an index. This could be because the column is, in fact, not an independent variable and would not be part of an index, or it could be because, for the sake of manipulation and transformation, we have a column that *could* be an index, or a level of an index, and is not. Both cases come up frequently. For example, with `indicators`, we could reasonably use either the country code or name as an index and (based on the specifics of the table we combine with) might need to combine using whichever is not the index.

Logically, the operation is like the `join()` we have examined above, but, without the extra knowledge of index levels, we need to call a different method, called `merge()`, and we need to provide additional information to make the table combination possible.

Consider the two data sets we have been working with in this chapter: the `topnames` has, by year and by sex, the name and count of the top application for the US social security card. In the full `indicators` data set, we have, among other things, the population for each of the countries, including the population for the USA. We might want to create a new data frame based primarily on the `topnames` data set, but where we augment each row with the US population for that year. That could be used, in an analysis, to divide the count of applicants by the population for that year to be able to see what percentage of the population is represented. This could give a more fair comparison as we try to compare between years, because an absolute count does not take into consideration the changes in the population over time.

We prepare the source data frames, and in `us_pop`, we have a data frame with columns of `year` and `pop`, the changing population of the USA from 1960 to 2017. We show the first five rows of `us_pop`:

```
us_pop.head()
```

```
|     year          pop
| 0   1960    180671000
| 1   1961    183691000
| 2   1962    186538000
| 3   1963    189242000
| 4   1964    191889000
```

Note that the units of the pop column are in persons, not in millions of persons, like our earlier examples.

Because we want to illustrate a table combination that does *not* use an index or index level, our second data frame is topnames0, which has only the default integer index, and columns year, sex, name, and count. We show the prefix:

```
topnames0.head()
```

```
|     year       sex   name   count
| 0   1880   Female   Mary   7065
| 1   1880     Male   John   9655
| 2   1881   Female   Mary   6919
| 3   1881     Male   John   8769
| 4   1882   Female   Mary   8148
```

The merge() function has, as its first two arguments, the two DataFrame objects to be combined, where the first argument is considered the "left" and the second argument is considered the "right." The on= named argument specifies the name of a column expected to exist in both data frames and to be used for matching values. The how= named parameter allows us to specify the logical equivalent of an inner join, a left join, or a right join.

In the following code, we show both a left join/merge and an inner join/merge. Since topnames0 is the "left," the first will give us all the rows of topnames0 and will add the pop column from matching rows in us_pop. For those rows and years where us_pop does not have population data (i.e., those years before 1960), the merge1 will have NaN indicating missing data.

The merge2, by contrast, does an "inner" join/merge. The result will only have rows where the year has common values from *both* data frames. So this result will effectively prune the topnames0 portion to the years after 1960.

```
merge1 = pd.merge(topnames0, us_pop, on='year', how='left')
merge2 = pd.merge(topnames0, us_pop, on='year', how='inner')
print("Rows in merge1:", len(merge1),
    "Rows in merge2:", len(merge2))
```

```
| Rows in merge1: 278 Rows in merge2: 116
```

Looking at the first six rows of merge2, we see the added column, and the data starting with 1960, not 1880. Also note that since topnames0 has two rows for each year the same population is repeated for each row, since both match the year value from us_pop.

```
merge2.head(6)
```

```
|     year       sex     name   count        pop
| 0   1960   Female     Mary   51475   180671000
| 1   1960     Male    David   85929   180671000
| 2   1961   Female     Mary   47680   183691000
| 3   1961     Male  Michael   86917   183691000
```

```
| 4   1962   Female     Lisa   46078   186538000
| 5   1962     Male   Michael  85034   186538000
```

The merge() function is quite versatile and can be used for almost all the variations of combining taken covered in this section. It can also be used for hybrid situations where the criteria for matching rows between two tables might involve an index or index level from one source frame, and a regular column from the other source frame. See the pandas documentation for all the parameters that make this possible.

8.3.4 Reading Questions

8.46 Give a real-world example of two data sets with the same set of rows, but with different columns, that you might want to combine.

8.47 Give a real-world example of two tables that are related to each other through the values of a particular column, where you might want to combine the tables based on matching values.

8.48 What could go wrong if you try to naively concatenate two data frames along the row dimension, when they *do* contain rows with the same row index? Give an example.

8.49 Consider the code that concatenates indicators1 and indicators2 by row (axis=0). Suppose that, when defining indicators1, we use the code
```
indicators1 = indicators.loc[['CHN', 'IND', 'USA'],
'life' ]
```
instead of
```
indicators1 = indicators.loc[['CHN', 'IND', 'USA'],
:'life' ]
```
Explain what happens in combined and why.

8.50 What happens if we use concat with ignore_index = True on a data frame that *does* have a meaningful index, like 'code' in indicators?

8.51 Consider the example of concatenating data frames along the column dimension. When you do this with a command like pd.concat([cols1, cols2], axis=1, sort=True), what field does it sort based on?

8.52 Give a real-world example, different from the one in the book, when you might want to combine two data sets with the same rows but different columns, along a column level. Does your example represent tidy data? Why or why not?

8.53 Is the inner join symmetric? In other words, does it matter if you do join1.join(join2, how="inner") or join2.join(join1, how="inner")?

8.54 Do you get the same data by doing `join1.join(join2, how="left")` versus by doing `join2.join(join1, how="right")`?

8.55 Give a real-world example, different from the one in the book, plus a reason you might want to do an inner join instead of a left or right join.

8.56 What happens if you use `join` when you meant to use `merge`, i.e., on a column that is not an index?

8.57 What happens if you use `merge` when you meant to use `join`, i.e., on a column that is an index?

8.3.5 Exercises

8.58 On the book web page, you will find two csv files, `educationTop.csv` and `educationBottom.csv`, both based on data hosted by www.census.gov. Both have the same columns, and each row is a US metropolitan area, with data on population, education, and unemployment. The first `csv` file contains metropolitan areas starting with A–K, and the second starting with L–Z. Read both into `pandas` data frames, using the column `GEO.id2` as an index. Concatenate these two data frames along the row dimension (with the top one on top), and call the result `educationDF`.

8.59 On the book web page, you will find two csv files, `educationLeft.csv` and `educationRight.csv`, both based on data hosted by www.census.gov. Both have the same rows, and each row is a US metropolitan area, with data on population, education, and unemployment. The first has information on individuals without a college degree, and the second has information on individuals with a college degree. Read both into `pandas` data frames. Concatenate these two data frames along the column dimension, and call the result `educationDF2`.

8.60 On the book web page, you will find two csv files, `educationLeftJ.csv` and `educationRightJ.csv`, both based on data hosted by www.census.gov. In both, rows represent US metropolitan area, with data on population, education, and unemployment. However, they do not have exactly the same set of rows, and the columns are totally different except for the index column `Geography. 0`. Read both into `pandas` data frames, with names `educationLeftJ` and `educationRightJ`. 1. Make a copy of `educationLeftJ` called `educationLeftOriginal`. 2. Starting with `educationLeftJ`, do a left join to bring in the data from `educationRightJ`, storing your answer as `dfJ`. 3. Starting with `educationLeftOriginal`, do an inner join to bring in the data from `educationRightJ`, storing your answer as `dfJ2`. 4. Now read the original csv files in as `eduLeft` and `eduRight` with no meaningful index. Then, starting from `eduLeft`, do an inner merge along the column `Geography`, storing your answer as `dfJ3`.

8.61 Finish the application from the end of the section. This means, merge together `topnames0` and `us_pop`, using whichever type of join is appropriate for this application, and then normalize the application counts by `pop`. Use the resulting data frame to find the highest and lowest percentage counts over the years, and think about why a social scientist might be interested in this kind of information.

8.4 Missing Data Handling

We conclude with a brief discussion of the issues that arise with missing data. As we have seen, both join and merge operations can easily produce missing data. So can operations that create a new column, e.g., $life/cell$ if a row had 0 for cellphones. Lastly, many data sets encountered in the wild are missing data, as the following examples illustrate. Missing data can occur when:

* Individuals do not respond to all questions on a survey.
* Countries fail to maintain or report all their data to the World Bank.
* An organization keeping personnel records does not know where everyone lives.
* Laws prevent healthcare providers from disclosing certain types of data.
* Different users of social media select different privacy settings, resulting in some individuals having only some of their data publicly visible.
* Many other situations analogous to these.

When encountered in the real world, there is no true standard by which data sources denote (or "code") missing data. For example, a provider of healthcare data may code missing white blood cell counts as 0, knowing that no reader versed in healthcare would ever think that an individual truly had no white blood cells. In this situation, blindly applying an aggregation function (like `mean()`) could result in a badly wrong result. Even worse, some providers code missing data using a string, like `"N/A"`, `"Not Applicable"`, `"Not applic."`, `"NA"`, `"N.A."`, etc. Such a situation can entirely break simple code, e.g., that reads from a CSV file into a native dictionary of lists and then invoking an aggregation function.

Thankfully, information on how missing data is coded is almost always contained in the *metadata* associated with the data set, sometimes called a *codebook*. Whenever investigating a new data source, the reader should seek out this metadata and learn how missing data is coded, before attempting to read the data into native data structures or `pandas`. If there is no associated metadata, then it is usually possible to figure out how missing data is coded by a combination of the following steps:

* Careful inspection of the data in its original format.
* Visualizations (e.g., histograms and scatterplots), to look for data values that stand out from the rest.
* Consultation with an expert, to learn about conventions in the field, and about which data values look unusual or impossible.

Once we know how missing data is denoted, we should replace that coding with a special type. In pandas, this type is denoted nan and is displayed as NaN. Importantly, it is NOT a string. This special type has certain rules, such as "anything plus nan is nan." We illustrate, using the numpy library to gain direct access to the nan type.[2]

```
from numpy import nan
x = nan
print(x+2)
```

```
| nan
```

This rule is also satisfied for aggregation operators, as we illustrate below. Note that the nan elements still count toward the length of data, which is important for making sure all columns in a DataFrame have the same length. The sum is nan because the accumulation pattern used to compute it involves adding nan elements.

```
data = [1,2,nan,6,nan]
print(len(data))
```

```
| 5
```

```
print(sum(data))
```

```
| nan
```

When we have a DataFrame with missing data, like the pop column of merge1 above, built-in Python aggregation operations again produce nan if they involve elementary operations (like addition and multiplication) with a nan. We illustrate first demonstrating that the type of the first entry in the 'pop' column really is nan:

```
merge1.iloc[0]['pop']
```

```
| nan
```

```
print(len(merge1))
```

```
| 278
```

```
print(sum(merge1['pop']))
```

```
| nan
```

---

[2]Some practitioners might choose to use import numpy, in which case, we refer to the nan literal as numpy.nan.

As we know, not all of merge1['pop'] is null. The null values are only those from before 1960, when a match was not obtained in the left merge from the previous section. It would be reasonable to compute aggregation operations, like count, sum, and mean, on just the non-missing data. Thankfully, this is precisely what the pandas methods do:

```
print(merge1['pop'].count())
```

```
| 116
```

```
print(merge1['pop'].sum())
```

```
| 29144237198.0
```

```
print(merge1['pop'].mean())
```

```
| 251243424.12068966
```

We can also use the method isna() to find which values are nan (or, we could use notna() to find which are not). Below, we apply this on the population Series, but we could have also applied it on the entire DataFrame, resulting in a new DataFrame where every value is a Boolean.

```
(merge1['pop'].isna()).tail()
```

```
| 273      False
| 274      False
| 275      False
| 276       True
| 277       True
| Name: pop, dtype: bool
```

If we are certain that we do not want rows with nan values, the DataFrame method dropna() can be used to drop all such rows. As usual, the parameter inplace= tells pandas whether we want to modify the DataFrame in place or not. Here we choose not to, and we see that the new DataFrame no longer contains years before 1960:

```
merge1clean = merge1.dropna(inplace=False)
merge1clean.head()
```

```
|       year      sex      name   count              pop
| 160   1960   Female      Mary   51475   180671000.0
| 161   1960     Male     David   85929   180671000.0
| 162   1961   Female      Mary   47680   183691000.0
| 163   1961     Male   Michael   86917   183691000.0
| 164   1962   Female      Lisa   46078   186538000.0
```

This can be confirmed with the attribute merge1clean.shape, or again using len(), to see that the new DataFrame is smaller than merge1:

```
print(len(merge1clean))
```

| 116

If we prefer to fill in nan values with some specific scalar, this can be done with the fillna option, but we will not need that functionality. It would be appropriate, for example, if we had survey data and one column asked if individuals were willing to be contacted for a follow-up survey. If a survey recipient left that question blank, we could probably safely fill in the answer "No."

Lastly, it is common for a CSV file to have blank cells for missing data, e.g., the line "Michael","Male",,"35" means the third column is missing. The pandas.read_csv() method will automatically read such missing cells in as nan in the relevant row–column location, and hence the methods of this section apply.

8.4.1 Reading Questions

8.62 Have you ever analyzed a data set with missing data? If so, how were missing values coded?

8.63 Have you ever analyzed a data set with an attached codebook? If so, describe the data set and codebook. What kind of information was in the codebook?

8.64 The developers of Python made a design decision that adding a number and nan should yield nan. The developers of pandas made a design decision that methods should ignore nan values. What considerations do you think went into those design decisions?

8.65 In the reading, we discussed several options for dealing with missing data, including replacing aggregates by nan when nan data is encountered, ignoring nan data, dropping nan data, and filling in a single scalar value for all nan data encountered. Can you think of other ways of dealing with missing data?

8.66 Imagine at a scatterplot with a regression line, based on a data set of individuals where you know the age and salary of each. Suppose you want to predict the salary of a 32-year old, but none of the individuals in your data set were 32. What technique could you use to predict the salary of a 32-year old based on the other data that you do have? Could an approach like this be used to fill in missing data? Would it always work?

8.67 The book web page has a file, indicators.csv, with our indicators data for many countries, from 1960 forward. It has tons of missing data (e.g., because there were no cell phones in the 1960s). Use pandas to read this data into a DataFrame, then illustrate each of the functions from the reading on that DataFrame. Please use the cell column when you illustrate the fillna() function, and remember that there were no cellphones before 1980.

Chapter 9
Tabular Model: Transformations and Constraints

Upon completion of this chapter, you should understand the following:

- The restructuring performed by transformation operations of transpose, pivot, and melt.
- How to recognize when data is not in tidy form and what transformations and operations should be used to normalize it.
- How to use transformations and operations to achieve common presentational goals for a data set or subset.

Upon completion of this chapter, you should be able to do the following:

- Perform transformations in pandas:

  - Transpose,
  - Pivot and Pivot Table,
  - Melt.

- Use a sequence of pandas operations to achieve normalization and presentational goals.

© Springer Nature Switzerland AG 2020
T. Bressoud, D. White, *Introduction to Data Systems*,
https://doi.org/10.1007/978-3-030-54371-6_9

9.1 Tabular Model Constraints

Recall that constraints are *limitations* on the organizational structure, the relationships, and the values entailed in a data set. For the tabular model in general, the structural constraints are articulated by the precepts of tidy data, repeated from Sect. 6.1 here:

1. Each column represents *exactly one variable* of the data set (*TidyData1*).
2. Each row represents *exactly one unique (relational) mapping* that maps from a set of *givens* (the values of the independent variables) to the values of the dependent variables (*TidyData2*).
3. *Exactly one table* is used for each set of mappings involving the same independent variables (*TidyData3*).

For a particular data set, we might, additionally, have data set specific constraints like the following examples:

- A column named `age` must consist of integer values.
- A column named `income` is a categorical variable and must be one of the strings, `"high"`, `"middle"`, or `"low"`.
- A column named `year` must not have any missing values.
- If an `income` column is `"low"`, a `salary` column must be a float with values less than or equal to 24000.0.

The first three of these are intra-column value constraints, and the last is an inter-column constraint, which, for this structure, says something about a relationship between two variables. These additional constraints might be required for the sound and correct interpretation of the data set for analysis.

When data is created by a provider source, it is often created either without adhering to the structural constraints of tidy data or not necessarily adhering to intra- and inter-column constraints. In the tabular model, there is neither checking nor enforcement to ensure any limitations on the data are met.

In order to provide a check to see if a given data set conformed to these kinds of limitation, there must be some means to specify the limitations themselves. However, in the tabular model, our data arrives in files or through network streams that are almost always in a "record per line" and "fields delimited by a special character" format, conveying the data itself. There is no additional information conveying limitations, as there is with the relational and hierarchical models. So in the tabular model, all constraints are by *convention*—a common understanding, or by being specified in some *outside* documentation, or both. Outside documentation describing the form and assumptions of the data is called a *data dictionary* and often conveys metadata.

Since any creator of data can choose how to take data and put it into a row and column tabular form, the biggest concern is whether or not the data conform to the structural limitations of the model. In other words, we must always be asking the question of whether or not the data violates any of TidyData1, TidyData2, or TidyData3. Any intra- and inter-column constraints are data set specific, and, without additional information, cannot be addressed in a general manner.

In Sect. 6.1, we illustrated, for the `topnames` data set, four different representations of the same set of data values, all of which violated one or more of the tidy data principles. It may be worth reviewing those examples before continuing on in this chapter. The main observation is this: just because data is organized in rows and columns does *not* mean that it conforms to the tabular model. Data can be organized in rows and columns in many different ways, and many of those ways violate the tidy data requirements.

Since intra- and inter-column constraints are data set specific and cannot be handled in a general manner, the goal of the current chapter is to better understand the three principles of tidy data, to help recognize when data does not conform to the model, and to give some tools and examples so that, in the eventuality of having data in an un-tidy form, we can transform it into its tidy data equivalent.

In Sect. 9.2, we discuss some advanced operations that allow us to transform one tabular structure into another tabular structure. These, along with the operations of Chaps. 7 and 8, will give us the tools we need in order to work with data in various un-tidy forms and to transform and restructure it. In Sect. 9.3, we present a series of vignettes that illustrate many commonly occurring un-tidy forms and give examples of the normalization process. Finally, in Sect. 9.4, we generalize some of the characteristics seen in the normalization examples and articulate some of the "red flag signs" that can help us to recognize when data is not tidy.

9.1.1 Reading Questions

9.1 Give 2–3 reasonable column constraints for the data frame `topnames`.

9.2 Give 2–3 reasonable column constraints for the data frame `indicators`.

9.3 Give an example from a previous time in your life when you have needed to reference information from a *data dictionary*.

9.4 Describe one of the examples of an un-tidy way to display the `topnames` data set, and explain what specifically makes it un-tidy.

9.1.2 Exercises

9.5 The data set `ratings.csv` has information on several individuals and the ratings they provided for each of two restaurants (called "A" and "B"). Is this data in tidy form? Explain.

9.6 Explore the data provided as `ratings.csv`, and then draw the functional dependency that it should have when in tidy form. Hint: there should be only one independent variable.

9.7 The data set `restaurants_gender.csv` has information on several individuals and the ratings they provided for each of two restaurants. Is this data in tidy form? Explain.

9.8 This question concerns popular songs, based on a public domain data set known as `billboard`:

   `https://www.kaggle.com/sausen7/billboards-dataset`

A portion of this data is hosted on the book web page, and another snippet is visible here:

   `https://github.com/hadley/tidy-data/blob/master/data/billboard.csv`

Is this data in tidy form? Justify your answer.

9.9 This question concerns the data set `matches.csv` hosted on the book web page and keeping data on sports match-ups. Each row is a team, and the columns tell how well a particular team did against each opponent. Does this data conform to the tidy data restrictions? Justify your answer.

9.10 This question concerns data from the World Bank about population per country from 2000 to 2018, including the total population, the population growth, the urban population, and the urban population growth, for each country and each year.

0. Read `world_bank_pop.csv` into a data frame `dfwb`.
1. Give a reason why `dfwb` fails to represent TidyData.

9.11 Explore the data provided as `world_bank_pop.csv`, and then draw the functional dependency.

9.12 Download the World Health Organization's "Global Tuberculosis Report" data set as a CSV file from

   `https://www.who.int/tb/country/data/download/en/`

Look at the CSV file. Is it in tidy data form?

9.13 Consider the data set `members.csv` on the book web page. Please list all steps that would be required to make this data tidy.

9.2 Tabular Transformations

Transformations are the process of converting from one tabular form into another tabular form, while maintaining the information and values represented. We should note that, while these are viewed as complex operations of the tabular data model, they are simply algorithms that are constructed from the more basic operations: selection, projection, iteration, and so forth, which we have already learned. They are well defined and, if we needed to, we could construct these operations for ourselves. But because of their common utility, packages such as `pandas` in Python, or `dplyr` in R, have written them for us and allow us to leverage their implementation.

While, in the context of this chapter on constraints, our focus will be on using these transformations to normalize a data set from an un-tidy form, these tools can also be used for some types of analysis. They could also be used when communicating about a data set, manipulating the data into a form for presentation, and allowing us to communicate effectively. Because of this, the discussion in this section will be general, conveying the effect of each transformation, while also highlighting data situations where each transformation should be used.

9.2.1 Transpose

In linear algebra, a *transpose* is described as reflecting a matrix over its main diagonal. The columns of the matrix become the rows, and the rows become the columns, and an element that was at row i and column j becomes the element at row j and column i. The same transformation can be used to swap the rows and columns of a data frame, in case they are provided incorrectly, with variables as the rows and individuals as the columns.

We construct an example data frame, `example1`, with row labels A, B, and C and four columns, I, 1, 2, and 3. The values within the table are constructed to incorporate elements of their row and column label. This will help us to understand where values "land" in a transformation result.

```
example1
```

```
|       I     1     2     3
| A   A.I   A.1   A.2   A.3
| B   B.I   B.1   B.2   B.3
| C   C.I   C.1   C.2   C.3
```

In `pandas`, the transpose is obtained using the `.T` attribute of the `DataFrame`, so:

```
result1 = example1.T
result1
```

```
|       A     B     C
| I   A.I   B.I   C.I
| 1   A.1   B.1   C.1
| 2   A.2   B.2   C.2
```

```
| 3   A.3   B.3   C.3
```

The columns are now A, B, and C, and the rows and values are as expected. Also, as we would expect, the transpose of a transpose yields the original data frame, as the second transpose swamps the columns and rows back:

```
result2 = result1.T
result2

|       I     1     2     3
| A   A.I   A.1   A.2   A.3
| B   B.I   B.1   B.2   B.3
| C   C.I   C.1   C.2   C.3
```

In pandas, the transpose operates so that the row labels become the column labels and vice versa, so the transpose is almost always used in situations where the row labels are a meaningful index.

9.2.2 Melt

The melt operation converts columns into rows, like the transpose, but in a very specialized way, shown in Fig. 9.1. The name is meant to evoke a candle lying across the top of the data frame, melting from column information into row information.

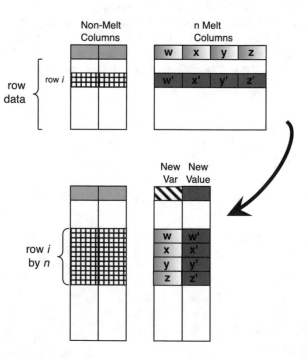

Fig. 9.1 Melt operation

We use this transformation when information stored in column names should actually be stored in the rows, as we saw in Sect. 8.3.2 when we had one column per year. We reproduce that example here, as this kind of presentational form is a dead giveaway that a melt is required:

```
females
```

	year	2013	2014	2015
	name	Sophia	Emma	Emma
	count	21223	20936	20455

The melt operation takes a subset of the columns, the *melt columns*, and uses the *column labels* of the melt columns as table entry *values* under a single new column, called the *new variable*. The operation uses the values from the original melt columns to become a single new *value column* in the transformation. This operation is particularly useful when the names of columns in a data set are, in fact, values in the data set, and we need to create a new column for these values.

Figure 9.1 highlights the transformation of row i in the original data frame, with column labels w, x, y, and z and values in row i of w', x', y', and z'. For example, in our un-tidy topnames data, i could be name, the melt columns could each be a year (e.g., $w = 2013$, $x = 2014$, etc.), and the red values could be the top names (e.g., $w' =$ "Sophia," $x' =$ "Emma," etc.).

In the transformed result, this single row i becomes four rows, corresponding to the original columns (w, x, y, and z) and the values of w', x', y', and z' are gathered in a single new column. The same transformation is applied to every row, e.g., a row of counts (with one column per year) is reshaped into a column named count with one row per year.

For an in-depth example, consider the columns of example1. For this example, we wish to *melt* the columns labeled 1, 2, and 3, and we wish not to melt column I. In the result, we will have three columns:

- the non-melt column, I, which remains and will repeat its values in the rows of the melted result,
- the new variable column, which we name K, whose values come from the column labels in the original, and
- the new value column, which we name V, whose values come from the values in the melt columns of the original.

The rows of the result are obtained by expanding, for each row in the original, n rows in the result, where n is the number of melt columns. The n rows will repeat the I value, will use exactly one of the column labels for the K column, and will use a lookup of I and the particular melt column to get the value for V.

We repeat example1 and note the $n = 3$ columns to be melted:

	I	1	2	3
A	A.I	A.1	A.2	A.3
B	B.I	B.1	B.2	B.3
C	C.I	C.1	C.2	C.3

We now show its melted result, where I is not melted, where columns 1, 2, and 3 are melted, and where we have specified column labels for the new variable name and for the new value column:

```
meltresult1 = example1.melt(id_vars=['I'],
                            value_vars=['1', '2','3'],
                            var_name='K', value_name='V')
meltresult1
```

```
|       I   K     V
| 0   A.I   1   A.1
| 1   B.I   1   B.1
| 2   C.I   1   C.1
| 3   A.I   2   A.2
| 4   B.I   2   B.2
| 5   C.I   2   C.2
| 6   A.I   3   A.3
| 7   B.I   3   B.3
| 8   C.I   3   C.3
```

The values of K are exactly the names of the melted columns. And the values of V are exactly the data values from inside table example1. The table meltresult1 has a default integer row index that we use to help describe the result of the melt operation. Rows 0, 3, and 6 in meltresult1 are transformations of row A in the original table. Note that the entry A.I appears $n = 3$ times in column I. In row 0, the column 1 from the original has become value 1 under new variable column K, and the value in row A, column 1 of the original, namely A.1, has become the value under column V for row 0. Similarly, row 3 is obtained from row A and column 2 of the original, and row 6 is obtained from row A and column 3 of the original.

When we consider the general information needed by a melt operation to perform its function, we see that its arguments must

- allow the function to partition the columns into the melt columns and the non-melt columns,
- provide a name for the *new variable* column, and
- provide a name for the *new value* column.

In the pandas melt() method of a DataFrame, the id_vars= named parameter allows us to specify the non-melt column or columns, the value_vars= named parameter allows us to specify the melt columns, the var_name= provides the name for the new variable column, and the value_name= provides the name for the new value column. Since we are partitioning the columns between those to be melted and those not melted, we only need to specify one of id_vars and value_vars, and the operation can determine the other. If we do not specify var_name or value_name, the melt() will use very generic defaults of var and value for the new column labels.

Table 9.1 Top baby names by year

Sex	2015	2016	2017
Female	Emma	Emma	Emma
Male	Noah	Noah	Liam

Table 9.2 Baby names melted

Sex	Year	Name
Female	2015	Emma
Male	2015	Noah
Female	2016	Emma
Male	2016	Noah
Female	2017	Emma
Male	2017	Liam

As a (non-abstract) example, consider a variation of one of the tables from Sect. 6.1, shown here as Table 9.1.

In this example, we would define sex as the non-melt column(s) and the columns 2014, 2015, and 2016 as the melt columns. The melt operation then introduces a new variable, year, composed from the original column names. Lastly, the entries in the original table are gathered in a new column, name. The desired result after the melt is given in Table 9.2.

In code, the Python variable names_by_year refers to the original data set:

```
         sex    2015    2016    2017
 0   Female    Emma    Emma    Emma
 1     Male    Noah    Noah    Liam
```

and we melt to transform into a data frame with columns of sex, year, and name:

```
melted_names = names_by_year.melt(id_vars=['sex'],
                        var_name='year',value_name='name')
melted_names
```

```
         sex   year    name
 0   Female   2015    Emma
 1     Male   2015    Noah
 2   Female   2016    Emma
 3     Male   2016    Noah
 4   Female   2017    Emma
 5     Male   2017    Liam
```

In this case, we specify only the non-melt column of sex as id_vars, and pandas uses the remaining columns for the melt columns. As usual, the values of the non-melt column have been repeated, explaining why we have three rows with "Female" and three with "Male."

The name of *melt* for the operation is intended to convey the idea that some of the columns are "melted" and become rows in the result, transforming a wider table into a table with fewer columns, but with more rows. In pandas, this operation is also performed by the `stack` method, which uses the row label index as the non-melt column and all other columns as the columns to be melted. Other languages, like R, have a `gather()` function to perform the same operation.

When there is more than one column that makes up the non-melt columns, all the values of the non-melt columns are repeated in each of the rows of the result generated from the same row or the source. We show this with the abstract `example2`, where we have two non-melt columns, `I` and `J`:

```
|       I     J     1     2     3
| A   A.I   A.J   A.1   A.2   A.3
| B   B.I   B.J   B.1   B.2   B.3
| C   C.I   C.J   C.1   C.2   C.3
```

```
meltresult2 = example2.melt(id_vars=['I', 'J'],
                            var_name='K', value_name='V')
meltresult2
```

```
|       I     J   K     V
| 0   A.I   A.J   1   A.1
| 1   B.I   B.J   1   B.1
| 2   C.I   C.J   1   C.1
| 3   A.I   A.J   2   A.2
| 4   B.I   B.J   2   B.2
| 5   C.I   C.J   2   C.2
| 6   A.I   A.J   3   A.3
| 7   B.I   B.J   3   B.3
| 8   C.I   C.J   3   C.3
```

Again, rows with indices 0, 3, and 6 are the melted result from row `A` in the original, and all of these have the same values for the non-melt columns, `I` and `J`. Note how this introduces the redundancy of having the `I` as well as `J` values repeated in the result.

In this section, we have focused on `melt()`, but if a data set and/or developer are skilled at using single- and multi-index with data sets, the `stack()` (and `unstack()`) can be useful for accomplishing the same goals. Additional information about `pandas` facilities for transformation/reshaping can be found in their user guide [45] and in the references for `melt()` [41] and `stack()` [44].

9.2.2.1   [Optional] Stack Examples

This optional section presents simple examples of the use of `stack()` to accomplish the same goal as explored with `melt()`.

In the first example, we define a `DataFrame` `df` with row labels as an `Index` with values `A.I`, `B.I`, and `C.I` and a name for the `Index` of `I`. The column labels are a single-level `Index` with values 1, 2, and 3 and a name for the index of `K`.

example3

```
| K       1    2    3
| I
| A.I   A.1  A.2  A.3
| B.I   B.1  B.2  B.3
| C.I   C.1  C.2  C.3
```

With row and column logical indices set up this way, a `stack()` operation is the same as a `melt()`. The `stack` uses the non-`Index` columns as the set of columns to be melted, in this case columns 1, 2, and 3. The column `Index` gets "stacked" and becomes a new level of the row index. So the row index transforms from a single-level (`I`) to a two-level index (`I` and `K`), with the values following in a fashion equivalent to the `melt()`. In this case, `stack()` does its job, by default, on the one level of the column `Index` (as the melt columns) and uses the row `Index` as the non-melt column.

example3.stack()

```
| I    K
| A.I  1    A.1
|      2    A.2
|      3    A.3
| B.I  1    B.1
|      2    B.2
|      3    B.3
| C.I  1    C.1
|      2    C.2
|      3    C.3
| dtype: object
```

In this case, the result is one-dimensional—we are left with only one non-index column, and so the result is a `Series` object, albeit one that has a two-level `Index`. We have seen this exact kind of data frame in the context of `topnames` in Sect. 6.3.2. If K consisted of year data (say, 2014, 2015, and 2016), and `I` consisted of sex data, then `A.I` might be Male (with the three years as the inner index) and `B.I` might be Female (again representing three rows, one per year). The values `A.1`, `A.2`, etc. would be the applicant numbers for top births by sex and year. The reader who is still confused is encouraged to pause and draw the table we have just described.

Our next example demonstrates `stack()` when there are multiple non-melt columns giving the row index. This works just like in the `melt()`; we end with the non-melt values being repeated, and we add a level to the row index based on the melted/stacked columns. So the result has a three-level row index.

```
example4
```

```
| K               1     2     3
| I   J
| A.I A.J   A.1   A.2   A.3
| B.I B.J   B.1   B.2   B.3
| C.I C.J   C.1   C.2   C.3
```

```
example4.stack()
```

```
| I     J     K
| A.I   A.J   1     A.1
|             2     A.2
|             3     A.3
| B.I   B.J   1     B.1
|             2     B.2
|             3     B.3
| C.I   C.J   1     C.1
|             2     C.2
|             3     C.3
| dtype: object
```

9.2.3   Pivot

The pivot() operation provides the dual operation to that of melt(). It is used to combine a *set of rows* into a *single row* by adding columns and moving the data from the multiple rows into the new columns of the transformed data set, as shown in Fig. 9.2.

For a pivot transformation, there must exist a column, called the "Pivot" column, where the set of unique values in that column *becomes the labels* for the newly created columns. The *values* to be populated in the newly created columns come from a single "Value" column from the original table, and the column where the value is placed in the new table is based on the value in the "Pivot" column. We most often use a pivot transformation when the values of the "Pivot" column are actually column names in our desired data presentation, e.g., pop, gdp, and life. We illustrate with an example, ind_to_pivot, featuring data from our indicators data set that would need a pivot transformation to be made tidy:

```
ind_to_pivot
```

```
|     code   ind       value
| 0   CAN    pop        36.26
| 1   CAN    gdp      1535.77
| 2   CAN    life       82.30
```

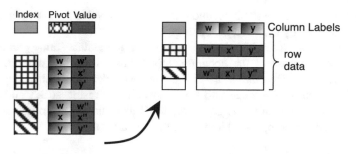

Fig. 9.2 Pivot operation

```
| 3   USA   pop      323.13
| 4   USA   gdp    18624.47
| 5   USA   life      76.25
```

Shortly, we will show how to make this data tidy using a pivot transformation. First, we describe the pivot transformation in general. In Fig. 9.2 we show, among many possible rows in an original table on the left, two sets of rows that, using matching values in the column annotated as "Index", will be combined into a single row of the result. In the column annotated as "Pivot," we see three unique values, x, y, and z that are repeated between each of the two sets of rows (just like pop, gdp, and life). So there are six values in the column annotated as "Value."[1] The values themselves are denoted with single and double prime marks (e.g., because the pop of Canada can be different from the pop of the USA) and must be included in the transformed result.

After the pivot, each of the different values from multiple rows of the source and with a common "Index" column will be represented in exactly one row of the result. The values of the "Pivot" column, x, y, and z, have become the column labels in the result, as can be seen in our corrected indicators data frame:

ind_corrected

```
| ind         gdp     life       pop
| code
| CAN     1535.77   82.30     36.26
| USA    18624.47   76.25    323.13
```

Furthermore, the values in the resulting table (after the pivot transformation) can be characterized as follows. At the intersection of an "Index" value and one of the values from the "Pivot," we find the corresponding values from the "Value" column of the source. For example, in the source (ind_to_pivot), the "Value" associated

---

[1]We are using double quoted terms of "Index," "Pivot," and "Value" to refer to columns by their use in the pivot operation and to distinguish from a pandas row index or from columns whose actual labels could be named in a variety of ways.

with Canada and `life` was 82.30, and that is the same value we see after the pivot, in the row indexed by `CAN` and the column `life`.

We now illustrate a more in-depth example. We define data frame `pivot_ex1`, which, for the purpose of the upcoming pivot, has "Index" as column `I`, pivot column as `K`, and value column as `V`. We use row-index labels that help indicate the set of rows to be combined into single rows in the result, so `A1` and `A2` rows become row `A` in the result after the pivot, etc. This is analogous to how three `USA` rows above became compressed into one `USA` row after the pivot.

`pivot_ex1`

```
|        I     K      V
| A1   A.I   K.1   A1.V
| A2   A.I   K.2   A2.V
| B1   B.I   K.1   B1.V
| B2   B.I   K.2   B2.V
| C1   C.I   K.1   C1.V
| C2   C.I   K.2   C2.V
```

So, in this example, there are six distinct values from column `V` that we want to represent in the final table, two per each of `A`, `B`, and `C`. For each set of rows to combine, the values in the pivot column (`K.1` and `K.2`) distinguish the rows from each other, and the same pivot values are repeated in each set. There is exactly one column, `I`, that uniquely identifies the set of rows to collapse. Think of `I` and `K` as two independent variables that yield a dependent variable `V`.

In the `pandas` `pivot` method, the named parameter `index=` allows us to specify the "Index" column and this becomes the row label/index of the result. The `columns=` named parameter allows us to specify the "Pivot" column. In this example, only one other column (V) exists, and is used as the "Value" column for the operation. We show an example:

`pivot_ex1.pivot(index='I', columns='K')`

```
|           V
| K       K.1     K.2
| I
| A.I   A1.V    A2.V
| B.I   B1.V    B2.V
| C.I   C1.V    C2.V
```

Note how, when called this way, the resulting column label/index has two levels, with the outside level labeled with the "Value" column label and the inside level giving the new column labels represented by the "Pivot" column. If one wishes to avoid the situation of a two-level column index, then all three of `index`, `columns`, and `values` should be specified to the `pivot` method:

```
pivot_ex1.pivot(index='I', columns='K', values='V')
```

```
| K     K.1    K.2
| I
| A.I   A1.V   A2.V
| B.I   B1.V   B2.V
| C.I   C1.V   C2.V
```

This is, in fact, the exact command we used to convert from ind_to_pivot to ind_corrected:

```
ind_corrected = ind_to_pivot.pivot(index="code",
                         columns="ind",values="value")
```

We will return to an analogous example in Sect. 9.3, where we will also learn how to drop a level of a multi-level index.

Extending the abstract example, we can see what happens if there is more than one "Value" column with the same "Index" and "Pivot" columns. In our indicators example, this would be like if we had pop data from each country in each of two years, so instead of value we might have value2016 and value2017 as columns in ind_to_pivot. For the abstract setting, consider data frame pivot_ex2:

```
pivot_ex2
```

```
|       I     K      V1       V2
| A1   A.I   K.1   A1.V1   A1.V2
| A2   A.I   K.2   A2.V1   A2.V2
| B1   B.I   K.1   B1.V1   B1.V2
| B2   B.I   K.2   B2.V1   B2.V2
| C1   C.I   K.1   C1.V1   C1.V2
| C2   C.I   K.2   C2.V1   C2.V2
```

Here, the "Index" column is still I, and the pivot column is still K, but now there are two "Value" columns, V1 and V2, each with unique values to be represented in a pivoted result. In this case, beyond the row label index, we expect four columns, with one for each pairing of the pivot column values K.1 and K.2 with the two value columns V1 and V2.

The named argument values= allows us to specify multiple value columns for the pivot, and we obtain our desired columns in the result:

```
pivot_ex2.pivot(index='I', columns='K', values=['V1', 'V2'])
```

```
|            V1                V2
| K        K.1     K.2      K.1     K.2
| I
| A.I   A1.V1   A2.V1   A1.V2   A2.V2
| B.I   B1.V1   B2.V1   B1.V2   B2.V2
| C.I   C1.V1   C2.V1   C1.V2   C2.V2
```

The two-level index of the result of a pivot is helpful here, as it allows a naming for the combinations of value column and pivot column in the result.

For a non-abstract example, suppose we start with the topnames data from the previous section after a melt has been performed and referred to as melted_names:

melted_names

```
|          sex   year   name
|  0   Female   2015   Emma
|  1     Male   2015   Noah
|  2   Female   2016   Emma
|  3     Male   2016   Noah
|  4   Female   2017   Emma
|  5     Male   2017   Liam
```

If we wanted a transformation with the values of year as the columns, and rows for the two values of sex (since pivot is dual to melt), the corresponding pivot is given by

melted_names.pivot(index='sex', columns='year', values='name')

```
|  year      2015   2016   2017
|  sex
|  Female   Emma   Emma   Emma
|  Male     Noah   Noah   Liam
```

If, on the other hand, we wanted a transformation with the two values of sex as the columns, and a row for each year, we just need to reverse the "Index" and "Pivot" columns:

melted_names.pivot(index='year', columns='sex', values='name')

```
|  sex   Female   Male
|  year
|  2015     Emma   Noah
|  2016     Emma   Noah
|  2017     Emma   Liam
```

As a non-abstract example where it makes sense to have more than one "Value" column to use in a pivot, let us return to our indicators data set. Say that we consider the population (pop) and GDP (gdp) indicators for the three countries of China (CHN), India (IND), and Great Britain (GBR) for the three years of 2005, 2010, and 2015, compiled into a DataFrame and referenced by indicators. For this example, we have removed the row label/index, with columns code, year, pop, and gdp all as non-index data columns.

```
indicators
```

```
|    code  year        pop          gdp
| 0  CHN   2005   1303.72      2285.97
| 1  CHN   2010   1337.70      6087.16
| 2  CHN   2015   1371.22     11015.54
| 3  GBR   2005     60.40      2525.01
| 4  GBR   2010     62.77      2452.90
| 5  GBR   2015     65.13      2896.42
| 6  IND   2005   1147.61       820.38
| 7  IND   2010   1234.28      1675.62
| 8  IND   2015   1310.15      2103.59
```

To present the data in a table with the country codes as the rows and with columns for each of the three years for each of the two indicators, we would employ the pivot transformation. This would be a pivot where `code` is the "Index" column and `year` is the "Pivot" column. The remaining two columns (`pop` and `gdp`) are the two "Value" columns, and we invoke the `pivot()` method as follows:

```
indicators.pivot(index='code', columns='year')
```

```
|             pop                            gdp
| year       2005      2010      2015       2005      2010       2015
| code
| CHN     1303.72   1337.70   1371.22    2285.97   6087.16   11015.54
| GBR       60.40     62.77     65.13    2525.01   2452.90    2896.42
| IND     1147.61   1234.28   1310.15     820.38   1675.62    2103.59
```

This results in the two-level column label/index and the six desired columns.

We could also use `pivot()` to transform the `indicators` data frame into one with years for the rows, and columns giving the combinations of country and indicator. In the invocation, we simply swap the `index=` and `columns=` named parameters.

```
indicators.pivot(index='year', columns='code')
```

```
|             pop                        gdp
| code       CHN     GBR      IND        CHN      GBR       IND
| year
| 2005   1303.72   60.40  1147.61    2285.97   2525.01    820.38
| 2010   1337.70   62.77  1234.28    6087.16   2452.90   1675.62
| 2015   1371.22   65.13  1310.15   11015.54   2896.42   2103.59
```

For many data sets, there may not always be exactly the same set of rows in an original to be combined into a single row in the result. Suppose, for example, that the indicators data set has years 2010, 2015, and 2018 for China, just years 2010 and 2015 for India and Great Britain.

The new `indicators` data frame would look like this:

```
|     code  year        pop         gdp
| 0   CHN   2010   1337.70     6087.16
| 1   CHN   2015   1371.22    11015.54
| 2   CHN   2018   1392.73    13608.15
| 3   GBR   2010     62.77     2452.90
| 4   GBR   2015     65.13     2896.42
| 5   IND   2010   1234.28     1675.62
| 6   IND   2015   1310.15     2103.59
```

Now, when we pivot, we will have a column for 2018 under both `pop` and `gdp`, but only data for the CHN row. The remaining slots will be filled as missing values.

```
indicators.pivot(index='code', columns='year')
```

```
|                  pop                            gdp
| year           2010     2015     2018       2010       2015       2018
| code
| CHN         1337.70  1371.22  1392.73    6087.16   11015.54   13608.15
| GBR           62.77    65.13      NaN    2452.90    2896.42        NaN
| IND         1234.28  1310.15      NaN    1675.62    2103.59        NaN
```

9.2.3.1   Pivot Table

There are times when a `pivot()` operation is unable to handle some more general situations. Consider the following table, evaluating it for performing a pivot operation, with I as the "Index," P as the "Pivot," and V1 and V2 as two "Value" columns:

```
table
```

```
|     I  P  V1      V2
| 0   A  w  15    23.5
| 1   A  x  10    42.5
| 2   A  x   5    18.0
| 3   B  w   8    10.2
| 4   B  w   4    14.3
| 5   B  x   6    12.5
```

Here, we would want and expect the result after a pivot to have two rows, one for A and the other for B. But, unlike in the prior examples, we have subsets of rows of the same index with the *same* value of the pivot variable. For the A rows, there are two rows with a P value of x. But, in the result of a pivot, we would have only one A row and only one x column (for each of V1 and V2). Similarly, there are two B rows with the w pivot.

The situation is resolved with an operation, `pivot_table()`, that goes beyond the `pivot()` and combines rows with duplicate (I,P) pairs into a single destination row, using the technique of aggregation that was discussed in Sect. 8.1.2. This means, when we invoke `pivot_table()`, we must specify an aggregation function to use to combine the values with duplicate `(I,P)` pairs. The default aggregation function is the mean, so that we would combine the `V1 = 8` and `V1 = 4` that appear in `(B,w)` rows into an output value of `V1 = 6`. Similarly, we average the two `V2` values in `(A,x)` rows (42.5 and 18) to get the single `V2` value in the output (30.25). In this way, we arrive at a single row per "Index." By default, `pivot_table()` will expect numeric values in the "Value" columns and computes the mean when it must combine, but other aggregation functions can be specified.

```
pivoted = table.pivot_table(index='I', columns='P')
pivoted
```

```
|        V1              V2
| P      w      x        w        x
| I
| A    15.0    7.5     23.50    30.25
| B     6.0    6.0     12.25    12.50
```

We can see that the value for A and x under `V1` is the mean of 10 and 5, and the value for B and w under `V2` is the mean of 10.2 and 14.3.

In the `pivot_table()` method, the `aggfunc=` named parameter can accept a dictionary, which maps from a value column to the desired aggregation function. Other than the change in the named parameter, this behaves the same as in the aggregation we have seen earlier. In this example, we perform an aggregation by mean for `V1` and the two x values but use an aggregation by sum for `V2` and the two w values.

```
pivoted = table.pivot_table(index='I', columns='P',
                            aggfunc={'V1': 'mean', 'V2': 'sum'})
pivoted
```

```
|        V1              V2
| P      w      x       w       x
| I
| A    15.0    7.5    23.5    60.5
| B     6.0    6.0    24.5    12.5
```

For example, the `(A,x)` value of `V2` is now $60.5 = 42.5 + 18$.

A common occurrence that also requires a `pivot_table()` operation is when we have more than one column that makes up the operation "Index." Consider the following table, and assume we wish to pivot on the `indicator` column, making `pop` and `gdp` into column labels, with values based on the values in column `value`.

```
ind2
```

```
|      code   year  indicator       value
| 0    IND    2005        pop     1147.61
```

```
| 1    IND   2010          pop    1234.28
| 2    IND   2015          pop    1310.15
| 3    USA   2005          pop     295.52
| 4    USA   2010          pop     309.33
| 5    USA   2015          pop     320.74
| 6    IND   2005          gdp     820.38
| 7    IND   2010          gdp    1675.62
| 8    IND   2015          gdp    2103.59
| 9    USA   2005          gdp   13036.64
| 10   USA   2010          gdp   14992.05
| 11   USA   2015          gdp   18219.30
```

In the desired result, we want the values of the independent variables (code and year) to uniquely determine each row, effectively combining each pair of rows in the original for the same code and year, putting the values in the two new columns for that row. This means we have two columns that, together, define the row-label index of the result. The pivot() method is limited and requires a single column for the index= named parameter. Instead, we need to use the more general pivot_table().

```
pivoted = ind2.pivot_table(index=['code', 'year'],
                           columns='indicator')
print(pivoted)
```

```
|                     value
| indicator             gdp          pop
| code year
| IND   2005          820.38      1147.61
|       2010         1675.62      1234.28
|       2015         2103.59      1310.15
| USA   2005        13036.64       295.52
|       2010        14992.05       309.33
|       2015        18219.30       320.74
```

The result builds a multi-level Index in both the row and the column dimensions. We may need to reset the row label/index or perhaps drop the outer level of the result, depending on what further operations are required.

Details for the pivot() and pivot_table() operations are given the full documentation of the methods of pandas DataFrame objects [42, 43].

9.2.4 Reading Questions

9.14 Give a real-world example where you might want to take the transpose. Hint: normally, the rows are in the individuals and columns are traits, so think of a situation where you might want to switch your focus to the traits rather than the individuals.

9.15 Does the repetition created by the *melt* operation mean that the resulting data frame is un-tidy?

9.16 Is it possible to do a melt with no `id_vars`?

9.17 The reading gives an example of a data frame that melts into our usual `topnames` example. Give an example of a data frame that melts into `indicators[['code','year','pop']]`. Hint: your example should have lots of columns.

9.18 Write a sentence describing the number of rows after a melt, based on the number before the melt and the number of melt columns. Does it matter how many non-melt columns there are?

9.19 Read the documentation for stack, including how to stack only a subset of the columns. Then, apply stack to the `pandas` data frame created from this list of lists:

```
headers = ['sex',    2015,    2016,    2017]
data = [['Female',    'Emma',    'Emma',    'Emma'],
        ['Male',    'Noah',    'Noah',    'Liam']]
```

9.20 Looking at the Pivot Operation Figure, how can you tell how many new columns and new rows will be present after a pivot operation, based on the columns `index`, `pivot`, and `value` in the original data frame?

9.21 In the data frame `pivot_ex1`, could you have done `pivot` with other choices of index columns? If so, explain in detail what would have happened.

9.22 When doing a pivot, do the arguments `index,columns,values` have to partition the original columns? Explain.

9.23 In the `indicators` example, does the line

```
indicators.pivot(index='code', columns='year')
```

yield the same as

```
indicators.pivot(index='code', columns='year',
                 values=['pop','gdp'])
```

Explain.

9.24 Why does it make sense that the default for `pivot_table` is to take the average of the repeated values?

9.25 Carry out the arithmetic to explain why the values in the `pivot_table` examples are different, i.e., explain where 30.25, 7.5, 6, and 12.25 come from in the first table and where 60.5 and 24.5 come from in the second table.

9.26 In the last example of `pivot_table`, explain what would happen if you chose `columns = 'value'` instead of `columns = 'indicator'`.

9.27 Consider the original `topnames` data frame, with `year,sex,name, count`. Show how to use `pivot_table` to get a version with `sex` and `year` as indices, columns given by `name`, and with counts as the entries in the data frame.

9.2.5   Exercises

9.28 Make the following into a `pandas` data frame.

```
{'A': {0: 'a', 1: 'b', 2: 'c'},
 'B': {0: 2, 1: 4, 2: 6},
 'C': {0: 1, 1: 3, 2: 5},
 'D': {0: 1, 1: 2, 2: 4}})
```

Suppose further that we have determined that columns B and D are really *values* of a *variable* called X. What transformation/reshaping operation should be used to obtain a tidy version of this data?

9.29 At a minimum, what parameter arguments would be needed for this operation to do its job?

9.30 Give the column headers of the transformed data set and give at least two rows.

9.31 Make the following into a `pandas` data frame.

```
{'foo': ['one','one','one','two','two','two'],
 'bar': ['A', 'B', 'C', 'A', 'B', 'C'],
 'baz': [1, 2, 3, 4, 5, 6]}
```

If the values `one` and `two` from column `foo` should head columns (so it takes more than one row to interpret a single observation), and the values themselves come from the `baz` column, what transformation/reshaping operation should be used to obtain a tidy version of this data?

9.32 What parameter arguments would be needed for this operation to do its job?

9.33 Give the resultant transformed data set, including column headers.

9.34 Refer back to `ratings.csv` and the un-tidy data frame you made in the previous section. Make the data frame tidy, storing your result as `ratings_tidy`.

9.35 Refer back to the `restaurants_gender.csv` and associated data frame from the previous section. Pivot the `restaurants_gender` data into a matrix presentation with restaurant down one axis and gender across the other axis. This makes it easy to aggregate female ratings and male ratings. Store the result as `rest_mat`.

9.36 This question concerns the `billboard` data introduced in the exercises of the previous section. Download this data as a CSV file, read it into a `DataFrame`, and melt the data frame to make it tidy.

9.37 This question concerns the data set `matches.csv` hosted in the data directory and keeps data on sports match-ups. Each row is a team, and the columns tell how well a particular team did against each opponent.

1. Read `matches.csv` into a data frame `df`.
2. Transform the data frame to make it tidy, storing the result as `df_tidy`. Your data frame should have no `NaN` values.

9.38 This question concerns data from the World Bank about population per country from 2000 to 2018, including the total population, the population growth, the urban population, and the urban population growth, for each country and each year. 0. Read `world_bank_pop.csv` into a data frame `dfwb`. 1. Transform the data frame to make it tidy and store the result as `dfwb_tidy`.

9.39 Consider the original `topnames` data frame, with `year,sex,name,count`. Show how to use `pivot_table` to get a version with `sex` and `year` as indices, columns given by `name`, and with counts as the entries in the data frame. Call the result `dfPivoted`.

9.40 With reference to the exercise above, is `dfPivoted` in tidy data form? Explain your answer.

9.3 Normalization: A Series of Vignettes

With the full suite of transformation operations to help, we are now ready to look at a primary objective of this chapter: given a data set that does *not* conform to the principles of tidy data, construct a sequence of operations, as an algorithm, to normalize the data into tidy data form.

Because data may be "messy" in an almost infinite variety of ways, the current section is presented as a series of *example vignettes*, where we show some of the more common violations of tidy data and discuss how our operations from Chaps. 7 and 8 and our transformations from Sect. 9.2 may be employed to resolve the violation(s).

Before any attempt at normalization for a data set, it is of critical importance to *understand your data*. You *must* be able to separate what are values from what are the variables. You should be able to argue that variables are *different* measures. Variables identified as categorical should be assessed on whether they indeed provide different categories that partition observations or whether the "categories" could be variables themselves. Variables should be identified as dependent or independent. Functional dependencies should be written down and examined critically to understand if dependent variables do, in fact, depend on *all* the independent variables. If one or more dependent variables are determined by a subset of the independent variables, these should give *separate* functional dependencies and therefore *separate tables*.

9.3.1 Column Values as Mashup

TidyData1 states that each column must represent *exactly one variable*. A violation occurs if, in a given data set, either a column is not actually a variable or if a column is in violation because of the *exactly one* stipulation. Any time the values in a column are some kind of *composition* of values of *more than one* variable, a violation has occurred. A composition might mean a mashup, or it could be the result of a column that is a combination of two other columns, or even an aggregation of multiple columns.

9.3.1.1 Example: Code and Country Mashup

Consider the data frame `df` below, with column `country` as an independent variable and dependent variables `land`, `pop`, `gdp`, and `life`. We see that `country` is a mashup of a country code and a country name in a single string value.

`df`

```
|                         land      pop       gdp    life
| country
| CHN--China            9388210.0  1386.40  12143.49  76.41
| GBR--United Kingdom    241930.0    66.06   2637.87  81.16
| IND--India            2973190.0  1338.66   2652.55  68.80
| USA--United States    9147420.0   325.15  19485.39  78.54
```

At an abstract level, we need to introduce new columns that have, parsed out, the variable values from the current `country` column and then replace that column with the two new columns. We write our solution as an algorithm.

Solution
1. If necessary, reset the index to allow manipulation of `country` as a column.
2. Define two functions that, given a value of `country`, yield the `code` part and the `country` part.
3. Apply the function to get a code to the current `country` column, yielding a `codes` vector.
4. Apply the function to get a country to the current `country` column, yielding a `countries` vector.
5. Drop the current `country` column.
6. Add `codes` as a column.
7. Add `countries` as a column

8. If needed, set index to be the new code column.
9. If needed, change the order by projecting columns in the desired order.

The corresponding code to realize this algorithm is as follows:

```
working = df.reset_index()

get_code = lambda s: s.split('--')[0]
get_country = lambda s: s.split('--')[1]

codes = working['country'].apply(get_code)
countries = working['country'].apply(get_country)

working.drop('country', axis=1, inplace=True)
working['country'] = countries
working['code'] = codes

working.set_index('code', inplace=True)
column_list = list(working.columns)
lastcol = column_list.pop()
column_list.insert(0, lastcol)

working = working[column_list]
working
```

```
|                   country       land       pop       gdp    life
| code
| CHN                 China  9388210.0   1386.40  12143.49   76.41
| GBR        United Kingdom   241930.0     66.06   2637.87   81.16
| IND                 India  2973190.0   1338.66   2652.55   68.80
| USA         United States  9147420.0    325.15  19485.39   78.54
```

In the final version of working, we see the normalized data set. In this example, we could also have made country the index and had code as a dependent variable.

9.3.1.2   Example: Year and Month Mashup

Suppose we are collecting data on tourism for a (fictitious) city, Metropolis. Consider the following prefix of a table of data. This table compiles, by month and year from January 2015 through June 2020, a number of measures associated with tourism, including the number of visitors, the number of hotel rooms in the city, the occupancy rate of the hotel rooms, and revenue (Table 9.3).

It is quite common in data sets to encode a date as a string, and like in this case, to make that string be a single variable composed of the year and the month (or sometimes, the year, the month, and the day). Such encoding of dates as strings is a violation of TidyData1 and gives us good examples of why this is problematic:

Table 9.3 Metropolis tourism

Date	Visitors	Rooms	Occupancy	Revenue
Jan-2015	508800	28693	0.857	26124000
Feb-2015	450500	27866	0.656	25618000
Mar-2015	783900	27717	0.940	27317000
Apr-2015	545100	27566	0.887	23254000
May-2015	602100	27656	0.891	20700000

1. When dates are strings, even a simple *sort* operation, which occurs as a lexicographic ordering of the string characters involved, can yield unexpected results. In this case, a sort would put `Apr-2017` as the first entry. Similar problems can occur even if numeric months and days are used, depending on the ordering.
2. When data is in a time series over multiple months and years, analysis can often want to GroupBy the years of the data set, to characterize year over year information, or to GroupBy the months of the data set, to characterize monthly or seasonal trends.

To make time series encoded as composite strings tidy and normalize a data set, we have two choices. First, if a package supports it, we could make use of a `DateTime` object type. These objects are designed to abstract the nature of years, months, days, hours, minutes, seconds into a single object that can be sorted and, in packages like `pandas`, can be used in a GroupBy which specifies the frequency (like year or month) of the group. Second, we can use the approach employed above, splitting the string into its component types and replacing the composite column with the individual columns.

9.3.2   One Relational Mapping per Row

TidyData2 states that each row represents *exactly one mapping* from the values of the independent variables to the values of the dependent variables. This means that if, for a representation, more than one row is employed to give the value of dependent variables for the *same* independent variable, the data is not in tidy form. In the next two examples, we will look at variations of having multiple rows used to represent a single relational mapping.

9.3.2.1   Example: One Value Column and One Index Column

Suppose our single year indicator data set was represented as shown in Table 9.4, with columns `code`, `indicator`, and `value`, giving data for the indicators of population, GDP, and exports:

Table 9.4 Indicator data

Code	Indicator	Value
CHN	Pop	1386.40
CHN	Gdp	12143.49
CHN	Exports	2280.09
GBR	Pop	66.06
GBR	Gdp	2637.87
GBR	Exports	441.11
IND	Pop	1338.66
IND	Gdp	2652.55
IND	Exports	296.21

We might, just looking at the table and not understanding our data, think that the functional dependency entailed was

$$\text{code, indicator} \rightarrow \text{value}$$

which says that `code` and `indicator` are independent variables and, given a `code` and an `indicator`, the `value` variable is determined. The problem is that `indicator`, which might look like a categorical variable, has the names of actual variables in its column: `pop`, `gdp`, and `exports` are *separate measures*. Further, unlike a categorical variable, they do not partition the space of values for each `code`. Because of this, we have *three rows* representing a *single mapping* of a country to indicators, given by the functional dependency

$$\text{code} \rightarrow \text{pop, gdp, exports}$$

Normalizing this particular data set is straightforward and, modulo manipulation of the row or column `Index` in the result, requires the single operation of a `pivot`. The "Index" of the pivot is the `code` column, the pivot column is `indicators` of the source data frame, and the value column is named `value`. If the data from Table 9.4 is a pandas `DataFrame` referred by Python variable `mult_rows`, the pivot is given as follows:

```
tidy = mult_rows.pivot(index='code', columns='indicator',
                       values='value')
tidy
```

```
| indicator    exports         gdp        pop
| code
| CHN          2280.09    12143.49    1386.40
| GBR           441.11     2637.87      66.06
| IND           296.21     2652.55    1338.66
```

This yields the desired tidy form of the data.

It is worth noting that, when TidyData2 was violated by having more than one row per mapping, TidyData1 was also violated: neither `indicator` nor `value` in the original data frame satisfied the condition of a column representing exactly one variable.

9.3.2.2 Example: One Value Column and Two Index Columns

If we extend the previous example, we can see a normalization use case for the `pivot_table` operation. Suppose our data set is the time-series version of the indicator data, so that we have indicators that are dependent on both a year *and* a country code. The same kind of representation, where we have an `indicator` and `value` column, might produce a data set as seen in Table 9.5, where we show data for indicators of `pop` and `gdp` for three countries and for years 2015 and 2017:

Like before, this representation might lead us to the incorrect functional dependency

$$\texttt{year}, \texttt{code}, \texttt{indicator} \to \texttt{value}$$

where, instead, by understanding our data and the separately measured variable nature of the indicators, the functional dependency should be

$$\texttt{year}, \texttt{code} \to \texttt{pop}, \texttt{gdp}$$

Table 9.5 Indicator data

Year	Code	Indicator	Value
2015	CHN	Pop	1371.22
2015	CHN	Gdp	11015.54
2015	GBR	Pop	65.13
2015	GBR	Gdp	2896.42
2015	IND	Pop	1310.15
2015	IND	Gdp	2103.59
2017	CHN	Pop	1386.40
2017	CHN	Gdp	12143.49
2017	GBR	Pop	66.06
2017	GBR	Gdp	2637.87
2017	IND	Pop	1338.66
2017	IND	Gdp	2652.55

Observe also how the numbers in the `value` column, because they really come from different variables, are not comparable. If we aggregated the column, we would get a nonsensical result.

Because we have two columns that make up the index in a pivoted result, we must use the `pivot_table` method. The pivot column of `indicator` and the value column of `value` are as before, and the normalization again proceeds in a single step:

```
tidy = mult_rows2.pivot_table(index=['year', 'code'],
                       columns='indicator', values='value')
tidy
```

```
| indicator        gdp       pop
| year code
| 2015 CHN    11015.54   1371.22
|      GBR     2896.42     65.13
|      IND     2103.59   1310.15
| 2017 CHN    12143.49   1386.40
|      GBR     2637.87     66.06
|      IND     2652.55   1338.66
```

9.3.3 Columns as Values and Mashups

Another frequent violation of TidyData1 occurs when the column labels represent *values of a variable* instead of a single variable of the data set. Sometimes this is easy to see, like when the column labels are, in fact, numeric values,[2] but sometimes it is less obvious, like when the column name incorporates a variable name along with a value in a mashup as the column name. The next two examples look at variations of this scenario.

9.3.3.1 Example: Single Variable with Multiple Years

Table 9.6 presents a table with columns for country code and population for the years 2014 through 2017. There is a single row per country code. From our experience with the time-series version of the indicators data set, we should be able to write the correct functional dependency

$$\text{code, year} \rightarrow \text{pop}$$

---

[2]This case can often be resolved to normal form with a single melt operation.

Table 9.6 Population by year

Code	pop2014	pop2015	pop2016	pop2017
CHN	1364.27	1371.22	1378.66	1386.40
GBR	64.61	65.13	65.60	66.06
IND	1295.60	1310.15	1324.51	1338.66
USA	318.39	320.74	323.07	325.15

where we have a single dependent variable, pop, determined by code and year, which should guide us as we normalize this data and transform Table 9.6 into a data frame with columns for code, year, and pop.

To normalize this data, we need to start with a melt of the table using all the "popXXXX" columns as the melt columns. The value column of the result will have collected all the values from the original table, as desired. But the "New Var" in the result will still have the mashup of the string "pop" composed with the value of the year, so this will need to be resolved to obtain the integer year value. We write our solution as an algorithm.

Solution
1. If necessary, reset the index to allow the use of the code index as a column.
2. Melt the data frame, so that all columns except the code column transform into rows, where the value in the column becomes the pop column, and the previous column name becomes a value in a column named colname.
3. Define a function that, given a value of colname yields the year value.
4. Apply the function to the colname column and assign to the data frame as a column named year.
5. Perform the cleanup of dropping the colname column and setting the index based on two independent variables (year and code); if desired, change the row order so that the index shows grouping in more friendly presentation.

With the original data from Table 9.6 as a data frame referred to by pop_columns, this algorithm translates into the following code and yields the normalized result named tidy:

```
working = pop_columns.reset_index(drop=True)
melted = working.melt(id_vars='code', value_name='pop',
                      var_name='colname')
getyear = lambda s: int(s[-4:])
melted['year'] = melted['colname'].apply(getyear)
tidy0 = melted.set_index(['year', 'code'])
tidy0.drop('colname', axis=1, inplace=True)
tidy = tidy0.sort_index()
tidy
```

```
|                  pop
| year code
| 2014 CHN    1364.27
|      GBR      64.61
|
```

```
|       IND    1295.60
|       USA     318.39
| 2015 CHN    1371.22
|       GBR      65.13
|       IND    1310.15
|       USA     320.74
| 2016 CHN    1378.66
|       GBR      65.60
|       IND    1324.51
|       USA     323.07
| 2017 CHN    1386.40
|       GBR      66.06
|       IND    1338.66
|       USA     325.15
```

An equivalent tidy representation could swap the levels of the row Index, and possibly perform a sort_index, if we desired code as the outer index and year as the inner index.

9.3.3.2 Example: Multiple Variables with Multiple Years

Our last example in this subsection extends the previous example by just a bit, adding columns that are compositions of a second variable (gdp) with the year values of the data set. Interestingly, the normalization becomes more involved—the solution presented here involves both a melt *and* a following pivot to transform the data set into tidy form. Consider the data and structure represented in Table 9.7, where, relative to Table 9.6, we have added three GDP columns, giving the GDP for each of the countries for the same years of 2015, 2016, and 2017.

When we extend our (correct) functional dependency to include the variable of gdp, we get

$$code, year \rightarrow pop, gdp$$

Like in the last example, we clearly need to start with a melt, which will transform the data set into a frame with columns for code, as the index, ind_year for the new variable column, and value for the value column. Also, like the

Table 9.7 Population and GDP by year

	pop2015	pop2016	pop2017	gdp2015	gdp2016	gdp2017
CHN	1371.22	1378.66	1386.40	11015.54	11137.95	12143.49
GBR	65.13	65.60	66.06	2896.42	2659.24	2637.87
IND	1310.15	1324.51	1338.66	2103.59	2290.43	2652.55
USA	320.74	323.07	325.15	18219.30	18707.19	19485.39

previous example, the values in the `ind_year` column will be strings composed of indicator and year. But now, we will have strings like both `'pop2015'` and `'gdp2015'`. So now when we separate, we will get two columns, one for the indicator part and the other for the year part.

We show the code for realizing this part of the algorithm, essentially extending our previous algorithm, and display a prefix of the result, named `part_tidy`, and assuming that the original data is in a `DataFrame` referenced by `popgdp_columns`:

```
working = popgdp_columns.reset_index()
melted = working.melt(id_vars='code',var_name='ind_year')
getind = lambda s: s[:-4]
getyear = lambda s: int(s[-4:])
melted['indicator'] = melted['ind_year'].apply(getind)
melted['year'] = melted['ind_year'].apply(getyear)
part_tidy = melted.drop('ind_year', axis=1)
part_tidy.sort_values(['year', 'code'], inplace=True)
part_tidy.head(6)
```

```
|      code     value indicator   year
|  0   CHN    1371.22       pop   2015
| 12   CHN   11015.54       gdp   2015
|  1   GBR      65.13       pop   2015
| 13   GBR    2896.42       gdp   2015
|  2   IND    1310.15       pop   2015
| 14   IND    2103.59       gdp   2015
```

This should look familiar. After melting and cleanup, we now have a representation that uses multiple rows to map from the independent variables of `code` and `year` to the two dependent indicator variables of `pop` and `gdp`. This is like our examples in Sect. 9.3.2. Fortunately, with two columns making up the index needed for a pivot, a `pivot_table` can transform into normal form, referenced by variable `tidy`.

```
tidy = part_tidy.pivot_table(index=['year', 'code'],
                             columns='indicator')
tidy
```

```
|                  value
| indicator          gdp        pop
| year code
| 2015 CHN      11015.54    1371.22
|      GBR       2896.42      65.13
|      IND       2103.59    1310.15
|      USA      18219.30     320.74
| 2016 CHN      11137.95    1378.66
|      GBR       2659.24      65.60
```

```
|        IND     2290.43   1324.51
|        USA    18707.19    323.07
| 2017  CHN    12143.49   1386.40
|        GBR     2637.87     66.06
|        IND     2652.55   1338.66
|        USA    19485.39    325.15
```

In this particular case, the column Index has, unnecessarily, two levels, with the outer level being the generic value. The pandas module supports dropping a level of an Index, which we can do if the remaining levels retain all the necessary information. So we want to drop the outer level, level 0, along the column dimension, which is axis 1:

```
tidy2 = tidy.droplevel(0, axis=1)
tidy2
```

```
| indicator        gdp       pop
| year code
| 2015  CHN   11015.54   1371.22
|        GBR    2896.42     65.13
|        IND    2103.59   1310.15
|        USA   18219.30    320.74
| 2016  CHN   11137.95   1378.66
|        GBR    2659.24     65.60
|        IND    2290.43   1324.51
|        USA   18707.19    323.07
| 2017  CHN   12143.49   1386.40
|        GBR    2637.87     66.06
|        IND    2652.55   1338.66
|        USA   19485.39    325.15
```

Alternatives

For this example, the original data set, as a data frame, could have been normalized in a couple of other ways than what was presented here.

First, we could have taken the original data frame and, using two different column projections, created a data frame with all the population data and another data frame with all the GDP data. Each of these could have been transformed into a partially tidy data set, with population or GDP indexed by code and year. Then, using the table combination techniques of Sect. 8.3, these partial data sets could be combined into a tidy whole.

A second alternative would involve the use of a two-level Index in the column dimension. If we transformed the data frame, through manipulation of the column Index, so that it had one level of the indicator and the other level of the year, then the pandas unstack() method, which generically is really transforming column Index levels into row Index levels, could be used to unstack just the

year level. This is a more advanced topic specific to pandas, so we have chosen
to omit it.

9.3.4 Exactly One Table per Logical Mapping

Up to now, the examples of messy data we have looked at have all involved
TidyData1 and/or TidyData2. The other constraint of this model is TidyData3,
which requires that *exactly one table* be used for a set of mappings (i.e., rows)
involving the same combination of independent variables. To better understand this,
it helps to think in terms of functional dependencies, where the combination of
independent variables is specified by the left-hand side of the relationship.

This means that if we have, for example,

$$\text{code}, \text{year} \to \text{pop}$$

and

$$\text{code}, \text{year} \to \text{gdp}$$

then these should be represented in a *single* table, not in two separate tables. If
they were in separate tables, the complete tables (like a pop table and a gdp table)
would each define the values for a variable, so this also violates TidyData1, because
the variable is not specified by a *column* but rather, in this case, by a *table*.

The dual, where two separate things are combined into a single table, must also
be upheld—we would not have a tidy data set if some of the dependent variables
in a table were related to one combination of independent variables, and some
other dependent variables in the same table were related to a *different* combination
of independent variables. To see this point, consider the dependent variables of
country and land in some of the variations of our indicators data set. If the
data set is *not* a time series, the functional dependency could be given by

$$\text{code} \to \text{country}, \text{land}, \text{pop}, \text{gdp}$$

where all four dependent variables are determined by code alone. However, in the
time-series version of the data set, we have, by definition of these indicators, that
pop and gdp are determined by *both* code and year:

$$\text{code}, \text{year} \to \text{pop}, \text{gdp}$$

But let us consider more carefully the dependent variables of country, the English
name of the country, and land, the land area of the country. If the value of these
variables *does not change* over time (i.e., with the year), then the year variable

is *not* part of their functional dependency. This means that the correct functional dependency for these dependent variables is given by

$$\text{code} \rightarrow \text{country}, \text{land}$$

So if we have a *single table* that combines these two separate logical mappings into one, we have violated TidyData3, and, to normalize the data set, we should separate the two logical mappings into two tables, one for each of these last two functional dependencies.

9.3.4.1 Example: Variable Values as Two Tables

We start with an example that uses the top baby names data set and was first presented as the first of the alternative representations illustrating tabular structure in Sect. 6.1. Tables 9.8 and 9.9 show a prefix of the data where we have separate tables for the top female and top male baby names and counts by year. Each of these tables has 139 rows, one per year of collected data.

Recall that Female and Male are *values* of a categorical variable, sex, which, along with year, are the independent variables of the data set and which determine the dependent variables name and count. To achieve the functional dependency of

$$\text{year}, \text{sex} \rightarrow \text{name}, \text{count}$$

we need to combine these tables together. Suppose the two source tables are referenced by variables topfemale and topmale. In the final result, we need

Table 9.8 Top female baby names

Year	Name	Count
2018	Emma	18688
2017	Emma	19800
2016	Emma	19496
2015	Emma	20455
2014	Emma	20936
2013	Sophia	21223

Table 9.9 Top male baby names

Year	Name	Count
2018	Liam	19837
2017	Liam	18798
2016	Noah	19117
2015	Noah	19635
2014	Noah	19305
2013	Noah	18257

all four columns of year, sex, name, and count, but there is currently no sex column, since that is encoded as part of the separate table representation.

When we have the same columns and what are, logically, different rows, we know from Sect. 8.3 that we can use the concat() function along the row dimension. But, before we combine, we should add the sex column to each of the source tables, setting values in the column to "Female" for the topfemale table and to "Male" for the topmale table.

```
topfemale['sex'] = "Female"
topmale['sex'] = "Male"
tidy = pd.concat([topfemale, topmale], axis=0)
```

At this point, tidy is a DataFrame with 278 rows, with no index, and with all the female rows preceding all the male rows. A more convenient situation would have year and sex as the row Index and use sort_index for the rows to be sorted in index order:

```
tidy.set_index(['year', 'sex'], inplace=True)
tidy.sort_index(inplace=True)
tidy.head(6)
```

```
|                    name   count
| year sex
| 1880 Female   Mary    7065
|      Male     John    9655
| 1881 Female   Mary    6919
|      Male     John    8769
| 1882 Female   Mary    8148
|      Male     John    9557
```

9.3.4.2   Example: Separate Logical Mappings in a Single Table

For our final example of this section, we look at the example described above, where we have a single combined table, but the two different functional mappings of

$$\text{code}, \text{year} \rightarrow \text{pop}, \text{gdp}$$

and

$$\text{code} \rightarrow \text{country}, \text{land}$$

Our source table, which will be referenced by the variable mixed_table is shown in Table 9.10. Note how, in different rows, the contents of code, country, and land always have the same values, even while the values in pop and gdp differ.

The strategy for normalizing this data involves using column projection twice on the original source to obtain the correct set of columns for each of the constituent

Table 9.10 Combined table

Year	Code	Country	Land	Pop	Gdp
2000	CHN	China	9388210	1262.64	1211.35
2000	IND	India	2973190	1056.58	468.39
2000	USA	United States	9147420	282.16	10252.35
2017	CHN	China	9388210	1386.40	12143.49
2017	IND	India	2973190	1338.66	2652.55
2017	USA	United States	9147420	325.15	19485.39

tables. The code below projects the columns code, country, and land into the
table countries and projects the columns code, pop, and gdp into the table
indicators, and we show the results for the latter table.

```
countries = mixed_table[['code', 'country', 'land']]
indicators = mixed_table[['code', 'pop', 'gdp']]
indicators
```

```
|     code       pop        gdp
| 0   CHN   1262.64    1211.35
| 1   IND   1056.58     468.39
| 2   USA    282.16   10252.35
| 3   CHN   1386.40   12143.49
| 4   IND   1338.66    2652.55
| 5   USA    325.15   19485.39
```

But when we look at the countries table, we see:

```
countries
```

```
|     code            country            land
| 0   CHN              China       9388210.0
| 1   IND              India       2973190.0
| 2   USA   United States       9147420.0
| 3   CHN              China       9388210.0
| 4   IND              India       2973190.0
| 5   USA   United States       9147420.0
```

This still violates TidyData2 since there are multiple rows (duplicates) conveying
the same mapping. While we could use iteration to obtain, row-by-row, the contents
of countries with duplicates and accumulate a new set of rows for a table
where we added a row only if it were not already present, pandas has a
drop_duplicates method that can perform this functionality for us.

```
countries = countries.drop_duplicates()
countries
```

```
|     code            country            land
```

```
| 0   CHN              China  9388210.0
| 1   IND              India  2973190.0
| 2   USA   United States  9147420.0
```

The resulting two tables of `countries` and `indicators` are now a tidy data set. We will explore these tools further in the exercises at the end of the chapter.

9.3.5   Reading Questions

9.41 In the `country` example, explain the 3 and 5 that appear in the lambda expression. Illustrate with an example.

9.42 Give an example of a way to report dates numerically, but where sorting gives a result different from chronological order.

9.43 Give an example of a real-world seasonal trend you might want to study by grouping data by month.

9.44 In the un-tidy example of `indicators` (with three rows for each mapping from a country to its indicators), explain why it is a bad idea to have a single column with three types of separate measures. Rather than speaking in generalities, give a specific example of something that could easily go wrong in such a representation.

9.45 Often in scientific inquiry, it is valuable to think about the *units* of various quantities, e.g., distance being measured in miles. Please write a sentence about how the consideration of units can be useful in determining if a given column satisfies **TidyData1**.

9.46 Consider the example of `Population by Year`, where normalization resulted in an index of `['year','code']`. If the index had been set as `['code','year']` instead, what would the result after sorting look like?

9.47 In the example `Population and GDP by Year`, the data was normalized with a melt followed by a pivot, and alternatives were given after. Could you have normalized it by pivoting first and then doing a melt? Why or why not?

9.48 In the `Top Baby Names` example, what is the difference between `pd.concat([topfemale, topmale], axis=0)` and `pd.concat([topmale, topfemale], axis=0)`?

9.49 In the `countries` example (with `pop,gdp,country,land`), is any information lost with the projections
```
    countries = mixed_table[['code', 'country', 'land']]
indicators = mixed_table[['code', 'pop', 'gdp']]?
```

9.50 Describe the relationship between `countries`, `indicators`, and `mixed_table` using the language of joins.

9.51 How do you think the `drop_duplicates()` command is implemented?

9.4 Recognizing Messy Data

This section attempts to arm the reader with some tools for recognizing violations of one or more of the structural constraints of tidy data (i.e., TidyData1, TidyData2, and TidyData3). We frame these as *red flags*—characteristics of the structure and the values of a given data set that *may* indicate a violation. Even to employ these red flags and to truly assess whether or not a given data set is in tidy data form, we must start by *understanding the data*, as emphasized at the start of Sect. 9.3.

9.4.1 Focus on Each Column as Exactly One Variable (TidyData1)

The following is a sequence of questions we must ask ourselves.

- If a table has columns whose value is derived by other columns, e.g., GDP per capita.
- If a table has columns where the column labels incorporate multiple parts. The examples in the chapter include cases where there are mashups of a variable name and the value of another variable
- If a table has column labels that are values of a variable.

  - Column labels that are the values of years, months, or days.
  - Column labels that are the various categories of a single categorical variable; we could imagine topnames with a column for Female and another column for Male.
  - In a table with a single independent variable, the values of the independent variable are the column labels; consider the country codes in indicators that could be structured as the column labels or the names of patients in a treatment data set.

- If the *values* in a column can be seen as a mashup, possibly using a hyphen (-) or slash (/) or comma (,) to divide parts of the mashup.
- If we have dates with multiple parts (year, month, and day) as *strings* in a single column. Note that a `DateTime` object in a column is allowed, as it allows proper sorting and GroupBy in the various dimensions of the time series (by year, month, and day of month).
- If it makes sense and we are inclined to aggregate values in *more than one column* for a single aggregate value, then this could indicate that multiple columns are, in fact, values for a single variable.
- If it does *not* make sense to aggregate in a column. This is particularly clear if the *units* of the values are different, but like units can still exhibit this red flag. Consider the open, close, high, and low stock price in a table of stocks. Taking an

average of these values makes no sense, and so a single column of stock prices
and another column of high/low/open/close would violate tidy data.

- – Another way to see this red flag is if we are treating as categorical something
  that does not give a partitioning and are really *independent measures*.

- If it is not treating as a variable something that gives a categorical partitioning of
  the data (e.g., Male/Female).

9.4.2   Focus on Each Row Giving Exactly One Mapping (TidyData2)

The following is a sequence of questions we must ask ourselves.

- If it makes sense to aggregate *across a row*, it may mean that the values are
  somehow compatible and could be values of the same variable, instead of a given
  row having values of *different* variables. This could be a subset of the values in a
  row, as opposed to the full row.

  - – In this case, the row labels would be the names of variables.

- If we can write down the functional dependencies and if we can see that a *group
  of rows* is needed to capture one mapping instance.
- If the *name of a dependent variable* is a value in a single column. Our "indicator"
  meta-variable with "values" of `'pop'`, `'gdp'`, `'cell'`, and `'life'` is an
  example where it would take multiple rows to capture a single logical mapping.
- If the name of a column is too generic, like "value" or "variable." This red flag
  often occurs at the same time as the previous point.
- The use of *missing data* (or some other encoding) to mean that there is not a
  value for this variable in this row because it is *not applicable*.
- Data rows that are dependent on other rows, like rows that are aggregations for a
  subset of the rows.

9.4.3   Focus on Each Table Representing One Data Set (TidyData3)

The following is a sequence of questions we must ask ourselves.

- If there are multiple tables and the *names* of the tables are values of a categorical
  variable, like we had with a `Female` and a `Male` table for the topnames data set.
- If there are multiple tables and the *names* of the tables name different *dependent
  variables*, particularly if they are dependent on the same independent variable.
  For instance, the indicators data set might have one table for population and

another for GDP, with both being dependent on the independent variables of the year and country code. In this case, if we write down the functional dependencies, the different tables would have the same left-hand side and differ in the right hand sides.

- If, when we understand the data and write down the functional dependencies, there are *different* functional dependencies with one having a subset of the variables on the left-hand side of another. Our example in the book had code, year -> pop,gdp but code->country, land.
- If there is repeated data and, upon examination, the redundant data is because, in a row (instance mapping), the repeated data is dependent on less than the full set of independent variables.

9.4.4   Reading Questions

9.52 The first red flag for TidyData1 states "If a table has columns whose value is derived by other columns." Does this description *always* lead to a violation of TidyData1? Illustrate with examples. This question hinges on how you interpret "derived," so please be clear in what this term means.

9.53 Give an example, different from any in the text, of a table where the column labels incorporate multiple parts.

9.54 Give an example, different from any in the text, of a table where we might be inclined to aggregate values in more than one column for a single aggregate value.

9.55 Give an example, different from any in the text, of a table with a numerical column that it does not make sense to aggregate, e.g., because it contains values in different units. Your example should involve "treating as categorical something that does not give a partitioning and are really independent values."

9.56 The last red flag listed for TidyData1 says "Not treating as a variable something that gives a categorical partitioning of the data (e.g., Male/Female)." Suppose you had a data set containing salary data for individuals of different hair colors. What would it look like to represent this data set in an un-tidy way, leading to this red flag? What would it look like to represent this data in a tidy way?

9.57 Give an example, different from any in the text, of a table where it makes sense to aggregate across a row, and give a specific TidyData assumption your example fails to satisfy.

9.58 Give an example, different from any in the text, of a table where a *group of rows* is needed to capture one mapping instance of the functional dependency.

9.59 Give an example, different from any in the text, of a table where the *name of a dependent variable* is a value in a single column.

9.60 Give an example, different from any in the text, of a table where missing data is used to mean that a variable is not applicable to a given row.

9.61 Referring to the language of the TidyData assumptions, why is it a violation to have "Data rows that are dependent on other rows, like rows that are aggregations for a subset of the rows"?

9.62 Give an example, different from any in the text, of a table where "there is repeated data and, upon examination, the redundant data is because, in a row (instance mapping), the repeated data is dependent on less than the full set of independent variables."

9.4.5 Exercises

9.63 The data set members.csv has a mashup column containing both first and last names. Download this data from the book web page, read it into a data frame, and correct this mashup column.

9.64 The data set members.csv has a date column. Correct this as shown in the reading, by splitting it into three columns.

9.65 Now correct the date column in members.csv by creating a column of DateTime types.

9.66 The data set members.csv has another mashup column. Find it and correct it.

9.67 Read us_rent_income.csv from the book web page into a data frame, and then look at the data. Is it in tidy data form? Explain your answer. Note: this data set contains estimated income and rent in each state, as well as the margin of error for each of these quantities. This data came from the US Census.

9.68 Explore the data provided as us_rent_income.csv, and then draw the functional dependency.

9.69 Read us_rent_income.csv into a data frame (with "GEOID" as the index), and then transform as needed to make it tidy. Store the result as df_rent.

For each of the next four exercises, assess the provided data against our Tabular-Tidy data model constraints. If the data conforms, explain. If the data does not conform, explain why and describe the transformation(s) needed to bring the data set into conformance. Be clear about the columns that would be in the resultant data set.

9.70 Below is some (fake) data about the connection between religion and salary. This was inspired by real research done by the Pew Research Center.

Religion	Under 20K	20–40K	40K–75K	Over 75K
Agnostic	62	140	173	132
Buddhist	49	68	84	61
Catholic	1025	1432	1328	1113
Protestant	1413	2024	1661	1476
Jewish	39	50	39	91

Is it tidy? Assess.

9.71 Below is (fake) weather data, giving the minimum and maximum temperature recorded at each station on each day.

Date	Element	Station	Value
01/01/2016	tempmin	TX1051	15.1
01/01/2016	tempmax	TX1051	26.9
01/01/2016	tempmin	TX1052	14.8
01/01/2016	tempmax	TX1052	27.5
01/02/2016	tempmin	TX1051	13.9
01/02/2016	tempmax	TX1051	26.4
01/02/2016	tempmin	TX1052	14.6
01/02/2016	tempmax	TX1052	30.7

Is it tidy? Assess.

9.72 The following is public transit weekday passenger counts, in thousands.
****

Mode	2007–01	2007–02	2007–03	2007–04	2007–05	2007–06
Boat	4	3.6	40	4.3	4.9	5.8
Bus	335.81	338.65	339.86	352.16	354.37	350.54
Commuter Rail	142.2	138.5	137.7	139.5	139	143
Heavy Rail	435.29	448.71	458.58	472.21	474.57	477.32
Light Rail	227.231	240.22	241.44	255.57	248.62	246.10

Is it tidy? Assess.

9.73 Here is some (fake) stock price data by date.

Date	Apple Stock Price	Yahoo Stock Price	Google Stock Price
2010-04-01	177.25	179.10	178.62
2010-04-02	176.11	175.39	175.84

Is it tidy? Assess.

9.74 Similarly to the above, another way that stock data is commonly presented is with one row per company and one column per month. Would that form be tidy? If not, what transformations would be needed to fix it? Justify your answer.

9.75 Explore the data from the previous exercise, and then draw the functional dependency. Note: this data came from a survey of individuals about their religion and their income.

9.76 Read the data from the previous exercise into a data frame, and then transform as needed to make it tidy.

9.77 Download the World Health Organization's "Global Tuberculosis Report" data set as a CSV file from

```
https://www.who.int/tb/country/data/download/en/
```

Then, wrangle this data into a tidy data frame. Hint: you will need to study the data dictionary carefully.

Chapter 10
Relational Model: Structure and Architecture

Chapter Goals

Upon completion of this chapter, you should understand the following:

- The relational model structural concepts of:

  - Tables defining a data relation,
  - Rows providing data records,
  - Columns defining the fields of a record, with specific data type,
  - Primary Key to uniquely identify records, and
  - Relationships between tables, from one-to-one, one-to-many, and many-to-many.

- The background motivation for relational databases, basic types of databases, and some of the advantageous characteristics.
- How the notion of clients and providers fits into a database system architecture.

The single most important form of a data system, grounded in nearly 50 years of development of both underlying theory of structure and efficient implementations, is the Relational Database system. In Sect. 5.3, we presented an overview of the relational model, describing it as composed of an inter-dependent set of tables,

© Springer Nature Switzerland AG 2020
T. Bressoud, D. White, *Introduction to Data Systems*,
https://doi.org/10.1007/978-3-030-54371-6_10

and distinguishing it from the tabular model by its database design prior to data population, and its mathematical foundation. We also gave a high-level definition of a database management system (DBMS). The relational model is extremely powerful, facilitates very efficient access to the data it stores, and is protected (via constraints) from a wide variety of errors. The trade-off is that careful thought is required prior to designing the rigid relational structure that the data must conform to. The current chapter describes and explores the relational model further.

We start by giving additional background on the requirements that a database system must meet and how relational databases, in particular, meet those requirements. We highlight the types of relational databases that are of particular interest in a broad treatment of data systems, the subject of this book. The chapter then presents more information on the structural details, and places the client and database system provider in the context of an architecture.

In subsequent chapters, we focus on operations to query data from a relational database. We introduce SQL query operations and progress through queries on a single table in Chap. 11 and then SQL queries involving multiple tables in Chap. 12. Because SQL is a declarative language, specifying what data is desired, and is agnostic to the programming language or system of the client, these chapters focus on just the SQL, and defer the programming aspects of using a particular package and API to Chap. 13.

Within the relational data model, many of the constraints are related to the *design* of a database. We thus group a discussion of constraints with advice regarding database design in Chap. 14. There, we also cover the SQL for realizing the design of a database through SQL CREATE, and populating the database through SQL INSERT.[1]

The treatment of relational databases, as well as the SQL for operating on the database, in this book is, necessarily, a small fraction of the subject. We are interested in understanding the relational data model as one of our data forms and the provider system as a source of data. Our perspective is that of a client interacting with this particular data system, as needed for acquiring data and preparing it for analysis. As such, we omit topics like updating database records, deep understanding of consistency issues, and transactional support. We omit the underlying mathematical set theory but note that it facilitates the relational algebra that makes relational databases so powerful and efficient. We also omit the discussion of the provider-side system design for the efficient storage and retrieval of data.

---

[1] The "commands" of SQL, by convention, are written in all uppercase, to help clarify the use as an SQL syntactic reference.

10.1   Background

The term *database* is used in many contexts, both in the real world, and as used in this book. We begin with a comparison of definitions. The definition from Wikipedia states [86]:

Definition 10.1 A *database* is an organized collection of data, generally stored and accessed electronically from a computer system.

The Wikipedia description of a database also states: "Where databases are more complex they are often developed using formal design and modeling techniques." This definition is perhaps too general. It might, with appropriate argument, be applied to the entire spectrum of data systems presented in this book. The Merriam-Webster definition [33] enhances this definition by adding that the collection is usually *large*, and that the organization allows for *"rapid search and retrieval"*. Within these definitions, we can focus on the *organized structure*, and allow that this might include a structure that is well-defined, and the idea of constraints that can be enforced so that the organized structure can be maintained. We can also focus on the *rapid search and retrieval*, and allow that this might include a standard language that allows specification for what data should be retrieved.

With this additional focus, we arrive at the notion of a *relational database*, that is, those databases based on the relational model of data. In practice, this means the database comes equipped with a schema relating the different tables within it, that the database is organized for rapid search and retrieval, that the database contains information about its own constraints, and that these constraints are enforced. The notion of a relational database proposed by [4], with its formal methods for determining tables and relationships, is a prime example. We will not have need for the overly general definition of database given above, and will instead often use the term *database* as a substitute for "relational database," often implemented on a dedicated system of the provider, called a *relational database management system*, or sometimes just a *database management system*. The use of the relational model achieves the goal of organized structure, and the use of SQL achieves the goal of rapid search and retrieval.

10.1.1   Motivation and Requirements

The primary motivation for any database system is to provide a repository of *persistent* data that centralizes the data and allows for the organization, search, and retrieval components of the above definitions. Often, a database system is hosted by a data system provider, and provides access to data system clients with appropriate permissions. However, it is also possible for the database to be hosted locally on the computer of the data system client, if the client desires the data in a more structured form, and with more efficient query access, relative to the tabular data model.

In our motivation, we employ the adjective of *centralized* (at least logically) and the notion of a *persistent repository* to guide us from a potential concept of a database as one or more files on a local system for just the client or application running on that local system[2] to the more general idea of a system where the same data set can be reliably accessed and queried by *multiple* clients, possibly executing applications on different machines, and to have only one place to go for the data, and for the clients to be able to do their work without any coordination between these clients, even though they may be executing at the same time.

Persistence means that the data repository maintains its information and its integrity across clients coming and going and even across the stopping and starting of provider system software and hardware. Even more beneficial is if the repository can maintain its information and integrity across unexpected failures, like power outages or logical errors in client applications.

A database system should allow us to reduce common sources of errors. Through the tabular model, we have seen three common sources of errors that we would like to prevent:

- Allowing data to violate structural constraints (violations of tidy data).
- Allowing data in a column to not have an appropriate data type, or be restricted to a set of acceptable values.
- Allowing some data to be repeated (i.e., redundant) in a way that allows changes to one copy not be reflected consistently across all copies.

In the tabular model, these errors could occur because constraints are a set of conventions, and no mechanism exists to check or enforce compliance. In the relational model, we can and will enforce such constraints.

Allied with our initial motivation, a requirement may be stated that separates the client or a set of clients from the provider, where the data is actually stored and maintained. This allows a single and consistent database to be available to multiple users and multiple client applications. And once we make the step of, at least potentially, a separation of client from provider and their server system, we open up the possibility of operations that might be permitted for some clients and not permitted for others.

Beyond the requirement of independent clients and their access to a single database, might be the requirement of *sharing*. In this requirement, when one client makes one or more updates to the database: be they the addition of records, the changing of existing records, or the deletion of records, the other clients will eventually see the update, and algorithms of clients may proceed along different paths, depending on this interaction.

The definition of database includes the idea of rapid search and retrieval. For ease of retrieval, and to complete the separation between the provider and the client, we require relieving the client of knowledge of the physical model entailed in how the

---

[2]For instance, under the tabular model, we might have a local directory structure that organizes a set of CSV files into an informal kind of database.

data is stored at the provider. We, instead, want the clients to employ a logical and programming language agnostic way of specifying the operations to be performed on the database. In other words, we want to *declare* what data should be retrieved or updated while abstracting away the details of how, given the physical model and reality, that is actually accomplished.

The final requirement to be discussed involves what, in distributed systems parlance, is called *consistency semantics*. This requirement is at the confluence of two earlier-cited requirements: we must maintain the integrity of the database, and we also must support multiple clients, executing concurrently. The requirement of consistency semantics arises because two (or more) clients might be accessing the same data in the database at nearly the same time, and the individual operations of the clients might be performed in some interleaved way, causing the database to become logically corrupted.

Using an illustrative example, first suppose that an individual operation, like a read of a record, or to change a value in a record, can be executed by a provider without interference. Then suppose that one client wants to perform the three steps and two database operations of:

- Read the value of field A out of record n, obtaining value x.
- Compute a new value, say $2 * x$.
- Update field A of record n with $2 * x$.

Another client might, at approximately the same time, be performing the three step sequence of:

- Read the value of field A out of record n, obtaining value y.
- Compute a new value, say $y/2$.
- Update field A of record n with $y/2$.

Say that, before execution, the A field of record n contains the value 32. If one client executed all three steps, and then the other client executed all three steps, then the result would be that the A field of record n again contains the value 32.

Suppose, instead, that the interleaving of the clients happens such that *both* take their first step before either takes their third step. That means that x is 32, and also that y is 32. If the first client then completes and then the second client completes, then afterward the A field of record n contains the value 16. If the second client performs their update followed by the first client, then afterward, the A field of record n contains the value 64.

So we have three possible outcomes, with final values of 32, 16, or 64. If these sets of operations involved deposits and withdrawals from a bank account whose data were maintained in a database, the logically correct solution would be the value of 32. That value would be the final result whether the first client executed, followed by the second client, or vice versa. But it would not be correct (or *consistent*) for any possible interleaving were to be observable, with final results of 16, or 64.

While much more complex examples exist, this relatively simple example should illustrate the requirement for consistency when we must maintain logical integrity in the presence of a shared database with concurrent clients.

10.1.2   The Relational Database Solution

Relational databases are the most common form of database, and, for the provider side, many commercial (Oracle^{TM}) and open source (PostgreSQL, MySQL, SQLite) realizations exist. This subsection describes how the relational model and the relational database realizations satisfy the requirements of Sect. 10.1.1.

To support, in part, the requirement of maintaining integrity and to support reduction of common errors, relational databases use a complete design and definition of the database ahead of time, prior to any population of the data or issuing of any operations. This design of the database is called the *database schema* and the design of any particular table is called a *table schema*. The schema defines data types and valid values for column fields so that the system can enforce that no invalid data in the column field is even possible. The database design and, ultimately, the schema, often involves formal techniques of modeling.

Database modeling and design also supports the requirement of eliminating redundant information. In the design process, the tables are determined through a process called normalization such that each one models separate information. The net result is a larger number of tables, but designed so that redundancy is eliminated. Through the design and normalization process, the relationships that relate one table to another are also explicitly established.

In the relational database solution, because the design result is a set of related tables, the operations also include powerful basic operations to combine tables. Further, the operations, both simpler single table operations and operations to combine and to query combinations of tables, satisfy properties of the underlying mathematics. The set theory and predicate logic basis allow the mathematics to model tables as sets and to prove that, whenever we operate on a table or a subtable, we get a coherent table as the result. This allows the operations to be composed and chained in powerful ways.

The realization of relational database management systems use a client–server architecture to satisfy the requirements of a persistent and logically centralized repository for the data, separate from the client set. Depending on the system, it can manage a multiplicity of users with possibly differing access to data and permissions for operations. These realizations also, then, support sharing and coordination for access to common data.

Relational database management systems provide a meta-operation, called a *transaction* that allows clients to define groupings of operations that must be completed as a unit. This allows the system to recognize possible interleavings that might lead to incorrect/inconsistent results, and can use them to ensure the logical consistency of the database. In many aspects of the relational database solution, the core idea is to make sure, through schema and initial population, that the database starts in a consistent (aka *sound*) state, and each subsequent operation or transaction only transitions a database from one sound state to the next sound state, even in the presence of concurrent clients.

In the relational database solution, we have separated the client from the provider, and placed the onus of the physical realization on the provider. Since clients being implemented in different platforms or using different programming languages, it is essential to give clients a way to *declare* what data is required, without specifying how. The satisfaction of this requirement in the relational database solution is the language SQL.

In an ever changing and data-centric world, to address needs of greater scale and access from clients with different characteristics, the relational database solution is not the end-all and be-all. In recent years, many types of databases have been designed to address many of the same requirements. These alternatives are advanced topics beyond the scope of this book.

10.1.3 Types of Relational Databases

Relational databases are often categorized based on how they are used. While such a categorization is not "pure," in that some databases are used in multiple ways, looking at the categorization can help us to understand what parts of the relational database solution pertain to the objectives of this book.

An *Operational Database* is one used in the operation of an enterprise, be it in for-profit or non-profit business, the education sector, or any other enterprise. That means that the database is dynamic and "online," being used for maintaining information about contacts, vendors, inventory, sales, budgets, and so forth. The operations on an operational database are a mix of queries and are undergoing constant update, with additions of records, deletions of records, and changes to existing records. These databases are used in the minute-by-minute processes, involving both human and data-centric processes, of the enterprise. This type of database is also called an online transactional processing (OLTP) system.

An *Analytical Database*, on the other hand, is used to support analysis. Most often, this analysis is by the enterprise for decision support. Other analytical databases might be open and accessible for analysis by outside users, and be used for data science and machine learning on databases maintained for the public good. These analytical databases are often used to track historical and time dependent data. Such databases are relatively static, populated initially with a collection of data and then, possibly added to as additional data becomes available. So the operations are primarily queries and then the addition of new data, but rarely are there deletions of records, nor changes to existing records.

In an analytical database, even one used in a shared client environment, the operations are primarily read-only and the operations of one client are independent of one another. The issue of consistency is avoided because there are, generally, no updates to existing records shared by two clients.

Our objectives in this book are to understand data systems from a client user/application perspective, and to understand the forms and sources of data for the purpose of data acquisition and as a prelude to analysis. Thus, we are primarily interested in understanding and using analytical databases. As such, in this book, we omit coverage of the mechanisms of transactions, and also omit coverage of the parts of SQL that deal with updating existing records and with deleting existing records.

10.1.4   Reading Questions

10.1 The reading discusses the difference between a "database" (as defined by Wikipedia) and a "relational database." Draw an analogy with our discussion of the difference between tabular data (that is, with rows and columns) and tidy data.

10.2 Give a real-world example, different from any in the reading, of a situation where it might be important for many different individuals to have access to the same database.

10.3 Please think of a time you shared an electronic resource or document with another person. When one of you made a change, would the other person eventually see that? How do you think this was managed by the system?

10.4 Give a real-world example, based on your experience, of how a system you have used has handled consistency. What happened if two users tried to update it at the same time? What if users made changes while offline, and syncing was only possible when both came back online?

10.5 The reading explains the importance of database transactions. Using the language from the reading, give an example from the field of banking, showing why database transactions are important (e.g., what could go wrong if we did not have such a notion). Hint: you can use the example in the text as inspiration.

10.6 Give a real-world example illustrating the difference between a database schema and a table schema.

10.7 Give a real-world example where you might want to give some users different rights than others, as regards access and modification privileges for a shared database.

10.8 Based on how *operational database* was defined, what are some special issues that could make working with an operational database harder than working with an analytical database?

10.2   Structure

In the relational database model, the structure is a set of inter-dependent tables, where each table is comprised of rows and columns. An individual table is also, in its mathematical context, called a *relation*. Note that it is this term, of a table as a relation, that is the origin of the model as a *relational* database. A common misconception is that the name has to do with the relationships that might exist between tables.

A *relational database*, then, is the collection of tables that are designed together in a single schema. The database includes information governing the relationships between individual tables, as well as information about the users/clients permitted to access the database, and what types of access are allowed. A single provider system may host many relational databases, each with different schemas and with different user/client permissions.

10.2.1   Single Table Characteristics

We begin by considering the structure-related characteristics of a single table in the relational model. Like in the tabular model, tables are composed of rows and columns. But unlike in the tabular model, a table cannot have non-conformant table representations, like values for columns, nor variables for rows—the limitless variety of messy data we saw in the tabular model.

In the relational model, the columns of a table are also called *fields*, and the rows are called *records*.[3] Each column/field has a distinct name within the table along with a data type, a data size, and possibly other constraints determined at design time. Field names are used like variables, and are *not* arbitrary strings, like we saw in the tabular model and in `pandas`. For instance, no spaces, other whitespace, nor special characters are allowed as part of field names. We often use the term *field* synonymously with field name. Like in tidy data, the fields in the relational model should capture the notion of *exactly one variable*.

10.2.1.1   Functional Dependencies

The relational model, more formally than the tabular model (at least at design time), uses functional dependencies to describe the mapping provided by a table. So, for a table, we can write a mapping, like before, as:

$$independent\ vars\ \rightarrow\ dependent\ vars$$

Consider the example of a table maintaining records for each of the courses offered at a college. The fields might include a `courseid`, used internally by the registrar to uniquely identify the course, and also might have `coursesubject` like ECON or MATH and a `coursenum`, like 201, or 350. Students would look in the

---

[3]In the mathematical model, rows/records are called *tuples* and columns/fields are called *attributes*. We will avoid using these terms to avoid additional confusion.

catalog for ECON-350 for the description of the course. Beyond these, we assume fields/variables of `coursetitle`, the full title of the course, and `coursehours`, an integer for the number of credit hours of the course. In this example we can write the functional dependency:

`courseid`→`coursesubject, coursenum, coursetitle, coursehours`

If we can argue that the combination of `coursesubject` and `coursenum` is unique, we can, with equal validity, write:

`coursesubject, coursenum`→`courseid, coursetitle, coursehours`

It is perfectly normal and valid for multiple functional dependencies to characterize the table design and the anticipated data that will populate the table. But consider a similar example with `studentid`, the unique identifier of a student at the college, `studentlast`, the last name of the student, and `studentfirst`, the first name of the student. If `studentid` is unique, it can serve as the independent variable side of the functional dependency of the `students` table:

`studentid`→`studentlast, studentfirst`, *other dependent vars*

It may be tempting to try to identify students by their first and last name, as shown in the following potential dependency:

`studentlast, studentfirst`→`studentid`, *other dependent vars*

However, if the possibility exists for the combination of last name and first name to *not* be unique, then this would be *invalid*, because it is *not* a functional dependency. This is true even if, in the current collection of student data, there are no pairs of students with the same first and last name, because there is still the *possibility* that future data (more students registering) could lead to a name duplication. In this case, the relational constraints would prevent us from selecting this functional dependency, or, if we had selected it already before future data arrived, would prevent us from adding new students breaking the uniqueness requirement.

10.2.1.2   Table Keys

A *key* is one or more fields that can potentially be used to *uniquely* identify an individual record. In the relational model, given a value for a key, there is *only one* record that corresponds to the given key. If we have correctly written all possible functional dependencies for the mapping of a table, then the set of independent variable fields on the left-hand side of each functional dependency would enumerate all the potential keys. This set of potential keys is also called the *candidate keys* of the table.

If a key is composed of *more than one field*, the key is called a *composite key*. When there is a composite key, it is the *combination* of field values that must uniquely identify a record. In our example above, a coursesubject of "ECON" does not necessarily uniquely identify a record; it is the combination of "ECON" for coursesubject and the value 350 for coursenum that should identify a record.

The design of a relational database table must identify, *for every table*, a *primary key*. This, clearly, must be one of the candidate keys, and will be used by the system to build auxiliary structures to allow efficient access to a record, given its primary key. Remember that, for persistence, these tables reside on disk storage, and access can be very slow. If we have no means to "narrow the search," finding a record by using a linear search of all the records would make such a system infeasible. Since a primary key for a table could potentially be any of the possible candidate keys, the primary key might well be a composite key.[4]

Even when another candidate key might exist for a table, often an "artificial" key is used for the primary key, where the key is created as the table is populated with data records. These artificial keys might be created by the enterprise (like the studentid for a college adding a student for the first time), or it could be generated automatically by the DBMS.

10.2.1.3 Illustrative Example

Let us take the example of a courses table to more fully illustrate the structure concepts described above. The design phase takes place first, where we start by describing fully all the fields of the table. In Table 10.1, we describe the fields of the table:

Given the functional dependencies for this table of

courseid → coursesubject, coursenum, coursetitle, coursehours

Table 10.1 Fields of the courses table

Field	Data type and size	Field description
courseid	Integer (4 bytes)	Unique number associated with a particular course
coursesubject	String of up to 4 chars	The identifier for the subject, be it department or program
coursenum	Integer (4 bytes)	The number of the course, as used in the catalog
coursetitle	String up to 128 chars	Full title of the course
coursehours	Integer (4 bytes)	Credit hours for course, from 1 to 4

---

[4]When we use the term "key," it is a singular noun. Even if a key is composite, the singular is still used, so do not get misled.

Fig. 10.1 Table schema for courses

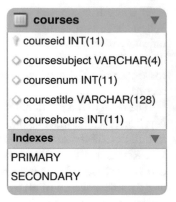

and

$$coursesubject, coursenum \rightarrow courseid, coursetitle, coursehours$$

we decide, in this design, to choose the non-composite key of `courseid` from the candidate keys of `courseid` and the composite key `coursesubject`, `coursenum`. Then, using a table schema tool, we arrive at the table schema in Fig. 10.1.

In the schema, a table is depicted in a rectangular box, with the name of the table in the bar at the top. Inside the box are each of the fields with the name of the field and the data type. In this case `INT` is used for the default integer size of 4 bytes. For strings, the SQL data type uses `VARCHAR` for variable length strings and the number in parentheses is the maximum number of characters to be used for this field.

The primary key is indicated in the schema with a small key icon. The icon next to the other fields is either filled in, or not, depending on whether we wish to enforce the constraint that the field is non-NULL or not. Here, `coursesubject` and `coursenum` must not be NULL (missing), but `coursetitle` and `coursehours` are allowed to be NULL. The `Indexes` section of the table schema refers to the auxiliary structures built by the DBMS to allow efficient access given a key. A `PRIMARY` index will always exist, automatically built for the primary key. In this case, we are also showing a secondary index that, while not shown in this picture, is the composite key of `coursesubject` and `coursenum`.

In the relational model, the tables of a database are created, one at a time, and we, in fact, use the SQL language itself to create the tables. Although details of table creation will be covered later, in Chap. 14, we show the SQL for this particular example to illustrate how a design and associated schema are realized:

```
CREATE TABLE `courses` (
    `courseid` INT NOT NULL,
    `coursesubject` VARCHAR(4) NOT NULL,
    `coursenum` INT NOT NULL,
    `coursetitle` VARCHAR(128) NULL,
```

```
'coursehours' INT NULL,
PRIMARY KEY ('courseid'),
INDEX 'SECONDARY' ('coursesubject' ASC, 'coursenum'
                                                 ASC))
```

A few notes on the above SQL:

- The capitalization of SQL keywords is by convention, but not required.
- For some database systems, the capitalization of field names is significant, so CourseID would be interpreted as a different field from courseid.
- The backticks around table and field names may, in many cases, be omitted.[5]
- For each of the fields, after the SQL data type, is the constraint of NULL or NOT NULL. Other constraints can also be included as part of the table creation.

Once a table is created within a database, it may be populated with data. The SQL INSERT command is used to add records to a table, and will also be covered in Chap. 14. Because the DBMS has the information from the CREATE shown above, it can *enforce* all constraints. So if we try to INSERT a record whose courseid already exists in the table, violating the uniqueness of a primary key, that SQL will result in an error. Likewise, if an INSERT tried to use a string that was longer than 4 characters for coursesubject, or a record that did not include an integer for coursenum, an error would result at insert time.

Once the population has occurred, we end with a table of rows and columns, very similar to the tidy data form of the tabular model. We can depict the courses table example using formatted output as shown in Table 10.2. Understanding the SQL operations for querying data from a single table, like this, will be covered in Chap. 11.

10.2.2   Multiple Table Characteristics

While the structural characteristics of any single table are described above, we now consider additional structural characteristics of the relational model that involve

Table 10.2 Populated courses table

courseid	coursesubject	coursenum	coursetitle	coursehours
1023	MATH	135	Calculus I	4
2055	CS	111	Discovering computer science	4
2099	CS	399	Independent study	4
3012	ECON	201	Micro economic theory	4

---

[5]Using delimiters allows more complex table and field names, such as incorporating spaces into the name. For simplicity and readability, the tables and fields used in the examples of this book will be simple, and delimiters omitted.

multiple tables. Given any two tables, we must provide a way for one table to have fields and field values that somehow correspond to the fields and field values in another table. In other words, we are seeking to characterize and provide for the *relationships* that may exist between two tables. And the tables, as a set of records for some mapping, are representing a set of "things," often with a correspondence in the real world.

Consider our `courses` table. Records of this table are intended to each represent exactly one course, like an item in the list of courses of a catalog for a college. The dependent fields then give all the information about that particular course. So given a course, the title for the course may be found, along with its course credits, and any other information we might wish to include. There might be multiple courses with the same title, say "Independent Study," one associated with CS-399, but the same title might be associated with HIST-299. But given a single course, we can find the one and only title associated with that course.

We could extend the database example and add a `classes` table. The intent of the classes table is to include a record for each *class instance* taught. So, in this terminology, a particular *course* might not be taught at all in a particular semester, but it might have multiple class sections. So we are using "class" to refer to an instance of a "course," taught as a particular section in a particular semester. We show a table schema for a `classes` table in Fig. 10.2. The `classterm` field will use values `"FALL"` or `"SPRING"`. Each class has a class registration number (CRN) that uniquely identifies a class instance within a semester, but is re-used in another semester, and this field is identified as `classcrn`, an integer. The class section is represented as a two character string, and would take on values like `"01"`, `"02"`, and so forth. Finally, and important in understanding table relationships, is the `courseid` field of the `classes` table. This field is intended to be used as a reference to the record in the `courses` table that corresponds to this class record.

We define the composite key of `classterm` and `classcrn` as the primary key of the `classes` table.

In this design, we are postulating a relationship where, given a record for a course, there may be multiple/many records in the `classes` table that correspond to that course. This is called a *many-to-one* relationship between two tables. Putting the two table schema together, we get the representation in Fig. 10.3. There are two differences of note beyond the schema for the individual tables. First, there is a

Fig. 10.2 Table schema for classes

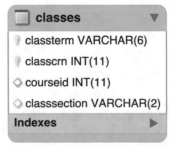

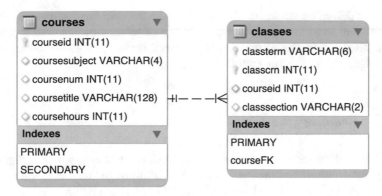

Fig. 10.3 Database schema for courses and classes

line denoting the relationship between the `classes` table and the `courses` table. The line has a multiplicity of connections, a "crows foot," on the `classes` side, indicating that this is the *many* side of a table relationship. On the `courses` side of the relationship line is a hash, which indicates this is the *one* side of a table relationship—so we have a one-to-many relationship between `courses` (on the one side) and `classes` (on the many side). The other aspect to note is the existence of an additional index for the `classes` table. This indicates the means by which we can efficiently use the `courseid` field in the `classes` table to reference the correct corresponding `courseid` in the `courses` table. This is called a *foreign key*, and it is a key in the table for the "many" side of a relationship that allows an index mapping to the primary key of the table on the "one" side of the same relationship.

When, by design, the fields of one table are used to establish relationships with another table through some corresponding field, we explicitly define a foreign key to allow for efficient table combination (i.e., a join). Again, this is part of the design of the database and part of the set of SQL used in the creation of the database, and before any population. This further enhances our ability to reduce common errors from being possible. In this case, an SQL `INSERT` into the `classes` table that included a `courseid` that *did not exist* in the `courses` table would be rejected with an error. Such inter-table constraints are called *referential integrity* constraints.

We could further expand our database of courses and classes and find table interactions that require a *many-to-many* relationship. For instance, we could add a `students` table. For a given student record, that student is generally taking multiple (many) classes. And it is certainly the case that, given a class, there are multiple (many) students taking the class.

In relational databases, many-to-many relationships are realized by adding an additional table, called a "joining" table or "linking" table, that allows us to define all the pairwise associations between the two original tables. By using this additional table and multiple one-to-many relationships, we can achieve the many-to-many relationship. We will explore this topic, from an operational standpoint, in Chap. 12, and from a design standpoint in Chap. 14. Because the many-to-many relationship is realized using an additional table and the one-to-many relationship (and appropriate foreign keys), it does not further complicate our discussion of the structural aspects of the relation model in this section.

10.2.3   Reading Questions

10.9 The reading mentions the phrase *mathematical relation*, which is a generalization of the notion of a function. Look up the definition of "mathematical relation" and try to explain how to think of an individual table as a relation.

10.10 When we were making data tidy in previous chapters, we often needed to move back and forth between integer and string representations of years by applying int() and str() to the column years. Would this be possible in a database? Why or why not?

10.11 Which of these are allowable field names and why?

1. Address/PO
2. num-cases
3. number of visits
4. pop_size
5. socialSecurity#
6. 4chanUsername

10.12 The reading gives several examples of functional dependencies. Write down a functional dependency for the classes table, whose fields are shown in the Figure.

10.13 Give an example of key other than student id that uniquely defines a student.

10.14 Make a list of 4–5 fields that could be present in a hypothetical student database (with information on students), then modify the CREATE TABLE command to fit this example. Do not forget to specify a key.

10.15 The reading mentions that, when a database is created, each column is given a name, data type, data size, and "possibly other constraints." Give a real-world example with a specific column and another constraint one might wish for it.

10.16 The reading discusses how one can connect the courses and classes tables by including courseid as a field in the classes table. The reading also discusses how courseid is a *primary index* for the courses table. Is it also a primary index for the classes table? Explain.

10.17 The reading explains the one-to-many relationship between courses and classes (since one course could be associated to many classes). Think of a database D such that `courses` would have a many-to-one relationship with D (that is, many courses associated to each record in D).

10.18 The reading discusses how one can connect the `courses` and `classes` tables by including `courseid` as a field in the `classes` table. Is it also possible to match up these two tables by adding a column to `courses` instead of adding a column to `classes`? Explain.

10.19 The reading provides examples of many-to-one and many-to-many relationships between database tables. Give an example of a one-to-one relationship between two tables.

10.3 Database Architecture

From a practical standpoint, when a relational database system is realized, at least in general, then we have shifted our *source* of data as well as our *form* of data. While we defer the general treatment of the various sources of data to Part III of this book, it helps to understand the relational model in the context of its architecture. We need to see how a client and a provider *co-exist*, often connected through a network, and how the provider responds to the needs of the client.

A client might be an application that we construct, but it also could be tool or utility that simply provides access to a database and allows us to issue SQL commands directly.[6] When we are in the relational model, and are considering the *provider* in this model, the provider and server are referred to as the Relational Database Management System (RDBMS), or just "database management system" (DBMS) in common parlance. With the dual roles of client and provider in mind, we return again to the question of formally defining what is meant by this term. The Wikipedia definition follows [86].

Definition 10.2 A *database management system* (DBMS) is the provider software and hardware that interacts with clients and client applications on one side, and manages the data and its storage organization on the other.

The DBMS software additionally encompasses the core facilities provided to *administer* the database. The sum total of the database, the DBMS, and the associated applications can be referred to as a "database system." Often the term "database" is also used to loosely refer to any of the DBMS, the database system or an application associated with the database.

Figure 10.4 depicts the general architecture possible for relational databases. On the left of the figure, we see the collection of clients. Client1 might be an

---

[6]A generic utility issuing SQL is possible because SQL is a language unto itself, agnostic to the programming language of the client, and using declarative commands to specify what data is desired.

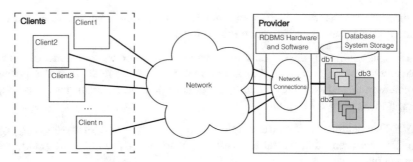

Fig. 10.4 Relational database architecture

application written in C++, while Client2 might be comprised of a web interface and JavaScript. Client3 might be in Python, or R, or even be executing using Notebook-style computing. While a particular client might be the only software executing on a given machine and interacting with the RDBMS, there might be multiple clients on one machine. If the execution software allows concurrency, there might be multiple client interactions in a single piece of software.

Each client establishes a connection over the network, and the target of that connection is within the software of the provider. The network, shown in the abstract as a bubble, could be as large as the Internet, and the provider could be one or more systems provisioned in the cloud, or the network could be the size of an enterprise and the provider could be a dedicated system maintained by an IT department. In particular, if the client is accessing the RDBMS over the Internet, and loses Internet connectivity, then access to the database(s) will be lost. On the smaller end of the spectrum, the network could be local, or even, virtually, within the confines of a single machine.

Each client must have a user identity that is part of the structure of the provider side, along with appropriate authentication and permissions. These are necessary so that different clients are not compromised, in a security sense, by the shared access.

On the provider side, we depict disk (or any other persistent) storage with a cylinder icon. In the figure we show that the RDBMS might be maintaining multiple databases, where an individual database (or schema) is defined, as above, as an interrelated set of tables. Here, we show db1, db2, and db3 as distinct databases. While we shall not concern ourselves with the physical model and structure, we should note that these are not simply files in the operating system. Database systems employ extraordinary efforts to manage raw disk storage and to provide for efficient access. Oracle, PostgreSQL, and MySQL are examples that realize the architecture of Fig. 10.4.

Figure 10.5 depicts a small-scale, single machine architecture for a relational database. Here, in a single client machine, we can support the relational model. In this depiction, there is no network, not even a virtual one. The client application communicates through a database package (like SQLite) which manages the database and its logical files. In this case, the database is maintained in a single

Fig. 10.5 Single machine database architecture

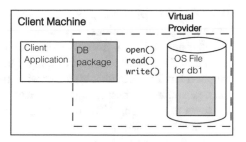

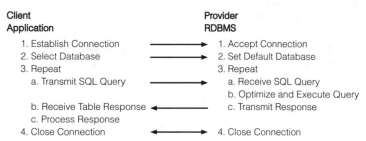

Fig. 10.6 Client/provider database interaction

regular file defined and accessible through the operating system (usually, a `.sql` file). Just as in the file I/O we have learned earlier, the file is `open()`'d and then the `read()` and `write()` provide the database package with the means to maintain all the required information about the database and table structure of a database, named `db1` here.

In this simplified architecture, there are no separate user identities maintained by the database package. The client application has its OS-provided identity, and permissions for the OS file for the database are the same mechanisms as with any other file in the file system.

The differences between these two architectures, while they might be apparent in the connection establishment and in the programming package(s) being used, are *not* apparent in SQL operations.

With the architecture in mind, it is useful to consider, at a high level, the interaction between a client and a provider, as shown in Fig. 10.6.

In the figure, the client is shown on the left and the provider on the right, and we use arrows to depict communication over the network, with the arrow head used to show the "net" direction of the communication.[7]

Before any queries or other SQL may be issued, the client first establishes a connection which, with appropriate authentication, is accepted by the provider. Depending on how the client established the connection and the defaults of the

---

[7]Reliable communication over a network always requires bi-directional communication.

database software, a client may next need to select the particular database to use for subsequent interactions.

From that point on, the client and provider coordinate to exchange requests and responses. The client constructs an SQL query and transmits it (over the network) to the provider. The provider, after receiving the query, must parse the query, optimize and determine how best to obtain the data, and execute the query. The provider then transmits the response. The response is then received by the client and processed. The whole request-response interaction may be repeated as many times as necessary to fulfill the needs of the client.

The reader should realize that this is an abstraction, and Chap. 13 will give many specific details. In particular, if the data is truly large, it may be inefficient for the response to be transmitted in a single network operation, so we will see how responses can be processed in more manageable sized units. With this architecture and the client-provider interaction in mind, we can now, usefully, explore SQL queries that involve single and multiple tables.

10.3.1   Reading Questions

10.20 The book says each client must have a user id, so that different clients are not compromised, in a security sense, by the shared access to the RDBMS. Describe a real-world scenario that might represent a security compromise if such protections were not in place.

10.21 To understand the Client/Provider Database Interaction, use a program like MySQLWorkbench to connect to the databases provided with the book. Describe your experience, as a user, of establishing a connection, selecting a database, and viewing the data inside it.

10.22 If you were designing a database management system from the provider side, and you knew many students would be accessing the databases, what permissions would you grant the students by default and why?

Chapter 11
Relational Model: Single Table Operations

Chapter Goals

Upon completion of this chapter, you should understand the following:

- SQL query operations on a single table to perform:
  - column projection of single and multiple columns,
  - selecting and filtering rows,
  - forming of columns of a result by computation,
  - aggregating results from multiple rows, and
  - partitioning results and combining with aggregation.

- The concept that every SQL query yields a *table* as a result.

Upon completion of this chapter, you should be able to do the following:

- Form syntactically correct SQL that can be issued to a RDBMS for all of the operations discussed for a single table.

The primary objective of this chapter and the next is to understand the *operations* of the relational data model. This chapter focuses on operations involving a single table, while Chap. 12 shifts the focus to operations involving multiple tables. Given this objective, we will defer details on aspects of connecting with a database, and the mechanisms of making requests and receiving replies needed for writing client

© Springer Nature Switzerland AG 2020
T. Bressoud, D. White, *Introduction to Data Systems*,
https://doi.org/10.1007/978-3-030-54371-6_11

applications in a particular programming language. Instead, we assume that the database is available and connection established, and that any tables within the database already exist and are populated.

As described in Chap. 10, requests to a database are specified using SQL, a language that declares what data is desired. Similar in some ways to the regular expression language of Chap. 4, this data request language is agnostic to client details, like the programming language or environment. For this reason, SQL queries are sometimes called *declarative queries*. In common with any language, SQL has the two dimensions of:

- **syntax**: the rules governing the required and optional keywords and elements, along with their sequence, needed to construct a syntactically correct SQL statement, and
- **semantics**: the meaning of the request (i.e., what data is requested) for a syntactically correct statement.

To encourage acceptance and adoption, particularly for users who want and need to request data from a database but might not be skilled at algorithmic problem solving, SQL was designed to be *readable* and, where possible, to mimic English sentences in its statements.[1]

Throughout the chapter, we will repeat the steps of describing the syntax and semantics for progressively more involved SQL statements, then demonstrate with one or more example SQL statements based on one of our database schema, and then show a table with the results of the query.

Because of their common use of a row–column table to represent observations or mapping instances of fields/variables, the operations of SQL have many parallels with the operations we learned in Chaps. 7 and 8. We will present the operations in a similar order, progressing from the projection of columns to the selection of rows and filtering based on condition, to computation with columns, and aggregating and grouping information.

A significant point of difference between the operations of the tabular model (using pandas) and the relational model (using SQL) involves the data type of the result. In pandas, depending on how we constructed the operation, the result might be a scalar, a one-dimensional column vector, or another two-dimensional data frame. By contrast, in SQL, the result of any SQL query is *always* another valid two-dimensional table. This property conveys great power, as we can compose results, and builds a query that, in fact, uses another complete query as a constituent part.

---

[1] In this regard, it is decidedly *not* similar to the language of regular expressions.

11.1   Example Data Sets

For use in illustrating operations in SQL, we will employ two different data sets (along with some variations thereof). These are the same data sets we used in Chaps. 7 and 8.

The indicators0 Table
This is a small subset of the full indicators data set. It contains the following variables, called *fields* or *attributes* in the relational model:

- code: the unique international three character code for the country.
- pop: the population of the country, in units of millions of persons.
- gdp: the gross domestic product of the country, measured in units of billions of US dollars,
- life: the life expectancy of the country, in years.
- cell: the number of cell phone subscribers in the country, in millions of subscribers.

For this data set, the functional dependency is:

$$code \rightarrow pop, gdp, life, cell$$

The table, with just code as the independent variable, is not a time series and simply captures a snapshot from year 2017 and only contains the rows for the countries with code 'CHN' (China), 'FRA' (France), 'GBR' (United Kingdom), 'IND' (India), and 'USA' (United States). Having a small table is helpful for illustration. The schema for this table is given in Fig. 11.1.

The indicators and countries Tables
These are the two tables that comprise the full indicators data. The fields of the indicators table include:

- year: the year associated with the measurement of the indicators for the given country.
- code: the unique international three character code for the country.
- pop: the population of the country, in units of millions of persons.

Fig. 11.1 Table schema for indicators0

Fig. 11.2 Table schema for
countries and indicators

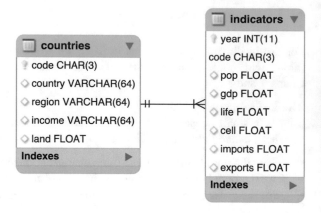

- gdp: the gross domestic product of the country, measured in units of billions of US dollars.
- life: the average life expectancy of persons in the country, in years.
- cell: the number of cell phone subscribers in the country, in millions of subscribers.
- imports: the gross dollar value amount of the products imported by the country, measured in units of millions of US dollars.
- exports: the gross dollar value amount of the products exported by the country, measured in units of millions of US dollars.

The fields of the countries table include:

- code: the unique international three character code for the country.
- region: the world region to which the given country belongs.
- income: the average income category for the persons in the given country.
- land: the land area of the given country, measured in square kilometers.

In Fig. 11.2, we depict the schema for countries and indicators, also showing their inter-table relationship.

The primary key of indicators is a composite key consisting of year and code, while the primary key of countries is code. For any given country record, there are many indicator records with the same code, so we have a one-to-many relationship from countries to indicators.

For the indicators table, the functional dependency is:

$$year, code \rightarrow pop, gdp, life, cell, imports, exports$$

For the countries table, the functional dependency is:

$$code \rightarrow region, income, land$$

Fig. 11.3 Table schema for
topnames

The topnames Table

As described previously, the topnames table gives, over the years 1880 through 2018, the top female and top male names in applications to the US Social Security administration, along with the count of those applications. The fields of topnames, then, consist of:

- year: the year associated with the US Social Security top applications.
- sex: the sex ('Female' or 'Male') of the social security application for this record.
- name: the name with the most social security applications in the given year for the given sex.
- count: the count of applications for the associated name in the given year for the given sex.

And the functional dependency of the table is given by:

$$year, sex \rightarrow name, count$$

The table schema for topnames is shown in Fig. 11.3, with integral data types for year and count and character data types for sex and year:

11.1.1 Reading Questions

11.1 Give an example line of Python code, and use it to explain the difference between *syntax* and *semantics* in the situation of your example. Then do the same with an SQL query.

11.2 Part of the power of SQL is that queries are composable, since the output of a query is always a valid two-dimensional table. Give an example of how this could fail for pandas.

11.3 Can you think of another table that the topnames table might have a relation to, analogous to the relationship between countries and indicators?

11.2   Projecting Column Fields

The goal in projecting columns from a table in the relational model, just like in the tabular model, is to specify either all or a subset of the possible columns contained in a relational table.

In SQL, queries of tables, from the most basic queries like requesting all fields (columns) of all records (rows) to complex requests involving multiple tables and aggregation and grouping, use the SQL SELECT statement. Since SQL is a language, it has rules and ordering that govern whether a statement in the language is syntactically valid. Throughout this chapter and Chap. 12, we will explore the syntactic complexity of the SQL SELECT statement, starting with simple forms and building to more complex ones. We begin with a limited form of SELECT.

The syntax specifications given in these chapters will follow a consistent notation. Each specification will consist of keywords (presented in uppercase and using a monotype font), punctuation/operators (also displayed in a monofont), and often placeholders, which might have their own syntax specification, or might be replaced by a table name or a field name. If a placeholder has its own syntax specification, we will use the $\models$ operator with the placeholder on the left and the syntax specification on the right as a way of providing these subordinate definitions. Placeholders are displayed in italics and use hyphenation to achieve multi-word placeholders to avoid the ambiguity of whether an actual space is required in the syntax. In syntax specification, we often need to denote when one or more elements are optional, or if a subsequence of the syntax can occur zero or more times. Following common convention, we use math-style brackets to denote optional syntax and use a superscript asterisk to denote element(s) that may be repeated zero or more times. This should be familiar from Chap. 4.

When composing SQL statements, the whitespace, outside of strings, *does not matter*, and we will often write more complex SQL statements using multiple lines and indentation to improve the readability of the statement.

The syntax for an SQL query to project column fields from a table is given by:

SELECT *result-field* FROM *table-spec*

where the *result-field* can be comprised of one or more *expression*s, separated by commas, as given in the syntax below.

result-field $\models$ *expression* [, *expression*]*

An *expression* can be as simple as the name of a field in a table. It can also be more complex, as we will see later, but an *expression* of a field name is the most common. The *table-spec* following the FROM, in its simplest form, names a single table from the current database. The *table-spec* also has more complex forms to be explored later.

Table 11.1 5 records

Pop
1386.40
66.87
66.06
1338.66
325.15

11.2.1   Single Column Field Projection

To project a single field from a table, we simply name the field after the SELECT keyword and name the table after the FROM keyword:

SELECT pop **FROM** indicators0

The result of this query, as with *all* SQL queries, is a *table*. It has, in this case, a single field, whose name comes from the projected column (Table 11.1). A valid result may have 0 records in the resulting table. In SQL, there is no notion of a column vector, or a scalar, as a result of a query.

Often, the result of a query can contain a very large number of records. As we explore a database and as we develop and test queries, we often want to restrict the result to a limited number of records, as we did with the head() function in pandas. In SQL, the LIMIT keyword, followed by an integer number of rows, is used to accomplish this. The LIMIT, if present, suffixes the rest of the SELECT query. For instance, in the topnames table, there are a total of 278 records, so to restrict a single column projection to the first five:

SELECT name **FROM** topnames **LIMIT** 5

Extending our syntax specification to denote the optional LIMIT clause, we have

SELECT *field-spec* FROM *table-spec*

[LIMIT *n*]

With no LIMIT clause, the result of SELECT name FROM topnames would yield all 278 rows. Also note, in contrast to the tabular model and pandas, even though there *is* a primary key, there is no Index displayed as part of a query. The result of an SQL query *only* gives fields (columns) for the *expression* sequence that is specified (Table 11.2).

11.2.2   Multiple Column Field Projection

When we want multiple columns projected in a single query, we compose the set of field names in a comma-separated list that follows the SELECT. So to get the fields year, sex, and name in the topnames table, limiting the result to the first six, we have:

Table 11.2 5 records

Name
Mary
John
Mary
John
Mary

Table 11.3 6 records

Year	Sex	Name
1880	Female	Mary
1880	Male	John
1881	Female	Mary
1881	Male	John
1882	Female	Mary
1882	Male	John

Table 11.4 5 records

Code	Population	GDP
CHN	1386.40	12143.50
FRA	66.87	2586.29
GBR	66.06	2637.87
IND	1338.66	2652.55
USA	325.15	19485.40

SELECT year, sex, name **FROM** topnames **LIMIT** 6

When we project multiple columns, the resultant table uses the expression ordering specified in the query, which can be different than the order of the table schema (Table 11.3).

In a slightly more complex form of *expression*, SQL allows us to provide an *alias* for any expression field of the SELECT by using a suffix of AS *new-name* after the rest of any given expression. This gives us the ability to rename columns of the result and will be convenient as our queries become more complex. So, for instance, if we wanted the pop field to be named Population and the gdp field to be named GDP, and to leave code unchanged, we could issue the query:

SELECT code, pop **AS** Population, gdp **AS** GDP
FROM indicators0

When we want *all* columns of a table to be projected, we use the special syntax of ⋆ as the field specifier of the SELECT (Table 11.4). Our updated syntax uses the math | to allow selection of alternatives:

$$\textit{field-spec} \models \textit{expression} \, [, \textit{expression} \,]^* \mid *$$

To project all columns, even if we do not know what they are, from indicators0, we could issue the following SQL:

Table 11.5 5 records

Code	Pop	Gdp	Life	Cell
CHN	1386.40	12143.50	76.4	1469.88
FRA	66.87	2586.29	82.5	69.02
GBR	66.06	2637.87	81.2	79.10
IND	1338.66	2652.55	68.8	1168.90
USA	325.15	19485.40	78.5	391.60

SELECT * FROM indicators0

Note how this wildcard is different from the same notation used in regular expressions (Table 11.5). There, the quantifier always followed a syntactic element, and it was that element that should appear zero or more times. Here, the same notation is used, also as a wildcard, but with different syntactic rules and semantics.

11.2.3   Simple Subquery

Since every time we specify a query, the result is a table, a query can, itself, be thought of as a table. One of the powers of SQL is its ability to compose a complex query that is comprised of an *outer* query, and an *inner* or *subordinate* query. Such a subordinate query is generally known as a *subquery*, and a subquery can, in general, be used in places where a *table* would be used.

More formally, our SELECT for projection was given by:

SELECT *field-spec* FROM *table-spec*

[*limit-clause*]

So if we are considering a subquery as a "table replacement," we could use a subquery as a valid *table-spec*, expanding our notion of *table-spec* to:

table-spec $\models$ *table* | (*subquery*) AS *new-name*

where a *subquery* is, itself, an SQL statement, and the AS clause gives a table name as an alias for the result of the subquery (previously, we saw AS to give an alias for a result field name). We introduce the idea of subquery at this point because our SQL has not yet become so complex as to obscure the relatively simple nature of a subquery just being a substitute for a table. Its utility will come later, as we try to construct solutions to more complex problems. In the current context, we can only use it to perform a column projection *on* a column projection, like in the following:

Table 11.6 5 records

Pop
1386.40
66.87
66.06
1338.66
325.15

```
SELECT pop FROM
      (SELECT code, pop, gdp FROM indicators0)
      AS select1
```

Here, the RDBMS first executed the inner query, obtaining a "virtual" table with the three-column projection of code, pop, and gdp, and named it select1. This was then used as the table for the outer SELECT, and just the pop column was projected from select1 (Table 11.6). When using subqueries, we use parentheses to delimit the subquery, as can also be seen from the syntax specification given above for *table-spec*.

11.2.4 Ordering Results

In the relational model, tables, as they reside in the database, do not have an a priori order to their records, or at least, an order we should rely on. Each record defines its own mapping and is part of the set of mappings that make up the table. In this, tables should be thought of like sets, which do not have an order. Once we begin building more complex queries, the order of a result is also influenced by how the RDBMS determines the execution of the query. But processing at a client or presentation of the results of a query may require the client to order the result.

SQL provides the capability to order (i.e., sort) the result of a query with an ORDER BY clause that follows all of the rest of the SQL query, and comes just before any LIMIT clause, if present. The ORDER BY is operating on the composed result, which we know will be a table, and that table will have column fields, or aliases thereof.

The syntax, extended with an optional *order-clause*, is given by

SELECT *field-spec* FROM *table-spec*

[*order-clause*]

[*limit-clause*]

For any field of the result, or expression involving the result fields, we can order by that expression and can order in either ascending or descending order. We can also order by more than one expression. When ordering by more than one expression, a sort is first performed on the first specified expression. If the result includes "ties" from the first sort, the second specified expression is used to sort just the subset of records involved in the tie. Additional order terms refine the overall sort in similar fashion.

$$order\text{-}clause \models \text{ORDER BY } order\text{-}term \text{ [, } order\text{-}term \text{]}^*$$

where

$$order\text{-}term \models expression \text{ [ASC | DESC]}$$

In other words, an order clause starts with the keyword sequence ORDER BY followed by a one or more order terms separated by commas. An order term gives an expression, typically the name of a field, and optionally specifies ASC for ascending order or DESC for descending order. If neither ASC nor DESC are specified, the default is an ascending sort.

Consider the topnames table of 278 records with records for two values of sex over 139 values of year. If we specify no ordering:

```
SELECT * FROM topnames
LIMIT 3
```

We see the default result begins with the two year 1880 records, with the Female record preceding the Male record and then proceeds to the year 1881 (Table 11.7).

If we wanted the most recent year to be first in the result, we could sort in descending order by the year field of the result:

```
SELECT * FROM topnames
ORDER BY year DESC
LIMIT 4
```

An effect of this descending sort is that, with each pair of records for the year, the Male record occurs before the Female record. If we wanted the opposite order, we could specify a second order term, requesting an ascending sort by sex *after* the descending sort by year (Table 11.8):

Table 11.7 3 records

Year	Sex	Name	Count
1880	Female	Mary	7065
1880	Male	John	9655
1881	Female	Mary	6919

Table 11.8 4 records

Year	Sex	Name	Count
2018	Male	Liam	19837
2018	Female	Emma	18688
2017	Male	Liam	18798
2017	Female	Emma	19800

Table 11.9 4 records

Year	Sex	Name	Count
2018	Female	Emma	18688
2018	Male	Liam	19837
2017	Female	Emma	19800
2017	Male	Liam	18798

Table 11.10 5 records

Year	Sex	Name	Count
1991	Female	Ashley	43478
1992	Female	Ashley	38453
1960	Male	David	85929
1996	Female	Emily	25151
1997	Female	Emily	25732

Table 11.11 8 records

Year	Sex	Name	Count
1947	Female	Linda	99689
1948	Female	Linda	96211
1949	Female	Linda	91016
1950	Female	Linda	80431
1921	Female	Mary	73985
1951	Female	Linda	73978
1924	Female	Mary	73534
1922	Female	Mary	72173

```
SELECT * FROM topnames
ORDER BY year DESC, sex ASC
LIMIT 4
```

We need not sort by one or more of the fields making up a primary key. For instance, we could sort by name or count. Here, we show the former. Note that we omit the ASC part of the order term, since we desire an ascending sort (Table 11.9).

```
SELECT * FROM topnames
ORDER BY name
LIMIT 5
```

In this final ORDER BY example, suppose we want to sort first by sex, to group Female and Male records together, and then to sort in descending order by count to see, within the sex category, the high-to-low progression of number of applications for the top-getting name (Tables 11.10 and 11.11). Our query to accomplish this:

```
SELECT * FROM topnames
ORDER BY sex, count DESC
LIMIT 8
```

11.2.5 Reading Questions

11.4 The reading mentions using multiple lines and indentation to write select statements in a more readable way, since whitespace does not matter. Give an example. You can peek ahead in the book if you want to.

11.5 Give an example of a *result-field* that might appear in a SELECT statement, taking the form *expression, expression, expression*, to illustrate the notation from the reading regarding optional elements in a result-field.

11.6 When using the SQL keyword LIMIT, is this analogous to using head, tail, or nlargest in pandas? Explain how you know.

11.7 When you use SQL to project onto multiple columns, does the order in your SELECT statement have to match the order of the columns in the table you are selecting from? Explain how you know.

11.8 The text reads "Our updated syntax uses the math | to allow selection of alternatives"—how do you pronounce the symbol | ? What does it mean?

11.9 Consider the example: SELECT pop FROM (SELECT code, pop, gdp FROM indicators0) AS select1

How could you modify this example to rename pop to Population in the final output?

11.10 In a SELECT query with both an ORDER BY and a LIMIT, what would happen if you put the LIMIT before the ORDER BY? Try it!

11.11 Where are you allowed to add in the AS keyword to a select statement of the form:

```
SELECT field-spec FROM table-spec  [ order-clause ] [
limit-clause ]
```

For each such place, please explain what precisely will be given an alias.

11.12 Consider the pandas code

```
topnames.sort_values(by = ['year','count'],
    ascending = False, inplace=False)
```

How can you replicate this in SQL using ORDER BY?

11.2.6 Exercises

11.13 Using the table countries, give the SQL query you would use to list all countries.

11.14 Select the year, code, population, and number of cell phones from `indicators`, then think about something interesting you could compute from this data.

11.15 Using the SQL table `countries`, produce a list of rows ordered list by landmass, from smallest to largest. Then ask yourself what might make the result more useful.

11.16 Use an SQL query to answer: which country had the most cell phones in any year?

11.17 Use an SQL query to answer: in which year was the max number of cell phones in any country achieved?

11.18 Is it easy to tell, based on the SQL you have learned so far, what country/year had the *smallest* non-zero number of cell phones? Justify your answer.

11.19 Use an SQL query to answer: which country had the most exports in any year?

11.20 Use a subquery to select the top ten entries in `indicators` by population, then select the bottom three of those by GDP.

11.21 Select `year`, `code`, `pop`, `gdp`, and `life` from `indicators` ordered by life expectancy from low to high.

11.22 Use an SQL query to find out: what is the most recent data in the `indicators` table?

11.23 Starting from `indicators`, select `year`, `gdp`, `imports`, `exports`, and then think of something interesting you could do with this data.

11.24 Use SQL to find the rows with the top 20 GDP values in `indicators`.

11.25 Use an SQL query to answer: what is the smallest count in `topnames`? Give the entire row.

11.26 Use an SQL query to answer: what is the newest data in `topnames`? Your results can include all the data as long as it is properly ordered.

11.27 How many column projections are possible from `topnames`? Explain your answer.

11.28 Consider the `enron` database. It has only one table, `emails`. Select every field from that table, to get a sense of what is in the data.

11.29 Select all users from `emails`, and think about what resulting table might be more useful.

11.30 Select all users and folders from `emails`, and think about what resulting table might be more useful.

11.31 Consider the `names` table inside the `imdb` database. Select all fields, ordered alphabetically by name. Then ask yourself what might be a more useful ordering.

11.32 Consider the `names` table inside the `imdb` database. Select birth and death year, and then ask yourself what might be a reasonable question to ask based on this.

11.33 Using SQL queries, investigate the `principles` table in the `imdb` database, and then explain what you think this table is about.

11.34 Using SQL queries, investigate the `ratings` table in the `imdb` database, and then explain what you think this table is about. You should include your observations about the possible range of `averageRating` and `numVotes`. Hint: ORDER BY statements can help you figure this kind of information out.

11.35 Using SQL queries, investigate the `titles` table in the `imdb` database, and then explain what you think this table is about. Specifically, please explain the range of `startYear` and `endYear`, as well as what you think `isAdult` is about.

11.36 Using the `nycflights13` database, find out how many airlines are present in the data.

11.3   Selecting and Filtering Rows

An important class of operations in any table-based model is the ability for *selection*, obtaining a subset of the rows (records). SQL provides a number of such operations, to be explored in this section.

Because, in SQL, such operations always still admit the SQL syntax for projecting column fields, there is no subsection for selection *without* projection. Most of the examples to follow will include examples of projection along with the selection being illustrated.

11.3.1   Uniqueness Filtering

Depending on a particular projection, the result of an SQL query could well include records that have the same value for all fields of the result (i.e., duplicate records). SQL extends its SELECT syntax to allow record filtering by omitting all duplicate records in the result. This filtering is achieved via the optional DISTINCT keyword immediately following the SELECT, giving the following syntax description:

SELECT [DISTINCT] *field-spec* FROM *table-spec*

[*order-clause*]

Table 11.12 Displaying
records 1–8

Income
High income
Low income
Lower middle income
Upper middle income
High income
High income
Upper middle income
Upper middle income

Table 11.13 4 records

Income
High income
Low income
Lower middle income
Upper middle income

Table 11.14 7 records

Region
Latin America & Caribbean
South Asia
Sub-Saharan Africa
Europe & Central Asia
Middle East & North Africa
East Asia & Pacific
North America

[*limit-clause*]

As an example, consider the `income` field of the `countries` table. There are 217 records in the table, and as a categorical variable, there are only four values of the `income` field. If we were to project the `income` field, we would get *many* duplicates in the 217 records returned, which we can see even in this display limited to eight records (Table 11.12):

`SELECT income FROM countries`

Using the `DISTINCT` keyword per the syntax given above, we can obtain the unique values in this single column projection (Table 11.13):

`SELECT DISTINCT income FROM countries`

In a similar fashion, we can use the `DISTINCT` qualifier in a query to discover the unique set of world regions used as values for the `region` field in the `countries` table (Table 11.14):

`SELECT DISTINCT region FROM countries`

Table 11.15 Displaying
records 1–7

Region	Income
Latin America & Caribbean	High income
South Asia	Low income
Sub-Saharan Africa	Lower middle income
Europe & Central Asia	Upper middle income
Europe & Central Asia	High income
Middle East & North Africa	High income
Latin America & Caribbean	Upper middle income

As with any SQL query, we can specify multiple expressions that follow the DISTINCT keyword. Records are omitted in the result if they are duplicates *on all* fields of the result. For instance, SELECT DISTINCT region, income FROM countries gives 24 results, as there are multiple pairings involved (Table 11.15).

SELECT DISTINCT region, income **FROM** countries

11.3.2   Row Selection by Filtering

An important category of row selection entails specifying a Boolean expression that, when true, indicates a row should be included, and, when false, indicates that a row should not be included. This type of row selection by filtering is performed with a WHERE clause in SQL.

The extended syntax of the SQL SELECT now becomes:

> SELECT [DISTINCT] *field-spec* FROM *table-spec*
>
> [WHERE [NOT] *filter-condition*]
>
> [*order-clause*]
>
> [*limit-clause*]

The WHERE clause follows the *table-spec* of the FROM source table specification and precedes the other clauses we have learned about so far.

The *filter-condition* behaves similarly to when we use a Boolean vector to filter rows in pandas, but it is *not* a Boolean vector. Rather, it specified as a Boolean predicate and can use the names of fields along with relational operators and scalar values, much like a condition of an if in a traditional programming language.

Suppose we want to filter for those records in indicators1 where the population is greater than 1000. We want to use the > relational operator to compare the field pop with the literal constant 1000, giving the following (Table 11.16):

Table 11.16 2 records

Code	Year	Pop	Gdp	Life
CHN	2007	1317.88	3550.34	74.6
CHN	2017	1386.40	12143.50	76.4

Table 11.17 4 records

Code	Pop	Gdp	Life	Cell
CHN	1386.40	12143.50	76.4	1469.88
FRA	66.87	2586.29	82.5	69.02
GBR	66.06	2637.87	81.2	79.10
IND	1338.66	2652.55	68.8	1168.90

```
SELECT * FROM indicators1 WHERE pop > 1000
```

Just like in our programming languages, we can build more complex Boolean expressions by using logical operators. In SQL, these operators are realized with the keywords of AND, OR, and NOT. For example, if we extend the prior query to also include those records where the life field is greater than 80 (Table 11.17):

```
SELECT * FROM indicators0 WHERE pop > 1000 OR life > 80
```

The set of SQL relational operators and the logical operators, specified with keywords, are summarized in Table 11.18.

These relational and logical operators, along with the names of fields, field aliases, literal constants, and even expressions involving arithmetic computation, are used to compose the *filter-condition* of a WHERE clause in SQL. As in conventional programming languages, we use parentheses to enforce a desired precedence order. When we are specifying constants, we must adhere to the SQL rules for constants, instead of the syntax and rules of our programming language. In particular:

- String constants are delimited by *single* quotes. To obtain a constant string that, itself, contains a single quote character, we use two single quotes in a row within the string to "escape" a single quote.
- Numeric constants are written as we would expect with integers and decimal numbers and can include scientific notation (like 3.7E-3) to give us greater range.
- A missing value in SQL is always specified with the keyword NULL, without quotes, and may be used in checks with relational operators (like = and <>).
- Boolean constant values are TRUE and FALSE in all caps, and with no string delimiter.

SQL syntax can be different than our programming language syntax in some other ways.

Notice that equality is with single =, and inequality is <>, and strings are delimited with single quotes. Logical operations are based on the keywords AND, OR, and NOT. It is important to remember that SQL is its own language and thus has embedded in it its own syntactic decisions and may differ from Python, R, or some other programming languages.

Table 11.18 Operators and keywords for filter conditions

Operator or keyword	Operation	Notes
=	Equal To	Binary operation to compare if two operands have the same value. For most systems, comparing strings is sensitive to case. Comparing operands of different data types can lead to unexpected results
<>	Not Equal To	Binary operation to determine if two operands are *not* equal in value
<	Less Than	Binary operation to determine if a first operand is less in value than a second operand. Comparing two operands of a string data type performs a lexicographic comparison that is based on a "collating sequence" that determines how characters are represented
>	Greater Than	Binary operation to determine if a first operand is greater in value than a second operand. Same string comments as above
<=	Less Than or Equal To	Binary operation to determine if a first operand is less than or equal in value to a second operand
>=	Greater Than or Equal To	Binary operation to determine if a first operand is greater than or equal in value to a second operand
NOT	Logical Negation	Unary operation performing the logical negation of the operand. If the operand is numeric, the NOT of 0 yields 1 and the NOT of any non-zero value yields 0. The NOT of a NULL value yields NULL
AND	Logical Conjunction	Binary operation that yields TRUE if both operands are TRUE and FALSE otherwise. An AND with a NULL as either operand yields FALSE
OR	Logical Disjunction	Binary operation that yields TRUE if either or both operands are TRUE and FALSE otherwise. An OR of NULL is only TRUE if the other operand is TRUE

Beyond the familiar composition of Boolean expression from relational and logical operators, SQL includes some additional constructs for *filter-condition* that involve additional SQL-specific keywords. We briefly cover a few of these Boolean conditions in the following.

Value in a Range: BETWEEN

When an SQL statement must specify that a field or expression lies within a bounded inclusive range, we use the BETWEEN keyword, as given in the following syntax description:

Table 11.19 1 records

Code	Year	Pop	Gdp	Life
CHN	2007	1317.88	3550.34	74.6

Table 11.20 4 records

Code	Pop	Gdp	Life	Cell
CHN	1386.40	12143.50	76.4	1469.88
FRA	66.87	2586.29	82.5	69.02
IND	1338.66	2652.55	68.8	1168.90
USA	325.15	19485.40	78.5	391.60

$$range \models expression\ [\ \text{NOT}\ ]\ \text{BETWEEN}\ low\ \text{AND}\ high$$

So *range*, as defined here, is one possible alternative for a Boolean *filter-condition* or can be used in a Boolean expression composed with logical operators in a composite Boolean expression. In a simple example like the one given below, the *expression* can simply name a field but could easily be a more complicated expression as well. Likewise, in simplest form, *low* and *high* could be constants of appropriate type for *expression* but can, themselves be more complex expressions as well (Table 11.19).

```
SELECT * FROM indicators1
WHERE life BETWEEN 65 AND 75
```

Set Inclusion: IN
SQL also provides for conditions that test whether a field or more complex expression is equal to one of a number of discrete values in a set. The negation of inclusion in a set can also be specified. The syntax specification *in-clause* yields, on evaluation of any record, a Boolean and can be used in a WHERE or in a compound statement in which a Boolean expression is appropriate.

$$in\text{-}clause \models expression\ [\ \text{NOT}\ ]\ \text{IN}\ (\ set\text{-}members\ )$$

For example, suppose we want `indicator` records for the set of country codes that are one of `'CHN'`, `'IND'`, `'FRA'`, `'USA'` and also for the year 2017 (Table 11.20). The SQL to accomplish this is given by:

```
SELECT code, pop, gdp, life, cell FROM indicators
WHERE year=2017 AND code IN ('CHN', 'IND', 'FRA', 'USA')
```

The IN clause also enables us to see the utility of a subquery. If we can define a query whose result is a single column and the rows define a set of elements of interest, the subquery can be used to define the *set* to be used with IN (Table 11.21). For instance, the query below determines the set of country codes for countries with GDP greater than 3500:

```
SELECT DISTINCT code FROM indicators WHERE gdp > 3500
```

Table 11.21 4 records

Code
USA
JPN
CHN
DEU

Table 11.22 4 records

Code	Pop	Gdp	Life	Cell
CHN	1386.40	12143.50	76.4	1469.88
DEU	82.66	3693.20	81.0	109.70
JPN	126.79	4859.95	84.1	172.79
USA	325.15	19485.40	78.5	391.60

The above query can be thought of as a set, and so we can define a nested query, just like our original IN example, but where the set is the SQL query just listed (Table 11.22):

```
SELECT code, pop, gdp, life, cell FROM indicators
WHERE year=2017 AND code IN
  (SELECT DISTINCT code FROM indicators WHERE gdp > 3500)
```

String Pattern Matching: LIKE

We can use the relational operator, =, to determine if a field exactly matches a literal string, but sometimes we need more flexible string matching capabilities. Like with regular expressions (Chap. 4), we want to be able to match string *patterns*. The SQL standard defines another Boolean-generating keyword, LIKE, that allows comparison of a field or expression to a pattern. Again, this can be used in a WHERE or in compound expressions requiring a Boolean. The syntax specification for a *like-clause* is given as follows:

$$like\text{-}clause \models expression\ [\ NOT\ ]\ LIKE\ pattern$$

The *pattern* is given as a constant string that consists of literal characters and metacharacters to allow wildcard matching. The two metacharacters supported by all SQL systems for pattern matching are:

- %: Match zero or more characters of a target at this point in the pattern, and
- _: Match exactly one character of a target at this point in the pattern.

In the example below, the first LIKE matches any country field that ends with 'Republic' as a suffix, and the second LIKE matches any three character code that ends with the literal characters 'ZA' (Table 11.23):

```
SELECT code, country, region FROM countries
WHERE country LIKE '%Republic' OR code LIKE '_ZA'
```

Table 11.23   8 records

Code	Country	Region
CAF	Central African Republic	Sub-Saharan Africa
CZE	Czech Republic	Europe & Central Asia
DOM	Dominican Republic	Latin America & Caribbean
DZA	Algeria	Middle East & North Africa
KGZ	Kyrgyz Republic	Europe & Central Asia
SVK	Slovak Republic	Europe & Central Asia
SYR	Syrian Arab Republic	Middle East & North Africa
TZA	Tanzania	Sub-Saharan Africa

Table 11.24   7 records

Code	Country	Land
CUW	Curacao	NA
MAF	St. Martin (French part)	NA
MCO	Monaco	NA
SDN	Sudan	NA
SSD	South Sudan	NA
SXM	Sint Maarten (Dutch part)	NA
XKX	Kosovo	NA

11.3.3   Missing Values

In SQL and in the relational tables stored at the RDBMS, the special value NULL
is used to represent missing data. NULL should not be mistaken for the value
0, or a blank, or some other data types but represents its own unique type and
value. If NULL is used consistently to represent missing or unknown values in
any dependent variable column, we want our facilities for filtering and for Boolean
logical operations to be able to inter-operate with NULL-valued fields (Table 11.24).

Obtaining a Boolean TRUE/FALSE to use in a WHERE clause or as part of
compound Boolean, perhaps with other logical operators, we use a *null-clause*
whose syntax is given by:

$$\textit{null-clause} \models \textit{expression} \text{ IS } [\text{ NOT }] \text{ NULL}$$

```
SELECT code, country, land FROM countries
WHERE land IS NULL
```

When we use relational operations and one of the fields is NULL, the result will
always be NULL (and so it is neither TRUE, nor is it FALSE). So one could imagine
a Boolean expression like land > 500, and then, for the rows shown above where
land is NULL, the Boolean expression yields NULL. This brings up the possibility
that a compound Boolean expression using AND, OR, and NOT could operate on

Table 11.25 1 records

Name
Noah

Table 11.26 4 records

Year	Count
2013	18257
2014	19305
2015	19635
2016	19117

NULL for one of its operands. This is why the definition of these logical operations takes NULL into account, as articulated in Table 11.18.

11.3.4   Additional Examples

Here we present just a few examples of other common patterns. First, we can easily construct a query that obtains a singleton result (meaning, only one), at the intersection of exactly one column and selecting exactly one row (Table 11.25). For instance, to get the name column in topnames for the unique row where year is 2016 and sex is 'Male':

```
SELECT name FROM topnames WHERE year = 2016 AND sex = 'Male'
```

This is still a table, but with exactly one row and exactly one column. We can also use such a query as a subquery as part of a WHERE clause as a comparison point (Table 11.26). For instance, if our question was to find the years and counts when the top male name was the same as that in 2016, we could execute the following:

```
SELECT year, count FROM topnames WHERE
   name = (SELECT name FROM topnames
           WHERE year = 2016 AND sex = 'Male')
```

If we want to filter rows by some number of the largest values in some field (like the pandas data frame nlargest() method), we combine an ORDER BY clause with a LIMIT clause (Table 11.27). The ORDER BY can use a descending order so that the largest is first in the result, and the LIMIT class determines the number of rows in the result:

```
SELECT year, sex, name, count FROM topnames
   ORDER BY count DESC LIMIT 5
```

In similar fashion, we can obtain the *n* smallest values for a field (Table 11.28):

```
SELECT code, life FROM indicators0 ORDER BY life ASC
   LIMIT 3
```

Table 11.27 5 records

Year	Sex	Name	Count
1947	Female	Linda	99689
1948	Female	Linda	96211
1947	Male	James	94757
1957	Male	Michael	92704
1949	Female	Linda	91016

Table 11.28 3 records

Code	Life
IND	68.8
CHN	76.4
USA	78.5

11.3.5   Reading Questions

11.37 How could you find all unique names that have ever appeared in `topnames`?

11.38 Suppose `table` is some generic table (i.e., any example you want to think about). What would it mean if `SELECT DISTINCT * FROM table` resulted in fewer rows than `SELECT * FROM table`?

11.39 The reading says `SELECT DISTINCT` is used to select distinct *rows*. Is there an analogous command to select distinct *columns*? Explain.

11.40 What happens if you put a `WHERE` clause after an `ORDER BY`? Try it and justify your answer.

11.41 In a query of the form

```
SELECT field FROM table WHERE condition
```

where can you put the keyword `AS`, and what does it do in each place?

11.42 When using Boolean operators in SQL, what happens if you forget to capitalize them? Try it, e.g., on

```
SELECT * FROM indicators0 WHERE pop > 1000 or life > 80
```

11.43 Write out example select statements to illustrate each of the Operators and Keywords (e.g., =, < >, etc.) from the reading.

11.44 Rewrite the SQL query

```
SELECT * FROM indicators1
    WHERE life BETWEEN 65 AND 75
```

to avoid `BETWEEN` (but you can use operators and keywords as needed). Hint: be careful with parentheses.

11.45 Consider the SQL query

```
SELECT code, country, region FROM countries
    WHERE country LIKE '%Republic' OR code LIKE '_ZA'
```

How could you test in Python if the country is "like" %Republic? How could you test if the code is "like" _ZA? Be precise.

11.46 Convert the following pandas selection into an SQL query (but one that will return a data frame instead of a scalar):

```
topnames.loc[(1961,'Female'),'count']
```

11.47 What SQL query would you use to select all columns for the top ten countries in the world by life expectancy?

11.48 If you wanted to select just the rows that are duplicates, could you do SELECT NOT DISTINCT * FROM table? Try it and explain.

11.49 Consider the following select query from the reading:

```
SELECT code, pop, gdp, life, cell FROM indicators
    WHERE year=2017 AND code IN
        (SELECT DISTINCT code FROM indicators WHERE
                                        gdp > 3500)
```

How is this different from using a WHERE clause looking for gdp > 3500?

11.3.6 Exercises

11.50 In reference to the table indicators, write a query to find all distinct years appear in the table.

11.51 In reference to the table indicators, write a query to find all rows with no missing data for gdp.

11.52 In reference to the table indicators, write a query to find all rows with no missing data.

11.53 In reference to the table indicators, write a query to find all rows where exports are higher than imports. Select all fields for such rows.

11.54 In reference to the table indicators, write a query to find the fraction of rows where exports are higher than imports (as a fraction of the number of rows where both of these fields are non-null).

11.55 In reference to the table indicators, write a query to find all distinct country codes that appear.

11.56 Is the number of distinct countries in `indicators` the same as the number of distinct countries in `countries`? Explain.

11.57 In reference to the table `indicators`, write a query to find the minimum non-zero number for `cell` that appears. Since we have not learned about `MIN`, you will have to use `ORDER BY` to arrange it so that the first row in your result has that minimum value.

11.58 Using the `enron` database, construct a query to select all distinct users from the `emails` table.

11.59 Select all messages from the `emails` table in the `enron` database for user `allen-p`.

11.60 In the `enron` database, construct a query to obtain all the `message` items from the `inbox` of user `allen-p`.

11.4   Column-Vector Operations

In the relational model, just like in the tabular model and `pandas`, we want to be able to compute vectors of values by operating on fields and using functions, operators, and scalars to obtain these new vectors. In `pandas`, we defined procedural methods by which we could compute vectors of values, and could then assign and work with the results in stepwise fashion. In a declarative language like SQL, we have no such steps. Instead, column-vector computations are specified using one or more comma-separated *expression*s following the `SELECT` keyword. These expressions can operate on tables and fields that are entailed by the *table-spec* following the `FROM` keyword.

In SQL, we compose our expressions in a *single row* perspective. So unlike `pandas`, we are not considering something like `pop` as a vector. Instead, it is a *field*, and on each row of the result, the field will have a value. A closer analogy would be how we might work with a set of variables inside a loop, where, in each iteration of the loop, the variables take on the values of the fields of the current row iteration.

Consider the following row-centric expressions:

- `pop * 1.15`—compute a 15% increase from the `pop` field.
- `cell / pop`—compute a per-capita measure of cell phone, subscription for the current field values of `cell` and `pop`.
- `country || ' (' || code || ')'`—using the SQL concatenation operator (`||`), create a string concatenation incorporating constant strings with the current field values of `country` and `code`.
- `life > 75`—compute Boolean TRUE/FALSE value giving an indicator of life span over 75; the interesting aspect of a Boolean calculation is that TRUE values yield a 1 and FALSE values yield a 0.

Table 11.29 5 records

Code	newpop	cellper	longlife
CHN	1594.3600	1.0602135	1
FRA	76.9005	1.0321519	1
GBR	75.9690	1.1973963	1
IND	1539.4590	0.8731866	0
USA	373.9225	1.2043672	1

The "language" of these expressions uses the scalar constants and operators with which we are familiar. To understand the "nouns" (i.e., the fields and tables) that are legal in these expressions, we should understand that, in a RDBMS, execution of an SQL statement follows a prescribed order:

1. First, determine the rows and columns for the table or subquery of the FROM clause. This could introduce alias names for tables.
2. Second, determine the subset of rows selected by the WHERE clause from the table (or table combination) from Step 1.
3. For each selected row, compute the set of expressions after the SELECT as the values of the fields of the result.

So this is different than the tabular model and pandas, although it achieves similar goals. In pandas, we compute Series vectors and then mutate an existing data frame. In SQL, we generate a result table, building it, at least logically, selected row by selected row.

When, in the one or more *expressions* following the SELECT, the *expression* consists of a field name, then that field name is used for the result table. When an *expression* is a computed value, the RDBMS generates its own field name. But these are often ungainly, as so an alias is frequently used to control the field name of the result.

In the following example, we incorporate the numeric and Boolean computation expressions given above and, for each, specify an alias to give a readable field name of the result (Table 11.29):

```
SELECT code, pop * 1.15 AS newpop,
       cell / pop AS cellper, life > 75 AS longlife
FROM indicators0
```

In the next example, we demonstrate the string concatenation expression example. Note how we can "mix and match" simple fields and computed expressions (Table 11.30).

```
SELECT country || ' (' || code || ')' AS country2,
  region
FROM countries
WHERE code IN ('FRA', 'GBR', 'IND', 'USA')
ORDER BY country2
```

Table 11.30 4 records

country2	Region
France (FRA)	Europe & Central Asia
India (IND)	South Asia
United Kingdom (GBR)	Europe & Central Asia
United States (USA)	North America

This example also illustrates how the ORDER BY clause, because it is evaluated after Steps 1, 2, and 3 above, can use the alias determined by an expression following the SELECT. More examples are explored in the exercises.

11.4.1   Reading Questions

11.61 Consider a select query yielding Boolean values, like

```
SELECT life > 75 FROM indicators
```

The reading points out that this statement yields a 0 for FALSE and a 1 for TRUE. What happens if life is NA?

11.62 In the following SQL query, what happens for a row where pop = 0?

```
SELECT code, pop * 1.15 AS newpop,
       cell / pop AS cellper, life > 75 AS longlife
FROM indicators0
```

11.4.2   Exercises

11.63 In reference to the table indicators, write a query to find the GDP per capita of each country, without changing the order of the rows.

11.64 In reference to the table indicators, select all columns, plus a new column named new that tells whether exports are larger than imports.

11.65 In reference to the table indicators, select all fields plus a new column, new, that tells, for each row, whether any data is missing.

11.66 In reference to the instructors table in the school database, select a new column full_name obtained as the last name, followed by a comma and a space, followed by the first name.

11.67 In reference to the courses table in the school database, select a new column catalog obtained as the course subject, followed by a space, followed by the course number, followed by a colon and a space, followed by the course title. Only include distinct catalog entries and do not include any NULL entries.

11.68 Using `indicators`, select all data including a new column `trade_deficit` that computes the difference between `imports` and `exports`.

11.5 Aggregation

In the context of a single table, we want the ability to aggregate one or more columns, possibly in conjunction with filtering so that the aggregation is over a subset of rows. These operation requirements are similar to those we saw in the tabular model and `pandas` in Sect. 8.1. In SQL, such aggregations are still part of the `SELECT` query statement, and the aggregation functions, along with the fields being aggregated, are specified in the comma-separated *expressions* following the `SELECT`. The filtering of rows can occur using the `WHERE` clause variations as seen throughout Sect. 11.3.

The next few subsections will illustrate some of the most useful aggregations.

11.5.1 Counting Rows for Fields

We use the `COUNT (expr)` aggregation to count the number of rows in a result. If *expr* is `*`, then all of the rows of the result are counted. If *expr* is a field or other computed quantity, then the count only incorporates rows for which *expression* is not `NULL`.

An example querying the number of records in the `countries` table (Table 11.31):

```
SELECT COUNT(*) FROM countries
```

Notice the name given by SQL to the resultant field.

We now show another example, where we use aliases to give field names in the result for the aggregations. This also shows how, in the presence of `NULL` field values, the total count of rows differs from the count on a field (Table 11.32):

```
SELECT COUNT(*) AS total, COUNT(land) AS has_area
FROM countries
```

Table 11.31 1 records

COUNT(*)
217

Table 11.32 1 records

Total	has_area
217	210

Table 11.33 1 records

numname
18

Table 11.34 Common SQL aggregation functions

Aggregation function	Description
AVG (*expr*)	Compute the average of the non-NULL values of *expr* from the rows of the result
MAX (*expr*)	Compute the maximum non-NULL value of *expr* from the rows of the result
MIN (*expr*)	Compute the minimum non-NULL value of *expr* from the rows of the result
SUM (*expr*)	Compute the sum of the non-NULL values of *expr* from the rows of the result

Table 11.35 1 records

numrec	maxgdp	avggdp	minlife
5	19485.4	7901.122	68.8

Table 11.36 1 records

numrec	Name	Year
139	John	1880

The count aggregation can optionally include, in its argument, the DISTINCT keyword and then will count the unique instances of a field, as exemplified here (Table 11.33):

```
SELECT COUNT(DISTINCT name) AS numname FROM topnames
```

Standard SQL includes many of the same aggregation functions we have seen from the tabular model. The four most common are listed in Table 11.34.

The next example shows how to use multiple aggregations in the same query, one counting all the rows in the result, two performing different aggregations on the same field, and one computing a different aggregation on another field (Table 11.35).

```
SELECT COUNT(*) AS numrec, MAX(gdp) AS maxgdp,
       AVG(gdp) AS avggdp, MIN(life) AS minlife
FROM indicators0
```

Whenever we use aggregations in a single table query, and with no partitioning/grouping, the result is always a single row. If we were to use a field name as one of the expressions following SELECT in the presence of other aggregations, the "aggregation" performed would be to select the *first* instance of the field (Table 11.36).

```
SELECT COUNT(*) AS numrec, name, year
FROM topnames
WHERE sex = 'Male'
```

Code	Pop	Gdp
CHN	1386.40	12143.5
USA	325.15	19485.4

Table 11.37 2 records

An aggregation can be a powerful tool when used as a subquery. In the following, we compute the average GDP and can use that in a WHERE clause to filter for the rows that exceed the average. This could not be done with a hard-coded number and would continue to "do the right thing" as the table is updated, making it a more robust solution (Table 11.37).

```
SELECT code, pop, gdp FROM indicators0
WHERE gdp > (SELECT AVG(gdp) FROM indicators0)
```

More examples will be explored in the exercises.

11.5.2 Reading Questions

11.69 In pandas when we did aggregate and groupby, it was permissible to choose different sets of aggregates for different variables, e.g., `df.agg({'population':['min','max'],'cases':['mean','max']})`
Can you do the same with SQL? Explain.

11.5.3 Exercises

11.70 Using the `topnames` table, find the number of distinct names in the data.

11.71 Using the `topnames` table, find the number of distinct female names in the data.

11.72 Using the `topnames` table, find the average count of female top name applications over all years.

11.73 Using the `indicators` table, find the average life expectancy, the minimum year, and the maximum number of cell phone subscribers.

11.74 Using the SQL table `indicators`, use a select query to answer the question: which country and year had the highest cell phone number? Hint: use a subquery and MAX to find the overall highest cell phone number, then find which row has that number.

11.75 Create an SQL query to determine the *number* of unique users in the `emails` table.

11.76 Create an SQL query to retrieve the *length* of each email `message` in the `emails` table, not changing the order. This will involve the `LENGTH` function.

11.77 Based on your answer to the previous problem, use Python to determine the maximum length of any of the email messages. Then find the same using an SQL query.

11.6   Partitioning and Aggregating

Aggregations that are based on the *partitioning* of a data table allow us to compare between different categories of the rows in our table. For instance, in our `countries` table, we may wish to compute and compare the sum of the land area for the portioning of the regions of the world. In the `topnames` table, we may wish to compare the minimum, maximum, and average number of applications between the Female top names and the Male top names.

This is the same functionality we saw previously in the tabular model and `pandas` with the `groupby()` method and the ability to then apply a set of aggregation functions on specified columns of the resultant group by object. See Sect. 8.1.4 for a review.

We repeat the figure from Sect. 8.1.4 so that we can explain grouping in the context of the relational model and SQL.

We start on the left side of the figure, which depicts a single table. In the example figure, field A provides the value to be used for partitioning. All rows with a matching value in A are gathered together to get to the second portion of the figure, where we have a *set of partitions*. Aggregation is then performed *per partition*, and the result is a single row, with fields for A, defining the partition, and the remaining fields are the computed aggregations for the partition. The operation of aggregation on a partition is the same as that discussed in Sect. 11.5, Finally, the set of one-row summaries of the partitions are combined, and we obtain the final result at the far right of the figure (Fig. 11.4).

The SQL adds a `GROUP BY` clause that follows the `WHERE` clause. After the `GROUP BY` keywords, we specify one or more *field-spec*, which are used to determine, based on their unique set of values, the partitions. We know that the `WHERE` clause is evaluated after the `FROM` table-spec is determined, but before any computations in the `SELECT` have been performed. But we often want to filter, not on the original *table-spec*, but after the partitioning and the aggregation values have been computed. The `HAVING` clause allows this to occur, and its *partition-condition* is a Boolean that can include the computed values of partition aggregation.

The extended syntax of the SQL `SELECT` now becomes:

`SELECT [ DISTINCT ]` *field-spec* `FROM` *table-spec*

`[ WHERE [ NOT ]` *filter-condition* `]`

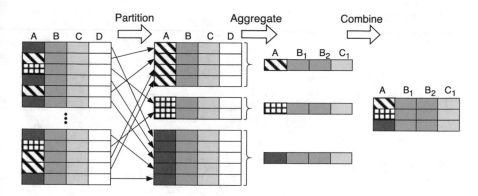

Fig. 11.4 Partition/aggregate/combine

Table 11.38 7 records

Region	region_area
East Asia & Pacific	24361338
Europe & Central Asia	27429255
Latin America & Caribbean	20038832
Middle East & North Africa	11223466
North America	18240984
South Asia	4771604
Sub-Saharan Africa	21242361

[GROUP BY *field-spec* [, *field-spec*]*]

[HAVING *partition-condition*]

[*order-clause*]

[*limit-clause*]

The existence of a GROUP BY does not necessitate a HAVING clause.

We now provide the SQL for the first example of this section, where we wanted, for the countries table, to compute and compare the sum of the land area for the partitioning of the regions of the world in the table (Table 11.38).

```
SELECT region, SUM(land) as region_area
FROM countries
GROUP BY region
```

If, after aggregation, we want to filter for the partitions where this new region_area is less than 20000000, we add the HAVING clause (Table 11.39):

```
SELECT region, SUM(land) as region_area
FROM countries
GROUP BY region
HAVING region_area < 20000000
```

Table 11.39 3 records

Region	region_area
Middle East & North Africa	11223466
North America	18240984
South Asia	4771604

Table 11.40 2 records

Sex	min_count	avg_count	max_count	num_name
Female	6919	42372.68	99689	10
Male	6900	47597.81	94757	8

Table 11.41 Displaying records 1–10

Name	Sex	num_name	avg_count
Mary	Female	76	40768.79
Michael	Male	44	70109.64
John	Male	44	19577.34
Robert	Male	17	61293.18
Jennifer	Female	15	57280.60
Jacob	Male	14	26484.21
James	Male	13	81248.31
Emily	Female	12	24542.33
Jessica	Female	9	44218.00
Lisa	Female	8	52571.50

Note that trying to specify this filter in a WHERE clause would result in an error because, at the time of evaluation of the WHERE, the region_area aggregation is not yet defined.

In our second example, we wanted to compare aggregations between male and female application counts in the topnames data table (Table 11.40):

```
SELECT sex, MIN(count) AS min_count, AVG(count)
  AS avg_count,
  MAX(count) AS max_count, COUNT(DISTINCT name)
  as num_name
FROM topnames
GROUP BY sex
```

As a final example, we partition by name in the topnames data set and compute aggregates (Table 11.41):

```
SELECT name, sex, COUNT(*) AS num_name, AVG(count)
AS avg_count
FROM topnames
GROUP BY name, sex
ORDER BY num_name DESC, avg_count DESC
```

If we consider the examples above, we should see some patterns. For each example, the expression list following the SELECT is always comprised of one or a few non-aggregate expressions and then our aggregate expressions. Unless we are intending to use a non-aggregate to get the value of the *first* field in the partition, each of our non-aggregate expressions should be included as field specifications in the GROUP BY. Some SQL systems will produce an error if this is *not* the case. More examples will be explored in the exercises.

11.6.1 Reading Questions

11.78 In the example below, what precisely is being computed, in real-world terms?

```
SELECT year, MAX(life) AS max_life
FROM indicators
GROUP BY year
ORDER BY max_life DESC
```

11.79 In reference to the question above, how would you achieve the same goal using pandas?

11.80 In the example below, what precisely is being computed?

```
SELECT sex, AVG(count) AS avg_count,
    COUNT(DISTINCT name) as num_name
FROM topnames
GROUP BY sex
```

11.81 In reference to the questions above, how would you achieve the same goal using pandas?

11.82 What SQL query would tell you the baby name that had the most births per year, over all years in the data?

11.83 What SQL query would tell you the number of applications associated with the most popular baby name of all time (i.e., largest total count over all years where this baby name was most popular)? Note that you do not yet have the tools to actually find this baby name, but you can find the number.

11.84 What SQL query would tell you the top ten countries in indicators by GDP per capita (i.e., gdp / pop)?

11.85 What SQL query would tell you the number of countries in each of the eight regions?

11.6.2 Exercises

11.86 Using the SQL table countries, use a select query to answer the question: how many countries are there in each region? Alias your new column as new.

11.87 Using the SQL table `countries`, use a select query to answer the question: how many countries are there for each starting letter of the alphabet? Please order your answer so it is easy to see the most common starting letter. For a challenge, make a histogram of counts for each letter. Hint: to get the first letter of `country`, use `LEFT(country,1)`. Please use the alias `count` for your new column.

11.88 Use the `indicators` database to find the total world population in each year (as the sum of country populations). Use the alias `total_pop` for your new column.

11.89 Treating your query above as a subquery, find the maximum for `total_pop` over all years. Use the alias `m` for the new column.

11.90 Not all countries are growing, so the largest population a country ever had might be in a previous year. For each country code in `indicators`, find the max population that country ever had. Alias your new column as `max_pop`. You should have one row per country. Do not change the order (i.e., your records should still be ordered by `code`).

11.91 With reference to the above, find all records where the max population is less than 1 (remember, this is measured in millions of people). Use a `HAVING` clause. Keep the original ordering of the data (alphabetically, by `code`).

11.92 Use the `indicators` database to find the total world population in each year (as the sum of country populations) and then return the rows where the total population is greater than 6000 (measured in millions of people). Use the alias `total_pop` for your new column.

11.93 Using the `indicators` table, create a `decade` column and then group records by both country and decade. Compute the resulting aggregate GDP in each group and think about what this kind of analysis tells you about the world.

Chapter 12
Relational Model: Multiple Tables Operations

Chapter Goals

Upon completion of this chapter, you should understand the following:

- Operations for combining multiple tables based on matching values in common between columns of the two tables, including the variations of:
  - inner join,
  - left/right join, and
  - outer join.
- Given combined tables using partitioning and aggregation to answer more complex questions.
- Using subqueries to compose more complex queries.

Upon completion of this chapter, you should be able to do the following:

- Form syntactically correct SQL that can be issued to a RDBMS for all of the operations discussed for multiple table queries.

Much power in SQL is enabled by the operations that allow us to query tables in combination. For instance, to be able to group information in the indicators data set by regions of the world, or by income category of the countries involved, we need the information from countries combined with information from

© Springer Nature Switzerland AG 2020
T. Bressoud, D. White, *Introduction to Data Systems*,
https://doi.org/10.1007/978-3-030-54371-6_12

indicators. With the combined result, we can then construct a "group by" that partitions by region or by income.

Conceptually, we have already seen the combination of multiple tables with the concat/merge/join operations in Sect. 8.3 during the discussion of advanced operations of the tabular model. In the tabular model, because the structural constraints of tidy data are by convention, many tables are not fully tidy, and, for instance, the indicators data set might well include the countries table information and avoid the need for table combination. In the relational model, however, database design and normalization is much more formal and strict, and violations are much more rare. In this model, we are driven to become proficient combining tables. We do so all the time to be able to construct the queries we need to get the data to answer various questions about the data.

12.1   Preliminaries and Example Data Set

To help illustrate some of the variations of multiple table operations, we will, in addition to the tables from Chap. 11, need a more sophisticated example database as well. The school database will be used for these more involved examples.

12.1.1   Data Set: The school Database Schema

Figure 12.1 depicts the full database schema for the school database. It consists of eight interrelated tables:

Table 12.1 describes the eight tables of the database. In keeping with SQL types, discussed in Chap. 13, we write "character id" (rather than "string id") to mean a field, id (which is also the primary key), made up of characters.

12.1.2   Table Relationships

The schema in Fig. 12.1 includes labeled lines indicating the relationships between tables. As described earlier in Chap. 10, the ends of these lines between tables indicate either a one-to-one relationship, which is rare, or a one-to-many relationship. Many-to-many relationships are realized in relational tables with a linking table and using two one-to-many relationships. In the figure, we label with a 1 the one-side of a one-to-many relationship and with a m the many-side of a one-to-many relationship.

We can understand the relationship notation/meaning by starting with the side labeled 1 and call that the *near table* and the other linked end (the "many" end) of a one-to-many link as the *far table*. The meaning is then, "Given exactly one record of the near table, there are potentially many/multiple corresponding records in the

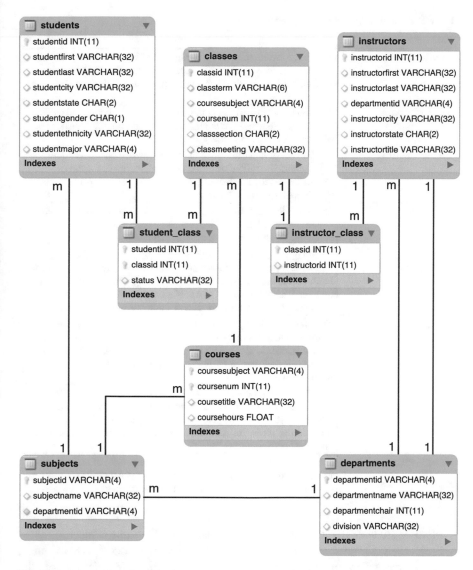

Fig. 12.1 Schema for school database

far table." Equivalently, starting from the perspective of the m side of the link and calling that the near table and the opposite end the far table, then the meaning of a many-to-one link is: "Multiple of the records of the near table can be linked to the same (single) record of the far table."

Sampling some of the relationships in the figure, we can see the following:

- Given a particular course, there can be multiple class instances of that course.
- Given a particular department, it can host multiple subjects.

Table 12.1 Tables in the `school` schema

Table name	Description
`students`	Maintains information about the full set of students in the data set. A generated id is used as the primary key, and the table includes fields for the student's name, home city/state, gender, ethnicity, and major
`instructors`	The full set of instructors at the school, with a generated id as primary key, name, id of the instructor's department, city/state part of their address, and a title
`departments`	The full set of departments at the school, with a character id for the department, the full name of the department, the id for the instructor serving as chair of the department, and a string giving the division to which the department belongs
`subjects`	This table lists the subjects that are taught at the school. A given department can teach multiple subjects. Each subject has a short character id, along with a longer textual name and a field giving the department, which is home to the subject
`courses`	The table represents the full catalog of courses taught at the school. Each course has an associated subject, along with an integer course number. The combination of a subject and number will be unique. A course also has a longer title and a number of credit hours
`classes`	The classes table represents an instance of a course, taught as one of the possibly multiple sections, and in a particular term. It has a unique integer id along with the course subject and number for the course it is an instance of. It also has a string representing its meeting days and times
`student_class`	This is a linking table, where entries consist of a student id and a class id. An entry in the table indicates that the given student is/was registered for the given class. A status entry gives information about, for a student/class pair whether the course was "normal" or "withdrawn"
`instructor_class`	This is another linking table, where entries consist of an instructor id and a class id. An entry in the table indicates that the given instructor was the teacher of record for the given class

- Multiple students (by their major) are associated with the same subject.
- Each class is present exactly once in the `instructor_class` table, but a particular instructor can be present in multiple records of the `instructor_class` table.
- Given a particular student, there can be multiple records in the `student_class` table.
- Given a particular class, there can be multiple records in the `student_class` table.

These last two relationships, taken together, are what give a "virtual" many-to-many relationship between students and classes—a single student takes multiple classes, and a single class is taken by multiple students.

A database schema, by its choices, can also limit what can be represented in a database. For instance, the following are not allowable by the current database design:

- One subject that is in multiple departments.
- A department with multiple chairs.
- A student with more than one major.

12.1.3 SQL Execution Plan

As our SQL becomes more complex, it is important to emphasize and understand the order in which an SQL query is processed. While our SQL often reads more like a natural language sentence, the processing does NOT occur in a left to right order. From a conceptual perspective, the execution of a query must begin by determining a full set of possible rows and columns of an originating data set, and then filtering down by particular rows, doing any partitioning and filtering again, and then performing column computations.

More precisely, the order of execution of a query is as follows:

1. Use the FROM clause to determine a source data set, defining all the possible columns and the full set of rows. In Chap. 11, this was a single table, and thus the possible columns were those of the specified table. Once we have multiple tables that can be combined together in a join, we have a set of possible columns that are from multiple tables, and we have to be able to unambiguously refer to columns from the multiple tables.

2. Even with a complex FROM clause, the result of Step 1 is a *table* (that we call the *source table*), with its full set of columns and rows. The second step of query execution uses the WHERE clause to filter rows, including only some of them in the result. Since, at this step, we only have the source table and have not considered the column projections, some versions of SQL will not recognize aliases that come from the SELECT in the WHERE clause. This is even more strict when it comes to fields that do not yet exist because we have not reached the execution step where an aggregation occurs. The order of clauses within the WHERE can be significant, and so row filtering that yields a smaller result in an early clause can speed up a query.

3. Once the rows have been filtered by the WHERE, execution *then* performs the partitioning defined by the GROUP BY. Until the GROUP BY partitioning has occurred, there can be no aggregations. So execution order is the reason that attempting to use an aggregate value as a condition in the WHERE is illegal for SQL.

4. Given the aggregated data from the GROUP BY, the HAVING clause is executed next, allowing us to filter rows that form the aggregates of the partitions, and can use aggregate values.

5. The final set of rows and columns resulting from the HAVING is now processed for the columns and column expressions based on the SELECT expressions, with one column for each expression for each row, and now named with any aliases specified in the SELECT. If a DISTINCT keyword is specified, the uniqueness filtering happens at this point in the execution, i.e., after the SELECT operation.
6. The computed columns and rows are then processed based on the ORDER BY to determine a row ordering of the result.
7. Finally, a LIMIT truncates the result as specified, and the final result becomes the response to go back to the client.

12.1.4   Reading Questions

12.1 Recall that, for a many-to-one relationship in a database schema, given a record of the *neartable*, there is exactly one corresponding record in the *fartable*, and given a record of the *neartable*, there are potentially multiple corresponding records in the *fartable*. With this in mind, in the classes table, why did the database designers choose to name coursesubject and coursenum with a course prefix, but named classsection with a different prefix?

12.2 Notice that the link between classes and instructor_class has a "1" on both sides. So for each record in class, there is a single corresponding record in instructor_class (identified by classid), and for each record in instructor_class, there is a single corresponding record in class (identified also by classid). What would be an alternative design that allows this database to determine which instructor is teaching which class instance?

12.3 If you answered the above, you may have concluded that instructor_ class is *not* required in this design. What real-world scenario (i.e., under what conditions) would require the instructor_class linking table? Put another way, what could a school using this database be able to do?

12.4 With reference to the previous question, name two changes to the schema that accomplish the ability of the new scenario. Hint: no new fields are required.

12.5 If you were designing the school database, what other fields might you maintain for students? For instructors? For classes? Give at least one new field you could add for each.

12.6 With reference to the student database, why is there a field classterm in classes but no courseterm in courses?

12.7 Why do you think the table student_class has a field status but instructor_class does not? What kinds of data would be stored in this field? Hint: first think about the purpose of the student_class table.

12.8 What is the connection between the subjects table and the students table?

12.1.5 Exercises

12.9 In reference to the `school` database, how many departments are there? Your query should result in a 1×1 table with just one number in it.

12.10 In reference to the `school` database, how many departments appear to cover distinct areas of study? Hint: this question is asking about the field `departmentname`, but your answer should be general enough that it would work even if some departments changed name (e.g., "Cinema and Culture") or if new departments were added (e.g., "Agriculture/Culinary studies").

12.11 In reference to the `school` database, how many classes have at least three sections? Your query should result in a 1×1 table.

12.12 In reference to the `school` database, how many students are from each state? Please order your results by state (alphabetically) and alias your new column as `count`. Please omit NULL entries. For a challenge, make a graph displaying this information.

12.13 In reference to the `school` database, what is the gender breakdown of the student body? Your query should return a 2×2 table, with `count` as the new column.

12.14 In reference to the `school` database, what is the breakdown of the student body by ethnicity? Your result should be 8×2. Please sort by your new column (`count`), from highest to lowest.

12.15 In reference to the `school` database, how many courses are missing a title? Your result should be 1×1.

12.16 In reference to the `school` database, how many classes are there with numbers 100–199, inclusive? Your answer should be 1×1.

12.17 In reference to the `school` database, how many courses are there with numbers 100–199, inclusive? Your answer should be 1×1.

12.18 In reference to the `school` database, make a data frame telling the number of courses at each level: the 400 level (i.e., 400–499), 300 level, etc. Hint: the FLOOR function might be useful. Please name your new columns `level` and `count` and include only these columns.

12.19 In reference to the `school` database, make a data frame telling the number of classes at each level: the 400 level (i.e., 400–499), 300 level, etc. Hint: the FLOOR function might be useful. Please name your new columns `level` and `count` and include only these columns.

12.20 In reference to the `school` database, how many courses are there worth zero credits?

12.21 In reference to the `school` database, make a data frame telling the number of courses at each credit hour level, e.g., 4 credits, 3 credits, 2 credits, etc. Please Call the new columns `credits` and `count`. It should be 7×2. Think about whether something is fishy and investigate.

12.22 In reference to the `school` database, what fraction of classes have a meeting time strictly before 10 am? This should be as a fraction of all rows where `classmeeting` is not null.

12.23 In reference to the `school` database, how many instructors are there?

12.24 In reference to the `school` database, please rank departments (highest to lowest) in terms of how many instructors there are in each department.

12.25 In reference to the `school` database, how many students are there?

12.26 In reference to the `school` database, how many subjects are there?

12.27 In reference to the `school` database, please rank departments (highest to lowest) based on how many subjects they offer?

12.28 In reference to the `school` database, how many students are missing a state?

12.29 In reference to the `school` database, please rank the majors (highest to lowest) in terms of number of students with each major.

12.30 Write an SQL query to retrieve all the course titles (from `school. courses`) that contain the words "Theory" or "Computer." List the results in reverse alphabetical order by the course title name. Please do not include any duplicate listings. Do not include more fields than that requested.

12.31 Write an SQL query to get the student id numbers, major, gender, and ethnicity of all students from Ohio where ethnicity and major data is not missing. Do not include any other columns.

12.2   Overview of Join Operations

Joins in the relational model have the same operational meaning as when we encountered them in the tabular model. Furthermore, the joins of the relational model subsume the tabular operations of a `concat` where we are combining in the column dimension, as well as the `join` and `merge` operations in `pandas`.

In general, a join allows us to combine two tables based on a *match condition* that involves comparing the values of one or more field in one table to the values of corresponding fields in another table. When there is a match, we yield a row in the result, and the values of the fields in that row are the values from the fields of the pair of records that produced a match. That is, each row has columns from both tables. The variations of the join operation (left join, inner join, etc.) come about

Table 12.2 pop_gdp

Code	Pop	Gdp
CHN	1386.40	12143.50
FRA	66.87	2586.29
GBR	66.06	2637.87
USA	325.15	19485.40

Table 12.3
country_land

Code	Country	Land
FRA	France	547557
GBR	United Kingdom	241930
IND	India	2973190
USA	United States	9147420
VNM	Vietnam	310070

when the value of fields used in the match condition in one table has *no* matching value in the second table, or vice versa.

Suppose, given Table 12.2 and 12.3, we want to use SQL to create a unified table with columns from both tables and to "match up" rows based on the code field:

SELECT * **FROM** pop_gdp

SELECT * **FROM** country_land

Notice that the pop_gdp table has code field values for CHN, FRA, GBR, and USA, while in country_land, we have code field values for FRA, GBR, IND, USA, and VNM.

In the join operation, we want a match condition of the code field in pop_gdp to equal the code field in country_land.

The different variations of the join operation can be distinguished in this example, and assuming that pop_gdp is the *left table* and country_land is the *right table*:

- **Inner Join**: include rows when *both* tables have corresponding fields based on the match condition. In this example, this occurs for the three codes of FRA, GBR, and USA.
- **Left Join**: include all rows of the pop_gdp table (the left table) and, for cases where there is no corresponding code value in the right table, fill in as *missing* the values for columns/rows in the right table that have no correspondence. In this example, the result would include rows with a code for CHN, FRA, GBR, and USA.
- **Right Join**: include all rows of the country_land table (the right table) and, for cases where there is no corresponding code value in the left table, fill in as *missing* the values for columns/rows in the left table that have no correspondence. In this example, the result would include rows with a code for FRA, GBR, IND, USA, and VNM.

- **Outer Join**: merge all rows from both tables, and fill in as *missing* for rows/columns in either table that lack a correspondence. In this example, the result would include rows with a `code` for CHN, FRA, GBR, IND, USA, and VNM.

As in Chap. 8, in a left join, we think of the left table as dominant, and taking information from the right table to enrich itself. We think of inner join as sitting between the two tables and drawing rows from both, when matches are found, and strictly insisting that a row actually be present in both tables before accepting it. We think of the outer join as being maximally permissive, still sitting between the two tables, but accepting every row, and trying to make sure no data is lost in the join.

In SQL, all of the various types of join syntactically occur in the form of the *table-spec*, which is the specification that comes after the `FROM` keyword and which generates the full set of possible rows and columns.

When our `FROM` clause is extended to join together two tables, our set of possible column fields has expanded, and thus our comma-separated expression list of field names after the `SELECT` must now be more specific. We must also understand, when dealing with two tables, that the field names between the tables might include duplicates.

In our example `SELECT` statements up to now, we have been using a shortcut when we simply name a field—because in our single table SQL queries, with only one table, the DBMS can unambiguously resolve a simple field name. In general, we can explicitly specify both a table name and a field name, using a dot notation, so we can be more specific with *table.field* instead of just *field*.

The possible fields that may be specified in the `SELECT` when a `JOIN` forms the `FROM` clause may include all fields from all joined tables.

12.3   Inner Joins

We use an inner join either when we *know* that the matches will be complete or when the problem we are trying to solve requires that the resulting table has meaningful values for all the columns to be included in the result, even at the price of eliminating the rows that fail to have matches in one direction or the other.

The operations that give us control on records to include in the result, beyond the form provided by the inner join, are the *outer* join and *left outer* joins and will be explored in Sect. 12.4.

12.3.1   Two Table SQL Inner Join

We extend our syntax description to allow a `FROM` to be described by an `INNER JOIN`:

Table 12.4 Joined table

Code	Pop	Gdp	code	Country	Land
FRA	66.87	2586.29	FRA	France	547557
GBR	66.06	2637.87	GBR	United Kingdom	241930
USA	325.15	19485.40	USA	United States	9147420

$$table\text{-}spec \models \quad table \mid (\ subquery\ ) \ \text{AS}\ new\text{-}name \mid join\text{-}table$$

where

$$join\text{-}table \models \quad table\text{-}spec\ \text{INNER JOIN}\ table\text{-}spec\ \text{ON}\ match\text{-}condition$$

The *match-condition* in this specification is specified as a Boolean condition that, when TRUE, indicates a match for the join and, when FALSE, means no match. The condition is typically one which checks for equality between a field of the left table and a comparable field of the right table. These fields need not be naming a key field, nor must they be fields that are named identically between the two tables.

In SQL, and when using an INNER JOIN, we will be explicit about naming the matching condition. Namely, for the current example, rows from two tables match when the code field in pop_gdp is equal to the code field in country_land. With this matching criterion, it does not matter whether the order of the rows in the two tables is the same.

Using full table-and-field naming for the fields of the SELECT and for the fields used in the *match-condition*, we can form our first inner join SQL operation (yielding Table 12.4):

```
SELECT pop_gdp.code, pop_gdp.pop, pop_gdp.gdp,
       country_land.code, country_land.country,
       country_land.land
FROM
       pop_gdp INNER JOIN country_land
               ON pop_gdp.code = country_land.code
```

There are three rows in the result, for the three row-cases where the code field in pop_gdp matches with the code filed in country_land. Notice also that the column/field names come from the field names of the tables, but without the table names. Each system must deal with the fact that some field names might be the same between the left and right tables. In the case of the MySQL software used in the generation of this book, the result maps a duplicate code to an alternate with a column number appended. Other systems may use other conventions for handling duplicate field names.

We can still use * as a wildcard to include all available fields for column projection to get *all* fields from the left table followed by all fields from the right table (yielding Table 12.5):

Table 12.5 The result of the SELECT * query

Code	Pop	Gdp	code	Country	Land
FRA	66.87	2586.29	FRA	France	547557
GBR	66.06	2637.87	GBR	United Kingdom	241930
USA	325.15	19485.40	USA	United States	9147420

Table 12.6 Selecting using an alias

Code	Pop	Gdp	cl_code	Country	Land
FRA	66.87	2586.29	FRA	France	547557
GBR	66.06	2637.87	GBR	United Kingdom	241930
USA	325.15	19485.40	USA	United States	9147420

Table 12.7 Selection with USING keyword

Code	Pop	Gdp	Country	Land
FRA	66.87	2586.29	France	547557
GBR	66.06	2637.87	United Kingdom	241930
USA	325.15	19485.40	United States	9147420

```
SELECT * FROM
        pop_gdp INNER JOIN country_land
        ON pop_gdp.code = country_land.code
```

We can use aliases for the tables involved in the join, and also for fields specified in the SELECT to both shorten the query and to resolve any repeated column names, as we do in the following example. By giving aliases for the tables involved in the inner join, we can shorten the *match-condition*. We use an alias in the SELECT to resolve the second field named code to be cl_code (yielding Table 12.6):

```
SELECT PG.code, PG.pop, PG.gdp,
        CL.code AS cl_code, CL.country, CL.land
FROM
        pop_gdp AS PG INNER JOIN country_land AS CL
             ON PG.code = CL.code
```

In many cases, an inner join *match-condition* involves *equality* between two fields that are *named the same* between the left and right tables. If both of these are satisfied, we can simplify the inner join syntax and instead of using the ON with a match condition, we use USING and put the one or more columns to be used for matching in a comma-separated list in parentheses following the USING keyword, as shown here (yielding Table 12.7):

```
SELECT *
FROM pop_gdp AS PG INNER JOIN country_land AS CL
     USING (code)
```

Table 12.8 Names table

Year	Sex	Name
2016	Female	Emma
2016	Male	Noah
2017	Female	Emma
2017	Male	Liam

Table 12.9 Counts table

Year	Sex	Count
2016	Female	19496
2016	Male	19117
2017	Female	19800
2017	Male	18798

Table 12.10 Join of names and counts

Year	Sex	Count	Name
2016	Female	19496	Emma
2016	Male	19117	Noah
2017	Female	19800	Emma
2017	Male	18798	Liam

Extending the syntax definition, we specify a second form of inner join:

join-table $\models$ *table-spec* INNER JOIN *table-spec* USING (*common-fields*)

This shortcut is available if both conditions are met:

1. the same field or fields is/are present in *both* tables,
2. the join is based on equality of the specified *common-fields*.

Note that, when using this syntax, the meaning of the $\star$ wildcard is changed slightly and does NOT project the duplicate of the *common-fields*.

More generally, a match might require the equality of *multiple* fields. Suppose, in the topnames data set, we have one table of top names, with primary key year and sex and a second table of corresponding counts, also with primary key year and sex (Tables 12.8 and 12.9).

```
SELECT * FROM names
```

```
SELECT * FROM counts
```

We want a match when both the year *and* the sex fields are the same between records in the left and right tables (Table 12.10). In the ON *match-condition* version of the JOIN, the query becomes:

```
SELECT C.year, C.sex, C.count, N.name
FROM counts AS C INNER JOIN names AS N
     ON C.year = N.year AND C.sex = N.sex
```

This problem also satisfies both of our conditions for the USING shortcut and the following query yields Table 12.10 again:

```
SELECT *
FROM counts INNER JOIN names
      USING (year, sex)
```

12.3.2 [Optional] Cartesian Product-Based Inner Join

The meaning of an inner join, and, in particular, what it means to "match" two records, can be defined more precisely using a Cartesian product. The Cartesian product of two sets is given by pairing all elements of the first set with all elements of the second set. This same idea can be applied to tables in the relational model, where each *record* of a table represents exactly one element of a set, and the table represents the set itself.

If the first table has n records, and the second table has m records, the Cartesian product will have $n \times m$ records. Each resultant record will contain *all* the column fields from both tables. So if the first table had x fields and the second table had y fields, the Cartesian product would have $x + y$ fields.

In SQL, we can obtain the Cartesian product by just naming both tables, comma separated, in the FROM clause. We use aliases in this example to help clarify which fields originated from which table (yielding Table 12.11):

```
SELECT   PG.code AS pg_code, PG.pop AS pg_pop,
         PG.gdp AS pg_gdp, CL.code AS cl_code,
         CL.country AS cl_country, CL.land AS cl_land
FROM pop_gdp AS PG, country_land AS CL
```

Table 12.11 Cartesian product

pg_code	pg_pop	pg_gdp	cl_code	cl_country	cl_land
CHN	1386.40	12143.50	FRA	France	547557
FRA	66.87	2586.29	FRA	France	547557
GBR	66.06	2637.87	FRA	France	547557
USA	325.15	19485.40	FRA	France	547557
CHN	1386.40	12143.50	GBR	United Kingdom	241930
FRA	66.87	2586.29	GBR	United Kingdom	241930
GBR	66.06	2637.87	GBR	United Kingdom	241930
USA	325.15	19485.40	GBR	United Kingdom	241930
CHN	1386.40	12143.50	IND	India	2973190
FRA	66.87	2586.29	IND	India	2973190

Table 12.12 Cartesian product filtered by a WHERE condition

pg_code	pg_pop	pg_gdp	cl_code	cl_country	cl_land
FRA	66.87	2586.29	FRA	France	547557
GBR	66.06	2637.87	GBR	United Kingdom	241930
USA	325.15	19485.40	USA	United States	9147420

The first four rows of the result have the FRA record of the second table paired with all four records of the first table, and so forth. Clearly, in the Cartesian product, there are many rows that we would not want for a table combination. Specifically, all rows where pg_code does *not* match cl_code should be eliminated. We can accomplish this in our SQL with a WHERE clause with an appropriate condition which, in fact, is the same as our *match-condition* (yielding Table 12.12).

```
SELECT   PG.code AS pg_code, PG.pop AS pg_pop,
         PG.gdp AS pg_gdp, CL.code AS cl_code,
         CL.country AS cl_country, CL.land AS cl_land
FROM pop_gdp AS PG, country_land AS CL
WHERE PG.code = CL.code
```

This is the semantic definition of an inner join, where the matching condition of the inner join is used to select the appropriate matching rows from a logical Cartesian product. Maintaining the same projected columns and aliases, the equivalent inner join is:

```
SELECT   PG.code AS pg_code, PG.pop AS pg_pop,
         PG.gdp AS pg_gdp, CL.code AS cl_code,
         CL.country AS cl_country, CL.land AS cl_land
FROM pop_gdp AS PG INNER JOIN country_land AS CL
ON pg_code = cl_code
```

The solution shown in this example, of building the Cartesian product and then filtering should rarely, if ever, be used, because it is exceedingly inefficient. When tables become of even modest size, the $n \times m$ rows of the Cartesian project grow quickly; if n and m are approximately the same size, the work to do a Cartesian product followed by the filtering of a WHERE is proportional to n^2. The inner join, on the other hand, is efficient and uses indexing and the matching condition as it is processing the tables, and the work is proportional to n.

12.3.3   Inner Join to Fill Redundant Fields

A common use for an inner join occurs when, by reducing redundancy in database design, we have split tables based on non-identical independent variables of the functional dependencies. For instance, we split our indicators data set into two tables, indicators and countries, because some fields, like the country

Table 12.13 Subset of indicators table

Code	Year	Pop	Gdp
CHN	2007	1317.88	3550.34
FRA	2007	64.02	2657.21
GBR	2007	61.32	3084.12
CHN	2017	1386.40	12143.50
FRA	2017	66.87	2586.29
GBR	2017	66.06	2637.87

Table 12.14 Inner join of indicators and countries

Code	Year	Pop	Gdp	Country	Region
CHN	2007	1317.88	3550.34	China	East Asia and Pacific
CHN	2017	1386.40	12143.50	China	East Asia and Pacific
FRA	2007	64.02	2657.21	France	Europe and Central Asia
FRA	2017	66.87	2586.29	France	Europe and Central Asia
GBR	2007	61.32	3084.12	United Kingdom	Europe and Central Asia
GBR	2017	66.06	2637.87	United Kingdom	Europe and Central Asia

name, the world region, and the land area, were only dependent on the country code, while other fields, like the population and GDP, were dependent on the country code *and* the year. So,

$$code \rightarrow country, region$$

$$code, year \rightarrow pop, gdp$$

A join allows us to readily build back the combined table. This, in turn, will allow us to perform operations on the combined result using partitioning and aggregation.

Suppose we have the following subset of `indicators` with data for country codes, CHN, FRA, and GBR and two years, 2007 and 2017 (Table 12.13):

```
SELECT * FROM indicators_subset
```

The goal is to use matching country code records from the countries table and fill in columns for country name and region redundantly in the resultant table (yielding Table 12.14):

```
SELECT I.*, C.country, C.region FROM indicators_subset AS I
        INNER JOIN countries AS C USING (code)
```

Note that we used another shortcut: SQL allows the * wildcard to be used with a particular table when we have a multiple table construction, and we wanted all the columns of the `indicators_subset` table and just two of the columns of the `countries` table.

Table 12.15 Inner join of instructor_class and instructors

classid	instructorid	instructorlast	instructorfirst
21014	9167	Foster	Helen
21088	9029	Hawkins	Grace
21256	9146	Garrett	Jason
21444	9050	Price	Taylor

Now, we have redundant data again in the `country` and `region` columns of this joined table result, but, in the source tables, the data appear only one time, so a change in the values for a `country` or a `region` would get reflected consistently for the joined result.

Consider another example from our `schools` database. Suppose a student has registered for four classes, whose `classid` fields are given by 21014, 21088, 21256, and 21444. If we want to find the instructors of those classes, we can use the linking field of `instructorid` to link to the `instructors` table, where `instructorid` is the primary field (intentionally named the same), and, by joining, can fill in a combined table starting with the `classid` fields and obtaining the instructor's last and first names (yielding Table 12.15).

```
SELECT classid, instructorid, instructorlast, instructorfirst
FROM instructor_class INNER JOIN instructors
     USING (instructorid)
WHERE classid IN (21014, 21088, 21256, 21444)
```

12.3.4 Three-Table Join

Our syntax of an SQL join uses *table-spec* for each of the tables to be joined. The astute reader will notice that this allows for a *table-spec* of a join to be, in fact, another *join-table*. This means that we can join two tables and then join that result to another table. We can, in fact, continue for as many joins as might be necessary to solve a particular problem.

For example, in the `school` database, suppose that we would like to see the subject, number, section, and meeting time for each of a set of classes combined with the instructor teaching the class using the `instructorlast` and `instructorfirst` fields.

To just see information about a particular set of classes, we would issue a single table query like so (yielding Table 12.16):

```
SELECT coursesubject || '-' || coursenum || '-' || classsection
    AS class, classmeeting
FROM classes
WHERE classid IN (21014, 21088, 21256, 21444)
```

Table 12.16 Class selection example

Class	classmeeting
PSYC-100-02	09:30–10:20 MWF
FYS-102-02	13:30–14:50 TR
ECON-240-01	11:30–13:20 MW
PHED-180-01	NA

Table 12.17 Inner join of classes and instructor_class

Class	classmeeting	instructorid
PSYC-100-02	09:30–10:20 MWF	9167
FYS-102-02	13:30–14:50 TR	9029
ECON-240-01	11:30–13:20 MW	9146
PHED-180-01	NA	9050

Table 12.18 Three way inner join (classes, instructor_class, instructors)

Class	classmeeting	instructorlast	instructorfirst
PSYC-100-02	09:30–10:20 MWF	Foster	Helen
FYS-102-02	13:30–14:50 TR	Hawkins	Grace
ECON-240-01	11:30–13:20 MW	Garrett	Jason
PHED-180-01	NA	Price	Taylor

The mapping of instructors to classes resides in a different table: `instructor_class`. So to retrieve the instructor for this class, we join on `classid` between the two tables (yielding Table 12.17):

```
SELECT coursesubject || '-' || coursenum || '-' || classsection
    AS class, classmeeting, instructorid
FROM classes INNER JOIN instructor_class USING (classid)
WHERE classid IN (21014, 21088, 21256, 21444)
```

But to get to the `instructorlast` and `instructorfirst`, we need to use the `instructorid` from this joined table result and join with the `instructors` table to then fill in the desired instructor fields (yielding Table 12.18):

```
SELECT coursesubject || '-' || coursenum || '-' || classsection
    AS class, classmeeting, instructorlast, instructorfirst
FROM classes INNER JOIN instructor_class USING (classid)
    INNER JOIN instructors USING (instructorid)
WHERE classid IN (21014, 21088, 21256, 21444)
```

In the FROM clause, inner joins are processed in the left to right order they appear. If we want a different order, we can use parentheses around a join clause to enforce a different precedence.

Note the order we used to stepwise build our desired query. We started with the table that has the most detailed information. It will almost always be on the "many"-side of a relationship. If we follow from the many-side of a relationship to the one-side of a relationship, we are logically going from the linking field (called a *foreign key*), which resides in the many-side, to the primary key on the one-side. It is *always* possible to perform a join using the primary key on the one-side. We can then repeat the process, based on the second table—if it has a relationship link and is the many-side of a many-to-one, we can follow to a one-side link to yet another table. We can think of the chain of joins as a table traversal. It is also an efficient way to go, as each of these primary key lookups on downstream tables is efficient.

12.3.5   Join Table from a Subquery

Another application of inner joins is to solve problems of the form "For every country, find the year in which that country's population hits its maximum." Such problems should be familiar from an introductory computer science course, as they are analogous to finding the index where the max occurs. However, solving such a problem with databases requires two steps. First, one needs a query to find the maximum population per country, over all possible years. This is accomplished with a GROUP BY clause (yielding Table 12.19).

```
SELECT year, code, MAX(pop) AS max_pop
    FROM indicators
    GROUP BY code
```

Note that this query selects the first year for each code, even if that year does not match the year in which the max_pop occurred. Hence, we must use an inner join to get rows of indicators where pop equals the max_pop computed by the query above, which we repeat as a subquery below (yielding Table 12.20):

```
SELECT I.code, I.year, maxTable.max_pop FROM
    ((SELECT code, MAX(pop) AS max_pop
        FROM indicators
        GROUP BY code) AS maxTable
    INNER JOIN indicators AS I
    ON (maxTable.max_pop = I.pop) AND (I.code = maxTable.code))
```

This query returns, for each code, all years in which the max_pop was achieved. Hence, it is allowed to have more results than the number of distinct

Table 12.19 GROUP BY example

Year	Code	max_pop
2017	ABW	0.11
2018	AFG	37.17
2018	AGO	30.81
1990	ALB	3.29
2004	AND	0.08
2018	ARE	9.63
2018	ARG	44.49
1989	ARM	3.54
1997	ASM	0.06
2017	ATG	0.10

Table 12.20 Codes along
with max_pop values, but
with duplicates

Code	Year	max_pop
KNA	1960	0.05
TUV	1960	0.01
KNA	1961	0.05
TUV	1961	0.01
KNA	1962	0.05
TUV	1962	0.01
KNA	1963	0.05
NRU	1963	0.01
TUV	1963	0.01
GIB	1964	0.03

Table 12.21 Countries with
the most recent year where
their max populations was
achieved

Code	Year	max_pop
ABW	2017	0.11
AFG	2018	37.17
AGO	2018	30.81
ALB	1990	3.29
AND	2004	0.08
ARE	2018	9.63
ARG	2018	44.49
ARM	1989	3.54
ASM	1997	0.06
ATG	2017	0.10

values of code (e.g., Gibraltar achieved its max population, of 0.03, in both 1964
and 1965). Changing the first SELECT to SELECT DISTINCT does not solve this
problem, because the year field will be different. If one desires only the first year
in which a max population was achieved, we could add another layer to the query
above, treating it as a subquery, and add a GROUP BY clause to only get one result
per code (yielding Table 12.21):

```
SELECT * FROM
    (SELECT I.code, I.year, maxTable.max_pop FROM
        ((SELECT code, MAX(pop) AS max_pop
        FROM indicators
        GROUP BY code) AS maxTable
        INNER JOIN indicators AS I
        ON (maxTable.max_pop = I.pop) AND
                (I.code = maxTable.code))) AS j
    GROUP BY j.code
    ORDER BY code
```

12.3.6 Reading Questions

12.32 The reading introduced dot notation (.) to identify a field with a specified table, e.g.,

```
SELECT pop_gdp.code FROM ...
```

How does SQL know when a dot is part of a column name and when it is meant to be interpreted as *table.field*?

12.33 In the general syntax for specifying a table:

$$table\text{-}spec \models table | (subquery) \text{ AS } new\text{-}name | join\text{-}table$$

Give an example where *table* is used, and give an example where *(subquery) AS new-name* is used.

12.34 In

```
SELECT IE.code, PG.pop, PG.gdp,
       PG.code AS ie_code, IE.imports, IE.exports
FROM pop_gdp AS PG INNER JOIN imports_exports AS IE
            ON PG.code = IE.code
```

How does SQL know which part each AS refers to?

12.35 The USING keyword is one way to solve the problem of a duplicate code column (at least, when the conditions for USING are met). Explain another way to avoid selecting a duplicate column when doing an INNER JOIN on PG.code = IE.code.

12.36 In the topnames join example, *what would happen and why*, if we only matched on year rather than year and sex, e.g., doing:

```
SELECT C.year, C.sex, C.count, N.name
FROM counts AS C INNER JOIN names AS N
     ON C.year = N.year
```

12.37 Using the example of a Cartesian product from the reading (regarding pop_gdp and imports_exports), show how to do a Cartesian product of the tables counts and names from the reading.

12.38 As another example to illustrate the use of INNER JOIN to link two data sets with different functional dependencies, imagine a table usa_pop with fields year and pop, containing the population of the USA for each year. Explain how to build this information into the topnames table using an INNER JOIN. It will help to draw the functional dependencies for each table.

12.39 As another example to illustrate a three-table join, imagine a table usa_births with fields year and births that gives the total number of new births each year. Explain how to build this information, and the information of the usa_pop table from the previous question, into the topnames table using a single query with multiple INNER JOIN commands.

12.40 Modify your answer to the previous question to use a WHERE command, to only yield rows with female births.

12.41 Suppose that, for one or more students, we wish to get the name of the department hosting their major. What is the chain of joins we would need, and on what criteria? Can we use the USING form throughout the chain?

12.3.7   Exercises

12.42 In reference to the school database, select all distinct course titles for classes offered in the fall semester.

12.43 Using school, select all students' first and last names, and the class ids and terms for the classes they are taking during the year. Only include students who are actually taking classes, and order your results alphabetically by students' last name.

12.44 Using school, use appropriate joins, and a select query to list instructors (first and last names) and the students they taught (first and last names). Only include instructors who were actually teaching. Please order your results by the instructors' last name.

12.45 From school, use a select and join to display all students (by first and last names) and the classes each student took (by classterm, coursesubject, coursenum, and classsection). Order your results by the student's last name. Only include students who were actually taking classes.

12.46 Using school, select departments (by name) and instructors in each department (first and last names). Order alphabetically by department name and then instructors' last name. Only include departments that have instructors and instructors who are part of departments.

12.47 Write a query to display students (last name and first name) and instructors (first name) who have the same last name, ordered by students' last name, then students' first name. Do not include duplicate results.

12.48 Write a query to display all the students (id, last name, and first name) who took math or computer science during the fall. Please order your results by studentid (lowest to highest). If a student took multiple math or CS courses, please include them multiple times.

12.49 In reference to the school database, which instructors (first and last names) were teaching in the spring semester? Your result should not include duplicates.

12.50 In reference to the `school` database, select all course titles for classes offered during the year, their class meeting times, and their terms. Keep the default ordering (by `coursetitle`). It is ok to include directed studies, but do not allow any NULL course titles or meeting times.

12.51 In reference to the `school` database, how many classes are offered by the economics department, not counting directed or independent studies? Please select all relevant queries.

12.52 In reference to the `school` database, what courses are offered by the econ department in the spring semester? Please list courses not classes, and do not count directed or independent studies.

12.53 Write a query to display all the students (id number, last name, and first name only) who were *registered* in a dance class during the spring. Please order your results by `studentid` (lowest to highest). If a student took multiple DANCE courses, please include them multiple times.

12.4 Outer Joins

An outer join allows us to control the join operation in the situation where records in one table do not have matching records in another table, beyond the inner join solution of throwing away any such cases. The basic ideas were discussed in Sect. 12.2, and we review the possibilities here. We begin by discussing the cases and showing the operation of the various forms of the outer join. We then consider some examples as a way of motivating why one might choose a particular join for a particular use case.

Let us assume that, in the binary operation of a join, the first table specified, which is positionally on the left side of the join operator, is called the *left table*, and the second table specified, which is positionally on the right side of the join operator, is called the *right table*.

There are three cases to consider in the situation where a record in one table has no match[1] in the other table:

1. Keep all information, so that all records from the left table, even those without a corresponding match in the right table, are included, along with all records in the right table, even those without a corresponding match in the left table. This is called a *full outer join* or just an *outer join*.
2. Keep all information from the left table, but restrict information from the right table. Here, we keep all records from the left table, even those without a

---

[1]A match is defined as before, with a *match condition* that specifies one or more columns from each table of the join where the values of a record using the column(s) in the left table are equal to the values of a record using the column(s) in the right table.

Table 12.22 ind0 table

Code	Year	Pop	Gdp
FRA	2007	64.02	2657.21
GBR	2007	61.32	3084.12
USA	2007	301.23	14451.90
FRA	2017	66.87	2586.29
GBR	2017	66.06	2637.87
USA	2017	325.15	19485.40

Table 12.23 countries0 table

Code	Country	Income
GBR	United Kingdom	High income
IND	India	Lower middle income
USA	United States	High income

corresponding match in the right table, but records in the right table that fail to match a record in the left are excluded from the result. This is called a *left outer join* or just a *left join*.

3. Keep all information from the right table, but restrict information from the left table. Here, we keep all records from the right table, even those without a corresponding match in the left table, but records in the left table that fail to match a record in the right are excluded from the result. This is called a *right outer join* or a *right join*.

Clearly, cases 2 and 3 are dual to each other, and we could transform a query from one to the other by swapping which is the left table and which is the right table.

12.4.1   Left and Right Joins

Let us start with a left join in a concrete example (Table 12.22). Assume that the ind0 table has records for FRA, GBR, and USA and for years 2007 and 2017:

SELECT * FROM ind0

and the countries0 table has records for GBR, IND, and USA, with code, country, and income (Table 12.23):

SELECT * FROM countries0

The union of *just the code field*, the field used for our matching condition, between the tables is FRA, GBR, IND, USA, and VNM, while the intersection is just GBR and USA. The inner join from Sect. 12.3 would include just the GBR and USA rows, with repetitions for each of the two years.

When we, instead, use a LEFT JOIN operation, we get Table 12.24:

Table 12.24 Left join of ind0 and countries0

Code	Year	Pop	Gdp	c_code	Country	Income
FRA	2007	64.02	2657.21	NA	NA	NA
GBR	2007	61.32	3084.12	GBR	United Kingdom	High income
USA	2007	301.23	14451.90	USA	United States	High income
FRA	2017	66.87	2586.29	NA	NA	NA
GBR	2017	66.06	2637.87	GBR	United Kingdom	High income
USA	2017	325.15	19485.40	USA	United States	High income

Table 12.25 Right join of ind0 and countries0

Code	Year	Pop	Gdp	c_code	Country	Income
GBR	2007	61.32	3084.12	GBR	United Kingdom	High income
GBR	2017	66.06	2637.87	GBR	United Kingdom	High income
NA	NA	NA	NA	IND	India	Lower middle income
USA	2007	301.23	14451.90	USA	United States	High income
USA	2017	325.15	19485.40	USA	United States	High income

```
SELECT I.*, C.code AS c_code, country, income
FROM ind0 AS I LEFT JOIN countries0 AS C
     ON (I.code = C.code)
```

We see that, while FRA was present in the left table, ind0, there was no match in the right table, countries0. Consequently, in the result, the record fills with NULL/missing values the corresponding columns (country and income) of the right table.

One way to think of this result is that we start with a corresponding inner join and then add records in the result for each non-matched record in the left table.

Similarly, a right join can be thought of as starting with an inner join and then adding records in the result for each non-matched record in the *right* table. In the following SQL, the only change was modifying the left join to a right join (yielding Table 12.25).

```
SELECT I.*, C.code AS c_code, country, income
FROM ind0 AS I RIGHT JOIN countries0 AS C
     ON (I.code = C.code)
```

Because of their equivalent power, some systems do not implement a RIGHT JOIN and only provide a LEFT JOIN.

Consider the following scenario as a use case for a left join over an inner join (Table 12.26). Let us again consider our school database. Using the departments table, we can find the set of divisions and the number of departments for each division:

```
SELECT division, COUNT(*) AS num_dept
FROM departments
GROUP BY division
```

Table 12.26 Department numbers in each division of the school database

Division	num_dept
Fine Arts	5
Humanities	6
Interdisciplinary	13
Natural Sciences	6
Social Sciences	6

Table 12.27   Inner join seeking all departments in the Natural Sciences division

departmentid	departmentname	chairname
CHEM	Chemistry and Biochemistry	Burton, Judith
MATH	Mathematics and Computer Science	Bradley, Betty
PHYS	Physics	Lee, Jasmine
PSYC	Psychology	Nair, Hemant

Note that the Natural Sciences division consists of six departments.

Suppose our goal is to list all the departments in a particular division, including the name of the department chair. We recognize the problem as needing a join between the departments table and the instructors table with a match condition of the departments.departmentchair equaling the instructors.instructorid. A query to generate this table using an inner join might look like the following (yielding Table 12.27):

```
SELECT D.departmentid, D.departmentname,
    I.instructorlast || ', ' || I.instructorfirst AS chairname
FROM departments AS D INNER JOIN instructors AS I
  ON (D.departmentchair = I.instructorid)
WHERE division = 'Natural Sciences'
```

The (possibly unexpected) result has fewer departments listed than the full set of departments in the sciences division. The problem is that, in the database, there can be valid reasons why a departmentchair value is missing, for instance, in a transition between chairs. But in our goal, we want to list *all* departments, not just ones with a current chair. Since we want all the records in the left table and then the information, where possible, from the right table, the left join is the correct operation to select here. The updated query is given below (yielding Table 12.28):

```
SELECT D.departmentid, D.departmentname,
       I.instructorlast || ', ' || I.instructorfirst
AS chairname
FROM departments AS D LEFT JOIN instructors AS I
     ON (D.departmentchair = I.instructorid)
WHERE division = 'Natural Sciences'
```

Another example occurs if our goal is to get the name of the instructors for a set of classes. If the classes, perhaps from the schedule of a particular student, are 21014,

Table 12.28 Left join getting all departments in the Natural Sciences division

departmentid	departmentname	chairname
BIOL	Biology	NA
CHEM	Chemistry and Biochemistry	Burton, Judith
GEOS	Geosciences	NA
MATH	Mathematics and Computer Science	Bradley, Betty
PHYS	Physics	Lee, Jasmine
PSYC	Psychology	Nair, Hemant

Table 12.29 Selecting from a list of class IDs, using inner join

classid	instructorid	instructorlast	instructorfirst
21014	9167	Foster	Helen
21088	9029	Hawkins	Grace
21444	9050	Price	Taylor

Table 12.30 Selecting from a list of class IDs using left join

classid	instructorid	instructorlast	instructorfirst
21014	9167	Foster	Helen
21088	9029	Hawkins	Grace
21132	NA	NA	NA
21444	9050	Price	Taylor

21088, 21132, and 21444, and we use an *inner join* to match the `instructorid` in the `instructor_class` table with the `instructors` table, we get the following (yielding Table 12.29):

```
SELECT classid, instructorid, instructorlast, instructorfirst
FROM instructor_class
    INNER JOIN instructors USING (instructorid)
WHERE classid IN (21014, 21088, 21132, 21444)
```

We see three records in the result with the appropriate instructor's last and first names. But our query sought information for *four* classes. Either one of the class identifiers specified in the IN clause did not exist in the table, or else the join, because of a missing value for `instructorid` in the `instructor_class` table failed to find a match in the `instructors` table, and the inner join excluded the record from the result.

This is the most common use case for a left join: a field used in a matching condition is (legitimately) NULL, and the record is dropped due to a missing match. If, instead, we use a left join (yielding Table 12.30):

```
SELECT classid, instructorid, instructorlast, instructorfirst
FROM instructor_class LEFT JOIN instructors USING (instructorid)
WHERE classid IN (21014, 21088, 21132, 21444)
```

we get our desired result and can see that an instructor has not yet been assigned to the 21132 course.

Table 12.31 Full outer join

Code	Year	Pop	Gdp	c_code	Country	Income
NA	NA	NA	NA	IND	India	Lower middle income
FRA	2007	64.02	2657.21	NA	NA	NA
FRA	2017	66.87	2586.29	NA	NA	NA
GBR	2007	61.32	3084.12	GBR	United Kingdom	High income
GBR	2017	66.06	2637.87	GBR	United Kingdom	High income
USA	2007	301.23	14451.90	USA	United States	High income
USA	2017	325.15	19485.40	USA	United States	High income

12.4.2   Full Outer Join

The SQL syntax for a full outer join parallels the syntax for the INNER JOIN, LEFT JOIN, and RIGHT JOIN by simply replacing with FULL OUTER JOIN. So, repeating our example using ind0 and countries0, we would have:

```
SELECT I.*, C.code AS c_code, country, income
FROM ind0 AS I FULL OUTER JOIN countries0 as C
     ON (I.code = C.code)
```

Many real systems do not provide a full outer join, but this can be achieved by taking the UNION of a left join and a right join. In the union, the rows obtained by the inner join part of a left join and a right join will be in common, and the union will include only one record for these in common. The other rows are the rows added based on non-matching (yielding Table 12.31).

```
SELECT I.*, C.code AS c_code, country, income
FROM ind0 AS I LEFT JOIN countries0 AS C
     ON (I.code = C.code)
UNION
SELECT I.*, C.code AS c_code, country, income
FROM ind0 AS I RIGHT JOIN countries0 AS C
     ON (I.code = C.code)
```

The takeaway is this: an outer join (left, right, or full) lets you see both the matches and the records that failed to match. With a single full outer join query, we can, for instance, count the instances of missing values using the match field from one table to get the number of non-matches in one direction and obtain the count of instances of missing values using the match field from the other table to get the number of non-matches in the other direction. In the departments/department chairs and instructors table, we could find out both what departments do not have a current chair, and which instructors are not a chair. We could also achieve this result with two queries, one a left outer join and the other a right outer join.

Consider another simple example: our students table contains information about all students registered at the college. If we were compiling a list of students

Table 12.32 All instructors who were teaching during the fall semester

instructorid	instructorfirst	instructorlast	coursesubject	coursenum	classterm
9204	Antoine	Arnaud	THTR	100	FALL
9114	Martha	Meyer	THTR	110	FALL
9122	Gary	Carlson	THTR	120	FALL
9267	Robert	Brooks	THTR	160	FALL
9122	Gary	Carlson	THTR	165	FALL
9267	Robert	Brooks	THTR	160	FALL
9114	Martha	Meyer	THTR	165	FALL
9204	Antoine	Arnaud	THTR	170	FALL
9204	Antoine	Arnaud	THTR	170	FALL
9242	Crystal	Boyd	THTR	230	FALL

based on some other criterion, and also wanted to include the students' major in the result, we would perform a join between students and subjects, using the studentmajor field of students to link with the subjectid of subjects. But many students may not declare their major until the end of the sophomore year. So their studentmajor field would be missing. If we employ an inner join, all those students who had not determined a major would be omitted from the result.

We can also use an outer join and then a WHERE clause that looks for a NULL that indicates that a match did not exist. This can be used to answer questions like "Find all the students who have not declared a major."

Another application of outer joins is to answer questions of the form "Find all instructors who were *not* teaching in the fall semester." We know that we can find all instructors who *were* teaching in the fall, by doing an INNER JOIN between the instructors and classes tables passing through the instructor_class table (yielding Table 12.32):

```
SELECT I.instructorid, I.instructorfirst, I.instructorlast,
    C.coursesubject, C.coursenum, C.classterm FROM
(instructors AS I
INNER JOIN instructor_class AS IC
ON I.instructorid = IC.instructorid)
INNER JOIN classes AS C
ON IC.classid = C.classid
WHERE classterm = 'FALL'
```

Simply changing classterm = 'FALL' to classterm <> 'FALL' has the result of giving us instructors who *were* teaching in the spring semester. What we actually want is all rows in instructors that are not present in the query above. This can be accomplished by a LEFT JOIN, starting from instructors and joining to the table resulting from the query above (yielding Table 12.33).

Table 12.33 Inner join used as an intermediate step to find all instructors who were not teaching during the fall semester

instructorid	instructorfirst	instructorlast	instructorid..4	coursesubject	coursenum
9000	Brandon	Santos	9000	ARTS	131
9000	Brandon	Santos	9000	ARTS	361
9000	Brandon	Santos	9000	ARTS	363
9000	Brandon	Santos	9000	ARTS	451
9001	Margaux	Gillet	9001	PHED	350
9002	Tom	Werner	9002	ENVS	102
9002	Tom	Werner	9002	ENVS	301
9003	Roger	Davidson	NA	NA	NA
9004	Sharon	Jimenez	9004	SPAN	361
9004	Sharon	Jimenez	9004	SPAN	451

```
SELECT I2.instructorid, I2.instructorfirst, I2.instructorlast,
    j.instructorid, j.coursesubject, j.coursenum
FROM
instructors as I2
LEFT JOIN
(SELECT I.instructorid, I.instructorfirst, I.instructorlast,
    C.coursesubject, C.coursenum,C.classterm FROM
(instructors AS I
INNER JOIN instructor_class AS IC
ON I.instructorid = IC.instructorid)
INNER JOIN classes AS C
ON IC.classid = C.classid
WHERE classterm = 'FALL') as j
ON I2.instructorid = j.instructorid
```

In this new table, having null values for the j fields means that the instructor did not appear in the table j of instructors who were teaching in the fall semester. We can use a WHERE clause to extract just these rows, and can also modify our SELECT statement to ignore the j fields (yielding Table 12.34):

```
SELECT I2.instructorid, I2.instructorfirst, I2.instructorlast
FROM
instructors as I2
LEFT JOIN
(SELECT I.instructorid, I.instructorfirst, I.instructorlast,
    C.coursesubject, C.coursenum,C.classterm FROM
(instructors AS I
INNER JOIN instructor_class AS IC
ON I.instructorid = IC.instructorid)
INNER JOIN classes AS C
ON IC.classid = C.classid
WHERE classterm = 'FALL') as j
ON I2.instructorid = j.instructorid
WHERE j.instructorid IS NULL
```

Table 12.34 All instructors
who were not teaching during
the fall semester

instructorid	instructorfirst	instructorlast
9003	Roger	Davidson
9011	Jennifer	Guzman
9033	Madison	Lawrence
9036	Sarah	Taylor
9041	Robert	Roberts
9045	Lori	Johnston
9050	Taylor	Price
9060	Gary	Gordon
9065	Rakesh	Mehta
9072	Rosie	Smith

This same pattern can be used whenever we wish to find all individuals in a *set difference*, i.e., individuals satisfying some condition (being in a table A) but not some other condition (being in a table B). In this generality, we seek individuals whose B-values are NULL in A LEFT JOIN B. To find individuals in B but not in A, we seek individuals whose A-values are NULL in B LEFT JOIN A. Many more examples are explored in the exercises.

12.4.3   Reading Questions

12.54 Rewrite the given left join as a semantically equivalent right join:

```
SELECT I.*, C.code AS c_code, country, income
FROM ind0 AS I LEFT JOIN countries0 AS C
     ON (I.code = C.code)
```

12.55 If the outer join is the same as the union of the left join with the right join, is the inner join the same as the intersection of the left join and the right join? Explain.

12.56 Give a real-world example of two tables and an application where you would want a full outer join.

12.57 Give a real-world example of two tables and an application where you would want a left join.

12.58 If the full outer join is made up of rows from an inner join, left join, and right join, then why is it that in forming the outer join, you do not need to UNION with the inner join, as well as the left and right joins?

12.59 Suppose you have reason to believe that there are typos in a shared column of two tables (e.g., if you believed the code in indicators had typos, but code in countries was correct). Explain how you could use the material from the section to identify the rows with typos.

12.60 Is it possible for a full outer join to result in entirely NULL rows? That is, rows where every field is NULL? Explain.

12.4.4   Exercises

12.61 In reference to the school database, please list all courses (subject and number) that were not taught as classes during the year. Please include directed studies.

12.62 Is there any difference between a left join and an inner join between the tables subjects and departments in the school database? Investigate and report your findings.

12.63 Recall that LEFT JOIN can be used to compute set differences. Find all English courses (subject, number, and title) that were not offered in either the fall or the spring semesters.

12.5   Partitioning and Grouping Information

When we combine the ability of combining tables through a JOIN with the partitioning and aggregation of a GROUP BY, we gain considerable power in our SQL queries. This section uses some examples to illustrate this point. The material here is not new. The capabilities of partitioning and aggregation were covered in Sects. 11.5 and 11.6, and the table combination operations were covered earlier in this chapter. Because the GROUP BY clause partitions and the aggregations following a SELECT perform their computations on the full table of rows and columns formed by the *table-spec* of a FROM, be it a simple table or a complex combination formed by multiple join operations, these operations perform the same way in either case. This is also enabled because the execution plan performs the FROM *before* these other steps.

For our first example, we consider the indicators and countries tables. It may help to recall the functional dependencies from these two tables. For indicators, we have

$$\text{code, year} \rightarrow \textit{indicator-variables}$$

and for countries, we have

$$\text{code} \rightarrow \text{country, region, income}$$

While region and income are dependent variables, with one per code in the countries table, once we perform a join with indicators, we essentially

Table 12.35 Average life expectancy in each region in the year 2017

Region	avg_life	num_country	num_life
North America	80.80	3	3
Europe and Central Asia	77.83	58	52
Latin America and Caribbean	75.10	42	36
Middle East and North Africa	74.74	21	21
East Asia and Pacific	74.63	37	31
South Asia	70.81	8	8
Sub-Saharan Africa	62.04	48	48
NA	NA	1	0

get every `code`, `year` pair augmented with these categorical variables. These are exactly the types of variables we desire to use for partitioning in a GROUP BY.

In this first example, we focus on `life`, the life expectancy indicator. Suppose we want to compare life expectancy in the year 2017 among the countries of the world organized by their region (Table 12.35). We create the left join of `indicators` and `countries` to achieve the augmentation discussed above, and the group by the `region` field available in the joined result. The aggregation computes the average for each partition, and, as well, counts the records in the partition and the number of non-missing values for the `life` field. This results in the following query:

```
SELECT region, ROUND(AVG(life),2) AS avg_life,
       COUNT(*) AS num_country, COUNT(life) AS num_life
FROM indicators LEFT JOIN countries
     USING (code)
WHERE year = 2017
GROUP BY region
ORDER BY avg_life DESC
```

In similar fashion, we can ask the same question across countries grouped by income categories (yielding Table 12.36):

```
SELECT income, ROUND(AVG(life),2) AS avg_life,
       COUNT(*) AS num_country, COUNT(life) AS num_life
FROM indicators LEFT JOIN countries
     USING (code)
WHERE year = 2017
GROUP BY income
ORDER BY avg_life DESC
```

Let us consider one more example from the `schools` database. In the `students` table, the field `studentmajor` links to the `subjects` table, and the `subjects` table links to the `departments` table. Suppose we want to compute the total number of majors per division (Table 12.37). This will require a GROUP BY on the division field, but only after we join tables from `students`, through

Table 12.36 Average life
expectancy by income
category grouping

Income	avg_life	num_country	num_life
High income	79.73	79	66
Upper middle income	73.69	60	55
Lower middle income	67.64	47	47
Low income	62.14	31	31
NA	NA	1	0

Table 12.37 Total number of
majors per division

Division	students_per_subject
Social Sciences	689
Natural Sciences	630
Humanities	435
Fine Arts	147
Interdisciplinary	121

subjects to departments. In this case, we can use an inner join since we do
not want to include the students for whom the major field is missing. With a three-
table join, we can then group by division and use the aggregation of COUNT(*) to
count the students in each partition.

The above described query is given in SQL by:

```
SELECT D.division, COUNT(*) AS students_per_subject
FROM (students ST INNER JOIN subjects SU
        ON ST.studentmajor = SU.subjectid)
     INNER JOIN departments AS D
             ON D.departmentid = SU.departmentid
GROUP BY division
ORDER BY students_per_subject DESC
```

12.5.1   Reading Questions

12.64 In the example query, what is the purpose of the ROUND command?

12.65 Would the example queries (joining indicators to countries) work
without restricting the year to be 2017? If so, what quantities would they be
computing?

12.66 How would you modify the example queries to avoid the NA row?

12.67 Please come up with another example of an interesting question that can be
answered with a JOIN and GROUP BY.

12.5.2 *Exercises*

12.68 Using the SQL database `school`, use appropriate joins, grouping, and a select query to list instructors (first and last names) and the *number of* students they taught during the year. Only include instructors who were actually teaching. Please name your new column `total_taught` and order your results from largest number to smallest.

12.69 Find the students (id only) who took more than 10 classes over the year. Include the number of classes they took as `count`.

12.70 In reference to the `school` database, please list, for every instructor, how many classes they were teaching. Zero should be an option (e.g., if the professor was on sabbatical). Include the first and last names of the instructor, and alias your new column as `num_classes`.

12.71 In reference to the `school` database, please list, for each course, the number of class sections being taught in the spring. Call your new column `numSections`.

12.72 For each class, find the number of students who took the class (call it `count`), along with the class id, course subject, and course number. Please order your results by `count`, from highest to lowest. Please only include students who were *registered* for the class (see the `status` column). Ignore the `classmeeting` column. Your resulting data frame will include classes that do not actually meet but are used to assign students AP credit.

12.73 For each class, find the number of students who took the class (call it `count`), along with the class id, course subject, and course number. Please order your results by `count`, from highest to lowest. Please only include students who were *registered* for the class (see the `status` column), and please do not include classes where the meeting time was `'None'`.

12.74 For each class meeting time, find the total number of students who took a class during that meeting time. Please do not include classes without a meeting time, call your new column `count`, and order your results by `count` (from most popular to least popular). For a challenge, graph this data in some way, to see which class meeting times are most popular.

12.75 For each course, find the number of sections offered over the whole year. Only include courses that were actually offered (so zero should not be an allowed result). Your fields should be `coursesubject`, `coursenum`, `numSections`, and you should keep the default ordering in the database (by subject and number).

12.76 In reference to the `school` database, please list, for each student (id only), the total number of credits they took during the year.

12.77 In reference to the `school` database, what is the average credit hour load of the students, including directed studies? Your answer should be 1×1.

12.78 In reference to the `school` database, what is the average credit hour load of the students, not including directed studies?

12.79 In reference to the `school` database, please list instructors (id, first name, and last name) and their teaching load in terms of total credit hours for the year. Alias your new column as `teaching_load`. Hint: `coursehours` in the `courses` table is how many credit hours a course is worth.

12.80 Find the departments (id only) that offer more than one subject. Keep the default ordering (alphabetical by `departmentid`).

12.81 For each course, find the total number of students who took that course (any section) over the whole year. Please order your results by `courseid` (lowest to highest).

12.6 Subqueries

A subquery is a fully formed SQL statement that is subordinate to another SQL query. We have seen a number of uses of subqueries in solving different types of query problems. The purpose of this section is to capture some of the most common forms of subquery.

The first thing to understand is, relative to an execution plan, when the subquery is performed. As one might expect, the subquery has precedence over the outer query and gets executed first. SQL also restricts the ability to perform a sort on subqueries, and so a subquery cannot include an ORDER BY clause. We can still, however, use an ORDER BY in the outer query.

Subquery with Singleton for Comparison in WHERE
The simplest pattern occurs when we want to filter, using WHERE, but the *filter-condition* depends on data within the database. One possibility in this space is when we want to perform a comparison with a singleton value. We first compose an SQL query that yields a one-row, one-column result. For instance, we could compute, in the `indicators` table, and for the year 2017, a value that is five times the average GDP (yielding Table 12.38):

```
SELECT AVG(gdp) * 5
FROM indicators
WHERE year=2017
```

As with all subqueries, we put the SQL within parentheses and then place it at the some point in an outer query where the comparison is made (yielding Table 12.39):

Table 12.38 Average GDP in 2017

AVG(gdp) * 5
1991.7035

Table 12.39 Countries with
above average GDP

Code	Gdp	Life
BRA	2053.59	75.7
CHN	12143.50	76.4
DEU	3693.20	81.0
FRA	2586.29	82.5
GBR	2637.87	81.2
IND	2652.55	68.8
JPN	4859.95	84.1
USA	19485.40	78.5

Table 12.40 Countries with
life expectancy over 83

Code
HKG
JPN
MAC
SMR
ESP
CHE
ITA

```
SELECT code, gdp, life FROM indicators
WHERE year=2017 AND
      gdp > (SELECT AVG(gdp) * 5
             FROM indicators WHERE year=2017)
```

The inner query could be as complex as necessary, as long as it results in a singleton table. This could occur through aggregation or by retrieving a value from a single column and a single row. Another SQL restriction is that the subquery must be on the right side of a comparison operator.

This type of subquery, with a singleton result used with a comparison operator, can also be used in the HAVING clause, for filtering a result after aggregation.

Subquery with Vector of Values for IN
We have also seen a similar use of a subquery to generate the inclusion set to be used for the IN condition in a WHERE. Development is best when we begin with writing and debugging the inner query, like the following where we find the countries which had a life expectancy over 83, at any year in the data set (yielding Table 12.40):

```
SELECT DISTINCT code
FROM indicators
WHERE life > 83
```

We can then embed this subquery in a query that calculates the average GDP for those countries in 2017 (yielding Table 12.41):

Table 12.41 Average GDP
for countries with life
expectancy over 83

longlife_gdp
1313.37714285714

Table 12.42 Number of
students in each major

studentmajor	nummajor
ARTH	17
ARTS	42
BCHM	44
BIOL	237
BLST	3
CHEM	39
CINE	31
CLAS	5

```
SELECT AVG(gdp) AS longlife_gdp
FROM indicators
WHERE year = 2017 AND
      code IN (SELECT DISTINCT code
               FROM indicators
               WHERE life > 83)
```

Subquery Providing Virtual Table in FROM

Since an SQL statement, including that of a subquery, results in a table, we can use
a subquery in various ways in the FROM clause. The subquery could be the only
operand of a FROM, or the subquery could generate a table to be used as part of a
join with another table, or even with another subquery.

Consider the following as a subquery (yielding Table 12.42):

```
SELECT studentmajor, COUNT(*) AS nummajor
FROM students
WHERE studentmajor IS NOT NULL
GROUP BY studentmajor
```

This could be placed within an outer query that projects the same fields and sorts
by order of the computed nummajor (yielding Table 12.43):

```
SELECT studentmajor, nummajor
FROM (SELECT studentmajor, COUNT(*) AS nummajor
      FROM students
      WHERE studentmajor IS NOT NULL
      GROUP BY studentmajor) AS majors
ORDER BY nummajor DESC
LIMIT 6
```

Table 12.43 Ordered results

studentmajor	nummajor
ECON	251
BIOL	237
PSYC	196
COMM	160
HIST	131
POSC	123

Table 12.44 Major numbers
with subjects and departments

studentmajor	nummajor	departmentid
ARTH	17	ART
ARTS	42	ART
BCHM	44	CHEM
BIOL	237	BIOL
BLST	3	BLST
CHEM	39	CHEM
CINE	31	CINE
CLAS	5	CLAS

One of the more common use cases is where we use a subquery to perform a summary, like we are doing with the `majors` subquery, and then to join with another table fill in fields from the second table corresponding to the partitions of the subquery. Here, we compute the summary of number of majors and join that with subjects to obtain the department id for each major (yielding Table 12.44):

```
SELECT studentmajor, nummajor, departmentid
FROM    (SELECT studentmajor,
         COUNT(*) AS nummajor
         FROM students
         WHERE studentmajor IS NOT NULL
         GROUP BY studentmajor) AS majors
     INNER JOIN
         subjects AS S
     ON S.subjectid = majors.studentmajor
LIMIT 8
```

12.6.1   Reading Questions

12.82 The reading points out that "the subquery must be on the right side of a comparison operator"—please give an example where this fails, and write down the error message you get.

12.83 The reading says you can use a subquery with a singleton result in a HAVING clause. Please give an example. You can modify the example from the text if you like.

12.84 Consider the example query that computes longlife_gdp. Do you know that all country codes selected by the subquery had life expectancy > 83 in 2017? Think carefully and explain your answer.

12.85 Consider the virtual table example for computing the number of majors. Why is the virtual table (the result of the subquery) aliased as majors? Do we ever use that alias (e.g., asking for majors.nummajor? What would happen if you did not use an alias?

12.86 In the last example, what is the point of joining with the subjects table? In other words, what new thing can we select after this join that we could not have selected without joining?

12.6.2   Exercises

12.87 For every country in indicators, find the year in which the population hits its minimum value. Alias your new column as min_year. You can have more than one row per country, if there is a tie. Hint: in a previous exercise, you learned how to find the min_pop, so now you just need all rows where pop = min_pop. But be careful that you do not match the min_pop of one country with the pop of another.

12.88 For every country in indicators, find the *first* year in which the population hits its minimum value. Alias your new column as min_year. Include a column for country code.

12.89 In reference to the school database, what is the average credit hour teaching load of faculty over the whole year? Your answer should be a 1×1 result. Hint: use your answer to the previous problem.

12.90 In reference to the school database, what is the average credit hour teaching load of faculty who are teaching in the fall semester?

12.91 In reference to the school database, how many departments have more than one subject? Your resulting data frame should consist of just a single cell.

12.92 In reference to the school database, which instructors were not teaching during the fall semester?

12.93 In reference to the school database, which instructors were not teaching at all during the year?

12.94 Which country codes appear in indicators but not in countries?

12.95 In reference to the `school` database, use SQL to find the department name with the most subjects, along with a count of the number of subjects. Your resulting data frame should have only one row.

12.96 In reference to the `school` database, use SQL to find the course subject and number with the most sections, along with a count of the number of sections. Your resulting data frame should have only one row. Please name the new column `numSections`. Hint: first figure out how to get the number of sections for each course, then figure out how to get the max number from the result, and then use appropriate joins to get the final answer. Unlike the example in the book, the table you are joining with is not one of the given tables in the database, but you do have a query that finds the table you need to join with.

12.97 For each class, find out how many different majors are represented among the students who took the class. Do not count `NULL` as a major. Please order your results by `classid`.

Chapter 13
Relational Model: Database Programming

Chapter Goals

Upon completion of this chapter, you should understand the following:

- Client–server architecture predominant in applications interacting with relational database systems.
- Request/response nature of issuing queries for data in this architecture.

Upon completion of this chapter, you should be able to do the following:

- Construct a connection string for accessing either a remote or a local database, and successfully establishing a connection.
- Build an SQL query to be issued over a connection, including:
  - String-only based query,
  - Query with bound variables.

- Issue a query and receive a result object.
- Use a result object and process it to obtain full results, including:
  - Fetching an entire table,
  - Fetch a result a line at a time,
  - Fetch a result as a set of chunks.

- Incorporate the results into pandas data frames.

© Springer Nature Switzerland AG 2020
T. Bressoud, D. White, *Introduction to Data Systems*,
https://doi.org/10.1007/978-3-030-54371-6_13

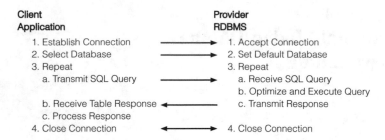

Fig. 13.1 Client/provider database interaction

Now that we have an understanding of SQL to build queries to request data, we must understand how to use our knowledge within a client application written in a programming language. We learned in Sect. 10.3 that the pattern of interaction between a client and a relational database provider proceeds as shown in Fig. 13.1.

To understand the programming associated with a client interacting with a database server, we must learn:

1. How to establish a connection to the provider and, if the provider supports multiple databases (schema), how to select a specific database.
2. How, in the programming language, to build an SQL query. These queries will often have to incorporate the values of variables of the executing program.
3. Given a constructed query, how to transmit/issue the query and send it to the provider.
4. How to retrieve the response from the provider.
5. Given the response, how to extract the table of data and incorporate it into data structures of the executing program.
6. How to close or terminate a connection with the provider.

From this list, 1 and 6 will be covered in Sect. 13.1, while 2 through 5 will be explored by starting with simple scenarios in Sect. 13.2, and proceeding to more involved techniques in Sects. 13.3 and 13.4.

In Fig. 13.2, we show some details of the software components involved. On the left is a representation of the client application. It is composed of the client-specific logic that interacts with a library that manages database interactions at a high level. This library, in turn, interacts with components called *drivers* that are specific to particular database management systems—interacting with an SQLite database requires different interactions than interacting with a MySQL database system than interacting with an SQL Server system.

A particular client might only use and interact through a single driver, but it is possible for a client to interact through multiple drivers, and also to have multiple connections through any given driver. The figure also shows, as a pipe icon, the connection from a specific driver to the provider. As discussed in Sect. 10.3, for many database systems, this connection may occur over a network, but for some, like SQLite, this connection is virtual and is realized through interaction

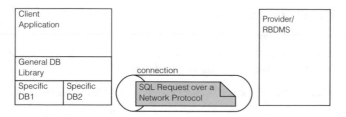

Fig. 13.2 Client software stack

with the local operating system. Over this connection, individual SQL queries are transmitted. Different database systems use different protocols, which is why it is necessary to use the correct driver software from the client.

Thus far, our descriptions apply to most client environments and programming languages. The interaction between a relational database client and a database management system proceeds in the same way, whether the client is written in R, in C++, or in Python. The client could be written using notebook-style computing in Python or R, with special cells for executing SQL with a connected database. But in any of these client-specific variations, the process is the same.

In the rest of this chapter, we will proceed using Python, and will be using the sqlalchemy package [3] for the generic database library. This particular package selection was made because:

- this library does an excellent job of unifying the interface to the client over many different database systems and their associated drivers,
- this library and the pandas library work well together, and
- this library has multiple higher levels of software providing higher levels of abstraction, like an object-oriented interface, which can provide benefits down the road.

13.1 Making Connections

The discussion here will be somewhat abstract, focusing on what needs to happen, but the specifics necessarily depend on both the client software environment and the particular database system. There must also be a step where the appropriate client libraries are incorporated, including sqlalchemy and a specific driver library, often called a *connector*.

13.1.1   The Connection String

For the general library, `sqlalchemy`, to make a connection to a database, it needs multiple pieces of information. It must determine:

- What protocol scheme and/or *kind* of database system is on the other end of the connection.
- Which lower-level database library (driver) will be required.
- If the database system is over a network, the network specifics for the machine and software process executing the provider-side software.
- If the database is local, like SQLite, the file system path of the database.
- If the database system supports multiple users in the same system, the credentials for the user associated with the currently executing client application connecting to this system.
- If the database system supports multiple databases, the specific database schema to select as the default database.

In `sqlalchemy`, as in many general libraries, this is accomplished with a single connection string, composed in the client program, and then passed to the library.[1]

Some of the information listed above could be sensitive, like the password of a user entity, and other information could change over time. For both sensitive information, and for information that changes, it is poor programming practice to include such information hard-coded in a client program. More strongly, passwords should *never* be included in the source code of a client program. Because of the frequent sharing of client program code, either directly or through maintenance in public-accessible repositories, such as by a `git` provider, this is a significant security risk.

For the purposes of the examples in this chapter, we employ a simple file, encoded as JSON, that can be read by our client programs and used to obtain the sensitive and changing information. Such an external file could be protected using operating system file system permissions, and avoids values hard-coded in our client programs. Refer back to Sect. 2.4 for information on basic reading and interpretation of JSON to populate Python variables with the values from the file.

For illustration, we will show the construction of connection strings for both a network example, using a MySQL provider on a non-local machine, and for a local example, using SQLite. For information on connection strings for other specific database systems, as well as other variations and parameters, the reader should refer to the database/engine connection documentation [62]. The general syntax for a connection string is:

$$protocol\text{-}scheme : / / resource\text{-}path$$

---

[1]Alternatives to a connection string for passing multiple pieces of information would include using a significant number of parameters for a connection function, or possibly using objects and inheritance for portions of the information.

In `sqlalchemy`, the connection string bears an intentional resemblance to a URL. It starts with a *protocol-scheme*, the character sequence `://`, then proceeds with a resource path that can include machine and user components. Unlike a protocol like `http` or `https` used in web requests, the sqlalchemy protocol specifies the kind of database along with the specific library, if needed. The kind of database is called a *dialect* and the specific library is called a *driver* and the protocol syntax is:

$$protocol\text{-}scheme \models dialect[+driver\,]$$

For an SQLite database, the protocol is just `sqlite`, but for a MySQL database using the `mysql-connector-python` library package, the protocol is `mysql+mysqlconnector`.

The *resource-path* portion of the connection string depends on whether the database is over the network, or is local. If over the network, the resource path has the syntax:

$$resource\text{-}path \models user:password@host/database$$

while for a local SQLite database, the resource path is given by:

$$resource\text{-}path \models /filesystem\text{-}path$$

Note the required forward slash before *filesystem-path*. This corresponds to the slash that follows the *host* in the network case; since there is no user and machine information, this portion is empty, and we proceed to the forward slash separator. If, in the local case, the path was given as a relative path, we see three slashes in a row. If the path was given as an absolute path, we would see four slashes in a row, two required to separate the protocol from the resource path, one to separate the empty user and machine information, and finally, the first slash that is part of an absolute path. The path should include any trailing extension that is part of the file name.

As mentioned above, information about a database location is changeable and may contain password information, so it is advisable to put such information into an external file. A JSON file with a dictionary mapping data sources to the information needed in an appropriate connection string is shown as follows:

```
{
    "mysql": {"protocol": "mysql+mysqlconnector",
              "host": "server.college.edu",
              "user": "user",
              "password": "pass",
              "database": "book"},
    "sqlite": {"protocol": "sqlite+pysqlite",
               "path": "datadir/book.db"}
}
```

Building a connection string is a matter of using Python string building and incorporating the information from the dictionary/ies. In the examples, we use the Python string format() [2] method with a pattern string to place the values into their corresponding places.

For our MySQL case:

```
credspath = os.path.join(datadir, "creds.json")
with open(credspath, 'r') as file:
    creds = json.load(file)

mysqlD = creds["mysql"]
pattern = "{}://{}:{}@{}/{}"
cstring = pattern.format(mysqlD["protocol"], mysqlD["user"],
                         mysqlD["password"], mysqlD["host"],
                         mysqlD["database"])
print(cstring)
```

  | "mysql+mysqlconnector://user:pass@server.college.edu/book"

For our SQLite case:

```
credspath = os.path.join(datadir, "creds.json")
with open(credspath, 'r') as file:
    creds = json.load(file)

sqliteD = creds["sqlite"]
pattern = "{}:///{}"
cstring = pattern.format(sqliteD["protocol"],
                         sqliteD["path"])
cstring
```

  | "sqlite+pysqlite:///datadir/book.db"

A typical client program would perform one or the other of these setup sequences, establishing the connection string in cstring.

13.1.2   Connecting and Closing

We import the sqlalchemy package and use an as clause to define a mnemonic (sa) to reference the functions and objects of the package. Online tutorials and references for the sqlalchemy package often use a

```
from sqlalchemy import xxx, yyy
```

---

[2]Recall that format() is a string method where the string object being operated upon has {} to indicate where values of variables are to be "filled in" by the arguments to format().

form of import to gain access to `sqlalchemy` functions and objects without having to use the mnemonic, but while we are learning, being explicit about what package a function belongs to is preferred.

A connection to a database through `sqlalchemy` proceeds in two steps. In the first, we create an `engine` object, using the connection string. This sets up all the internal data structures and gives the mechanism by which one or more connections to the same provider and database may be established. This means that, if desired, we can create a single engine and then establish multiple connections. If we require multiple connections to the same database, this allows the package to be more efficient in managing interactions with the database.

The second step uses the returned engine object and creates a *connection object* using the `connect()` method of the engine. No additional parameters are required.

With most of the work being accomplished by building the connection string, establishing the connection can be realized with

```
import sqlalchemy as sa

engine = sa.create_engine(cstring)
connection = engine.connect()
```

Just as with other external communication, such as with the file system, a connection *must* be closed when its use is complete, and the engine should be deleted. Because a connection to a provider is consuming local resources and is consuming remote resources, with the provider using memory and execution threads for each client interaction, we need to be especially careful to do our proper cleanup.[3]

```
try:
    connection.close()
except:
    pass
del engine
```

Similarly to opening and using files, we can use a `with` construct that, at the end of the `with`, will automatically perform a `close()`:

```
engine = sa.create_engine(cstring)
with engine.connect() as connection:
    # Perform database requests and process replies
    pass
del engine
```

---

[3] Some notebook-style computing systems maintain kernels that can keep a connection active, even if the browser window showing the notebook is closed. So these systems can "leak" resources and impact provider performance if we do not close our connections.

13.1.3   Reading Questions

13.1 To better understand "the client application," download a program such as `MySQLWorkbench` or `TablePlus`. Use it to establish a connection to a database, and then describe how it displays its database library and connection to the provider.

13.2 The drivers mentioned in the reading are special cases of the more general term *device drivers*. Look up this term and then explain the use of the word "driver." This will help you remember the term.

13.3 The combination of Jupyter notebooks and `sqlalchemy` and the ability to perform queries through cells in the notebook is known as "magic." Looking up the word "alchemy" if necessary, explain the connection to magic.

13.4 Regarding connections to SQLite, follow the discussion in the reading to explain why, if the path was given as a relative path, we would see three slashes in a row.

13.5 The reading explains a generalized MySQL connection string, and a generalized SQLite construction string:

```
"mysql+mysqlconnector://user:pass@server.college.edu/book"
"sqlite+pysqlite:///datadir/book.db"
```

1. Make up values, for your own network in the MySQL case, and for your own computer in the SQLite case, that would be appropriate for:

   - your user based on a login name
   - a made-up password based on your first pet
   - an actual machine on your network
   - a relative directory from your home folder where your course work is kept

2. Given this set of values, write down MySQL and SQLite connection strings that are specific.

13.6 Why is it that MySQL connection strings need a username and password but SQLite strings do not?

13.7 In the discussion of importing `sqlalchemy`, the reading describes the pattern of importing the package, then creating an engine object, then using an engine-object method (`connect`). Give an example of another time you have worked with an analogous pattern in Python.

13.8 The section ends with a discussion of closing a connection, e.g., using `connection.close()`. Is there also a command `connection.open()`? Look into the online documentation.

13.1.4   Exercises

These exercises require access to actual databases. On the book website (https://datasystems.denison.edu), you can find SQLite versions of the book and the school databases, in files named book.db and school.db, respectively. If this is the case, you should download these files and place them in an appropriate *data directory*, whose path should be accessible from your executing Python programs written for these exercises.

   In other learning environments, you may have access to a remote database that has been set up for you, in which case, you should have a username, a password, and a host machine specification, that you will need to access the remote database.

13.9 Create and edit a text file, named creds.json file, that follows the pattern in the book, which has a dictionary, and in that dictionary, there are entries for each of the remote or local database connections. Populate this file with the information specific to your environment. If you will only have a single database connection, your creds.json, at the outer level, will only have one element. Your file should be in a "known" location so that you can use it subsequently, like in a data directory for your work, or in the directory/folder where you will be executing your Python exercise code.

13.10 Write a function

```
mysql_connection_str(datadir, credsfile="creds.json")
```

that creates a connection string valid for a MySQL connection. Your function should use the credentials file named credsfile in a data directory named datadir, and, by reading the JSON and accessing the mysql dictionary therein, obtains the set of information needed to create a valid connection string, returning this string.

13.11 Write a function

```
sqlite_connection_str(datadir, credsfile="creds.json")
```

that creates a connection string valid for an SQLite connection. Your function should use the credentials file named credsfile in a data directory named datadir, and, by reading the JSON and accessing the sqlite dictionary therein, obtains the set of information needed to create a valid connection string, returning this string.

13.12 Write a function db_shutdown(engine, connection) that closes a given connection and deletes the engine.

13.13 Write code to create an engine named engine that is based on a connection string returned from either sqlite_connection_str or mysql_connection_str and, from that engine, create a connection object named conn. The database should be to the book database for correct testing. You can test your connection by executing the following:

```
print(conn.execute("SELECT * FROM ind0").fetchall())
```

13.2   Executing Queries and Basic Retrieval of Results

In the following, we will assume that the connection has been created and the connection object is referred to by the Python variable, `connection` and that the `indicators0`, `indicators`, `countries`, and `topnames` tables are available in the database specified by the connection string.

A connection object has a method called `execute()` that transmits an SQL query request. The SQL query itself is composed as a Python string, and the simplest form of `execute()` has that string as its sole parameter. Upon successful completion, the `execute()` method returns an object known as a *result proxy*, which can then be used in subsequent processing to obtain the rows and columns of the result, as well as additional information about the result. On an error, `sqlachemy` will generate a Python exception. It is not an error to get zero records as the result of a valid SQL query.

This use of an *object* returned by a request, that is then used to obtain both the data and information about the result, is similar to what we saw when studying regular expression programming in Chap. 4 with the *match object* return result. We will also see this same pattern in Part III of this book when working with general network requests.

13.2.1   Basic Query and Fetching Results

Making a request always requires the two step process of building the query into a Python string, and then calling the `execute()` method on the connection with the built string.

Any valid SQL query, as covered in the previous two chapters, could make up the value of the Python query string. In the simplest case, we could project all columns and select all rows from a table:

```
query = "SELECT * FROM indicators0"
result_proxy = connection.execute(query)
```

The variable `result_proxy` will be used to obtain the records themselves in one or more subsequent operations. If there was a syntax or other error in our SQL, an exception would be raised, with information about the error. This information should be examined carefully to help determine the cause of the error.

13.2.1.1   Result Data

The result proxy object provides numerous ways of obtaining the table resulting from an executed query. Because the architecture is, in general, a client/provider system over a network, the motivation behind this variety is to enable reduction of load on the provider and managing the quantity of data transmitted over the network at any given point in time. This becomes especially important when the resultant data is large.

At completion of the `execute()` method, the provider has completed its work in parsing and optimizing the query and executing the query at the provider machine, but the provider holds the data on the remote side until it is *fetch'd* by the client.

If the size of the resultant table of data is manageable, the client can request the entirety of the data from the provider through the `fetchall()` method call of the result proxy:

```
result_list = result_proxy.fetchall()
print("Length of result list:", len(result_list))
```

```
| Length of result list: 5
```

The result is a *list* of the records of the resultant table. If we want the number of records resulting from a query, we just take the `len()` of the result list. Each list element is a *tuple* consisting of the field values, in schema order, for the record. We can iterate over the list and print the tuple records:

```
for record in result_list:
    print(record)
```

```
| ('CHN', 1386.4, 12143.5, 76.4, 1469.88)
| ('FRA', 66.87, 2586.29, 82.5, 69.02)
| ('GBR', 66.06, 2637.87, 81.2, 79.1)
| ('IND', 1338.66, 2652.55, 68.8, 1168.9)
| ('USA', 325.15, 19485.4, 78.5, 391.6)
```

The tuples resulting from a query have special additional functionality. Say that one of the rows is referred to by the variable `row`. Here we set `row` to the tuple associated with the `'IND'` record:

```
row = result_list[3]
```

Like any other tuple, we can use an accessor with an integer position to get one of the values in the tuple, like `row[2]` to get the GDP value. But these "enhanced" tuples can also use the string name of the field in an accessor to specify which element is desired:

```
print('By position:', row[2], '\nBy field:', row['gdp'])
```

```
| By position: 2652.55
| By field: 2652.55
```

While we can now work with the data in its native Python data structure form, the next subsection takes it to the next logical step for our client applications.

13.2.1.2   Native Data Structure to pandas

For working with the table resulting from a query, we would like to build a pandas
DataFrame object, as this enables the power of the tabular operations. Thus far, we
have the data as a native Python data structure, as a list of row tuples, structurally
similar to our list of row lists from Chap. 3.

We begin by simply using the DataFrame constructor passing the result:

```
ind0 = pd.DataFrame(result_list)
ind0
```

	0	1	2	3	4
0	CHN	1386.40	12143.50	76.4	1469.88
1	FRA	66.87	2586.29	82.5	69.02
2	GBR	66.06	2637.87	81.2	79.10
3	IND	1338.66	2652.55	68.8	1168.90
4	USA	325.15	19485.40	78.5	391.60

The data in the table is as desired, but we want to provide pandas with the
column/field names. The keys() method of the result proxy retrieves these as a
list of strings, which we show below, followed by the construction of the data frame
passing both the data and the column names:

```
fields = result_proxy.keys()
ind0 = pd.DataFrame(result_list, columns=fields)
ind0
```

	code	pop	gdp	life	cell
0	CHN	1386.40	12143.50	76.4	1469.88
1	FRA	66.87	2586.29	82.5	69.02
2	GBR	66.06	2637.87	81.2	79.10
3	IND	1338.66	2652.55	68.8	1168.90
4	USA	325.15	19485.40	78.5	391.60

We can, if desired, set thecode field as a pandas index of row names. This is
accomplished by setting the index while dropping the integer index generated by
default at DataFrame creation:

```
ind0.set_index('code', drop=True, inplace=True)
ind0
```

	pop	gdp	life	cell
code				
CHN	1386.40	12143.50	76.4	1469.88
FRA	66.87	2586.29	82.5	69.02
GBR	66.06	2637.87	81.2	79.10
IND	1338.66	2652.55	68.8	1168.90
USA	325.15	19485.40	78.5	391.60

In summary, this section has shown a four step process for acquiring data from an SQL database into our client program as a data frame:

1. Compose the SQL query as a Python string.
2. Execute the query.
3. Fetch all records of the result.
4. Construct a `pandas` DataFrame.

13.2.1.3 Database Requests Directly Through `pandas`

The `pandas` module understands and can use `sqlalchemy` connection objects and provides two useful functions for streamlining the above process. There are two cases:

1. If we desire an *entire* table, we can use the `read_sql_table()` function and the result is a `pandas` DataFrame, compressing all four steps into one.
2. If we are not requesting an entire table, and need to compose our own, arbitrarily complex, SQL query, we can use the `read_sql_query()` function. Here we compose the query and then invoke the function and the result is a `pandas` data frame.

Here, we show an example of using the `read_sql_table()` function. The desired table is given, as a string, in the first argument, and the active `sqlalchemy` connection is given as the `con=` named parameter:

```
ind0 = pd.read_sql_table("indicators0", con=connection)
ind0
```

```
|     code      pop       gdp  life      cell
| 0   CHN   1386.40  12143.50  76.4   1469.88
| 1   FRA     66.87   2586.29  82.5     69.02
| 2   GBR     66.06   2637.87  81.2     79.10
| 3   IND   1338.66   2652.55  68.8   1168.90
| 4   USA    325.15  19485.40  78.5    391.60
```

When we need an SQL result that is not a full table, we construct the query as a Python string, and then pass that query as the first argument to the `read_sql_query()` function, along with the active `sqlalchemy` connection:

```
query = """
SELECT code, pop, gdp
FROM indicators0
WHERE life > 78
"""

ind1 = pd.read_sql_query(query, con=connection)
ind1
```

```
|     code       pop          gdp
```

```
|  0   FRA    66.87    2586.29
|  1   GBR    66.06    2637.87
|  2   USA   325.15   19485.40
```

Since queries themselves are Python strings, we can define a query string using single quotes, double quotes, triple-single quotes, and triple-double quotes. The least desirable of these would be single quotes, since SQL itself uses single quotes for string literals that are part of the SQL language. Triple-double quotes have the advantage of being able to compose the SQL query over multiple lines for readability, since, for SQL, the whitespace and line breaks are ignored in the parsing of a valid query.

The example below shows the use of a triple-double quoted string with single quotes within to follow the SQL language specification for its string literals.

```
query = """
SELECT DISTINCT sex, name
FROM topnames
WHERE name LIKE 'M%'
"""

M_names = pd.read_sql_query(query, con=connection)
M_names
```

```
|          sex        name
|  0   Female        Mary
|  1     Male     Michael
```

13.2.2   Reading Questions

13.14 When using execute(), how can you tell when a bug is in your Python code vs. in your SQL query? Do some testing and record what you learned.

13.15 Using the online documentation if necessary, describe precisely what you get as a result of an execute() command. For example, the reading shows that if result_proxy = connection.execute(query), then you can do result_proxy.fetchall() and result_proxy.keys(). This question is asking what else you can do with result_proxy.

13.16 With respect to information extracted from result_proxy, is len(result_list) the same as len(result_proxy.keys())? Explain.

13.17 What type of object does read_sql_query return? How can you tell?

13.2.3   Exercises

For the following exercises, you will either be using the book database or the nycflights13 database. Based on your learning environment, for a given exercise you will need to be able to:

- create a connection string, call it cstring
- create an engine object, call it engine
- create a connection, call it conn

It will probably be helpful to incorporate code and techniques from the last section.

13.18 Establish a connection to the book database. Then, write the code to issue an SQL query for all the rows and all the columns in the countries table. Retrieve *all* the rows from the result and use that data structure to print the land area of Zimbabwe (ZWE, the last record in countries). Make sure the database connection is closed upon completion.

13.19 Repeat the previous exercise using the pandas read_sql_table function, along with any appropriate access operations on the data frame. Use the documentation to see how to specify a column of the table as the row index.

13.20 Suppose we want to find the country with the largest area. While we could retrieve all the rows of countries and perform the work of looking through rows finding the maximum and finding the country, we can also use the power of SQL and retrieve the countries sorted by land area, and then we know the first entry will be the desired one. Compose such a query, named query, and use read_sql_query() and assign to the variable country the code of the country with the largest land area. Include all needed supporting code.

For the next set of questions, you will need a connection string (cstring) an engine (engine) for the nycflights13 database. Perform both Python assignment statements prior to answering the following sequence of questions.

13.21 Use sqlalchemy (without pandas) to open a connection and select all from the airlines table inside nycflights13. Call the proxy result of this query result. Because we will use this result proxy in future questions, **do not** close the database connection.

13.22 Using the results proxy, assign to columns the names of the columns in the resultant table.

13.23 Using the results proxy, retrieve all the rows and assign to table.

13.24 Use the table object and the list of columns to create a pandas data frame df with the results.

13.25 Use `read_sql_query()` to create a `pandas` data frame `df` from the `planes` table of `nycflights13`, by selecting all planes with more than 200 seats. Retrieve columns of the tail number, the year, and the manufacturer

13.3   More Advanced Techniques

As the number of rows retrieved in a result from a relational database system grows larger, we need greater control over how, after a query has been submitted, that we retrieve the data itself. This control is one of the reasons that, on a query, the object returned is a *result proxy*. The result proxy is a relatively small object that can be transmitted from the provider to the client application, and then, using the proxy, we can fetch rows of the result. In this way, the application can choose to retrieve subsets of the data, down to processing the result a record at a time.

The next few subsections explore some of these choices, and then we finish by thinking about how to write applications that need to work with multiple databases within the single application.

To illustrate the different techniques for obtaining record results based on a result proxy, we will use the following query:

```
query="""
SELECT departmentid AS DeptID, departmentname AS Name,
        division AS Division
FROM departments
WHERE division = 'Fine Arts'
"""
```

While this query only yields five records in the result, it can still be illustrative of the different retrieval patterns.

13.3.1   Record at a Time

In the limit, and given a result proxy, we may wish to process the results a single record at a time.

13.3.1.1   Result Proxy as an Iterator

One alternative for getting the data a record at a time is through direct iteration with the result proxy. A result proxy can act as an *iterator*, and so we can use it in a `for` loop and obtain each of the records in the result, one at a time:

```
result_proxy = connection.execute(query)

for record in result_proxy:
    print(record)
```

```
| ('ART', 'Art History and Visual Culture', 'Fine Arts')
| ('CINE', 'Cinema', 'Fine Arts')
| ('DANC', 'Dance', 'Fine Arts')
| ('MUS', 'Music', 'Fine Arts')
| ('THTR', 'Theatre', 'Fine Arts')
```

At each iteration of the loop, a subordinate acquisition of a single record occurs, sent from the provider to the client application.

From the printing in the example above, we can see that the Python variable named record, yielded by the proxy on each iteration of the for loop, is a *tuple*. This is a tuple that can use the name of a column with the access operator to retrieve individual values from the row. We repeat the prior example using this technique to print, for each record, the DeptID and the Division. Note how, in processing the results, we are using the aliases defined for the projected columns.

```
for record in result_proxy:
    print(record['DeptID'], record['Division'])
```

```
| ART Fine Arts
| CINE Fine Arts
| DANC Fine Arts
| MUS Fine Arts
| THTR Fine Arts
```

13.3.1.2   Fetch One

A result proxy also has a method named fetchone() that, in a manner similar to readline() for a file, obtains the *next* record, keeping track of a current position in the set of result records. This can be useful when we want to construct an indefinite loop using a while and invoking fetchone() both initially and at the end of the loop processing and immediately prior to the next iteration. Notice that the result, row, evaluates to False when there are no more results to be processed, and allows termination of the while loop.

```
row = result_proxy.fetchone()
while (row):

    print(row)
    # Process a single row

    row = result_proxy.fetchone()
```

```
| ('ART', 'Art History and Visual Culture', 'Fine Arts')
| ('CINE', 'Cinema', 'Fine Arts')
| ('DANC', 'Dance', 'Fine Arts')
| ('MUS', 'Music', 'Fine Arts')
| ('THTR', 'Theatre', 'Fine Arts')
```

13.3.2   Chunks

Processing records one at a time involves significantly more network processing. On each `fetchone()` or next item of the iterator, there are at least two network messages, one sending a "next record" request from the application to the provider, and then a reply message with the data of the record sent from the provider back to the application.

We may want a compromise between fetching all the records, which for some tables may be too large as a single result, and fetching one at a time, which add network latency and delay for each record. Logically, we want the control to specify a *chunk*, which is an application-specified number of records to retrieve at a time.

13.3.2.1   Fetch Many

The result proxy has a `fetchmany()` method that provides the capability of specifying the maximum number of records to retrieve in one operation. The method takes a single argument of an integer number of records to attempt to fetch.

To use a chunk size, we need our processing to operate at an outer level—where at each iteration we obtain a chunk-sized collection of result records, and an inner level—where at each iteration, we process the records *within* the chunk.

Consider the following example, where we process with a chunk size of, at most, two records per `fetch`. The outer loop is an indefinite loop that terminates when a `fetchmany()` returns a None result. This will happen when there are no more chunks to be retrieved. The inner loop uses the chunk returned by `fetchmany()` as an iterator that, on each iteration, yields a single record of the chunk result.

```
chunk_size = 2
rowset = result_proxy.fetchmany(chunk_size)
while (rowset):
    print("Next chunk @ length", len(rowset))
    for row in rowset:
        print(row)
        # Process a single row

    rowset = result_proxy.fetchmany(chunk_size)
```

```
| Next chunk @ length 2
```

```
| ('ART', 'Art History and Visual Culture', 'Fine Arts')
| ('CINE', 'Cinema', 'Fine Arts')
| Next chunk @ length 2
| ('DANC', 'Dance', 'Fine Arts')
| ('MUS', 'Music', 'Fine Arts')
| Next chunk @ length 1
| ('THTR', 'Theatre', 'Fine Arts')
```

We use print () to help understand the transitions from one chunk to the next. The row variable is an enhanced tuple object, similar to the earlier cases.

13.3.2.2 Using pandas with Chunk Size

When we are using the pandas package and its ability to interact with an SQL database, we also need the ability to control the number of records returned. To illustrate, we use the following query from our topnames data set that gives names that start with the letter 'J':

```
query = """
SELECT DISTINCT name, sex
FROM topnames
WHERE name LIKE 'J%'
"""
J_names = pd.read_sql_query(query, con=connection)
J_names
```

```
|          name      sex
| 0        John     Male
| 1       James     Male
| 2    Jennifer   Female
| 3     Jessica   Female
| 4       Jacob     Male
```

The pandas read_sql_query method has a named parameter, chunksize=, that we can use to control the number of results fetched at a time. The method returns an iterator. This iterator yields a complete data frame on each *next item* iteration of the for loop. So in the following example, we see that, in iterating over the valued returned from read_sql_query(), we get a DataFrame whose rows have the data of the chunk:

```
chunk_size = 3
iterator = pd.read_sql_query(query, con=connection,
                             chunksize=chunk_size)
for df in iterator:
    print(type(df))
    print(str(df))
```

```
| <class 'pandas.core.frame.DataFrame'>
|         name      sex
| 0       John     Male
| 1      James     Male
| 2  Jennifer   Female
| <class 'pandas.core.frame.DataFrame'>
|         name      sex
| 0   Jessica   Female
| 1     Jacob     Male
```

In our final example, we again use the chunk ability of the read_sql_query()
method, but this time, our goal is to create a single unified pandas DataFrame. We
can think of this as repeated instances of wanting to combine data frames with the
same columns along the row dimension. Our solution uses the same concat in an
accumulator pattern to build an aggregate result:

```
chunk_size = 3
iterator = pd.read_sql_query(query, con=connection,
                                    chunksize=chunk_size)

try:
    result_df = pd.DataFrame()
    while (True):
        df = next(iterator)
        print(type(df))
        print(str(df))
        print()
        result_df = pd.concat([result_df, df], axis=0,
                                    ignore_index=True)
except Exception as e:
    pass
```

```
| <class 'pandas.core.frame.DataFrame'>
|         name      sex
| 0       John     Male
| 1      James     Male
| 2  Jennifer   Female
|
| <class 'pandas.core.frame.DataFrame'>
|         name      sex
| 0   Jessica   Female
| 1     Jacob     Male
```

```
print("Aggregated result:\n{}".format(str(result_df)))
```

```
| Aggregated result:
|         name      sex
```

```
|  0         John      Male
|  1        James      Male
|  2     Jennifer    Female
|  3      Jessica    Female
|  4        Jacob      Male
```

13.3.3   Working with Multiple Databases

Sometimes the need arises for a client to work with multiple databases within a single application. There are few options:

- *Option 1*: Explicitly create two engine objects and two connection objects, using different Python variable names. Then each connection execute() or the connection passed to pandas would need to decide which connection to use for which query.
- *Option 2*: If a particular DBMS supports multiple database schema served by the same provider (MySQL, for instance), then we have some alternatives for selecting among multiple databases over the *same* connection. It is these options that will be described in the remainder of this subsection.

First, in composing the connection string, we must omit the last component (the database) of the resource path. Below we show the altered formatting of the connection string for a MySQL DBMS:

```
credspath = os.path.join(datadir, "creds.json")
with open(credspath, 'r') as file:
    creds = json.load(file)

mysqlD = creds["mysql"]
pattern = "{}://{}:{}@{}/"
cstring = pattern.format(mysqlD["protocol"], mysqlD["user"],
                         mysqlD["password"], mysqlD["host"])
cstring
```

```
|  "mysql+mysqlconnector://user:pass@server.college.edu/"
```

The engine creation and connection proceed as before:

```
engine = sa.create_engine(cstring)
connection = engine.connect()
```

Some databases (MySQL, Microsoft SQL Server) support an SQL statement, USE that takes the name of a database schema and changes the default database to the specified database schema name. The execute method is invoked with the appropriate USE statement prior to a sequence that needs to use a different database.

First, we change to the school database and then issue a query:

```
use_school = "USE school"
proxy0 = connection.execute(use_school)
query="""
SELECT departmentid AS DeptID, departmentname AS Name,
       division AS Division
FROM departments
WHERE division = 'Humanities'
"""
dept = pd.read_sql_query(query, con=connection)
dept
```

```
|    DeptID                 Name     Division
| 0    CLAS   Classical Studies   Humanities
| 1    ENGL             English   Humanities
| 2    HIST             History   Humanities
| 3    LANG     Modern Language   Humanities
| 4    PHIL          Philosophy   Humanities
| 5     REL            Religion   Humanities
```

Next, we change to the book database and issue a query to a joined table:

```
use_book = "USE book"
proxy0 = connection.execute(use_book)
query="""
SELECT I.code, country, region, pop, gdp
FROM indicators as I INNER JOIN countries USING (code)
WHERE year = 2017 AND code BETWEEN 'B' AND 'Bz'
      AND region LIKE 'S%'
"""
df = pd.read_sql_query(query, con=connection)
df
```

```
|    code       country                region      pop      gdp
| 0   BDI       Burundi   Sub-Saharan Africa    10.83     3.17
| 1   BEN         Benin   Sub-Saharan Africa    11.18     9.25
| 2   BFA  Burkina Faso   Sub-Saharan Africa    19.19    12.32
| 3   BGD    Bangladesh           South Asia   159.67   249.72
| 4   BTN        Bhutan           South Asia     0.75     2.53
| 5   BWA      Botswana   Sub-Saharan Africa     2.21    17.41
```

We can avoid the USE and explicitly specify the database schema along with the table.

```
query="""
SELECT departmentid AS DeptID, departmentname AS Name,
       division AS Division
```

```
FROM school.departments
WHERE division = 'Humanities'
"""
dept = pd.read_sql_query(query, con=connection)
dept
```

	DeptID	Name	Division
0	CLAS	Classical Studies	Humanities
1	ENGL	English	Humanities
2	HIST	History	Humanities
3	LANG	Modern Language	Humanities
4	PHIL	Philosophy	Humanities
5	REL	Religion	Humanities

```
query="""
SELECT I.code, country, region, pop, gdp
FROM book.indicators as I INNER JOIN book.countries
     USING (code)
WHERE year = 2017 AND code BETWEEN 'B' AND 'Bz' AND
      region LIKE 'S%'
"""
df = pd.read_sql_query(query, con=connection)
df
```

	code	country	region	pop	gdp
0	BDI	Burundi	Sub-Saharan Africa	10.83	3.17
1	BEN	Benin	Sub-Saharan Africa	11.18	9.25
2	BFA	Burkina Faso	Sub-Saharan Africa	19.19	12.32
3	BGD	Bangladesh	South Asia	159.67	249.72
4	BTN	Bhutan	South Asia	0.75	2.53
5	BWA	Botswana	Sub-Saharan Africa	2.21	17.41

13.3.4 Reading Questions

13.26 Why might you prefer to use `result_proxy` as an iterator, rather than iterating through `result_list = result_proxy.fetchall()`?

13.27 Consider the code

```
rowset = result_proxy.fetchmany(7)
```

What type of object is `rowset`? What is the significance of the 7?

13.28 Give a real-world application where you would use `fetchmany()`, and please be clear about the real-world meaning of `chunk_size` in your application.

13.29 When using `fetchmany()`, what happens if the `chunk_size` is larger than the number of records?

13.30 What type of object is returned by a function call like the following?
`pd.read_sql_query(query, con=connection, chunksize=chunk_size)`

13.31 What are the input and output types of the function `next()`?

13.32 List two pros and two cons of the `USE` keyword.

13.3.5 Exercises

13.33 Use `sqlalchemy` to open a connection and fetch just the first row from the `airlines` table inside `nycflights13`. Call the resulting object `row`.

13.34 The `flights` table in `nycflights13` is quite large with hundreds of thousands of rows. Suppose, to save memory, we only want to retrieve a single row at a time, and to "sample" the data by building a data set with one row out of each ten. Use an accumulation pattern and a `while` loop executing `fetchone()` and an integer counter to build a list of tuples with the sampled data. When finished, create a `pandas` data frame named `df` with the data and the appropriate column names.

13.35 Use `fetchmany()` to achieve the same result as the previous exercise, fetching results in groups of ten, and sampling and adding to the data set only the *first* row in each group.

13.36 Use the `pandas` ability to specify a chunk size and retrieve the flights data, sorted by ascending departure time, and retrieved in groups of 50. Create a list of the *average distance* in each group of 50, naming it `distances`.

13.4   Incorporating Variables

The programming of a client that acquires data through a relational database often requires variables in the executing program to be incorporated and to become part of the SQL queries requested of the database. For instance, a program working with the indicators data set may, at one part of the analysis, wish to request records with a life expectancy greater than 70 years, and at another point in the application, request records with a higher life expectancy of 80 years. Or in the `school` database, we may want a function that filters the students table based on last names in a range, for instance, retrieve all student records whose last name is between `'A'` and `'D'`. In such circumstances, a constant Python string containing the SQL query is insufficient to the task.

One way to look at this problem is as one of abstraction. We wish to generalize a specific (constant) query to abstract some part of it and be able to use a common sequence of code and generalize—dynamically creating and executing the query based on parameters. This perspective of abstraction drives us to define a *function* as the appropriate mechanism to generalize an SQL query using parameters. The examples in this section will all be developing a function to perform this work.

13.4.1 Python String Composition

Using mechanisms we already know, we can compose strings and incorporate the values of program variables using string methods, operations, and functions like format(), string +, and str() with our Python variables. Here, the goal is to substitute the values into a template of the SQL query to achieve a resulting string that is syntactically correct SQL, which we can then execute().

We start by writing the SQL query as a string, and use { } in the places of the query where the value of a variable should be inserted. This choice is in anticipation of using format() as a good tool for incorporating values into a string.

Suppose we often request data from the indicators table for a specific year and based on a life expectancy threshold. We wish the value of the year and the value of the threshold to come from variables. The generic form of the query, then, is given by:

```
template = """
SELECT code, pop, gdp, life
FROM indicators
WHERE year = {} AND life > {}
"""
```

With this in mind, and using a functional approach, we design a function, indicators_life() to perform the query. The function needs parameters for the year value, the life threshold value, and for the database connection. The result from the function should be a table, and for versatility, we choose to construct a pandas DataFrame result.

The algorithmic steps of the function are the same as we have seen before, with just a bit of additional complexity in the "Compose the SQL query" step.

```
def indicators_life(dbcon, year=2017, threshold=0):
    template = """SELECT I.code, country, pop, gdp, life
                FROM indicators AS I INNER JOIN
                    countries AS C USING (code)
                WHERE year = {} AND life > {}"""
    query = template.format(year, threshold)

    result_proxy = dbcon.execute(query)
    data = result_proxy.fetchall()
    df = pd.DataFrame(data, columns=result_proxy.keys())
    return df
```

Notice the use of parameters with default values, allowing this function to also be used with just the connection argument and obtain all records of the default year from the table.

Function invocation yields the desired table, issuing the query parameterized by the actual parameter values of our application.

```
indicators_life(connection, year=2000, threshold=80)
```

```
|    code                       country       pop        gdp   life
|  0  HKG   Hong Kong SAR, China        6.66    171.67   80.9
|  1  JPN                       Japan  126.84   4887.52   81.1
|  2  MAC       Macao SAR, China        0.43      6.72   80.4
```

Numeric values incorporated into SQL queries can generally be handled this way. But we must be aware of the data types of both the Python variables and the SQL data types of the fields used in the integrated SQL query. Suppose, for instance, that we are interested in incorporating Python string values into an SQL query. If those string values are used where we would have used SQL string literals in the non-generalized query, then the string values *must* be delimited, as required by SQL, by single quotes.

Consider the example of writing a function to obtain a subset of the students table based on a string range that uses the studentlast field. We want to select those records where the student last name is between an inclusive low value and an exclusive high value. We might write our SQL template like so:

```
template = """
SELECT studentid,
       studentlast || ', ' || studentfirst AS studentname,
       studentmajor
FROM students
WHERE studentlast >= {} AND studentlast < {}"""
```

Consider the result of invoking format(), using an example low and high as shown:

```
low = "A"
high = "Am"
query = template.format(low, high)
print(query)

|
| SELECT studentid,
|        studentlast || ', ' || studentfirst AS studentname,
|        studentmajor
| FROM students
| WHERE studentlast >= A AND studentlast < Am
```

Can you see the logical error? SQL would interpret the above query as asking if field studentlast is greater than or equal to *field* A, and if field studentlast is less than *field* Am. Without the single quotes to inform SQL that these are string literals, an unadorned sequence of characters is interpreted as a field. The syntax is perfectly valid, but query execution would raise an exception with an error message about missing field values for A and Am.

The solution is straightforward; we include the single quote marks within the template string so that the result gives an SQL query with the incorporated values as string literals. The correct template would be:

```
template = """
SELECT studentid,
       studentlast || ', ' || studentfirst AS studentname,
       studentmajor
FROM students
WHERE studentlast >= '{}' AND studentlast < '{}'
"""

low = "A"
high = "Am"
query = template.format(low, high)
print(query)
```

```
|
| SELECT studentid,
|        studentlast || ', ' || studentfirst AS studentname,
|        studentmajor
| FROM students
| WHERE studentlast >= 'A' AND studentlast < 'Am'
```

When we need to incorporate Python variables into an SQL query and we use the solution strategy of Python string methods, operators, and functions, we must take great care and understand the types of the Python variables and the corresponding data types required by the fields of the SQL.

13.4.2   Binding Variables

Incorporating the values of program variables into the ongoing interaction between a client and provider is a frequent occurrence, both for queries, as exemplified above, and particularly when we are *populating* data into a table. In the latter case, the row and columns data are all values originating in the executing program and need to be incorporated into the SQL from client to provider.

The libraries, both the generic library and the lower-level driver, provide mechanisms that work in concert with the protocols and software of the RDBMS to make incorporating variable values easier and more efficient. In general, the steps include:

1. Construct an object that has a combination of SQL syntax and elements designating the places where the value of a variable should be substituted. This is

analogous to our template string from Sect. 13.4.1. When this step is completed in an object-oriented language, the operation is performed on an existing *statement object*, and, in general, the step is called the *prepare* step.

2. When the variable values are available, which may be close in time to the prepare, or might be further removed in time, a *bind* operation is performed, that associates a set of values with their corresponding locations in the prepared SQL text. This step constructs a new object or modifies an existing object to obtain a *bound statement*. A bound statement object is more than the analogous result of string formatting, as it maintains the values and their associated data types separately from the prepared template. This enables type-correct use of the values based on their data type, as well as re-use of the object.

3. The bound statement is then conveyed as a request to the database. In our setting, this could be by using the `execute()` method, or we can also use `pandas` SQL facilities, which properly handle a bound statement as the query.

We will demonstrate each of these steps in turn:

13.4.2.1   Prepare

Within the `sqlalchemy` package, there is a general `sql` class. Within the `sql` class of `sqlalchemy` is a `text()` constructor that builds a prepared object, consistent with step 1 above. Using the same example from Sect. 13.4.1, we create a Python string using the syntax needed for preparing a statement:

```
import sqlalchemy as sa

pyquery = """SELECT I.code, country, pop, gdp, life
                FROM indicators AS I INNER JOIN
                    countries AS C USING (code)
                WHERE year = :yr AND life > :threshold"""
prepare_stmt = sa.sql.text(pyquery)
```

The `pyquery` variable references a Python string that uses the syntax for binding, namely that positions in the statement that require the binding of variables are designated with a colon, : immediately followed by a programmer selected name that should obey Python variable naming syntax. In this example, we have a binding location for the year, naming it `yr` and for the life expectancy threshold and name it `threshold`. This SQL-specific template is then prepared by calling `text()` and assigning to the `prepare_stmt` variable for later reference. One of the reasons for doing this is that the single `prepare_stmt` can be used multiple times in the future, once for each set of values to be bound to the same underlying statement.

13.4.2.2   Bind

The bind step invokes the `bindparams()` method of a prepared statement yielding a bound statement. The `bindparams()` method associates values by simply using *named parameters* as its arguments, where the variable name for the named parameters is exactly the same as the template statement. In this case, the named parameters are `yr` and `threshold`, and creation of the bound object is the single step:

```
bound_stmt = prepare_stmt.bindparams(yr = 2000, threshold = 80)
```

13.4.2.3   Execute

The execute step is performed, as before, but using the bound statement instead of a string for a query. We can use the `execute()` method and `fetchall()` to obtain the results, or we can use the `read_sql_query()` function of pandas to combine those steps and return a data frame:

```
df = pd.read_sql_query(bound_stmt, connection)
df
```

	code	country	pop	gdp	life
0	HKG	Hong Kong SAR, China	6.66	171.67	80.9
1	JPN	Japan	126.84	4887.52	81.1
2	MAC	Macao SAR, China	0.43	6.72	80.4

The `execute()` method, if given named parameters, will combine the bind and execute steps on our behalf, allowing us to skip an explicit bind:

```
result_proxy = connection.execute(prepare_stmt, yr=1999,
                                  threshold=70)
data = result_proxy.fetchall()
print('Number of records returned:', len(data))
```

```
Number of records returned: 96
```

Two of the advantages of using SQL variable binding over string-based composition of queries are reusability and the carrying of type information. Recall the problem in the previous section where we had to use our knowledge of the string type of the Python variable and/or the SQL string field in order to create a logically correct SQL query. With bound variables, and with the value and types packaged as part of the bound statement, the query is executed at the server with knowledge of the data types involved, and no single quotes are required because there are no longer string literals in the query.

Variable binding in our SQL queries can also give us a clean way to build functional abstractions that nicely generalize operations we wish to perform against a database.

```
def students_byname ( dbcon, name_min, name_max ):
    pyquery = """SELECT studentid, studentlast || ', '
                        || studentfirst AS studentname,
                   studentmajor
                 FROM students
                 WHERE studentlast >= :low AND
                       studentlast < :high
                 ORDER BY studentname"""
    prepare_stmt = sa.sql.text(pyquery)

    bound_stmt = prepare_stmt.bindparams(low=name_min,
                                         high=name_max)
    df = pd.read_sql_query(bound_stmt, con=dbcon)
    return df
```

Invocation of our new function can then be performed any time we want this functionality and can use different arguments to specify different name ranges.

```
students_byname(connection, "E", "Eh")
```

	studentid	studentname	studentmajor
0	62787	Edwards, Alisha	BIOL
1	63723	Edwards, Billy	None
2	63019	Edwards, Carolyn	BCHM
3	62222	Edwards, Eloise	ECON
4	61659	Edwards, Eva	FREN
5	62940	Edwards, Gary	MATH
6	62094	Edwards, Nancy	HIST
7	62265	Edwards, Stephanie	ARTS

We will employ variable binding in Chap. 14 when we are creating database tables and populating them with data.

13.4.3   Reading Questions

13.37 The key line of code in the indicators_life function is query = template.format(year, threshold), which fills in the { } spots with the given parameters. Show how to modify the function so that it selects records of countries with population greater than x and GDP per capita greater than y. Test your function with x=100 and y=30.

13.38 The reading explains how to use the pattern '{ }' for making queries regarding a string variable, e.g., low = "A". Is there any way to make queries with string variables using the pattern { } and modifying the string parameters (low and high in the example)? Explain.

13.39 In a "prepare" step, how does Python know when a colon signifies a position that requires binding, and when a colon is part of a variable or column name?

13.40 Consider the function call

```
result_proxy=connection.execute(prepare_stmt, yr=1999,
                                threshold=70)
```

How many parameters can the `execute` function take? How do you know?

13.41 Referring to the online documentation if necessary, determine if `read_sql_query` can also combine the bind and execute steps, like `execute()` can.

13.4.4   Exercises

The exercises in this section will employ techniques from the entire chapter, bringing together the programmatic execution of SQL using multiple databases and building functions with parameters for variables in the SQL, sometimes returning `pandas` data frames and sometimes returning native Python data structures.

The exercises will use both the `school` database along with a new database whose contents include a sampling of *emails* obtained during investigation of fraud of the Enron company. This data, now in the public domain, was part of the evidence of intentional actions by Enron employees based on email messages exchanged. The database is simple, with a single table, named `emails`. Each row represents exactly one email, with the following columns:

* `user`: the login id of the Enron employee
* `folder`: the folder in which the email was stored, with values `"inbox"` or `"sent_items"`
* `emailID"`: an integer id, unique per user and folder, for the email in this row
* `message`: text string containing the email itself. This can be long and includes all the "meta" information as *header lines* of To, From, Subject, Date, etc., along with the body of the message.

This database is called `enron` and its SQLite version is in the file `enron.db`. Before completing exercises here, ensure you have access to both databases and know the information to build connection strings to each.

13.42 Write code to construct connection strings and create engines for *both* databases, using separate connection strings for each. Then create connection objects with variable names `school` and `enron`, for the two databases.

13.43 Write a function

```
deparmentsByDivision(connection, division)
```

that queries the departments table for the set of departments in the given division. Use Python string composition and the format() method for building the query. Return all rows as a list of tuples.

13.44 Write a function

```
deparmentsByDivision(connection, division)
```

that queries the departments table for the set of departments in the given division. Use *variable binding* for building the query. Return all rows in a pandas data frame.

13.45 Write a function

```
subjectCourses(connection, subject, low, high)
```

that queries the courses table for the courses in the given subject whose course number is between low and high, inclusive. Use Python string composition and the format() method for building the query. Return a pandas data frame.

13.46 Write a function

```
subjectCourses(connection, subject, low, high, count)
```

that queries the courses table for the courses in the given subject whose course number is between low and high, inclusive. Use *variable binding* for building the query. In this case, only fetch count records, and return the list of tuples obtained.

Exercises now transition to the enron database

13.47 Write a function

```
getUsers(connection)
```

that returns a Python list containing the set of unique user names from the user field of the emails table. Be careful, as we do not want a list of tuples, but rather a simple list of string-based user names.

13.48 Write a function

```
userFolders(connection, user)
```

that uses a select query to select all distinct folders from the emails table, where the user is user. Return the resulting pandas data frame.

13.49 Write a function

```
userFolderIds(connectiuon, user, folder)
```

that uses a select query to select all email IDs from the emails table, where the user is user and the folder is folder. Return your answer as a Python list of integers (not tuples).

13.50 Write a function

```
selectMessage(connection, user,
              folder, emailId)
```

that uses a query to select the message from the `emails` table, where the user is `user`, the folder is `folder`, and the `emailid` is `emailId`. Return your answer as a Python string, and return None if no result with the given argument was found.

Chapter 14
Relational Model: Design, Constraints, and Creation

Chapter Goals

Upon completion of this chapter, you should understand the following:

- Characteristics that make a sound database design:
  - Tables that represent exactly one entity, without redundancy,
  - Fields that are properly defined and constrained to prevent error.

- Relationships between entities/tables, including:
  - Use of linking fields to allow for many-to-one relationships,
  - Use of intermediate linking tables for logical many-to-many relationships.

Upon completion of this chapter, you should be able to do the following:

- Given a set of interrelated data, design a database schema for representing that data.
- Given a database schema, construct SQL CREATE statements to define the tables and fields of the database.
- Given data for the records of one or more tables, use SQL INSERT statements to populate the tables with their data.

© Springer Nature Switzerland AG 2020   425
T. Bressoud, D. White, *Introduction to Data Systems*,
https://doi.org/10.1007/978-3-030-54371-6_14

The primary goal of this chapter is to understand the principles behind good database design and to give us the tools to be able to design, create, and populate our own databases. We begin in Sect. 14.1 by motivating the benefits of good database design and an overview of the process involved. The next three sections, Sect. 14.2, 14.3, and 14.4 discuss the process through the three primary areas of database design, the tables themselves, fields within tables, and the relationships between tables. The final sections cover the SQL and programming required to interact with a DBMS to create and populate database tables. Since our focus is on analytical databases, we omit coverage of SQL to perform operational updates and modifications to an existing database.

14.1 Motivation and Process

When a database is designed well and consistent with the principles of the relational model, we say that it is *sound*. A sound database design uses the relational model to its best advantage, eliminating redundancy and imposing constraints that help keep errors from occurring. A database is a long-lived entity, and inattention to the principles of sound database design can allow errors and confound the ability to retrieve the desired data, or the efficiency of such retrieval.

Understanding sound database design conveys the following benefits:

- When working with an existing database, it allows us to assess the design of a given database and can help us to know how to construct queries and/or make up for the deficiencies of a poor database design.
- We may need to design a database from scratch, or to participate in such a design.
- Database design allows us to understand the constraints of the relational model.
- Sound design means that the SQL will be simpler and will be able to be composed, and this gives greater power in our queries.
- Proper design can include the definition of index structures that can make common types of queries orders of magnitude more efficient.

Database design can also prove useful relative to our objectives in this book. Figure 14.1 depicts the data processing pipeline originally discussed in Chap. 1. In the data acquisition phase of the pipeline, we may acquire data from a myriad of sources: in tabular and hierarchical form from files, from API services, from web servers, and from databases. We then normalize the data, making it tidy and forming what generally are a set of in-memory tables, perhaps as `pandas` data frames. Once the data is in a normalized form, it makes sense to save this data in a local database repository to allow consistency and reuse for downstream operations like exploration, visualization, and analysis. A relational database, which can be shared and protected, is a logical choice for storage of our normalized data.

Separation of data acquisition and normalization from analysis through a relational database can also help our goal of reproducible and defensible research and results.

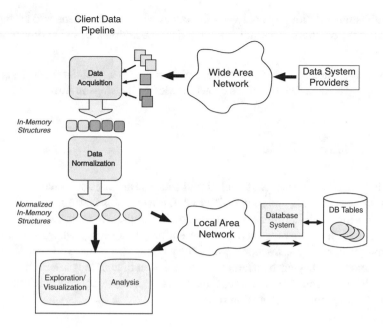

Fig. 14.1 Data pipeline

From the inception of the relational data model, there has built up a tremendous design foundation based on the theoretical and mathematical aspects of the model. Database textbooks rightly devote significant material to the properties and forms of normalization, and the modeling involved in design of a database. As stated at the onset, this book is not intended to replace a database systems course and textbook. Our approach, instead, is to build a basic understanding of what constitutes a sound database design, and to use an intuitive approach and examples to illustrate the most important concepts. With this information, we should be able to design a database that avoids some of the most common pitfalls.

The outcome of the database design process should be a set of interrelated tables, with appropriate fields per table and with explicit understanding of the relationships between tables. In other words, the outcome should include a database schema akin to the `school` database schema shown originally in Chap. 12.

The design process almost always starts with some existing data for a data set, and often with the data in existing row–column (though not necessarily tidy) form. From this starting point, design proceeds as a process of design and refinement, iterating through the process until the principles of a sound database design are satisfied. The three steps of design and refinement include:

1. Determining and designing tables for the database.
2. Refining the column fields of the defined tables, including appropriate constraints. This step can often derive additional tables for the database.

3. Establishing relationships between the tables based on the needs and anticipated uses of the database. This step can also derive additional tables and/or fields for the database.

These three steps are repeated to iteratively refine to a final database design and its schema. The next three sections focus on each of the steps in turn.

14.2   Designing Tables

At its highest level, the tables of a relational database are consistent with our view of the *entities* in a data set, and so we wish to define a table for each entity. This process starts by using our intuition as we think about the entities in a problem domain and then the fields/attributes for each of these entities. In a setting where we might be designing a database for a school, we start by noting that a student is different than an instructor, and they have different attributes, even though they are both people. A class is different from a course, because a course exists as a conceptual entity, while a class has a section and a meeting time and place.

14.2.1   Functional Dependencies

Ultimately, functional dependencies are fundamental in separating one table/entity from another table/entity as we design a database. Functional dependencies are about the *real* relationships between the data and variables for a database under design. In other words, the *data and variables* define the functional dependencies, and these are used to then create *tables*. It is a common trap to think that we define a *table* and that the table definition is what gives us the functional dependency.

Using functional dependencies to guide database table design implies that we understand the values and variables of the data set. The variables give us the columns/fields of tables, also known as *attributes* in database parlance.

When we understand our data and distinguish values from variables, then we can, for a data set, write down a set of functional dependencies. These are consistent with our understanding of tidy data in the tabular model. In particular, we write

$$independent\text{-}vars \rightarrow dependent\text{-}vars$$

An individual table should be defined for each unique combination of our independent variables. The table will then contain fields for each variable in the functional dependency, comprised of the union of the independent and dependent variables.

For instance, using the example from prior chapters, we know *from the data* in the indicators data set, that we have

$$\text{code} \rightarrow \text{country, land, region, income}$$

$$\text{code, year} \rightarrow \text{pop, gdp, imports, exports}$$

So our database design would have two tables, one for the first functional dependency and the other for the second. If we call the first table `countries`, we would define it with fields for `code`, `country`, `land`, `region`, and `income`. We could call the second table `indicators` and define it with fields for `code`, `year`, `pop`, `gdp`, `imports`, and `exports`. If our analysis of the data also yielded the functional dependency of

$$\text{code, year} \rightarrow \text{life, cell}$$

then, because this functional dependency has the same left-hand side, these are two functional dependencies that are defining the *same entity*. We could, equivalently, write

$$\text{code, year} \rightarrow \text{pop, gdp, imports, exports, life, cell}$$

So our database design should only have *one* table based on the independent variables of `code` and `year`. This notion of a table as a separate entity based on its independent variables is the same constraint that we called TidyData3 in the tabular model. And like in our tabular model case, the rows/records that will populate the table each represent a *unique* mapping (relation instance) from independent variables to dependent variables.

14.2.2   Table Design: Advice and Best Practices

A table is a collection of records corresponding to an entity. A good table name will use the name of the entity it represents and use the plural form to convey the "set of records" notion. For instance, in the `school` database, we name our primary entity tables `courses`, `students`, `departments`, etc. On the contrary, when we link two tables together with a separate intermediate table, it is good practice to name that table in a way that conveys its linkage between tables. Again, looking at the school database as an example, the table that links students with classes is named `student_class` and the table that links instructors with classes is named `instructor_class`. We use the singular version of the two entities to help convey the notion that the table consists of pairs of individual values. We will look at the design of tables that serve to link records between two entities in Sect. 14.4.

Databases, once designed and populated, become shared repositories. Therefore, we should consider the perspective of a non-expert who might use our database. This means that we should avoid abbreviations and acronyms that, while they might make sense to the designer, might give an external user trouble.

While we are considering tables and their primary attribute-based fields, we should take note of any fields where the name of the field is in a plural form, or where the field name is a mashup. These are indications of a field whose values would be multi-valued, which should not occur in a properly designed database. In a poor database design, a multi-valued field should be redesigned as a proper one-to-many relationship between two different entities. The further design of column fields are covered in Sect. 14.3.

14.2.3   Table Primary Key

We should determine a primary key at the time we define a table. A primary key can be a singleton or can be a composite key, consisting of multiple fields, but it *must* be the case that the primary key uniquely determines a record of the table. For a table, there might be multiple possibilities for the primary key. This is common if we have a synthetic identifier that we define for a table and also have one or more fields that can uniquely determine a record. In such a case, we can write multiple *equivalent* functional dependencies for the table, and each minimal combination of independent variables on the left-hand side of the functional dependencies gives us a possible key, called a *candidate key*.

When deciding on a primary key, it is important to think about future data and use of the database. For instance, given an existing data set, it might be that a particular field or set of fields is unique per row. We must ask if that uniqueness will hold in the future. Consider the combination of last name and first name for a student's table, which might be unique given current data values. If these two fields were chosen as a (composite) primary key, it would preclude ever having another student with the same combination of last name and first name in the future in such a database design.

Some other properties to consider when choosing a primary key:

- No field in the primary key can ever be NULL or optional.
- No primary key field should be a mashup or otherwise violate the single-variable-per-field principle.
- No field of a primary key should ever be possible to be changed/updated.

When choosing between candidate keys to select a primary key, it is generally better to select a simpler key (one with fewer fields, be it a singleton or a composite with just a few fields). Some database designs take this to an extreme and always create a synthetic, often integer, id field that is generated on each record insertion and is guaranteed to be unique. This can be done if truly necessary, but often a singleton or two-field key using existing fields of the table can make the table easier to deal with. Consider our `countries` table and using the three-character country code as the primary key versus generating integer ids; a synthetic id number like 642 means very little in the result of a query, while CHN conveys meaning.

14.2.4 Reading Questions

14.1 Explain why "sound" is a good word to describe a well-designed database. Look the word up in the dictionary if necessary.

14.2 In the `school.courses` table, what is the functional dependency? Notice that this is easy to tell from the figure where this table was first introduced.

14.3 Is it possible to have fields that are part of the set of independent variables but not part of a primary key? Explain.

14.4 Consider the primary key of the `indicators` data set. Will this key determine rows uniquely in the future? Explain, by thinking through what it would mean for the world if the answer was "yes" and if the answer was "no."

14.5 The `topnames` table has fields `name`, `count`, `year`, and `sex`. Considering the *data*, what field or fields would you select to be the primary key of the table?

14.6 Which fields of `school.students` might get changed or updated? Would the primary key every be changed?

14.7 Distinguish between the three terms of *primary key*, *composite key*, and *candidate key*, including examples.

14.2.5 Exercises

The exercises for the next few sections will consider designing a database "from scratch." In this section, we will start by planning for the *tables* that correspond to entities. The next section will focus on the *fields*, and then we will turn to the *relationships* between tables. Unlike coding exercises, the problem solving associated with designing a database from scratch rarely has a (single) *right answer* and can be quite open ended.

You should use the example of the `school` database to help your thinking.

Suppose you have been assigned to create a database for a team-based sports league. You can select your favorite domain, but it is wise, on a first database design, to keep your goals manageable.

Possible team-oriented, league-capable sports might include bowling, volleyball, or basketball. They could be made up of players who are of a single gender, or multiple genders. It would probably be better if we do not consider coaching allied with the teams, but the teams should have a single captain or possibly multiple captains. Contact information like a phone number and/or an address should be considered for each player. A player might also have a position, but this might not be the case (like in bowling). There should be some notion of games played between teams, with scores and win/loss/tie outcomes. There might be a notion of individual statistics or performance for a player within the context of a game.

14.8 In database design, we start with identifying the *entities* that are the collections of *things* that make up the database. Even before relationships add other tables, our first goal is to identify the entities and to give a *table name* for each entity. For your sports league database, give a list and description of each of the entities.

14.9 Check your table names. In this question, list any instances of where the *name of a table is not plural* or the *table is defining a relationship between entities* and has a name that is composed of two entities. Design is an iterative refinement process, so it is ok to find things that need to be improved.

14.10 For each entity table, list the functional dependency that gives the independent variable(s) and a first cut at the dependent variables for the entity.

14.11 Identify any cases where two entity tables have the *same set* of independent variables on the left-hand side of a functional dependency, and combine into a single table.

14.12 For each functional dependency, go through the dependent variables and verify whether or not the variable is dependent on *only a subset* of the independent variables. If found, break these out into *different* functional dependencies.

14.13 Identify the *primary key* for each table, and state yes or no whether it is a *composite key* or not.

14.3   Table Fields

Once tables have been defined, the process proceeds with a careful examination of the fields defined for each table. We must assess each field to see if it violates the principles of sound database design and, if so, take actions to remedy. These sound database design principles for fields are consistent with our tabular normalized data form of tidy data. Analogous to TidyData1, each column field should correspond to *exactly one* variable, and the variable should be entailed as an independent or dependent variable in the functional dependency for the table.

14.3.1   Single Field Issues

Once we have what we believe to be the set of fields/attributes for each table, we should examine each field and adjust our design if we find any of the following indications of a field violating our sound database principles.

- *Value-named attributes*: Column fields that partition data based on categories are the most common culprit of this violation (like Male/Female, or age/income categories, etc.). The remedy in the design is analogous to a "melt" where the "bad" columns become values within a categorical column, and table values align into a new column field.

- *Multi-part fields*: This violation occurs when a column, in the column name and/or the values for that column, has somehow combined multiple separable fields/attributes of the entity. The remedy is to create multiple columns based on the mashup and eliminate the mashup column.

- *Redundant information*: If we include a column that, in its data, has repeated information from the same table and/or information that should be in another table, then we have violated the "exactly one table per entity" principle. This is a sign of a many-to-one relationship between this table and another table. The remedy for this violation is to eliminate the redundant data column and make sure this table (the many-side) table has a field or fields that allow it to be joined with the one-side table. Consider if we were designing an `indicators` table with a composite primary key of `code` and `year`. If, in an initial design, we included an `income` category, table population would have the same value of, say, `'Upper middle income'` for all rows with code `BRA` (Brazil). The `income` categorical variable should be in a separate table (`countries`), and we should ensure that our design allows us to join `indicators` with `countries` using the `code` field.

- *Derived information*: If one field can be calculated from two more fundamental fields, it should not have its own field. The remedy here is to eliminate the derived field, since we can, through computation on columns, always generate the derived column.

- *Repeated fields between tables*: The same data-carrying column should not be repeated between tables. This is often a sign of the violation of "Redundant information" discussed above. To assess, we need to do more than just compare field names between two tables, as a field that allows linking/joining (rather than a data-carrying column) is often named the same between tables to help identify appropriate field(s) for a join. Further, sometimes a field is named the same even though it is carrying logically different data. For example, `lastname` might be in the students table and the instructors table, but the same last name does not indicate that the instructor and student are the same person. By good practice (see below), these should be named differently.

14.3.2 Field Relationship Issues

When we assess fields, there are two common cases of field choice that are a sign of a relationship problem in the table. That is, we have a one-to-many or a many-to-many relationship that the design is forcing into the structure of the current table.

14.3.2.1   List of Values in a Single Field

When a field name and the corresponding field values mean that a single field contains a *list of items*, it means that there is a one-to-many relationship between the table record and some different entity. At database design, it can seem simpler to encode a list of values into a single field. For instance, a `student` table could have a `phone-numbers` field, which could be defined as a string containing a comma, space, or other separator-based list of phone numbers. Similarly, a `movies` table might have a field named `genres`, which contains a list of the different genres to which the movie belongs, like "Adventure," "SciFi," or "Drama." While most well-named fields use *singular* names for the fields, a field anticipated to hold a list is often plural, and an indicator that a non-sound database design might be occurring.

When a field is plural and would contain a list, that means that a single record in the *"current"* table (the table under assessment) is related to many instances of some other entity. For example, each student can have (possibly) many phone numbers, or each movie can have (possibly) many genres. The phone numbers or genres should be in a separate table of like entities.

Two sub-cases exist. First, the current table might be in a one-to-many relationship with the new entity. The remedy is to create a new table for the new entity with a linking field back to the current table with a value to uniquely identify the one-side. In our example, if a phone number can belong to at most one student, but each student can have multiple phone numbers, we create a phone numbers table with a field for the student id to whom the phone number belongs.

Second, the current table might be part of a many-to-many relationship with the new entity. Consider our genres example; a given movie can belong to multiple genres, and a given genre can encompass multiple movies. In this case, a linking table is introduced to provide the pairwise associations between the two tables. Representing a many-to-many relationship will be discussed further in Sect. 14.4.

14.3.2.2   Using Multiple Fields Instead of List of Values

Another common mistake in database design occurs when we realize that we have an attribute with a many relationship with the current table, and we perhaps want to avoid a single field with list values, like in the prior case, and so the design introduces *multiple* columns for the same attribute. Consider a field for tracking the `major` of a student. If we define only one column field for major, we would preclude double (or triple) majors. If we want our database design to support double or triple majors, we might make `major` a field with a list of values or we could introduce `major1`, `major2`, and possibly `major3` in separate column fields.

This solution suffers from many of the problems of using a list in a field:

- Aggregating and grouping information based on the field(s) in question becomes very difficult.

- Even simple SQL queries become more complex; for instance, `WHERE` clauses have to incorporate more fields and use logical operators, or have to check for more possibilities.

The remedy for this problem is again based on whether the current table is in a one-to-many or a many-to-many relationship with the other entity. The remedies are the same as for the list-value violation.

14.3.3 Field Data Types

An important part of the proper design of fields in a table is determining the best data type to use for the field. On the DBMS provider side of a relational database, the system uses the sizes of each data type to allocate disk storage and to build efficient structures that satisfy the data type storage needs of the schema but allow for very fast determination of where to find both records and the fields within a record. This is why, the database systems define such a variety of data types that go well beyond the basic data types of many of our programming languages. Further, because the SQL and DBMS are agnostic to the programming language, the terminology does not always match up between what is expected in the SQL and the names of data types in the programming language.

The responsibility of the database designer is to select a data type for each field that:

- satisfies the data type needs of the field—integers versus floating or fixed point real numbers versus textual data,
- is sufficient for any current data in terms of representing the range of values and size of their representation,
- is sufficient for anticipated future use, and
- is as constrained as possible while first satisfying the above needs.

Data types for fields are a *constraint* of the relational model and, once defined, are enforced for any data placed in the database. As such, once they have been selected, it is significant work to have to change them, and such changes might not be backwardly compatible with software written to use the database.

14.3.4 Field Design: Advice and Best Practices

When designing the fields of tables as part of developing a database schema, we, of course, must avoid aspects that violate the principles of sound database design. But beyond that, there are some best practices that should also be followed.

First and foremost is to use the best SQL data type for the job. We want the power of enforced constraints based on data type to help us avoid errors. If we know a string should *always* use a fixed number of characters, we should use a CHAR (n) data type, not a VARCHAR (n) or TEXT. This gives us both better efficiency and will reject an attempt to populate with an invalid value. When we use a VARCHAR (n), we should think about the maximum number of characters such a string might have over the long-term use of the database. We should not use a floating point data type when an integral data type is appropriate.

The second principle of best practices can be summarized by advising to use *good naming*. There are a number of related points/advice on good naming:

- *Consistency* by using defined naming conventions is very important:

  - Sometimes single words do not convey enough meaning, and names use camel case or underscore. Use such conventions consistently.

- Avoid acronyms and abbreviations.
- When possible, use the *same name* when a field is a linking-type field and referring to a field in another table.
- Use *different names* for fields that do not refer to the same entity.

- For generic fields, make their names specific to the table; for instance, studentlast is better than lastname, if the latter might also be part of another table.

14.3.5   Reading Questions

14.14 Give an example of a table with a field that is derived from other fundamental fields. What does the reading recommend for such situations?

14.15 The reading recommends avoiding "repeated fields between tables." Does that mean that in the school database we should avoid having studentid in both students and student_class? Explain.

14.16 Give an example of two tables with a repeated field between them, which is not part of any primary key. What does the reading recommend to do in such cases?

14.17 Please think of an example, different from any in the book, of a field containing a list of values, and then explain the remedy.

14.18 Suppose our student table had a field, major, containing a list of values. So a joint major in biology and chemistry would have value ['BIO', 'CHEM']. Explain how to find all joint MATH and CS majors under this schema.

14.19 Suppose that, instead of a list of majors, the student table had columns major1, major2, major3. How could you find all joint MATH and CS majors now?

14.20 How, specifically, would you remedy the problem of students having multiple majors? That is: what new table(s) would you create and how would you link the new information to the previous schema?

14.21 In the case of `majors`, it is reasonable to assume that no student has more than three majors. Consider a database of pet owners, with a table `owners` where the entities are people (and various fields like `address`, `age`, `gender`, etc.). Imagine a column `cats` whose value for a given row (owner) is a list of all cats that person owns. Could this table be represented using columns `cat1`, `cat2`, etc.? Explain.

14.22 How would you remedy the problem of owners owning multiple cats, using database design?

14.23 Give an example to illustrate "the variety of data types" that database systems employ.

14.24 Give an example of a data type field constraint, and its impact in real-world terms.

14.3.6   Exercises

These exercises continue the database design process and the exercises from the previous section.

14.25 Where the initial table design included functional dependencies and the division into independent and dependent variables, defining fields for each table should strive to be complete and consider the use of the database. For each table and all variables associated with that table:

- Define the *field name* to be used for each field.
- Using the table at the end of the chapter, list the data type to be used for the field.
- Indicate whether the field can have a value that is *missing* (i.e., NULL).
- Write down 3 or 4 *example values* that you expect the field to have in different rows of the table.

14.26 Go back through your field list and identify any of the following violations:

- A field name that is really a *value* for a variable.
- A field that can be derived from other fields.
- A field whose example values are either a mash-up or a list of values.
- A set of fields that are different versions of the same thing (like position1, position2 to support a player who plays multiple positions).
- Any other poor naming or other violations from the chapter.

Any violations that indicate table relationship issues, set aside (like the third and fourth in the above list), but any that can be resolved by improving the design within the table, please fix.

14.4   Relationships Between Tables

The next step in a sound database design involves examining the entity-based tables that we have thus far, and, after assessing the fields in those tables, determine the set of relationships between the tables. Recall that the possible types of relationship are:

1. *one-to-one*: Each individual entity in the first table corresponds to exactly one entity in the second table, and each individual entity in the second table corresponds to at most one entity in the first table. This can have two sub-cases:

   a. The correspondence between is fully one-to-one, and the records in the two tables are uniquely determined by the same primary/candidate keys. In this case, unless the tables are separate for a permissions/security reason, the tables should be combined.
   b. A record in the second table *may* be associated with a record in the first table but need not be. Consider, from our `school` example, the `departments` as the first table and `instructors` as the second table. Each department has exactly one department chair, given by a record in the instructors table. But each record in the instructors table has *at most* one record in the departments table *for the department chair* relationship. We will not discuss this design separately, as the solution of adding a field to the first table, as we do in the design of many-to-one, can be employed here as well.

2. *many-to-one*: Given an entity instance in the first (the many) table, there is a correspondence to a single entity in the second table; given an individual entity in the second table, there may be a correspondence to multiple entities in the first table.
3. *many-to-many*: Given an entity instance in the first table, there may be a correspondence to many entities in the second table; given an individual entity in the second table, there may be a correspondence to multiple entities in the first table.

14.4.1   Designing for Many-to-One Relationships

If two tables have a relationship of many-to-one, where we will denote as the *first table* the many-side, and the *to-one* as the *second table*, it should be clear that, for a

Fig. 14.2 Student and housing tables

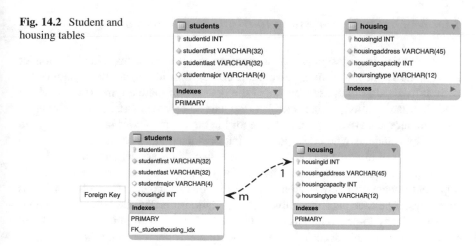

Fig. 14.3 Student and housing many-to-one

record in the first table, there needs to be a mechanism to uniquely identify the *one* record in the second table. The easiest mechanism is to add to the first table a field or fields that correspond to the *primary key* of the second table. If the primary key of the second table is a singleton, we add a single field; if the primary key of the second table is composite, we add multiple fields.

Suppose we were designing the school database and, for high-level entities, envisioned a table for buildings used for student housing (dorms, student apartments, or other forms of student housing). Suppose we call this new entity housing. A given record in housing would correspond to many students, and each student is associated with a single record in housing. So, prior to modification to account for this relationship, we would have the situation depicted in Fig. 14.2. We can see the basics of the students table and, for housing, have an integer primary key, housingid, and other attributes for tracking the address, the number of students who are capable of being housed, and a type field to allow distinguishing of dorms from apartment-style housing from a house.

To create the relationship-necessary ability to join these tables together, we add the field, housingid, to the many-side, which is students. This corresponds to the primary key in the housing table, and so, for any student, we can link to the record in housing that corresponds to that student (Fig. 14.3).

The figure also labels this new field as a *foreign key*, and we see that the database schema, under the *Indexes* part of the table schema for students, has a foreign key (FK) defined for this relationship.

14.4.2 Designing for Many-to-Many Relationships

If two entities in a design are in a many-to-many relationship, the solution of Sect. 14.4.1 is not sufficient to solve the problem. Consider again the design of the school database. Suppose we wanted the database to maintain information about student organizations and their membership, and so organizations becomes an entity in our database design. Given a record in the organizations table, there would be multiple students who correspond to that organization. A particular student can be a member of multiple organizations. These two tables are depicted in Fig. 14.4.

As we saw by example in Chap. 12 with the student_class table, to represent a many-to-many relationship we introduce a *linking table* whose records uniquely define each pairing involved in the relationship. This table is on the many-side of a many-to-one relationship with each of the two original tables. Like in Sect. 14.4.1, this table needs a field to uniquely identify a record on the one-side of these two relationships, typically using a field corresponding to the primary key on the one-side.

This solution is shown in Fig. 14.5. There is a foreign key named studentid that links to the students table, and a foreign key named organizationid

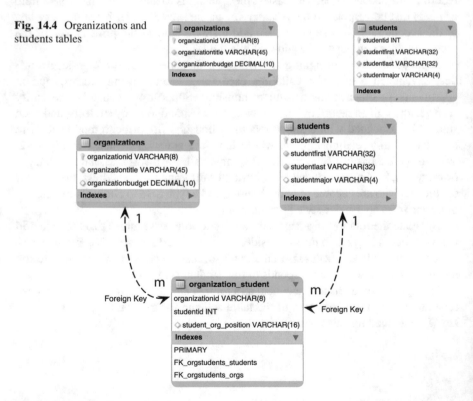

Fig. 14.4 Organizations and students tables

Fig. 14.5 Organizations and students many-to-many

that links to the `organizations` table. The primary key of this linking table is defined as a *composite key*, consisting of all the fields involved in the two foreign keys. In this example, the primary key is composed of both the `studentid` *and* the `organizationid`. Sometimes, a linking table has no more fields than those that give the foreign key to each of the one-side tables and that make up the primary key. On the contrary, if there is information that is dependent on a *pairing*, it can be added to this table. In the current example, we show the addition of the `student_org_postion` field, which could be used to record a particular student holding a position (like president, vice-president, treasurer, etc.) with a given organization. This would allow a student to hold a position with multiple organizations, for instance.

14.4.3   Reading Questions

14.27 At what point in the design process, do you determine the *primary key*, and when do you determine a *foreign key*?

14.28 Give an example, different from any in the book, of two tables linked by a foreign key. Your answer should include exactly two tables; identify a specific field that is the foreign key in one table, and the field in the second table that the foreign key links to.

14.29 Consider a sales order database, with entity tables for orders and for stock items. An order, as consisting of a set of stock items ordered, is related to many records in the stock items table, and a single stock item record is related to many orders. Give small table examples of both orders and stock items, and an example set of records in an `order_stockitem` linking table.

14.30 Give an example of a set of tables in a *recipes* database, in which two entity/tables have a many-to-many relationship, and a relationship where two entity/tables have a many-to-one relationship.

14.31 For the many-to-many relationship of the previous question, identify the *primary keys* of the two entity tables and draw a linking table with foreign keys linking to each of the entity tables.

14.4.4   Exercises

These exercises continue the database design process and the exercises from the previous section.

14.32 Answer the following basic questions in your particular sports league domain:

1. Can a player belong to more than one team?
2. Can a team have more than one captain?
3. Does a contest between two teams consist of more than one game?
4. Can a contest between two teams be played more than once?
5. Are all the contests played part of a single "season" or are there multiple tournaments that make up the season?

As you answer these questions, remember that with greater flexibility comes greater complexity, with more "linking tables" needed. So best to start small and then extend the design once the basic purposes are taken care of.

14.33 The answers to some of the above questions may have caused you to *add* tables that correspond to entities that you did not originally foresee. For each such new table, proceed through the table and field design steps.

14.34 Write down the relationships between any of your entity tables, listing each as:

• one-to-one,
• one-to-many,
• many-to-one, and
• many-to-many.

Make sure you properly understand the "one"-side and the "many"-side.
This step is best done on a graphical representation of the tables in your database.

14.35 Per the discussion in this section, add any needed linking fields to allow for one-to-many or many-to-one and add a linking table with an appropriate name for any many-to-many.

14.36 Determine and annotate the foreign keys. For the many-side of a one-to-many, this will be the added field that uniquely identifies the associated record on the "one"-side. Also, determine and annotate the primary key for any linking tables. This is typically the union of the two foreign keys that allow the linking table to reference the two entity tables.

14.5   Table and Schema Creation

Much of the benefit of the relational model in enforcing constraints occurs because every table, with all its field information, and each foreign key happen when tables are created and before any population. The SQL syntax for table creation is covered in this section, and the forms by which we can populate data into tables is covered in Sect. 14.6.

The general form of the SQL syntax is

CREATE TABLE [IF NOT EXISTS] *table-name* (

> *field-name data-type constraints*
>
> [, *field-name data-type constraints*]*
>
> [, *table-constraint*]*
>
>)

The *table-name* can simply be a valid SQL name for the table but could also, using dot notation, specify a specific database within which the table is to be created. As with column field names, SQL allows names that include spaces if they are delimited with back ticks. We can use the optional IF NOT EXISTS if we wish to generate an error and stop execution in an attempt to create a table that already exists. A table creation results in a table with **no** rows, so using this clause can protect against data loss. The *body* of the table definition, which includes the definition of each of the fields, their data types and field constraints, and constraints that apply to the whole table, is enclosed within required parentheses.

The body consists of a comma-separated list of field specifications, each consisting of a *field-name* and a *data-type* followed by zero or more constraints. All of the elements of a field specification are space separated, so that when a comma is encountered, SQL proceeds to the next field specification. At least one field must be specified in the body of a table definition. Following the set of field specifications are zero or more comma-separated table constraints.

14.5.1   Fields

Focusing on the definition of the individual fields within a table creation statement, we note:

- The *field-name* has similar syntax as *table-name*. It allows the table name to prefix the field name with a dot notation and can use back-ticks to surround the name itself, which is useful if embedding spaces in the name of the field.
- For supported values for *data-type*, refer to Table 14.4 for the most common data types in SQL. Supported data types and their syntax is the area of greatest divergence between various relational database systems, and the information given is meant to be illustrative and is by no means a complete documentation. See your own database system documentation for the full list and syntax of the data types supported by your particular system.
- Zero or more space-separated field constraints follow the data type in a field specification. The most common of these constraints include:

  - NULL or NOT NULL specifies whether a field is allowed or not allowed to be NULL.
  - UNIQUE ensures that no value in the column field is repeated among the records in the table.

- PRIMARY KEY, when used as a column constraint, specifies this as the one and only field making up the primary key for this table. If more than one field is used as parts of a composite primary key, the primary key must be specified as a table constraint following the set of field specifications.
- DEFAULT *val* specifies a default value for this column field when a record is added and an explicit value is not specified.

14.5.2   Table Constraints

For the *table-constraints*, which are *part of* the CREATE TABLE SQL, we list three, the PRIMARY KEY, the FOREIGN KEY, and the CHECK, constraints. While we list these as table constraints, some database systems allow these to be part of column constraints, if the constraint *only* involves the column being defined. Since the table constraint variation is more ubiquitous, we only cover that variation.

14.5.2.1   Primary Key

When we include a primary key specification within a table creation, the constraint is as follows:

Syntax

PRIMARY KEY (*key-fields*)

Description
The *key-fields* is a comma-separated list naming the fields, defined earlier in the set of field specifications, that make up the primary key, and may be a singleton or a set of fields for a composite key.

Examples
In the subjects table, the primary key is the singleton field of subjectid. Within the CREATE TABLE subjects (...), after the field specification, we would have the clause:

```
PRIMARY KEY (subjectid)
```

and the entire create could be given by

```
CREATE TABLE subjects (
   subjectid     VARCHAR(4)   NOT NULL,
   subjectname   VARCHAR(32)  NULL,
   departmentid  VARCHAR(4)   NOT NULL,
   PRIMARY KEY (subjectid)
)
```

In the `courses` table, where the primary key is composite, and given by the two fields of `coursesubject` and `coursenum`, the table constraint would look like the following:

`PRIMARY KEY (coursesubject, coursenum)`

Some systems use an extended syntax for primary keys as table constraints, and this extended syntax is given by

CONSTRAINT *constraint-name* PRIMARY KEY (*key-fields*)

In this case, the prior example would result in a clause that would look like the following:

`CONSTRAINT PK_course PRIMARY KEY (coursesubject, coursenum)`

where the programmer-selected name of the constraint often, by convention, includes the characters "PK" for a primary key, along with the table or fields involved.

Primary keys may be manipulated after a table is already created and may be added or dropped using the `ALTER TABLE` SQL statement.

14.5.2.2   Foreign Key

Recall that a foreign key is defined on the "many"-side table and it uses one or more fields that correspond to the primary key on the "one"-side table. If the "one"-side table has a singleton primary key, the foreign key in the "many"-side uses a single field; if the "one"-side table has a composite primary key, the foreign key in the "many"-side uses multiple fields.

Syntax

CONSTRAINT *constraint-name* FOREIGN KEY (*local-key-fields*)

REFERENCES *one-table* (*remote-key-fields*)

Description

When we define many types of constraints and indices, we use a programmer-selected name, in this case called *constraint-name*. For a foreign key, this name often includes the character sequence "FK" along with a composition of the two ends of the relationship.

The *local-key-fields* are the fields in the table being created (the many-side) that serve as the linking fields to the one-side table. The fields should correspond to the set of fields specified in *remote-key-fields*, which are naming the singleton or composite key fields that form the primary key in the remote one-side table, whose table name is specified by *one-table*.

We should note that *not* defining a foreign key in a table for a relationship that logically exists will not create an error situation. The JOIN operation in SQL will still perform the correct operation. However, the ability to enforce constraints such as referential integrity will be lost. Further, these JOIN operations become *much* more expensive to perform.

Example 1

In our school database, each record of courses has a particular subject, like MATH or BIOL for the course. For a given subject, like MATH, there are many records in the courses table that have the subject MATH. So we have a many-to-one relationship with courses on the many-side and subjects on the one-side. In creation of the many-side table, courses, we include a foreign key constraint that makes explicit the correspondence between the coursesubject field of courses with the one-side remote table of subjects and its subjectid field. Note that the corresponding fields need not be named identically, although often they are.[1]

Within the CREATE for courses, this foreign key constraint might be given by

```
CONSTRAINT FK_coursesubject FOREIGN KEY (coursesubject)
   REFERENCES subjects(subjectid)
```

This table constraint clause would be positioned in the CREATE TABLE statement after the set of field specifications, would be comma separated from the field specifications and any other table constraints, and would occur before the closing parenthesis of the CREATE.

Because of the reference to the subjects table, that table must already have been created in the database, or a reference error would occur.

Example 2

Between classes and courses, we have another many-to-one relationship, this time with classes on the many-side and courses on the one-side. In this case, the primary key of the one-side table is a composite key consisting of coursesubject and coursenum. The foreign key is defined in the many-side table, classes, and, in this case, the corresponding fields are named identically to the primary key in courses. The resulting foreign key constraint in the CREATE for classes might be given by

```
CONSTRAINT FK_classcourse
            FOREIGN KEY (coursesubject, coursenum)
            REFERENCES courses(coursesubject, coursenum)
```

Foreign keys may be manipulated after a table is already created and may be added or dropped using the ALTER TABLE SQL statement.

---

[1] When the fields on both sides of a foreign key to primary key are named the same, we enable the use of the shortcut USING clause instead of ON in a table join.

14.5.2.3   CHECK Constraint

A *check constraint* is used to specify, at least at the time of record insertion, that some (Boolean) predicate is satisfied. This type of constraint is used to help ensure data integrity, so that, at some time in the future, an unanticipated value of a field does not find its way into the table. For instance, we might want to ensure that the values of coursenum for all inserted records are greater than or equal to 100 and less than 500 (or be NULL). These kind of constraints are very much like the predicates of a WHERE clause. One difference is that an unknown value because of a missing (NULL) value is deemed to satisfy a check constraint predicate.

Syntax

CONSTRAINT *constraint-name* CHECK (*predicate*)

For some systems, the shortened syntax is allowed:

CHECK (*predicate*)

Description

The programmer-selected name of the constraint is *constraint-name*. This name might include the character sequence "CHK" along with a portion of the name that helps identify what is being checked. If an error occurs on the insertion of a record, the database system will report this name as the reason for the failure.

The *predicate*, for most systems, allows most of Boolean composition abilities from WHERE clauses and can involve more than one field of the current record, along with relational operators, logical operators, IN, BETWEEN, and so forth.

Examples

Example check constraint in creation of the classes table:

```
CONSTRAINT CHK_classterm
          CHECK (classterm IN ('FALL', 'SPRING'))
```

Example check constraint in creation of the courses table:

```
CONSTRAINT CHK_coursenum
          CHECK (coursenum >= 100 AND coursenum < 500)
```

14.5.3   Programming and Development Advice

In the following, we enumerate some good practices involved in the design and creation of a database and before population of data into the tables of the database.

1. The first goal is for a completed database schema and a sound database design. Toward this end, follow and complete the iterative design process described in Sects. 14.2, 14.3, and 14.4. This should result in an outcome of a documented schema for the database you wish to create.
2. Create the database/schema itself within which the tables of the database will reside. This step is database system dependent.
3. Within the programming language used for the client application, write functions for *each* table creation, abstracting the specifics of the operation.
4. Start with defining each of the tables with their appropriate fields and data types.
5. Expand table definitions to include the primary key.
6. Create one-side tables first and then move to the many-side tables that depend on them. In our `school` database, we would create the `departments` table first, then the `instructors` and `subjects` tables could be created, as they are many-side tables relative to `departments`, but are one-side relative to other relationships. Next, `courses` and `students` can be created. Then, `classes` can be created. Finally, the joining tables of `student_class` and `instructor_class` can be created.
7. Create a "drop" function per table and collect them together in a function that drops the many-side tables first and then the one-side tables (this order is the opposite order from creation).
8. Modify your creation functions to add in constraints:

    • column constraints,
    • then foreign key and check constraints.

Most database systems support scripts that consist of multiple SQL commands gathered together in a text file. Some developers prefer to use this mechanism for maintaining the operations for defining/creating/re-creating a database schema.

If our client program is responsible for creating tables prior to population, the mechanisms for that programming are the same as we covered in Chap. 13. In particular, we start by creating an engine object using a connection string and then establishing a connection. We then invoke the `execute()` method where the argument is a valid SQL CREATE (or DROP) statement.

In the following programming example sequence, we assume the engine has been created and that we have a connection object referenced by the Python variable `movie`, as our examples involve creating tables for a movie database. In the following, we walk through the design development process, in this case writing our function in Python to abstract the support of the `DROP` and the `CREATE` SQL statements.

Suppose we wish to create a table of records for movies and call the table `movies`. Further suppose we wish to include the following fields: - `movieid`: an integer with a unique identifier per movie. - `title`: the title of the movie (in English) with a string data type. Suppose the longest title we need in the database is less than 64 characters. - `release`: the release date of the movie, given to us as a string in YYYY-MM-DD format, but that we wish to be a DATE or DATETIME field in the database. - `rating`: a real-valued public rating of the movie in the range from 0.0 to 10.0.

A simple drop function that handles errors (exceptions) could be defined as

```python
from sqlalchemy import exc

def dropMovieTable(connection):
    drop_stmt = "DROP TABLE IF EXISTS movies"
    try:
        connection.execute(drop_stmt)
    except exc.SQLALchemyError as err:
        print("DROP of movies failed:", str(err))
```

The `exc` import from the `sqlalchemy` module defines the exceptions that can be raised when errors occur during operations defined in the module. This allows us to code appropriately in case the `execute()` was not successful.

A create function that optionally drops the table before creating and handles exceptions is shown next. Because the process of getting the SQL syntax correct can result in errors, handling errors as illustrated here is important when developing your code.

```python
def createMovieTable(connection, drop=True):
    create_stmt = """
    CREATE TABLE IF NOT EXISTS movies (
        movieid INT NOT NULL,
        title   VARCHAR(64) NOT NULL,
        release DATE NULL,
        rating  FLOAT DEFAULT 0.0,
        PRIMARY KEY (movieid)
    )
    """
    try:
        if drop:
            dropMovieTable(connection)
        connection.execute(create_stmt)
    except exc.SQLAlchemyError as err:
        print("CREATE of movies failed:", str(err))

createMovieTable(movie)
```

This process would then continue for other tables in the database until all tables and constraints had been defined.

14.5.4 Reading Questions

14.37 What are some examples of table constraints? When would they be useful in the real world?

14.38 Write down an SQL statement to create the countries table, including data type constraints, and using the UNIQUE keyword on one field. Allow some fields to be NULL.

14.39 Write down an SQL statement to create the indicators table including data type constraints and allowing certain fields to be NULL (e.g., in the case of missing data).

14.40 Revise your answer to the prior question to use the CONSTRAINT syntax to declare a primary key for indicators.

14.41 Show the CREATE SQL statement for countries and indicators, wherein the FOREIGN KEY that relates the two tables is part of the SQL in *one* of the two CREATE statements.

14.42 Use ALTER TABLE to set up a foreign key relationship between instructors and classes, assuming you had already created those tables.

14.43 Give an example, motivated by a real-world application and different from those in the reading, of a situation where you would want a CHECK constraint.

14.44 Explain how to use a CHECK constraint to ensure that, when inserting birth numbers into topnames, it is impossible to insert a negative number.

14.45 If you were creating a database to include countries and indicators, which would you create first and why?

14.46 If you were dropping tables from a database including countries and indicators, what order would you drop in and why?

14.5.5 Exercises

The goal of the exercises in this section is to build programming skills in the creation of tables of a database. To do this, you need to establish an engine and connection to a database where you have *write access*, so the setup might not be the same as in prior assignments. If you are using SQLite, then the database(s) that you create will be files on your local system, and permissions should not be an issue. If you are using a remote database, then we assume you, an instructor, or a database administrator has created a database for you to use that is empty and for which you have permissions to modify the database and create tables.

Before beginning these exercises, create an engine object and a connection object appropriate to your environment. In the first set of exercises, we will use sqlalchemy and execute() method. For checking your work, it is advisable to use a separate client tool, like `TablePlus` or `MySQLWorkbench` to view your database and the tables you create.

14.47 For the next several exercises, imagine you are working for a travel agency. Use `sqlalchemy` and your personal database to create a table `cities` with fields `city_id` (a primary key), `name`, and `country`. If there is already such a table, drop it first.

14.48 Use `sqlalchemy` and your personal database to create a table `hostels` with fields `hostel_id` (a primary key), `city_id`, `name`, `address`, `beds` (the number of beds), and `rate`. If there is already such a table, drop it first.

14.49 Repeat the previous problem, but add a foreign key linking the `city_id` field in this table to the `city_id` field in the `cities` table. Use the technique of dropping the `hostels` table, if it exists, before creating this new version.

14.50 Use `sqlalchemy` and your personal database space to create a table `stations` with fields `station_id` (a primary key), `city_id` (a foreign key), `city_name`, `country`, `latitude`, and `longitude`. If there is already such a table, drop it first.

14.51 Use `sqlalchemy` and your personal database space to create a table `weather` with fields `city_id`, `month`, `ave_temp`, and `rain_inches`, where `city_id` and `month` are primary keys. If there is already such a table, drop it first. Develop incrementally, starting by creating the table with no constraints, and then add constraints to check that `month` is between 1 and 12, that `ave_temp` is between −50 and 120, and that `rain_inches` is between 0 and 400.

14.52 The following is simulated data representing forecasters' weather predictions.

ID	Name	Prediction	Actual	Location
0	Merle	64	67	Hibbing
1	Jillian	68	63	Duluth
2	Julia	62	71	Elk River
3	Paul	76	76	Eden Praire
4	Rob	70	74	Bloomington
5	Andrew	65	71	Park Rapids
6	Maya	65	67	Eagan

Without worrying about populating the table, use Python to programmatically create a table appropriate for holding this data. Be sure and define a primary key and add constraints appropriate for the fields holding temperatures. Consider what fields may be allowed to be missing.

14.53 Consider the following table of baseball information.

Without worrying about populating the table, use Python to programmatically create a table appropriate for holding this data. Be sure and define a primary key and add constraints appropriate for the fields. Consider what fields may be allowed to be missing.

14.6   Table Population

We now proceed to defining the SQL for populating data in an existing (i.e., already created) table. We focus on SQL that populates a single record at a time, which uses the SQL INSERT INTO statement, described in this section.

Player	Team	Position	Age	Division	BattingAv
George	Wildcats	First base	50	West	0.316
Rob	Wildcats	Second base	42	West	0.197
Eric	Wildcats	Third base	28	West	0.388
Jim	Penguins	Pitcher	45	East	0.455
Greg	Penguins	Shortstop	38	East	0.215
Francis	Royals	Outfield	41	East	0.121
Moe	Raptors	Catcher	35	West	0.132
Edmund	Raptors	Pitcher	47	West	0.494
Curtis	Royals	First base	32	East	0.261
Dion	Wildcats	Catcher	46	West	0.207
Cecil	Royals	Second base	27	East	0.369

Syntax

INSERT INTO *table-name* [(*column-list*)]

VALUES (*value-list*)

Description

The INSERT INTO is used in two distinct ways. One way of using the statement occurs when, for the insertion of a given record, we have *all* the defined fields for a new record to be inserted into a given table. The other way of using the statement occurs when, for the insertion of a given record, we have a *subset* of the fields for that record in the given table.

When we are inserting values for a record and providing all the defined fields at the time of the insert, we omit the optional part of the syntax specification, where we have a parenthesized *column-list*. In this case, the *value-list* must consist of a comma-separated set of values, with the following requirements:

- the list contains a value *for every field*,
- the values are listed in the same order as the fields appear in the field specification portion of the table's CREATE TABLE, and
- the values are all of compatible data types for the data types specified for the fields at the time of the table creation.

The INSERT will fail if any of the requirements are not met.

If we have only a subset of the fields and wish to insert a record, we use the full syntactic form, which includes the parenthesized *column-list*. Here, the *column-list* consists of a comma-separated list of non-repeated field names of the table. While the order of the column list can be arbitrary, the requirements of this form of the insert include:

- the values specified in the *value-list* must be in the same order as the columns specified in the *column-list*,
- the values are all of compatible data types for the data types specified for the fields at the time of the table creation, and
- the primary key field(s) and all NOT NULL fields must be present in the *column-list* with corresponding values in the *value-list*.

The INSERT will fail if any of the requirements are not met. If the requirements are met, the record will be inserted and any fields that are *not* present in the *column-list* will have their DEFAULT value, if specified for the field, and NULL otherwise.

14.6.1 Examples

Here, we show the first form, where we are inserting all the defined fields into the movies table:

```
INSERT INTO movies
VALUES (109445, 'Frozen', '2013-11-27', 7.3)
```

We next show the second form, where we specify an explicit set of fields to update, along with their order, and can then omit any fields whose values are allowed to be NULL or have a DEFAULT value specified.

```
INSERT INTO movies (title, movieid)
VALUES ('Guardians of the Galaxy', 118340)
```

Table 14.1 Movies table
after two insertions

movieid	Title	Release	Rating
109445	Frozen	2013-11-27	7.3
118340	Guardians of the Galaxy	NA	0.0

If we then query all fields and all rows from the table where we inserted records, we get the following (yielding Table 14.1):

`SELECT * FROM` movies

Note how the rating field for `Guardians of the Galaxy` is filled with the default value specified at table creation. We note that, after data population is complete, records can be updated using the `UPDATE` statement, but we will not have the need of this functionality.

14.6.2   Programming for Table Population

We are next interested in applying the SQL for populating data in the context of our Python application programs. Conceptually, this is not different than the programmatic use of `sqlalchemy` to issue SQL from an application to a database through the `execute()` method, and after establishing a connection, as covered in Chap. 13. It can, however, be illustrative to see examples of such Python code, so in this section we walk through a few examples where we have in our program a structure with the data to be inserted into the `movies` table.

14.6.2.1   Example 1: Table Population from Python List of Row Lists

First suppose that our data is in a Python list-of-row-lists data structure referenced by LoL:

`print(LoL)`

```
[ ['109445', 'Frozen', '2013-11-27', '7.3'],
  ['118340', 'Guardians of the Galaxy', '2014-08-01', '7.9'],
  ['299536', 'Avengers: Infinity War', '2018-04-27', '8.3'],
  ['301528', 'Toy Story 4', '2019-06-21', '7.6'],
  ['420818', 'The Lion King', '2019-07-19', '7.1'],
  ['424694', 'Bohemian Rhapsody', '2018-11-02', '8.0'] ]
```

Using the technique for incorporating variables into our SQL through binding, the example below creates the movie table, then constructs and prepares the `INSERT` statement we wish to use, specifying all the fields and using binding variable names that match the field name in the SQL create.

The data is in a list of row-lists so, at each iteration of the list, variable `row` refers to the list of field values, and the order of the field values has the `movieid` at index 0, `title` at index 1, and so forth. We create a bound version of the prepared insert statement using named parameters for each of the binding variable names, and passing the appropriate row element based on the fields' index. Finally, we invoke the `execute` method with the bound statement, causing the insert to occur.

```
createMovieTable(movie)

insert_stmt = """
INSERT INTO movies
VALUES (:movieid, :title, :release, :rating)
"""
prepare_insert = sa.sql.text(insert_stmt)

for row in LoL:
    bound_insert = prepare_insert.bindparams(movieid=row[0],
                                             title=row[1],
                                             release=row[2],
                                             rating=row[3])
    rp = movie.execute(bound_insert)
```

The result is as expected (Table 14.2):

```
SELECT * FROM movies
```

14.6.2.2   Example 2: Table Population Using Python CSV `DictReader`

If our content is in a CSV file, and we employ a `DictReader` object from the Python `csv` module, we can improve on the previous solution and make it less error prone. Assume the CSV file has the same data as above and has a header line with column names as follows:

```
movieid,title,release,rating
```

These column names need not match the field names in the created table, but they *do* need to match the names we use in our `INSERT` template as the binding variable names. In the code below, after we create the movie table, we again construct and prepare the `INSERT` statement. We use a `with` construct to open the CSV file and associate it with the `csvfile` variable. This file is then used in the creation of the

Table 14.2 Movies table after insertions from the LoL

movieid	Title	Release	Rating
109445	Frozen	2013-11-27	7.3
118340	Guardians of the Galaxy	2014-08-01	7.9
299536	Avengers: Infinity War	2018-04-27	8.3
301528	Toy Story 4	2019-06-21	7.6
420818	The Lion King	2019-07-19	7.1
424694	Bohemian Rhapsody	2018-11-02	8.0

CSV `DictReader` object. This object, instead of returning a simple list of field values at each iteration of a loop, returns a *dictionary* whose keys are the column names found in the header, and whose values are the values from the CSV file at each row iteration.

We know that `bindparams` expects a set of named parameters in its parameter list. Python provides a mechanism that, if you have a *dictionary* whose keys are the same as a set of named parameters, it can take a single dictionary argument, prepended by `**`, and it will automatically convert the dictionary into the required set of named parameters. This is what is occurring in the `bindparams(**row)` invocation below. If `row` were a simple list instead of a dictionary, this technique would not work.

```python
import csv

movie_csv = "movie_example.csv"
path = os.path.join(datadir, movie_csv)
assert os.path.isfile(path)

createMovieTable(movie)

insert_stmt = """
INSERT INTO movies
VALUES (:movieid, :title, :release, :rating)
"""
prepare_insert = sa.sql.text(insert_stmt)

with open(path, 'r', newline='') as csvfile:
    reader = csv.DictReader(csvfile)
    for row in reader:
        bound_insert = prepare_insert.bindparams(**row)
        rp = movie.execute(bound_insert)
```

The reason this solution is more elegant and less error prone is that, even if a CSV file re-ordered the columns, the code will still work correctly. It is also less tedious, as you do not have to create a potentially long argument list for the binding step. Finally, the same basic code should work for inserting into any table based on an appropriate CSV file of data.

14.6.2.3   Example 3: Table Population from a `pandas` `DataFrame`

Suppose we have the data to be used to populate the `movies` table in a `pandas` `DataFrame` named `movies`.

movies

```
|     movieid                  title      release   rating
| 0    109445                 Frozen   2013-11-27     7.3
| 1    118340  Guardians of the Galaxy  2014-08-01    7.9
| 2    299536  Avengers: Infinity War   2018-04-27    8.3
| 3    301528            Toy Story 4   2019-06-21     7.6
| 4    420818         The Lion King   2019-07-19     7.1
| 5    424694      Bohemian Rhapsody   2018-11-02     8.0
```

The SQL INSERT statement works on a row at a time, so we need to iterate over the *rows* of the DataFrame. The facility to do this is the iterrows() method of a DataFrame, discussed in Sect. 7.3.6, which gives us an iterator that we can use in a for statement. At each iteration, we obtain the label for the row and a Series giving the values in the row. The following code snippet shows the use of the iterator in this example, printing the movieid and title for each iteration over the rows in the movies DataFrame:

```python
for rowlabel, rowseries in movies.iterrows():
    print(str(rowlabel) + ':', rowseries['movieid'],
          rowseries['title'])
```

```
| 0: 109445 Frozen
| 1: 118340 Guardians of the Galaxy
| 2: 299536 Avengers: Infinity War
| 3: 301528 Toy Story 4
| 4: 420818 The Lion King
| 5: 424694 Bohemian Rhapsody
```

The pandas Series, as shown in the example above, acts as a dictionary, allowing access to fields based on column name. This means that we can use the technique from the last example, where we convert a dictionary into a set of named parameters using the ** prefix on the dictionary as we are passing it to bindparams. This gives us the straightforward solution for this example:

```python
createMovieTable(movie)

insert_stmt = """
INSERT INTO movies
VALUES (:movieid, :title, :release, :rating)
"""
prepare_insert = sa.sql.text(insert_stmt)

for rowlabel, rowseries in movies.iterrows():
    bound_insert = prepare_insert.bindparams(**rowseries)
    rp = movie.execute(bound_insert)
```

14.6.2.4   Example 4: Table Population Using `pandas` Method

The `pandas` module provides a `to_sql()` method of its `DataFrame` class
that allows a data frame to be used to populate a table over an SQL connection.
The advantage of this mechanism for table population is its great convenience. If
a data frame in our client application has the needed columns and rows, a single
method call can both create and populate a table. The disadvantage is that we lose
the control in specifying primary and foreign keys, constraints on field values, or
default values. By default, the SQL data types are determined automatically based
on the Python data types. This can be controlled but requires deeper understanding
of `sqlalchemy` types, which is beyond the scope of the current text. And there
simply is not the control to specify the other field and table attributes/constraints
provided by a native SQL CREATE TABLE.

The most relevant `to_sql()` method parameters:

- name: a Python string with the name of the table to be created/populated,
- con: a reference to an `sqlalchemy` connection object,
- `if_exists`: a Python string from the set `'fail'`, `'replace'`, or
  `'append'`. The default is `'fail'`, so if the destination table already exists, the
  method invocation raises an error. If `'replace'`, an existing table is dropped
  and the operation creates its own new version of the table. If `'append'`, the
  existing table (including definitions of keys and constraints) and the rows of the
  data frame are added to the table,
- `'index'`: a Boolean determining whether or not to include the DataFrame
  `index` as one of the columns of data.

If we are using the database as storage for normalized data and are performing
full table population, perhaps with subsequent full table retrieval, we would do the
following:

```
engine = sa.create_engine(cstringmovie0)
connection = engine.connect()

movies.to_sql("movies", con=connection, if_exists='replace',
              index=False)

connection.close()
del engine
```

Table 14.3 Newer movies table

movieid	Title	Release	Rating
109445	Frozen	2013-11-27	7.3
118340	Guardians of the Galaxy	2014-08-01	7.9
299536	Avengers: Infinity War	2018-04-27	8.3
301528	Toy Story 4	2019-06-21	7.6
420818	The Lion King	2019-07-19	7.1
424694	Bohemian Rhapsody	2018-11-02	8.0

This replaced the old `movies` table and populated it as expected. In so doing, the primary key was lost and the strings became the SQL TEXT data type, with a maximum string length of 255 characters (Table 14.3).

If we want to maintain control of the constraints, an alternative would be, under our own control, drop the table followed by our own CREATE. This would create an empty version of the table. We would then invoke the `to_sql()`, but this time use `'append'` for `if_exists`, and then we maintain our structure:

```
engine = sa.create_engine(cstringmovie0)
connection = engine.connect()

createMovieTable(connection, drop=True)
movies.to_sql("movies", con=connection, if_exists='append',
              index=False)

connection.close()
del engine
```

We conclude this chapter with a table summarizing the common SQL data types that can appear in constraints (Table 14.4).

Table 14.4 Common SQL data types

SQL data type	Description
CHAR (n)	Used for *fixed* length strings, where the number of characters in the string is *always* the same. The n specifies the number of characters and defaults to 1
VARCHAR (n)	Used for *variable* length strings, where the number of characters in the string may range from 0 characters up to the maximum, given by n
TEXT	Used for variable length strings. On some systems, this is equivalent to VARCHAR, while on others, the maximum length of the string may be greater than that allowed by VARCHAR
INT (n)	Used for integers. On most systems, the n specifies the number of decimal digit characters plus a possible negative sign, for display/print of the integer. Some systems have variations or replacements, like an INTEGER synonym, or variants based on the needed size, like SMALLINT or BIGINT
FLOAT (p)	Used for floating point numbers. Depending on the value of p, this data type can be used for both FLOAT represented by 4 bytes when $p \leq 24$ and an 8-byte DOUBLE, when $p \geq 25$. Some systems use REAL for a 4-byte floating point value. On other systems, DOUBLE or DOUBLE PRECISION with optional size and precision parameters gives other options for floating point real-valued numbers

(continued)

Table 14.4 (continued)

SQL data type	Description
DECIMAL(s, d)	Used for fixed point numbers when explicit precision is required, with s giving the total number of digits, and p giving the digits after the decimal point. Common synonyms are NUMERIC and DEC
BLOB(n)	An acronym for Binary Large OBjects, this is provided by most systems for holding arbitrary binary data. Some systems support larger-sized BLOBs with data types MEDIUMBLOB and LONGBLOB, and smaller binary data with TINYBLOB and BINARY
DATE	Used for holding date values, generally specified in the format YYYY-MM-DD, and allowing conversion from strings in the same format
DATETIME	Used for the combination of a date and a time, specified in a format of YYYY-MM-DD hh:mm:ss and often using a 0 to 23 for hours. Some systems just use DATE and, when doing so, default the time to 00:00:00
ENUM$(vals)$	Used for an enumerated type, but supported by fewer database systems. The $vals$ are a comma-separated list of literal constants, most often string types

14.6.3　Reading Questions

14.54 Please give an example of a field, other than a primary key, where you might want to insist that NULL values are not allowed. Justify your answer.

14.55 With respect to the movies table, what are the pros and cons of making the default value for rating be 0.0 instead of NULL?

14.56 Consider the LoL example. How would you modify this example to work even if the LoL has columns in the wrong order, e.g., putting rating as the third field and release as the fourth? This might happen if the LoL was created based on an old version of the database schema.

14.57 How could you modify the code to work even if movieid is not provided in the LoL? Speculate on an answer.

14.58 How would the example code for inserting from a csv file change if the headers in the csv file were ID, name, released, stars?

14.59 If the csv file in example 2 had the same data as the LoL in example 1, what would be the first value of row in the for loop in example 2?

14.6.4　Exercises

14.60 In an exercise from the last section, you created the table cities. Using execute() method invocations, populate that table with the following data. Given that a database does not allow duplicate keys (or any other unique field), you may need to drop and recreate the table if you need to perform the population more than once.

city_id	name	country
1	Chicago	USA
2	Granville	USA
3	Budapest	Hungary

14.61 In an exercise from the last section, you created the table `hostels`. Using `execute()` method invocations, populate that table with the following data.

hostel_id	city_id	name	address	beds	rate
1	1	The Windy Hostel	400 E. Montrose Ave.	77	30.00
2	1	The Cubby Bear	5000 N. State St.	92	32.50
3	2	Buzzards	2000 College St.	80	17.25
4	3	The Lucky Lizard	500 Parliament Ave.	240	7.99

14.62 Repeat the previous problem, but this time, first put the table data into a CSV file and then use the book's CSV `DictReader` solution for populating the table from a CSV file.

14.63 In an exercise from the last section, you created the table `stations`. Using `execute()` method invocations, populate that table with the following data.

station_id	city_id	city_name	country	latitude	longitude
21	1	Chicago	USA	41.9	−87.6
32	1	Chicago	USA	41.92	−87.61
36	2	Granville	USA	40	−82.5
45	3	Budapest	Hungary	47.5071	19.0457

14.64 Put the data from the last exercise into a CSV file. Then, use `pandas` to read the CSV file into a data frame, and then use the book's technique of iterating over the rows of the data frame to populate the table data.

14.65 In an exercise from the last section, you created the table `weather`. If created properly, this table explicitly constrained data in `month` to be in the range 1 to 12, as well as constrained `ave_temp` and `rain_inches`. Note that, in the table below, the second to last row has an illegal value for month. Using any of the techniques (except to_sql()), insert the data below row by row. Note what happens when a constraint is violated. Did *any* of the rows get inserted?

city_id	month	ave_temp	rain_inches
1	1	28	2.4
1	7	82	2.2
2	5	69	2.5
2	10	55	1.9
3	−1	37	2.1
3	12	26	1.8

14.66 Improve your solution from the last question so that, when an INSERT raises an error, your code reports the problem in an informative way but then *continues*, so that *all* legal rows get inserted into the 'weather' table from question 14.65.

Chapter 15
Hierarchical Model: Structure and Formats

Chapter Goals

Upon completion of this chapter, you should understand the following:

- The structure of a graph-based and tree-based data model.
- How conceptually we can use nesting to represent such a model in Python native data structures.
- With JSON and XML as representative format instances of hierarchical models, understand correct syntax and structure to represent data within this model.
- The use of paths in JSON and XML to identify individual elements within the tree.

Upon completion of this chapter, you should be able to do the following:

- Construct in-memory structures from local files and strings containing JSON and XML formatted data.
- Carry out simple extraction of data from JSON or XML format.

In Sect. 5.4, we presented an overview of the hierarchical model, briefly describing its tree-based structure, and providing example formats such as JSON, XML, and XHTML. The current chapter deepens our discussion of the structure of

© Springer Nature Switzerland AG 2020
T. Bressoud, D. White, *Introduction to Data Systems*,
https://doi.org/10.1007/978-3-030-54371-6_15

hierarchical data, beginning with a formal treatment of trees, how to represent them in Python, and common terminology we will need for subsequent chapters. We then discuss the syntactic details of the JSON and XML formats.

The current chapter details more fully the structure and formats associated with the hierarchical data model. In Chaps. 16 and 17, we study the operations and constraints of the hierarchical model. This includes

- procedural programming operations for JSON and XML,
- declarative data query operation for XML using XPath, and
- encoding and enforcing constraints using JSON Schema, Document Type Definition (DTD), and XML Schema.

Operational examples will include access of data from a hierarchical structure, and storing it in appropriate data structures, learning how to modify data in a hierarchical structure, and using constraints to guarantee that a structure we build or receive is a well-formed tree, containing the data we need in the location we expect.

15.1   Motivation

Examples of hierarchical data abound in data science. For instance, we have already seen that file systems are naturally hierarchical. Date/time data is hierarchical. We will see that data obtained from web scraping and from APIs is also often naturally hierarchical. Data from emails has a hierarchical structure. Even linguistic data can be viewed as having a hierarchical structure, as can be seen from sentence diagramming and grammatical rules. Furthermore, many topics in statistics can be viewed more clearly through a hierarchical lens, including time series analysis (with a level for time), clustering (with a level telling which cluster an individual belongs in), multiple linear regression, ANOVA, longitudinal data analysis, multi-level modeling, and statistical methods for filling in missing data in a data set (with one level for modeling the missing data, and another level for modeling the response variable). For the moment, we will limit our discussion to local files with JSON and XML formats, but the reader is encouraged to remember that these files can store data from a wide variety of application areas.

It is worth taking a moment to compare the hierarchical model with the relational and tabular data models. The relational model is the most structured (e.g., constraints are built-in), followed by the tabular model (where constraints are "by convention" to achieve tidy data). We will see that the hierarchical model is the least structured of the three. This will create added complexity when we engage our discussion of hierarchical constraints.

Despite the additional complexity, the hierarchical model has a number of benefits over the previous two data models:

- First, as we have already described, the hierarchical model has tremendous expressive power. Furthermore, anything that can be expressed in tabular form can also be expressed in hierarchical form, as we will see below.
- Second, hierarchical models are excellent for storing large and complex data sets in such a way as to make the data efficiently searchable. We have already seen this with our example of the file-structure, where it is possible to get from the root directory to each file in a small number of steps, even though there might be a large number of files in total.
- Third, the hierarchical model is very flexible. We will see in the next chapter that it is much easier to add data in the hierarchical model (indeed, anywhere in the tree) than it is to add data to the tabular or relational models. This also makes it easier to including missing data in our model. Instead of requiring a special symbol like NA, we can represent missing data simply by the absence of a leaf in the tree.

15.2 Representation of Trees

Hierarchical data refers to data with the underlying structure of a tree, so we must pursue a careful discussion of trees, and begin this discussion with the more general concept of a graph.

A graph is a mathematical model and is common across much of the field of computer science. A *graph* is a set V of nodes, and a set $E \subset V \times V$ of edges between nodes. It is helpful to think of an edge $v_1 v_2$ as signifying a relationship between nodes v_1 and v_2, and when such an edge is present we say v_1 and v_2 are *adjacent*. Graphs model entities and the relationships between them, and are ubiquitous in the study of systems. For example, the web can be thought of as a graph, where the nodes are web pages, and there is an edge between two web pages if one of the pages contains a hyperlink to the other. We will see in Chap. 19 that graphs can also be used to model the Internet, as shown in Fig. 15.1.

Here the nodes represent devices (either routers or hosts), and an edge denotes a medium through which information can flow (e.g., a physical wire, a fiber optic cable, etc.). A *path* is a sequence of edges $v_1 v_2 \ldots v_k$ from some initial node v_1 (the *source*) to some node v_k (the *destination*). When we think of edges as bidirectional, we sometimes say the path goes *between* v_1 and v_k. The reader is encouraged to trace out a path (following the edges, one by one) from the "Cloud Server," a host in the lower right of the figure, to the one of the "Desktop" host nodes on the far left of the figure.

If $v_k = v_1$, the path is called a *cycle*. The graph above has several cycles, including triangles, quadrilaterals, and one hexagon. We say a graph G is *connected* if, for any two nodes in G, there exists at least one path between them.

A graph with no cycles is called a *forest*. A graph that is *connected* and has no cycles is called a *tree*. In this way, a forest is made up of one or more disconnected

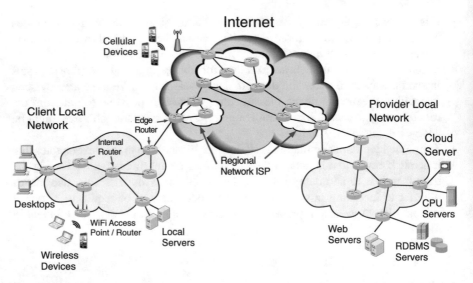

Fig. 15.1 Network architecture

trees. Both trees and forests are common in data mining algorithms. We recall the following picture of a tree, from Sect. 5.4.

In this figure, nodes are labeled by strings of text. The edge between author and "Tolstoy" denotes that the author of "War and Peace" was Tolstoy. The edges beneath book [id="bk102"] denote that author, title, and price are constituent pieces of book bk102. The edge above book bk102 signifies that bk102 is part of the catalog. It is not difficult to show that if a tree has n nodes, then it must have $n - 1$ edges. With any fewer edges, the tree would not be connected. With any more edges, it would have a cycle. We note that it is possible to extend beyond hierarchical data to graph-structured data, and that the resulting model has even more expressive power and flexibility, but we will not delve into it here.

15.2.1 Terminology

For the purposes of this book, we will always display our trees as shown in Fig. 15.2, with a single node at the top, and the rest of the tree displayed in *levels* below. We summarize the relevant terminology:

- The node at the top is called the *root*.
- For a given node, the nodes directly below it are called its *children*. For example, the children of the root, catalog, are bk101, bk102, ..., bk200.
- For a given node, the node directly above it is called its *parent*. For example, the parent of bk102 is catalog.

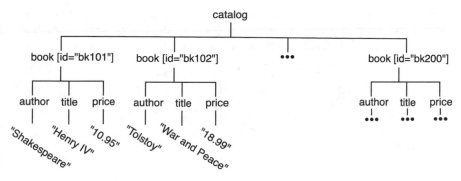

Fig. 15.2 Book catalog tree

Fig. 15.3 Example graph

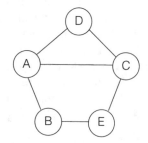

- For a given node X, the nodes along the path from the root to X are the *ancestors* of X. Nodes below X, whose paths from the root contain X, are the *descendants* of X.
- If two nodes have the same parent, we call them *siblings*.
- We allow nodes to have a symbolic name, sometimes called a *tag* or *label*.
- We allow two different nodes to have the same label (e.g., there are 100 nodes labeled `author`), and we use their location in the tree to distinguish them.
- A node with no children is called a *leaf* node.

We note that there is a unique path between any two nodes in a given tree. The length of this unique path (that is, the number of edges in the path) is called the *distance* between the two nodes. In Fig. 15.2, all leaves are distance three from the root. In such a case, we say the tree has *depth* three.

15.2.2   *Python Native Data Structures and Nesting*

15.2.2.1   **Representing Graphs**

The essential data of a graph is its adjacency structure, i.e., which nodes are adjacent to each other. There are several ways to represent graphs and trees naively in Python, that we will illustrate using the graph in Fig. 15.3.

First, a graph can be represented as a dictionary, where the keys are nodes and the value associated with any node v is the list of nodes adjacent to v. For example, the graph in Fig. 15.3 would be represented by:

```
graphDict = {
   'A':['B','C','D'],
   'B':['A','E'],
   'C':['A','D','E'],
   'D':['A','C'],
   'E':['B','C']
}
```

This representation makes it easy to find out which nodes a given node is adjacent to, but it duplicates information (representing each edge twice). In the case of a tree, this representation could be modified so that each node (key) is associated with its list of children. However, such a representation suffers from the drawback that it is challenging to determine the parent of a given node.

Another way to represent a graph is via its *adjacency matrix*, where each row and each column correspond to a node. An entry a_{ij} is 0 if there is no edge between nodes i and j, and is 1 if there is an edge. We represent this in Python as a list of lists. For the graph of Fig. 15.3, we get:

```
adjMatrix = [
   [0,1,1,1,0],
   [1,0,0,0,1],
   [1,0,0,1,1],
   [1,0,1,0,0],
   [0,1,1,0,0]
]
```

We read the first row as saying node A has edges to nodes B, C, D (the three middle 1s in the row) and does not have edges to A or E. Like the previous way of representing a graph, this representation duplicates information. Note that if you reflect the matrix over its diagonal (the line from the upper left to the lower right), you do not change the matrix at all, because the information that A has an edge to C is the same as the information that C has an edge to A. Furthermore, this representation contains a great deal of unnecessary information in the form of 0s. Many graphs in the world have a huge number of nodes, but not very many edges. Such graphs are called *sparse*, and examples include the web, the Internet, and social network graphs. When using an adjacency matrix, the number of bits required to represent a graph only depends on the number of nodes, not on how sparse the graph is.

15.2.2.2 Representing Trees

Because trees have more structure than graphs (namely, a chosen root, and representation in levels), it is easier to represent trees than it is to represent general graphs.

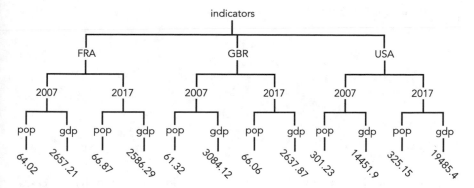

Fig. 15.4 Indicators0 tree

For consider the tree shown in Fig. 15.4, which represents our data set ind0 of country indicators in 2007 and 2017 (Fig. 15.4).

The essential structure here is the information that `indicators` has children `"FRA"`, `"GBR"`, and `"USA"`, that each country has children for years `"2007"` and `"2017"`, that each year has children `pop` and `gdp`, and that each of these has children consisting of a real-valued number. As we did for graphs, we can represent this as either a dictionary or a list, but now with radically different interpretations. In both cases, we use *nesting* of data structures to correspond to the levels in the tree, something we could not do for a general graph. Each additional level of nesting corresponds to an additional level as we descend the tree.

We can represent ind0 as dictionary whose name corresponds to the root note (indicators) and whose keys correspond to the children of the root node (here, "FRA," "GBR," and "USA"). The value associated with each of these is a dictionary representing the tree below it, as we now demonstrate:

```
indicatorsDict = {
  "FRA": {
    "2007": {
       "pop": 64.02,
       "gdp": 2657.21},
    "2017": {
       "pop": 66.87,
       "gdp": 2586.29}
     },
   "GBR": {
     "2007": {
       "pop": 61.32,
       "gdp": 3084.12},
     "2017": {
       "pop": 66.06,
       "gdp": 2637.87}
```

```
    },
  "USA": {
    "2007": {
       "pop": 301.23,
       "gdp": 14451.9},
    "2017": {
       "pop": 325.15,
       "gdp": 19485.4}
    }
  }
```

Note that we have used horizontal white space (indentation) to make it easier to "see" the tree structure, though of course Python does not care about this white space. For example, we can see that the children of the node "USA" are "2007" and "2017" (the keys of the dictionary associated with "USA"), because these are below "USA" and indented one level farther to the right. Similarly, we can see that the children of "2017" at the bottom are "pop" and "gdp." In this particular example, leaf nodes have no siblings. If a tree had leaf nodes with siblings, e.g., if there were two children of the final "gdp" instead of one, then we could use a list for the innermost layer of this nested data structure. But in all of our examples, we will have leaf nodes without any siblings, because a leaf will represent a unit of data, and the path from the root to a leaf will represent the meaning of that datum. We will not belabor this point.

When we use *dictionaries* as the primary structure used in the nesting to achieve a tree, we will call this a *Dictionary of Dictionaries* (DoD). We use this terminology even when there are more than two levels of nesting, e.g., the previous example is technically a dictionary of dictionaries of dictionaries but we do not call it a DoDoD.

We can also represent a tree as a Python list, again using the root as the name of the list, and children as the elements in the list. Here, each child may itself be a list, representing the tree below it. This means we would represent ind0 as a list indicators = [FRA, GBR, USA] where each of FRA, GBR, USA are themselves lists. For example, FRA could be the list [F07, F17], where F07 is the list [F07P,F07G], and F07P = 64.02, F07G = 2657.21. Similarly, F17 is a list (consisting of the population and gdp in France in 2017), and GBR and USA are defined as nested lists analogously to FRA. This means indicators is a 3D lists, because the tree ind0 has depth three.

When we use *lists* as the primary structure used in the nesting to achieve a tree, we will use the same List of Lists (LoL) term introduced in Chap. 3. Again, a particular tree can have additional levels that are also lists, but we will still use the LoL designation.

One benefit of the list representation is that we do not need explicit names of nodes as we gave above, but instead use integer indices into the lists. We could simply have written the 3D list as follows, where the top row corresponds to France, the second row corresponds to GBR, and the third row corresponds to the USA:

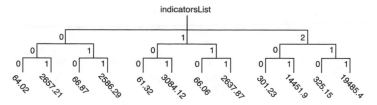

Fig. 15.5 Nested lists as a tree

```
indicatorsList = [[[64.02,2657.21],[66.87,2586.29]],
                  [[61.32,3084.12],[6606,2637.87]],
                  [[301.23,14451.9],[325.15,19485.4]]]
```

This allows integer indices to "stand in" for node names when we access data. For example, indicators[0][0][0] represents the population, in 2007, of France. Similarly, indicators[2][1][1] represents the gdp, in 2017, of the USA. In Fig. 15.5, we depict this nested list structure as its corresponding tree, annotating each child edge with their list index. So, from the root, index 1 is associated with the subtree representing GBR, and its children, annotated 0 and 1, correspond to the years 2007 and 2017, etc.

As a corollary of the nested list representation of a tree, we see that every tabular data set can be represented as a tree. In particular, given a tabular structure, stored as a list of lists $L = [L_0, L_1, \ldots, L_r]$, we can represent it as a tree in the following way:

- The root is named L.
- The children of the root are labeled 0, 1, 2, ..., r, and each represents one of the lists L_i.
- Each i has c children, where c is the number of columns, i.e., c is len(L_0).
- The jth child of node i represents the data stored at $L[i][j]$.

This procedure is essentially what happened with ind0, except we ended up with a third layer, because ind0 has a two-level index. This helps justify our use of the term "level" in Chap. 7. Note that not every tree can be represented as a list of lists, e.g., trees of many levels, or irregular trees (e.g., where not all leaves are the same distance from the root).

Both the dictionary representation and the list representation of trees have the property that compositions allow multiple levels of the tree. That is, to build on the tree to make it one level deeper, we can simply nest our data structures one level deeper, rather than having to build a new representation from scratch.

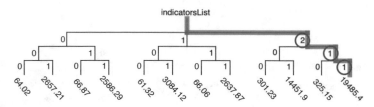

Fig. 15.6 Nested lists path traversal

15.2.3   Traversals and Paths

With either of these representations of trees through nested Python data structures, it is easy to build paths in the tree. For instance, a path in the example indicators data set, stored in `indicatorsDict`, from the root to the datum of the gdp, in 2017, of the USA, can be thought of as `indicatorsDict["USA"]["2017"]["gdp"]`. Python reads this from left to right, first moving inside the dictionary `indicatorsDict`, to get the value (itself a dictionary) associated with the key `"USA"`, then moving inside this dictionary to get the value associated with the key "2017," etc. We can write paths in `indicatorsList` in the analogous way, e.g., writing `indicatorsList[2][1][1]` to access the datum from above, where index 2 maps to the `"USA"` subtree, index 1 maps to the data associated with 2017, and index 1 at the final level maps to `"gdp"`. This concept of a path, starting at the root, and traversing along integer indices is depicted in Fig. 15.6, where we highlight in purple the path, and circle in red the indices along the path.

This way of thinking–that moving through the nested data structures one by one is equivalent to stepping through the tree one level at a time–will come up again in Chap. 16, when we discuss XPath.

15.2.4   Reading Questions

15.1 You may have seen trees before, when learning about binary search. A list of numbers can be represented as a tree, with one node for each item. The tree structure determines how quickly we can search for a given item. For example, let

    L = [1,3,4,6,7,8,9,10,11,13,15,16,17,18,20].

If you represent L as a *linear tree*, this means the root is 1, and there are edges 1->3, 3->4, 4->6, etc. Searching this tree for `item = 17` requires 13 steps, i.e., a path of length 13. We say item 17 is stored at *depth* 13 in the tree.

Please use the language of trees (root, paths, depth, etc.) to explain what makes binary search faster than linear search. Start by thinking about how to represent L in a way that lines up with binary search, analogously to how we just did it for linear search. Please also give a binary search path in L to the item 17, and include the depth of 17 in your answer. If you are stumped, you could search for "Binary Search Tree" online. Be sure to explain what tree property makes binary search take time O(log(n)) while linear search takes time O(n).

15.2 Building on the question above, why is the tree structure on your computer's file system important? What sorts of things does it make more efficient? Please illustrate with a concrete example, then give another real-world example where a tree structure makes something more efficient.

15.3 Please discuss the trade-offs between structuring data into a tree form, to facilitate efficient retrieval, versus leaving data in an unstructured form. It might help to think about the cost involved with sorting a list, to facilitate the use of binary search, versus leaving it unsorted. At what point does it become worthwhile to build the additional structure?

15.4 Consider a data set consisting of a university web site, with news, academics, admissions, events, research, etc. Would this be best to represent in relational form or hierarchical form? Justify your answer.

15.5 Consider a data set consisting of university records: students, courses, grades, etc. Would this be best to represent in relational form or hierarchical form? Justify your answer.

15.6 Please make an analogy between how paths are given when we use a native Python dictionary to represent a tree, and your experience with URLs on the web, and with using the Terminal or Command Prompt on your machine. Please give a concrete example illustrating the analogy.

15.3 JSON

As discussed in Sect. 2.4, the JavaScript Object Notation (JSON) format is a programming language independent, straightforward, way to move between data stored in a text file, and data stored in a data structure. This makes JSON a very convenient format for hierarchical data, as both of the Python native data structures of dictionaries and lists, discussed above, can be easily converted into JSON format [9, 21, 28].

It is important to note that the only scalars allowed for JSON are Booleans, numbers, strings, and a special data type called *null*. Thus, a JSON file cannot store a tree of file objects, but can store a tree of strings (including file names). Furthermore, strings *must* be represented by double quotes, instead of single quotes, e.g., "Chris" is an acceptable string, but 'Lane' is not.

In JSON, when data is stored inside square brackets, in an ordered list, such as [1,"one",True] or our indicatorsList is example, the data type is known as an *array*. When data is stored inside curly brackets, with key:value pairs, like

{"Noah":2, "Becky":1, "Evan":1}, or our indicatorsDict exam-
ple, the data type is known as an *object*. These are the only two container data
types allowed by JSON, though nesting is possible (that is, we can have an array of
numbers, an array of arrays, or even an array of objects). These data structures are
universal, and so connect easily with Python, C++, Java, and many other languages.

If we have the data structures indicatorsDict and indicatorsList in
a Python program, we can write them to a JSON file using the dump function:

```
import json
pathD = os.path.join(datadir, "ind0Dict.json")
pathL = os.path.join(datadir, "ind0List.json")

f = open(pathD, 'w')
json.dump(indicatorsDict, f)
f.close()

g = open(pathL, 'w')
json.dump(indicatorsList, g)
g.close()
```

The result of this code is to create and fill two files, ind0Dict.json and
ind0List.json in the data directory. Sometimes it is advantageous to have
a JSON-formatted string. We demonstrate for indicatorsDict (the case of
indicatorsList is analogous):

```
import json
mystring = json.dumps(indicatorsDict)
```

The resulting string, mystring, could easily be passed as a parameter to a
function. This is one way to avoid concerns about mutability of a Python native data
structure.

In addition to dumping Python data structures into JSON files, we can also load
a data structure from a JSON file. For example, the code above will result in a JSON
file or string containing the following text on one line (which we have truncated):

```
{"FRA": {"2007": {"pop": 64.02, "gdp": 2657.21},
         "2017": {"pop": 66.87, "gdp": 2586.29}}, ...}
```

To load this into a Python native dictionary, we use the load function:

```
import json
pathD = os.path.join(datadir, "ind0Dict.json")

f = open(pathD, 'r')
indicatorsDict = json.load(f)
f.close()
```

This code assumes there is a file "ind0Dict.json" in the data directory. We open the file with the intent to read, then load the data from the file into indicatorsDict, then close the file. If the data was given to us as a json-formatted string, s, instead of a file, we would load it using the line indicatorsDict = json.loads(s).

We will demonstrate programming with JSON in the next chapter, followed by JSON Schema and constraints in Chap. 17. We will also see JSON files in Chaps. 20 and 21, in the context of data obtained from Internet sources and web scraping.

15.3.1 Reading Questions

15.7 Explain the difference between JSON, which is *textual* (i.e., in a text file, or in a *string* in our programs), and the *data structure* that we get from performing a load() or loads() based on the JSON.

15.8 What is the name for the JSON equivalent of a *Python list*? What about a *Python dictionary*?

15.9 Go to https://www.json.org/json-en.html and look up the allowed syntax for a *JSON string*. Based on that syntax, write down three examples of *Python strings* that would be *illegal* as JSON strings.

15.10 Look carefully at the syntax diagrams (aka train-track diagrams) for JSON at https://www.json.org/json-en.html. Consider the following nested structure as JSON:

```
{"a": 152,
 "b": ["foo", "bar", true],
 }
```

Using the syntax diagrams, trace through the example. Then describe *why* JSON supports trees of information through nesting.

15.11 Suppose you have the following small subsets of data from the subjects and courses tables of the school database. Write a single JSON text string (or file) whose contents, in a single tree, represent the same information.

subject	name	department
CS	Computer Science	MATH
MATH	Mathematics	MATH
ENGL	English Literature	ENGL

15.12 Please write down a portion of your family tree with depth at least three, and with at least 8 nodes. You will be the root. Please give your tree in the form of a

subject	coursenum	title
CS	110	Computing with Digital Media
CS	372	Operating Systems
MATH	210	Proof Techniques
ENGL	213	Early British Literature

JSON *object*, following the model in the reading. Please use horizontal whitespace to make the tree structure easy to follow, and represent one node per line.

15.3.2   Exercises

15.13 Consider the table of data:

subject	name	department
CS	Computer Science	MATH
MATH	Mathematics	MATH
ENGL	English Literature	ENGL

This could be represented in a Python program as a *dictionary* mapping from a key, `"subjects"`, to a *list* of dictionaries, one per subject. These inner dictionaries map from `subject`, `name`, and `department` to the respective value for the row.

Write Python code to construct the above representation as a Python in-memory data structure called `ds`, and then have your code encode the data structure into a JSON-formatted string, and print the string. Then take `ds` and encode it as a JSON-formatted string, but include `indent=2` as an argument in the conversion, and print it.

15.14 As in the last question, create the in-memory data structure, but now create a local file named `school.json` with the JSON-formatted data. Use the `indent=2` argument when encoding as JSON.

15.15 This question assumes that you successfully accomplished the previous question, and have a file `school.json` with the JSON-formatted data structure originally specified in the first exercise.

Write code to read and decode the JSON-formatted `school.json` file into a data structure, assigning to variable `ds2`. Once created, manipulate the `ds2` data structure by appending a row of data to the `subjects` list with value:

```
{
  "subject": "PHYS",
  "name": "Physics",
  "department": "PHYS"
}
```

15.4 XML

More powerful, general, and widespread than the JSON format, the eXtensible Markup Language (XML) is a general structure for modeling hierarchical data, with the power to create a wide variety of new formats [90]. XML was created initially for the Internet, and can be viewed as a super-set of the HyperText Markup Language commonly seen in web design (that is, HTML is a special case of XML). The XML format is a common format widely used for data sharing and data transport, e.g., when data is provided by an API, in web pages, and in local files. XML data is stored as plain text, so an XML file can be opened with any text editor (ideally, one that is "syntax aware"). Like JSON, XML is designed to be programming language agnostic, that is, to work with any application that seeks to use it. Because XML data is simple text data, it is easy to communicate between operating systems, data systems, applications, etc. In web scraping, one often comes across XML pages as well as HTML and XHTML pages [73]. XML files are also commonly used as a format for data hosted by the US Government, e.g., [69].

Despite the word "language" appearing in its name, XML does not represent a *programming language*. When given an XML file, there is no code to execute. Instead, it is a language for communicating hierarchical data with great power and generality, with certain grammatical rules that we will discuss shortly.

15.4.1 XML Structure

An XML document represents a tree, and we will provide an example shortly displaying the indicators data set in XML format. As a tree, an XML document must have exactly one root. There are several types of nodes in an XML tree, including text nodes, attribute nodes, and internal nodes. Each internal node represents an XML *element*, which starts with a *start tag* (e.g., <indicators> for the root) and ends with the corresponding *end tag* (in this case, </indicators>). All data between the start tag and end tag represents data that is structurally subordinate to the node in question. Since the tree contains internal nodes under the root, there will be more start and end tags under <indicators> and above </indicators>. This is analogous to the use of curly brackets and nested dictionaries in JSON. To validly represent a tree, an XML document must *properly nest* these start and end tags.

Basic XML consists of three things:

- Tagged elements, which may be nested within one another.
- Attributes on elements, representing metadata associated with the element.
- Text, often representing essential data.

For example, to represent the tree of Fig. 15.2 in XML, we would start with the element catalog as the root, and the entire tree would be nested under this element. We would have 100 elements tagged book, all children of catalog. Each book element has a single attribute, mapping the key id to the value consisting of the id for the book in question. In general, attributes are stored in a dictionary, though in this example each attribute dictionary only has one key:value pair. In this example structure, each book has three elements nested under it: author, title, and price. The first author element has one child, a text node with the value "Shakespeare".

We will discuss attributes and text in a moment, but first we wish to explain the meanings of tags. In HTML, tags denote formatting, e.g., <title>, <i>, <table>, etc. To get a taste of HTML tags, you can open your favorite website in Google Chrome, then click View -> Developer -> View Source, to see the underlying HTML structure of the web page. There is a predefined set of HTML tags, with specific interpretations that affect how web pages are displayed. So when we consider HTML, there are two separable aspects:

- the tree structure, with a specific organization that constrains what tags are legal children of other tags, etc.
- the *interpretation* of those tags, which assigns *meaning* that, for instance, <h1> is a top level header, etc.

XML, on the other hand, only governs the legal structure–the tags and XML do not, in of themselves, have interpretations

In XML, we use the tags to denote the *meaning* of data, e.g. <Student>, <Book_Title>, etc. Tags can be any strings (case sensitive), subject to some restrictions on the characters allowed. Tags are chosen by the person creating the data set, and are part of the data, analogous to column headings in the tabular model. The tag structure means an XML document describes its own syntax. The structure of XML makes it both human and machine readable. As in our examples from the previous section, the tree structure is indicated both by start and end tags (which, for JSON, were curly brackets), and via the horizontal white space on each line (to clearly display the nesting structure). This structure can be seen below where we display the indicators0 data set in XML format.

In the example below, the first line is an optional XML prolog, and is *not* part of the XML tree itself. This prolog is almost always present when the XML is in a file. The second line and last line are the start and end tags for indicators discussed above. We also see the same structure for each of the three children of the root (countries). Each country element starts and ends with a tag. For example, the start tag for France occurs on line 3 and the corresponding end tag on line 12, denoting that all data associated with France is on these lines in the XML file. Similarly, the timedata start and end tags denote portions of the tree for 2007 (under FRA), for 2017 (under FRA), for 2007 (under GBR), etc. Lastly, under each timedata we have the population and gdp, each containing a single number (in text form, as is required by XML), wrapped in start and end tags.

```
<?xml version='1.0' encoding='UTF-8'?>
<indicators>
```

```
<country code="FRA" name="France">
  <timedata year="2007">
    <pop>64.02</pop>
    <gdp>2657.21</gdp>
  </timedata>
  <timedata year="2017">
    <pop>66.87</pop>
    <gdp>2586.29</gdp>
  </timedata>
</country>
<country code="GBR" name="United Kingdom">
  <timedata year="2007">
    <pop>61.32</pop>
    <gdp>3084.12</gdp>
  </timedata>
  <timedata year="2017">
    <pop>66.06</pop>
    <gdp>2637.87</gdp>
  </timedata>
</country>
<country code="USA" name="United States">
  <timedata year="2007">
    <pop>301.23</pop>
    <gdp>14451.9</gdp>
  </timedata>
  <timedata year="2017">
    <pop>325.15</pop>
    <gdp>19485.4</gdp>
  </timedata>
</country>
</indicators>
```

Because of this logical structure, it is easy to write code to identify sections of the tree, by searching for an end tag corresponding to a given start tag. The nested structure also allows us to build more complicated trees using composition, or to quickly identify locations to add new data to the tree. For example, if we receive data from a fourth country, we could find an end tag </country> and then add the data we receive starting on the next line with a new start tag <country> corresponding to the new data. Additionally, the information about how to decode the XML file is present in the top level tag, which also specifies the XML version used.

As we have mentioned, XML attributes allow us to store metadata in the tags. Specifically, every XML element comes with a set of *attributes* of the form key="value". For example, we see this in tags above like <country code="FRA" name="France"> and <timedata year="2007">. We see that when an element has multiple attributes, they are separated by a space. For a given element, attribute keys must be unique (e.g., one could not have `<country code="FRA" name="France" code="fra">`). Note that some authors write "name" where we have written "key," but we chose to avoid confusion with the explicit key `name` in the indicators example. Attributes are common in HTML, e.g., the `href` attribute that we will discuss in Chap. 22.

One special attribute is the *identifier* attribute *id*, which assigns an identifier to an element, unique across all elements in the XML tree. Including this special attribute is analogous to including an index in the tabular or relational data models.

Lastly, text nodes are encoded in XML as strings of text that appear between a start tag and an end tag, e.g., `<pop>301.23</pop>` tells us that the internal node `pop` has one child, `"301.23"`. It is possible for an element to have both element children and text children (e.g., if the text `"Home of the Free"` was added after the start tag `<country code="USA" name="United States">` and before the first `timedata` child in `indicators0` above). Such an element is said to have *mixed content*. We shall not consider such elements, as they violate our policy that text leaf nodes should not have siblings. As XML allows a great deal of flexibility, defining and communicating such constraints on XML trees is extremely important in practice. This is the topic of Chap. 17.

The person creating the XML data decides whether data appears as an attribute or as an element. For example, in `indicators`, instead of the element

```
<timedata year="2007">
  <pop>64.02</pop>
  <gdp>2657.21</gdp>
</timedata>
```

we could provide the same data in the form

```
<timedata>
  <year>2007</year>
  <pop>64.02</pop>
  <gdp>2657.21</gdp>
</timedata>
```

It is always possible to translate an XML that uses attributes into one that does not by making each attribute a child of the element in question.

This decision is made based on whether `year` is thought of as metadata or not. Of course, to extract XML data into another form (e.g., tabular), one needs to know ahead of time whether the data is encoded via attributes or text nodes. Since attributes cannot contain tree structure or multiple values, there is an argument for avoiding attributes as a way to store anything other than metadata. For instance, choosing to store all the data in a single attribute dictionary undermines all the values of XML data.

XML allows the user to define their own tags, elements, or attributes, and this explains the way in which XML is "extensible" (i.e., it is possible to extend it to meet your needs). Applications built to work for an old version of an XML file will continue to work for a new version, even if some tags are deleted and others added. Applications should be designed to work with the tags that are present at any moment.

XML files are required to satisfy a "grammar" with rules such as "there must be exactly one root element," "every start-tag must have a corresponding end-tag," and "elements must nest properly" (e.g., in indicators, we could not close a country element before closing one of its timedata children). Tag names cannot contain whitespace, colons, or the sequence xml, and attribute values cannot use the symbols &, <, or >, to avoid confusion with the symbols reserved for starting and ending tags. XML has reference characters, such as > for > and & for &, that allow us to include these reserved symbols in a way that does not cause confusion.

When an XML file satisfies the grammatical rules, we call it *well-formed*. When an XML file is well-formed, it can be read in an automated way, the relationships between elements (i.e., the edge structure) can be mapped, and the hierarchical data represented can be drawn as a unique tree. This automated process is known as *parsing*, and we return to it in Chap. 17. Often, data system providers will create constraints, so that only a certain set of tags can be used, and only in certain ways. Such constraints streamline the process of transmitting XML data between different applications. Examples include XMLNews (for XML data in news media applications), XML for national weather data applications, and, of course, HTML.

For the purposes of this book, we will always include an end tag associated with every start tag. However, the reader should be aware that there are XML files that omit end tags, by using a special syntax for start tags. Specifically, if a start tag ends with the symbol /, this means no end tag will appear. This sometimes happens when a node has no children. For example, if we wanted to add a fourth country, China, for which we had no data on population or gdp in 2007 or 2017, then we could add the China node as a child of the root, and without any children of its own. We could do so with a single self-ending tag <country code="CHN" name="China"/>, e.g., placed in the second-to-last line of the XML file above.

We also feel obliged to mention that XML data can be formatted using CSS (Cascading Style Sheets) or XSL (Extensible Stylesheet Language) to translate XML to HTML. However, we will not have need of this fact. We also mention that XML files can contain comments, and we will return to this point in Chap. 17.

15.4.2 Extracting Data from an XML File

When given an XML file, we use the Python library lxml to extract data from the file [31]. This abstracts the process of parsing the file to map out the tree structure, and allows us to access individual elements in the tree. The name etree stands for "Element tree", and the etree module is designed to parse a given XML file into the associated Element tree. In the code below, stripparser is a custom parser that removes blank text nodes. In Chap. 22, we will use a custom parser that knows about the particular structure of HTML files. Once we have the `Element` tree, `indtree`, the `getroot()` function returns its root `Element`, `indroot`. This `Element` contains the XML data of the entire tree.

```
from lxml import etree

pathInd = os.path.join(datadir, "ind0.xml")

stripparser = etree.XMLParser(remove_blank_text=True)
indtree = etree.parse(pathInd, stripparser)
indroot = indtree.getroot()
```

Once we have extracted the root, we can look at summary information about it. For example, the type of `indroot` is `Element`, and the *length* of an element is the number of children it has. We can also extract the tag associated with an element, or its attributes (which are empty for the root of `indicators`).

```
print("Type:", type(indroot), "Length:", len(indroot))
```

```
| Type: <class 'lxml.etree._Element'> Length: 3
```

```
print("Root Tag:", indroot.tag)
```

```
| Root Tag: indicators
```

```
print("Root Element Attributes", indroot.attrib)
```

```
| Root Element Attributes {}
```

Indeed, it is possible to print the entire tree, and doing so we recover the XML structure we displayed earlier. Note that the option `pretty_print=True` guarantees that we display one start tag per line, and use horizontal whitespace to display the nesting structure.

```
s = etree.tostring(indroot, pretty_print=True)
.decode('utf-8')print(s)
```

```
| <indicators>
|   <country code="FRA" name="France">
|     <timedata year="2007">
|       <pop>64.02</pop>
|       <gdp>2657.21</gdp>
|     </timedata>
|     <timedata year="2017">
|       <pop>66.87</pop>
|       <gdp>2586.29</gdp>
|     </timedata>
|   </country>
|   <country code="GBR" name="United Kingdom">
```

```
|      <timedata year="2007">
|        <pop>61.32</pop>
|        <gdp>3084.12</gdp>
|      </timedata>
|      <timedata year="2017">
|        <pop>66.06</pop>
|        <gdp>2637.87</gdp>
|      </timedata>
|    </country>
|    <country code="USA" name="United States">
|      <timedata year="2007">
|        <pop>301.23</pop>
|        <gdp>14451.9</gdp>
|      </timedata>
|      <timedata year="2017">
|        <pop>325.15</pop>
|        <gdp>19485.4</gdp>
|      </timedata>
|    </country>
| </indicators>
```

In addition to finding information about the root element, we can easily find information about other elements. For example, if X were the middle country node, then X.attrib would be the dictionary {"code":"GBR", "name":"United Kingdom"}, and X.tag would be "country". Furthermore, the find function will locate children with a given tag, one at a time. This function returns the first Element found with the given tag, searching from top to bottom in the XML file (equivalently, searching the children from left to right in the associated tree).

```
findCountry = indroot.find('country')
findCountry.attrib
```

```
| {'code': 'FRA', 'name': 'France'}
```

We see that, as expected, the find function finds France first. We can use the getnext() function to find the next match, i.e., the sibling just to the right of the current Element.

```
nextCountry = findCountry.getnext()
nextCountry.attrib
```

```
| {'code': 'GBR', 'name': 'United Kingdom'}
```

As expected, the second country under the root is Great Britain. We can also use the findall function to locate all children. This returns a list of Elements.

```
countrylist = indroot.findall('country')
```

We have already seen in Chap. 2, the importance of being able to provide a path from the root of a file system tree to any individual element. Relative paths, such as from a node representing a folder to a file stored in some sub-folder, are also important. Finding such paths in XML trees is easy, and forms the basis for XPath, which we will discuss in the next chapter.

15.4.3   Reading Questions

15.16 Please write down a portion of your family tree with depth at least three, and with at least 8 nodes. You will be the root. Please give your tree in the form of a XML format, with one node per line, and using tags and attributes as illustrated in the reading. Use horizontal whitespace to make the tree structure clear to follow.

15.17 Please describe three kinds of attributes you might include in your family tree, and why it would be valuable to store this data.

15.18 When we use the `lxml` library, why is it that doing `getroot()` then printing the result, has the effect of printing the entire tree?

15.19 Is it possible to use the `find()` function within the `etree` module to determine the depth of an element you are interested in?

15.4.4   Exercises

15.20 Consider the table of `subjects` data:

subject	name	department
CS	Computer Science	MATH
MATH	Mathematics	MATH
ENGL	English Literature	ENGL

Using a *text editor*, edit and create a file named `subjects.xml` in the current directory that creates a legal XML representation of this data. Once created, write a Python code sequence to read and parse the file, and then, using the technique from this section, print the entire tree.

15.21 Building on the last question, using a text editor, edit and create `school.xml` that contains a single XML tree with both the `subjects` table from the previous question, and the `courses` table presented here:

subject	coursenum	title
CS	110	Computing with Digital Media
CS	372	Operating Systems
MATH	210	Proof Techniques
ENGL	213	Early British Literature

Once created, write a Python code sequence to read and parse the file, and then, using the technique from this section, print the entire tree.

Further Exploration

There are many online lessons and tutorials that can be used to delve more deeply into JSON and XML [72, 65, 77, 66].

In addition, there are online resources for formatting and validating JSON and XML [12, 27, 13, 5].

Chapter 16
Hierarchical Model: Operations and Programming

Chapter Goals

Upon completion of this chapter, you should understand the following:

- Procedural mechanisms (using Python and `lxml`) as operations to operate on and retrieve data from tree structures.
- The constraints on when tree data can be read into a `pandas` data frame.
- Differences between the hierarchical model and tabular/relational models in terms of how missing data is handled.

Upon completion of this chapter, you should be able to do the following:

- Extract hierarchical data from a JSON object into a `pandas` data frame, when this is possible.
- Add data to a JSON-formatted hierarchical structure, and update data that is already present, by specifying a path within the tree.
- Extract hierarchical data from an XML file into a `pandas` data frame when this is possible.
- Create an XML file from a tree stored as a JSON object
- Create and structure declarative queries to get desired data out of a tree (i.e., XPath).

© Springer Nature Switzerland AG 2020
T. Bressoud, D. White, *Introduction to Data Systems*,
https://doi.org/10.1007/978-3-030-54371-6_16

Chapter 15 focused on hierarchical structure, and included a discussion of the use of JSON and XML for representing tree data. This chapter will focus on the operations involved in *traversing* such structures and formats, and explore both declarative and programmatic means to process the data and build data frames in our programs to allow for analysis. We assume throughout that our XML files are well-formed, as discussed in the next chapter. We also discuss how to *create* hierarchical data structures, both JSON and well-formed XML, given Python native data structures.

16.1   Operations Overview

When we have a tree, and that tree contains data, we have to change our view of the operations that are appropriate to this new structure. For both the tabular and relational model, we started with operations to access rows and columns of the structure, as well as to access individual elements. We also partitioned our operations into those that were explicit in our programming—the operators and functions in `pandas`, from the operations that specified what we wanted, but let the language and systems determine how to get it—SQL.

In the hierarchical model we will look at procedural operations, both for trees that come from JSON-formatted data, and for `lxml etree` structures that come from XML-formatted data. In the space of XML, we will also see the declarative type of operation, in the form of a language called *XPath*, so called because its declarative statements come in the form of paths that can be used to specify sets of tree elements for the desired data.

We start with describing, at a logical level, the basic operations that may be required when our structure is a tree. Our goal is to, in this structure, access the data that makes up the values of variables for a data set contained in the tree.

Access Operations
We saw in the last chapter that the data in a JSON-based tree are the scalars within lists and mapped to in dictionaries. When our tree is XML-based, we talk about *elements* or *nodes*, and the data could be entailed in the *tag*, the *attributes*, and the *text* of those element nodes, depending on how the creator designed the structure to hold the data set.

Traversal Operations
When we are explicitly writing procedural algorithms to access data from multiple elements in a tree, we must have operations that, given an element, give us the ability to *traverse* other elements that are treewise related to that element. These operations could include:

- Given an element, traverse the *set of children* of that element.
- Given an element, traverse to the *parent* of that element.
- Given an element, traverse to either the left- or right-*sibling* of that element.

In JSON-based trees, the first of these traversals, over children, is simply based on iterating over indices or list items when the current node is a list, and iterating over keys or items when the current node is a dictionary. These trees do not have explicit operations for traversing to parents or to siblings from a node.

In XML-based trees, all three types of traversal are supported.

Declarative Operations

Declarative operations specify what data we want extracted from a tree, without explicit iteration and traversal over the nodes of the tree. Declarative operations only apply to XML-based trees. Analogous to SQL in the relational model, we use an independent language (XPath and/or FLOWR) to specify, using the paths that are inherent in the structure of a tree, the *set of data* we wish to extract from the tree, and the language allows us to incorporate wildcards and predicates/conditionals to select the data of interest.

Update Operations

At times, our goals in building data systems clients might entail update and creation—write-style operations, as opposed to traversal and access—read-style operations. Given an existing tree and its element nodes, we require operations to *update values*. For JSON-based trees, this means changing values in lists and dictionaries, and for XML-based trees, this means changing the tag, the text, or the attributes associated with an element node. If we are creating a tree where one did not exist previously, or are augmenting an existing tree, we must be able to create/construct the new element nodes in a tree, and specify values for their tag, text, and attributes.

The remainder of the chapter will proceed through the presentation of examples that illustrate many of the most useful operations applicable to JSON- and XML-based tree structures. Given our data systems focus, we are particularly interested in how we start from a tree containing a data set and map the data contained therein into one or more tabular structures. In data analytics and data science, this work of understanding and transforming between different forms of data is known as *wrangling*, and we also employ that term here.

We will also illustrate instances of going in the other direction, and *creating* JSON- and XML-based tree structures, starting from one or more tabular structures.

We rely on the JSON and XML associated software packages available for Python, and so presume, beyond the standard imports, the appropriate imports for json and for the lxml etree have been issued.

```
import json
from lxml import etree
import util
```

Also in our examples, we will be making use of the custom util module (documented in Sect. A.1) to access convenience functions for printing JSON- and XML-based tree structures, in particular a print_data() and print_xml().

These functions have commonality in allowing specification of the width of the output and the number or lines to print, and are the analogue to the `pandas` `head()` method and to `LIMIT` in SQL.

16.1.1   Reading Questions

The following set of examples ask you to consider the book catalog tree from the previous chapter and ask you to think of concrete examples, using this tree, of the kinds of operations covered in this section.

16.1 Based on the book catalog tree, give two concrete examples of *Access Operations*.

16.2 Based on the book catalog tree, give concrete examples of *Traversal Operations* involving:

- given a "current" node, traversing the *children* of that node,
- given a "current" node, traversing to the *parent* of the node, and
- given a "current" node, repeated traversing to the right sibling of that node.

16.3 To get an idea of what a declarative operation might look like, try the following:

Write an English sentence that conveys, in as specific a way as you can, but without using terms like "loop" or "iterate," state that you want the set of text associated with the price nodes in the tree.

16.4 Based on the book catalog tree, give a concrete example of a set of *Update Operations* on the tree.

16.2   JSON Procedural Programming

We begin with hierarchical data based on the JSON format. Recall the Python native data structures we introduced in Chap. 15 for representing hierarchical data. The given JSON file can provide us hierarchical data in a number of forms. We will focus on the case of a dictionary of dictionaries (DoD), which is far more common in practice, and has the benefit of containing name information for the data stored. Note that the list of lists (LoL) `indicatorsList` from Chap. 15 does not include information about the meaning of the data it stores, e.g., that some leaves represent population data and the others represent gdp data.

We begin with illustrating how to read and traverse DoDs, thus focusing on our access operations and traversal operation. One advantage of working with JSON data is that numbers really are numbers, so we do not have to convert from string type [55].

16.2.1 Access and Traversal Operations Example

In this section, we illustrate how to extract (wrangle) data from a JSON-based data structure into a pandas data frame. Note that not every tree can be put into tabular form in a meaningful way. We will focus first on trees of depth two, which readily yield a row and column field tabular structure. Then we will turn to trees of depth three, we will need a tabular structure with a two-level index.

When our goal is a pandas data frame, we must choose between the various native Python representations for row–column data sets that will be used in the construction of the DataFrame object. We originally described these representations in Sect. 3.4. When starting from our example tree, the traverse will proceed, at the outer level, row by row. For each row, we will be extracting the names and values for each of the column fields within the row. Because of the row-centric outer processing, we want a row-centric data structure, with the outer object being a list of rows. Because, for each row, we will have the name of the column along with the value for the column, and to make our software robust against possible ordering differences in the tree, we want each row to be represented by a dictionary.

Summarizing the processing plan:

- Wrangle the data from the tree into a list of dictionaries (LoD), where

  - the items in the list represent rows, and
  - the dictionary maps column names to field values.

- This data structure can be passed to the DataFrame constructor and, without additional work, result in the desired pandas data frame.

The process of carrying out this plan gives us practice with the procedural operations one usually needs for hierarchical data. Specifically, we will illustrate the following operations:

- Given a node in the JSON-based DoD structure, traverse over its children:

  - For a Python dictionary D, D.items() yields an iterator of tuples T where T[0] is a key in D and T[1] is the associated value (in this case, itself a dictionary).
  - We iterate over D.items() and extract the data one row at a time.

- Restructure a DoD tree of depth = 2 into a LoD or LoL to pass to pandas.
- Given a DoD tree of depth > 2, restructure it into a pandas data frame with a multi-index.

  - When a leaf is a dictionary, use the dictionary access operator to get the value associated with a key.
  - To build the rows of a list of lists, we sometimes have to access individual row dictionary elements.

Problem Solving Strategy Takeaways

- We first need to understand the *structure* of the JSON-based data and to visualize the table or tables that are desired.
- We need to have a clear picture of the native Python data structure that we are processing into that will allow a data frame creation.

On the positive side, with regard to this processing, it is *easy* to traverse over the lists and dictionaries from JSON-based data set—these are simple (but nested) `for`-loops. On the negative side, because the design of each JSON-based tree of data can be different, `pandas` cannot automatically help us here, like it could with CSV-formatted files or in obtaining data from SQL queries.

16.2.1.1   Example: Simple Table in JSON

In this section, we work with a data set of economic indicators from three countries in 2016. This data set can naturally be structured either in tabular or hierarchical form. We receive the data as a hierarchically structured JSON object, and produce a `pandas` data frame for the data. We begin by reading from the JSON file into a Python native DoD, and print the first set of lines to help us understand the structure.

```
path = os.path.join(datadir, "ind2016.json")
with open(path) as f:
    ind2016 = json.load(f)


util.print_data(ind2016, nlines=15)
```

```
| {
|    "CAN": {
|       "country": "Canada",
|       "pop": 36.26,
|       "gdp": 1535.77,
|       "life": 82.3,
|       "cell": 30.75
|    },
|    "CHN": {
|       "country": "China",
|       "pop": 1378.66,
|       "gdp": 11199.15,
|       "life": 76.25,
|       "cell": 1364.93
|    },
```

Looking at this data, it is clear that the internal dictionaries represent the rows we want in our resulting `pandas` data frame. That is, we should have rows corresponding to country codes, and columns for `country`, `pop`, etc. Our plan will be to read this data into a list of dictionaries (one per country), where each of the dictionaries contains a key:value pair for code, that we can use as index in the resulting pandas data frame.

Using ind2016.items(), we can gain access to the key:value pairs in ind2016. Here each key is a code and the value is the dictionary of indicators. Because we want a column for the country codes, we add one more key:value pair to each interior dictionary, representing the code. Note that, because dictionaries are mutable, we make a copy() of each of the interior dictionaries so as not to mutate the originals in ind2016.

```
LoD = []
for code, rowD in ind2016.items():
    rowD = rowD.copy()
    rowD['code'] = code
    LoD.append(rowD)
df = pd.DataFrame(LoD)
df.set_index('code', inplace=True)
df.head()
```

```
|               country        pop        gdp    life      cell
| code
| CAN            Canada      36.26    1535.77   82.30     30.75
| CHN             China    1378.66   11199.15   76.25   1364.93
| IND             India    1324.17    2263.79   68.56   1127.81
| USA     United States     323.13   18624.47   78.69    395.88
```

In the solution above, in the first iteration we have a dictionary in rowD that is already close in structure and content to what we want as the first dictionary in LoD, which is {'code':'CAN','country':'Canada',...,'cell':30.75}. Making a copy of rowD and adding the code mapping to complete the row translation. From the perspective of traversals, this means it was enough for Python to only consider paths of length one in ind2016. In general, we will have more work to do, looking at the subtree that represents each row. As the data is inherently two-dimensional, we should be ready to use nested loops to access the data. When we move to trees of depth three, we will need triply nested loops. To build up to this, we present an alternative approach to the problem above that wrangles the data into a list of lists instead of a list of dictionaries, illustrating the use of nested loops.

When we wrangle ind2016 into a list of lists, we will also need a list of column headers. These come from the header name code plus the keys for one of the rowD dictionaries of ind2016. The code below employs a Boolean, first that allow us to record these column names on just the first iteration of the outer loop. We next need to iteratively fill rows with the items from the interior dictionaries of ind2016, and then append each row to the LoL we are building. We illustrate:

```
LoL = []
column_names = []
first = True
for code, rowD in ind2016.items():
```

```
        if first:
            column_names = ['code'] + list(rowD.keys())
            first = False
        row = [code]
        for field, value in rowD.items():
            row.append(value)
        LoL.append(row)
util.print_data(LoL, nlines=17)
```

```
 | [
 |    [
 |        "CAN",
 |        "Canada",
 |        36.26,
 |        1535.77,
 |        82.3,
 |        30.75
 |    ],
 |    [
 |        "CHN",
 |        "China",
 |        1378.66,
 |        11199.15,
 |        76.25,
 |        1364.93
 |    ],
```

The first list, row, in LoL is ['CAN', 'Canada', ..., 30.75]. From the perspective of traversals, the line row.append(value) occurs at depth two, i.e., at the end of a path of length two in the tree. Note that the use of rowD.items() in the code above ensures that each value is inserted into row in the correct order, matching the header list obtained from rowD.keys(). Compare the header order listed below with the column order above.

column_names

```
 | ['code', 'country', 'pop', 'gdp', 'life', 'cell']
```

Once we have a LoL and list of headers, pandas is able to create a data frame.

```
df = pd.DataFrame(LoL, columns=column_names)
df.set_index('code', inplace=True)
df
```

	country	pop	gdp	life	cell
code					
CAN	Canada	36.26	1535.77	82.30	30.75
CHN	China	1378.66	11199.15	76.25	1364.93

```
| IND          India   1324.17    2263.79  68.56  1127.81
| USA United States     323.13   18624.47  78.69   395.88
```

16.2.1.2   Single Table from JSON with Additional Level

We turn now to a tree of depth three, representing the indicators data from Chap. 15, which has, for each country, both multiple years and, for each year, multiple indicators. We parse and print a prefix of the data to help understand its structure:

```
path = os.path.join(datadir, "ind0.json")
with open(path) as f:
    ind0 = json.load(f)
util.print_data(ind0, nlines=20)
```

```
| {
|    "FRA": {
|       "2007": {
|          "pop": 64.02,
|          "gdp": 2657.21
|       },
|       "2017": {
|          "pop": 66.87,
|          "gdp": 2586.29
|       }
|    },
|    "GBR": {
|       "2007": {
|          "pop": 61.32,
|          "gdp": 3084.12
|       },
|       "2017": {
|          "pop": 66.06,
|          "gdp": 2637.87
|       }
```

Here, we see that we have children of the root representing countries, and then children of the country nodes representing years, and children of the years representing the individual indicators.

One solution strategy is to wrangle this data into a pandas data frame with a two-level column index, with one level being year, and another level being the indicators within the year, like we saw originally in Chap. 6. A row in our target data frame will contain all the information for one country, and there will be four data-carrying columns, one each for pop in 2007, gdp in 2007, pop in 2017, and gdp in 2017.

Our outer loop iterates over the children of the root (that is, country codes) and should generate exactly one row in our accumulation of rows per root. We cannot directly mimic the LoD approach from above, as it would result in a list of DoDs, which pandas cannot handle. Instead, we have to "look inside" the internal dictionaries in ind0, and traverse each branch of the tree. Because each leaf is at depth three, this requires a triply nested loop.

Normally, a triply nested loop implies three-dimensional data. Of course, for a pandas data frame, we want two-dimensional data. We can reduce from three-dimensional data to two-dimensional data by inserting a composite value into our LoD, representing an indicator value and the year in which the data occurred. The solution below carries out this plan, using nested loops (by year, and then by indicator), and, within the inner loop, creating columns for each year/indicator combination and inserting data into these columns. We use string manipulation to combine the key of the year loop with the key of the indicator-loop to get our column names.

```
LoD = []
for country, country_subtree in ind0.items():
    countryD = {'code': country}
    for year, indicators in country_subtree.items():
        for indicator, value in indicators.items():
            col_name = "{}-{}".format(year, indicator)
            countryD[col_name] = value
    LoD.append(countryD)
df = pd.DataFrame(LoD)
df.set_index('code', inplace=True)
df
```

	2007-pop	2007-gdp	2017-pop	2017-gdp
code				
FRA	64.02	2657.21	66.87	2586.29
GBR	61.32	3084.12	66.06	2637.87
USA	301.23	14451.90	325.15	19485.40

This sort of data frame should look familiar from Chap. 8. In order to make it conform to tidy data standards, we can convert the column names into a two-level index and then stack the year in pandas.

An alternative approach would be to read ind0 directly into a pandas data frame with a two-level index. For this, we first identify the index we want (namely code and year), then for each value of the index, we create a dictionary holding the data of that row. Our list of dictionaries consists of these dictionaries, and each becomes a row in the resulting data frame. In the code below, the country loop and year loop give us all values of the index, we create a dictionary for each (code, year) pair, and the innermost loop fills this dictionary, with key:value pairs like "pop": 64.02 and "gdp": 2657.21.

```
LoD = []
for country, country_subtree in ind0.items():
    for year, indicators in country_subtree.items():
        code_yearD = {'code': country, 'year': year}
        for indicator, value in indicators.items():
            code_yearD[indicator] = value
        LoD.append(code_yearD)
df = pd.DataFrame(LoD)
df.set_index(['code', 'year'], inplace=True)
df
```

		pop	gdp
code	year		
FRA	2007	64.02	2657.21
	2017	66.87	2586.29
GBR	2007	61.32	3084.12
	2017	66.06	2637.87
USA	2007	301.23	14451.90
	2017	325.15	19485.40

16.2.2   Node Creation

Having discussed how to read a JSON-formatted tree into a pandas data frame, by traversing each edge in the tree, we turn now to the problem of how to create new nodes in the tree. This is a realm where hierarchical structure behaves very differently from tabular or relational structure. With tabular data, we would need to add an entire row to represent new data (and would need to include cells with value NA to represent missing data). In the hierarchical model, we can add additional information *anywhere in the tree* whenever we decide we want to. This means we do not have to add a value for every column/field, but rather only the ones that need it.

When the JSON data comes to us as a dictionary, we use the usual syntax for adding key:value pairs to a dictionary, namely D[key] = value. Here, the key is a country code, and the value is a tree, represented as a dictionary. For the example of ind2016, the equivalent of adding an entire row would be adding a subtree at depth one, e.g.:

```
ind2016["CHN"] = {"country": "China",
    "pop": 1378.66,
    "gdp": 11199.15,
    "life": 76.25,
    "cell": 1364.93}
```

We could also add data for countries where we have incomplete information, by simply adding a dictionary with fewer than five entries, e.g.:

```
ind2016["CUB"] = {"country": "Cuba",
    "pop": 11.48,
    "life": 79.74,
    "cell": 3.99}
```

We could even just add a single indicator, e.g.:

```
ind2016["IMN"] = {"country": "Isle of Man",
    "pop": 0.08}
```

Sometimes, data arrives at different times. In such situations, it can be added to the tree as it arrives. The first step is to create a subtree for the country in question, e.g.:

```
ind2016["SXM"] = {"country": "Sint Maarten (Dutch part)"}
```

Later, when a key:value pair associated to "SXM" in 2016 arrives (e.g., "pop":0.4), we can add it under the node "SXM" in our tree, i.e., add it to the dictionary associated to "SXM". To do so, we recall how paths are defined in JSON tree data, and we simply provide the path and the datum we wish to add:

```
ind2016["SXM"]["pop"] = 0.4
```

Note that this only works once a dictionary ind2016["SXM"] has been established. For example, to update ind0 to include the datum above, we would need three steps:

```
ind0["SXM"] = {}
ind0["SXM"]["2016"] = {}
ind0["SXM"]["2016"]["pop"] = 0.4
```

Alternatively, we could do this in one step by specifying a nested dictionary:

```
ind0["SXM"] = {"2016": {"pop":0.4}}
```

In this case, we have specified a new path of length three from the root to the new leaf.

16.2.3   Node Attribute Updates

In the last examples above, we saw how to add a single leaf to an existing tree. The procedure for updating a leaf to take a different value is completely analogous, thanks to the simplicity of Python dictionaries. For instance, if we learned that the data had an error, and in fact the population of Sint Maarten in 2016 was 40,000 instead of 400,000, then we could update the value by specifying the path and then specifying the new value. We illustrate for each of our two example trees:

```
ind2016["SXM"]["pop"] = 0.04
ind0["SXM"]["2016"]["pop"] = 0.04
```

If we wanted to change the units of all population leaves in the tree (to put each into units of people, for instance), we could do so by specifying paths to each and updating them. We demonstrate for ind2016:

```
for code in ind2016:
    oldPop = ind2016[code]["pop"]
    ind2016[code]["pop"] = oldPop*1000000
util.print_data(ind2016, nlines=16)

    | {
    |    "CAN": {
    |       "country": "Canada",
    |       "pop": 36260000.0,
    |       "gdp": 1535.77,
    |       "life": 82.3,
    |       "cell": 30.75
    |    },
    |    "CHN": {
    |       "country": "China",
    |       "pop": 1378660000.0,
    |       "gdp": 11199.15,
    |       "life": 76.25,
    |       "cell": 1364.93
    |    },
    |    "IND": {
```

The process for ind0 is very similar, but we must specify paths of length three to get the old value of population and to update the new value of population.

```
for code in ind0:
    for year in ind0[code]:
        oldPop = ind0[code][year]["pop"]
        ind0[code][year]["pop"] = int(oldPop*1000000)
util.print_data(ind0, nlines=22)

    | {
    |    "FRA": {
    |       "2007": {
    |          "pop": 64019999,
    |          "gdp": 2657.21
    |       },
    |       "2017": {
    |          "pop": 66870000,
    |          "gdp": 2586.29
```

```
|        }
|      },
|      "GBR": {
|        "2007": {
|          "pop": 61320000,
|          "gdp": 3084.12
|        },
|        "2017": {
|          "pop": 66060000,
|          "gdp": 2637.87
|        }
|      },
|      "USA": {
```

16.2.4   Reading Questions

16.5 The data in ind0 can be structured into a tree in multiple ways. For instance, the children of the root can be country codes, and their children can be years, as we saw in the reading. Alternatively, the children of the root can be years, and their children can be country codes. These two representations represent two ways of thinking about the data, and what is considered "top level" information. Recalling that iterating over children is easy, which of these two representations would you use if your goal was to produce a graphic with time on the x-axis, population on the y-axis, and one curve per country (showing how the population changes over time)? Justify your answer, using a discussion of nested loops and append operations that would be required. You may assume that in either representation, the years are increasing as you move from left to right in the tree (i.e., from top to bottom in the XML file).

16.6 Now suppose the data was given to you with the opposite hierarchical structure from the question above (swapping the top two levels). Describe a graphic you could easily make using this structure. Hint: think about what esthetic is at the "top level."

16.7 Based on ind0.json, show how to modify the GDP of France in 2017 to make it 5000.

16.8 Show how to set the GDP of the USA in 2018 to be $20.50 trillion (in the correct units to match the rest of the tree). Note that the GDP of the USA in 2017 was $19.39 trillion. You may assume the current status of ind0 is as in the reading.

16.9 The reading uses a DoD representation of a tree to read from JSON into pandas, and describes how to insert or update data into such a tree. The alternative representation for a tree is a list of lists (e.g., see indicatorsList from the previous chapter). What kind of header information would be required to uniquely identify data within this LoL structured JSON tree? Hint: start with a concrete question, like, how to update the GDP of France in 2017 (e.g., to set it to 5000).

16.10 Consider the `topnames` data set from the tabular data unit. Please give an example of how this data might logically come to you in JSON. Include at least 2 years (so, four "sex" subtrees). If you are stumped, you can look below to see it in XML.

16.11 Describe how you would read the JSON-formatted data you just created into `pandas`. You can also answer this question by providing the code. At the very least, be clear about whether this can be done with just one loop, or with doubly nested loops. Justify your answer.

16.2.5 Exercises

The first set of exercises involves processing a file, `slac.json`, that contains JSON-formatted data for the course catalog at a Small Liberal Arts College (slac).
Inspect the file and understand its structure:

- the top level is a dictionary, with only one key, `course` that maps to a list of dictionaries,
- each of these dictionaries describes exactly one course offering, with fields about the course, including `reg_num`, `title`, etc.,
- two of the entries within these dictionaries are dictionaries themselves.

16.12 Write "prefix" code to read in the JSON-formatted data from the file and generate a data structure called `slac` that contains the parsed data. Then use the utility function, `print_data()` to print the first 38 lines of the data structure.

16.13 Write code to iterate over the first 5 elements of the `course` list and print out the `reg_num` and `title` fields.

16.14 Write a function

```
slacDataframe(data)
```

that creates and returns a `pandas` data frame from the `slac` data. There should be a row per course, and columns, named as they are in the dictionary used to represent each course, but skipping the "time" and "place" sub-dictionaries. The data frame should have `reg_num` as the row index.

16.15 Rewrite function

```
slacDataframe(data)
```

that creates and returns a `pandas` data frame from the `slac` data from the last exercise. But in this case, traverse the "time" and "place" sub-dictionaries to populate columns `start_time`, `end_time`, `building`, and `room` in the built data frame.

The last exercise in this section extends our JSON-based data processing for the file `school_small.json`. While the file itself is not large, the design of this file incorporates some new aspects. For instance, the top level of the file is a dictionary with mappings to `departments`, `courses`, and `instructors`. The data is derived from the `school` database from the relational database unit, and this structure is a means to incorporate *three* tables into a single tree.

In addition, the `departments` entry in the JSON combines `subjects` with `departments`, so that, if a department supports more than one subject, a department node has a `subject` list describing the multiple subjects. If, on the other hand, a department has only one subject, it is assumed to be the same id and the same name as the department itself.

16.16 Write a function

```
processSchool(data)
```

that processes a JSON-based data structure, `data`, whose tree is that shown in `school_small.json`. The processing should produce a tuple of *three* pandas data frames, one for `departments`, one for `courses`, and one for `instructors`. This question is open-ended and the solution should determine the specific columns of the solution and determine how to handle multiple-subject departments in the design of the departments table.

16.3   XML Procedural Operations

We turn now to trees provided in XML format. We begin with a brief reminder of several relevant observations about XML, from Chap. 15:

- It is based on a Hierarchical/Tree Model.

  - It can better match a data or application domain.

- XML itself is a "meta-specification," different *grammars* or *vocabularies* use the same framework to define domain or application-specific specifications, to which instances can conform.
- It does not require predefined, fixed schema.
- It has a flexible hierarchical structure (also arbitrary graphs).
- The XML XPath declarative query can be more complex to understand:

  - An alternative is to use procedural methods to extract the desired information.
  - Contrast this complexity with the relational model and its declarative language of SQL, which is easier to understand.

As in the previous section, we first discuss reading from XML into `pandas`, and then we turn to creating and modifying XML-formatted data.

16.3.1 Reading and Traversing XML Data

Given an XML file, we can use `lxml` and `etree` to represent the data in terms of its root `Element`, as we saw in Chap. 15. Such an `Element` has many important attributes:

- The `.attrib` attribute provides a dictionary of the XML attributes associated to the `Element`.

  - The usual Python methods and tools, `.keys()`, `.values()`, `.items()`, `[key]`, `in`, etc., work on the `.attrib` dictionary.
  - The `get(<attribute>)` method of an `Element` node; a `get()` for an attribute that does not exist yields `None`.
  - The `set(<attribute>)` method works similarly.

- The `.text` attribute provides the text of the `Element`.

  - There is also `.tail` to get text that follows the `Element` and appears before the next `Element`.

- The `.tag` attribute gets the tag of the `Element`.

  An `Element` also has a number of useful methods we will need:

- `find(<tag-name>)` can be used to find the first matching subelement by tag name.

  - If there is no match as a child, the function reaches down further into the subtree.
  - This is a design decision, but in practice is better than counting on the order of data in the XML document to always give the right position for `find()`.

- `iterchildren(tag=<tagname>)` iterates over the children with the given tag.

We will demonstrate these operations presently. There are many, many, more attributes and methods associated with objects of type `Element`, and we encourage the reader to consult the online documentation [77]. In addition, we will demonstrate how, given a node, we can iterate over the node's children. This is *easy*, as any node, itself, acts as an iterator over its children, and thus can be "sequence" of a `for` loop.

One important place where care is required is in extracting XML `Element` attributes, because they are separate notions from Python language attributes of an object. We rely on the dictionary produced by `.attrib`, for this task.

16.3.1.1   Indicators Example

Throughout this section, we will work with the example of ind0, and we recall its structure and constraints:

- The root is indicators.
- The children of indicators are all country nodes.
- Each country node has attributes of code and name.
- The children of country nodes are all timedata nodes.
- Each timedata node has an attribute of year.
- The children of timedata are individual indicators.
- The indicator uses its tag to name the indicator, and its text property to give the (string) value for that indicator.

We read in the data just as in Chap. 15, and represent the tree in terms of its root Element, root0:

```
xmlparser = etree.XMLParser(remove_blank_text=True)
ind0_path = os.path.join(datadir, "ind0.xml")
try:
    tree0 = etree.parse(ind0_path)
    root0 = tree0.getroot()
except:
    print("Exception in parsing XML")
```

Because the data set ind0 has a natural tabular structure (with a two-level index, as discussed in the previous section), it makes sense to attempt to wrangle the tree above into a pandas data frame. To do this, we need to create a list of dictionaries as in the previous section. We use a triply nested loop, where the outermost loop iterates over the children of the root (extracting the country code and name from each), the middle loop iterates over depth two nodes (extracting the year from each), and the innermost loop iterates over the depth three nodes.

As discussed above, iterating over the children of a node is easy: the syntax for child in node provides an iterator for the children of the node in question. For each depth 1 and depth 2 nodes we consider, we use node.get(<attribute>) to get the value associated with the given attribute, e.g., country_node.get('code') gives the value associated with the attribute code, e.g., FRA. Recall that this node is represented in XML with the start tag <country code="FRA" name="France">, and the attributes are represented in Python as the dictionary {'code':'FRA','name':'France'}. The tag of this node is country, and the text of the node is everything between the start tag and the end tag (i.e., all data associated to France).

When we get to the depth three nodes, we extract the indicator as the tag of the node, e.g., for France in 2007, the child <pop>64.02</pop> tells us that the population in that year was 64.02 million people. If this node is stored in Python as indicator_node, then indicator_node.tag is "pop" and `indicator_node.text` is `"64.02"`. We cast the text as a float, and insert the corresponding pair into our dictionary representing the row for France in 2007.

For each row of the data frame we are trying to create (i.e., for each country and year), we build a dictionary, `rowD`, as we traverse the tree, and we append to the LoD at the end of the middle loop, before iterating to the next year or to the next country. Each time we append `rowD` to LoD, it is a dictionary of five mappings, one for each column in the resulting data frame. Pulling this solution together, we achieve the following code:

```python
LoD = []
for country_node in root0:
    code = country_node.get('code')
    name = country_node.get('name')
    for timedata_node in country_node:
        year = int(timedata_node.get('year'))
        rowD = {'code': code, 'name': name, 'year': year}
        for indicator_node in timedata_node:
            indicator = indicator_node.tag
            value = float(indicator_node.text)
            rowD[indicator] = value
        LoD.append(rowD)
df = pd.DataFrame(LoD)
df.set_index(['code', 'year'], inplace=True)
df
```

```
|                            name        pop        gdp
| code year
| FRA  2007           France      64.02     2657.21
|      2017           France      66.87     2586.29
| GBR  2007   United Kingdom      61.32     3084.12
|      2017   United Kingdom      66.06     2637.87
| USA  2007   United States     301.23    14451.90
|      2017   United States     325.15    19485.40
```

Now that we understand the structure of our solution above, we wish to apply it to the full `indicators` data set, stored as `indicators.xml`. Because the full data set has the same structure and constraints as `ind0`, it can be processed identically. Thus, we abstract our solution above into a function, and then apply it to the root `Element` obtained from parsing the full data set using `etree`.

```python
def xml_indicators(root_node):
    LoD = []
    for country_node in root_node:
        code = country_node.get('code')
        name = country_node.get('name')
        for timedata_node in country_node:
            year = int(timedata_node.get('year'))
```

```
            rowD = {'code': code, 'name': name, 'year': year}
            for indicator_node in timedata_node:
                indicator = indicator_node.tag
                value = float(indicator_node.text)
                rowD[indicator] = value
            LoD.append(rowD)
    df = pd.DataFrame(LoD)
    df.set_index(['code', 'year'], inplace=True)
    return df
```

When we call this function on the `Element` `root0` obtained from parsing the file `ind0.xml`, we get the same data frame we just saw:

```
xml_indicators(root0)
```

```
|                          name      pop        gdp
| code year
| FRA  2007              France    64.02    2657.21
|      2017              France    66.87    2586.29
| GBR  2007      United Kingdom    61.32    3084.12
|      2017      United Kingdom    66.06    2637.87
| USA  2007       United States   301.23   14451.90
|      2017       United States   325.15   19485.40
```

We are now ready to parse the file `indicators.xml` into a tree and then call our function on the root `Element`. We then look at an arbitrary subset of the rows and columns.

```
ind_path = os.path.join(datadir, "indicators.xml")
try:
    tree = etree.parse(ind_path)
    root = tree.getroot()
except:
    print("Exception in parsing XML")

df = xml_indicators(root)
df.iloc[200:205,0:4]
```

```
|                name   pop  life  cell
| code year
| ALB  1983   Albania  2.84  70.9   0.0
|      1984   Albania  2.90  71.1   0.0
|      1985   Albania  2.96  71.4   0.0
|      1986   Albania  3.02  71.6   0.0
|      1987   Albania  3.08  71.8   0.0
```

As we can see, this yields the expected data frame. It is worth noting that our solution yields values of NA in the pandas data frame when there is missing data in the xml file. Our solution never attempts to access data that is missing from the tree (note that a list of headers is not provided along with the xml file). If, for a given country code and year, there is missing data, then the corresponding dictionary in the LoD has fewer entries than the other dictionaries in the LoD. When the LoD is passed to pandas, those missing entries are automatically filled in with value NA.

As a consequence, this implies that if the xml file has a typo, e.g., if for the USA in 2015, "gdp" is entered erroneously as "gpd," then one of the dictionaries in the LoD will have a key:value pair where the key is "gpd," and pandas will create a column called "gpd" which takes value NA for all entries except the USA in 2015. One should always check the resulting pandas data frames for errors of this sort.

16.3.1.2 School Example

We turn now to a more complicated example, based on our school database from Chap. 12. This database has a natural hierarchical structure:

- school is the root, with three children: departments, courses, and instructors.
- Each child of departments represents a single department, and can have children such as name, division, chair, subject, etc.

  - Each department has an id attribute, e.g., <department id="MATH">, that we can use to distinguish it from its siblings (which each have a tag of department).
  - We allow chair to have children, for first and last name of the current department chair.

- Each child of courses represents a single course, and can have children such as title, hours, and class.

  - Each course has attributes including subject and num to distinguish it from its siblings.
  - Each class has an id to distinguish it from other class offerings of its parent course.
  - A class node can have children such as term, section, meeting (time), and instructorid.

- Each child of instructors represents an individual instructor, and can have children such as first, last, departmentid, city, state, title, etc.

  - Each instructor has an id attribute, to distinguish it from its siblings and for use in table joins.

From a statistical perspective, thinking about classes nested in courses, and courses nested in schools (even though this data set represents only one university) is a valuable way to account for variation between different classes, e.g., across a large data set of transcripts. Many statistical models leverage hierarchical structure to better explain variation.

We store the data described above in a file, `school.xml`, and as in the previous section we work with a snippet, `school0.xml`, with identical structure. We use functional abstraction to apply solutions from the snippet to the full data set. The data is fairly large, but we can visualize it in pieces:

```xml
<school>
  <instructors>
    <instructor id="9000">
      <first>Brandon</first>
      <last>Santos</last>
      <departmentid>ART</departmentid>
      <city>Granville</city>
      <state>OH</state>
      <title>Associate Professor, Art</title>
    </instructor>
    ...
  </instructors>
  <departments>
    <department id="LANG">
      <name>Modern Language</name>
      <division>Humanities</division>
      <subject id="ARAB" name="Arabic"/>
      ...
      <subject id="SPAN" name="Spanish"/>
    </department>
    <department id="CINE">
      <name>Cinema</name>
      <division>Fine Arts</division>
      <chair id="9042">
        <lastname>Rice</lastname>
        <firstname>Theresa</firstname>
      </chair>
    </department>
    ...
  </departments>
  <courses>
    <course subject="ARAB" num="111">
      <title>Beginning Arabic I</title>
      <hours>4.0</hours>
      <class id="40184">
        <term>FALL</term>
        <section>01</section>
        <meeting>09:30-10:20 MWRF</meeting>
        <instructorid>9216.0</instructorid>
      </class>
```

```
        </course>
          . . .
      </courses>
  </school>
```

As we have seen in the previous section, we should be careful in our solution to avoid assuming that each subtree contains some fixed set of tag names. It is much better to infer what data we have from the XML file we are provided.

We begin by reading in the file and parsing it, and using the access operator to obtain the three primary subtrees.

```
school0_path = os.path.join(datadir, "school0.xml")
try:
    school0_tree = etree.parse(school0_path)
    school0_root = school0_tree.getroot()
except:
    print("Exception in parsing XML")

departments = school0_root[0]
courses = school0_root[1]
instructors = school0_root[2]
```

We will work through wrangling school0.xml into pandas in pieces, starting with each child of the root.

16.3.1.3 Wrangling Instructors

The simplest child of school is instructors, as it represents a subtree of depth one (that is, all nodes are of distance 2 or less from the root, school). We can easily iterate over all instructors with a loop of the form for inst_node in instructors_root. We begin by looking at an example instructor node:

```
<instructor id="9106">
  <first>Kathleen</first>
  <last>Campbell</last>
  <departmentid>MATH</departmentid>
  <city>Granville</city>
  <state>OH</state>
  <title>Associate Professor, Math &amp; CS</title>
</instructor>
```

An instructor Element, inst_node, is presented with start tag of the form <instructor id="9106">. The first step is to extract the id, using inst_node.get('id'). The next step is to get the data from all the children, which we can do with child_node.text. It is tempting to iterate over the children of an instructor node, e.g., with code like the following:

```
headers = ['id', 'firstname', 'lastname', 'departmentid',
           'city', 'state', 'title']
LoL = []
for inst_node in instructors:
        innerL = [inst_node.get('id')]
        for child_node in inst_node:
                innerL.append(child_node.text)
        LoL.append(innerL)
```

An even more concise version of the code above might use a list comprehension for the inner loop. While concise, this code assumes that the children will be provided in the same order (reading the XML from the top to the bottom) as the row structure we are trying to build. It is unwise to make such assumptions about hierarchical data, since data might have been added to the tree at different times for different instructors, and possibly by different data curators. The freedom that XML files provide (e.g., for the data provider to define their own tags or put nodes in any order) means that we cannot make assumptions that XML files will interface naively with pandas data frames. Furthermore, if data on a particular instructor was missing (e.g., 'city'), then the code above would result in a LoL where the internal lists might have different lengths. Such a LoL is sometimes called "ragged" to differentiate it from rectangular LoLs. Ragged LoLs do not read easily into pandas.

A better approach is to build a list of dictionaries, which as we have seen is protected from issues of missing data upon being read into pandas. To handle the issue of child nodes appearing in different orders, we will use the find function to locate a child node. This means that, for every tag that appears, we find the corresponding datum, extract its text, and add a tag:text pair to the dictionary. Because we use this functionality so often, we have abstracted it into a function:

```
def child_value(node, tag):
    first_find = node.find(tag)
    if first_find != None:
        return first_find.text
```

This function returns None if the given node does not have a descendant with the given tag. Otherwise it returns the text of the first descendant found. In the school.xml file, there is always a unique descendant with the given tag, when this function is called. With this function in hand, we can carry out our plan from above: to extract the 'id' as an attribute and to extract the text data from each child, storing this data inside a dictionary, one for each child of the instructor_root.

```
def xml_instructors(instructors_root):
    LoD = []
    for inst_node in instructors_root:
        rowD = {'id': inst_node.get('id')}
        rowD['firstname'] = child_value(inst_node, 'first')
        rowD['lastname'] = child_value(inst_node, 'last')
```

```
        rowD['departmentid'] = child_value(inst_node,
                                          'departmentid')
        rowD['city'] = child_value(inst_node, 'city')
        rowD['state'] = child_value(inst_node, 'state')
        rowD['title'] = child_value(inst_node, 'title')
        LoD.append(rowD)
    df = pd.DataFrame(LoD)
    df.set_index(['id'], inplace=True)
    return df
```

We call this function with the specific node, `instructors`, from our earlier parsing. This node is a child of `school` and is the parent of all instructor nodes.

```
instr_df = xml_instructors(instructors)
instr_df.head().loc[:, ['firstname','lastname','title']]
```

```
|       firstname   lastname                                    title
| id
| 9024        Lori   Williams             Professor, Math & CS
| 9031      Ankita  Jayaraman  Associate Professor, Math & CS
| 9059      Arthur    Chapman     Vis. Instructor, Math & CS
| 9077      Jordan        Ray         Professor, Mathematics
| 9106    Kathleen   Campbell  Associate Professor, Math & CS
```

Note that the title field allows arbitrary string values, which can include typos, as shown above.

16.3.1.4   Wrangling Departments

Slightly more complicated than `instructors` is `departments`. First, it has more missing data. Secondly, it can have grandchildren, rather than only children. We look at an example:

```
<department id="MATH">
    <name>Mathematics &amp; Computer Science</name>
    <division>Natural Sciences</division>
    <chair id="9140">
      <lastname>Bradley</lastname>
      <firstname>Betty</firstname>
    </chair>
    <subject id="CS" name="Computer Science"/>
    <subject id="MATH" name="Mathematics"/>
</department>
```

We note that some departments have no chair listed, and some have no subjects listed. We therefore need to attempt to `find()` the chair node, and only try to extract its id and children if the chair node is successfully found. Next, because a department could, a priori, have many subjects (indeed, Modern Languages has eight!), we extract the subjects into a list. In order to iterate over all children of `dept_node` with the tag `subject`, we use the `Element` method `iterchildren(tag='subject')`. We now show the code, and discuss it further below.

```python
def xml_departments(departments_root):
    LoD = []
    for dept_node in departments_root:
        rowD = {'id': dept_node.get('id')}
        rowD['name'] = child_value(dept_node, 'name')
        rowD['division'] = child_value(dept_node, 'division')
        chair_node = dept_node.find('chair')
        if chair_node is not None:
            rowD['chair_id'] = chair_node.get('id')
            rowD['chair_last'] = child_value(chair_node,
                                              'lastname')
            rowD['chair_first'] = child_value(chair_node,
                                               'firstname')
        subjectlist = []
        for subject in dept_node.iterchildren(tag='subject'):
            subjectlist.append(
                "{}/{}".format(subject.get('id'),
                               subject.get('name')))
        if len(subjectlist) == 0:
            spec = "{}/{}".format(rowD['id'], rowD['name'])
            subjectlist = [spec]
        rowD['subjectlist'] = ",".join(subjectlist)
        LoD.append(rowD)
    df = pd.DataFrame(LoD)
    df.set_index(['id'], inplace=True)
    return df
```

Most of the code above proceeds like our `instructors` example, with the added wrinkle of the if-statement. However, the `subjectlist` deserves more attention. First, since each subject has both an id and a name, we fill this `subjectlist` with mashup data containing both the id and the name. We do this to avoid a list of tuples or a list of lists. Next, rather than include `subjectlist` in the dictionary `rowD` that we are building (which would result in a `pandas` data frame where one column is allowed to have list values), we use `join` to turn `subjectlist` into a string. Recall that `join` is the inverse operator to `split`, so in the code above it produces a comma-separated string representing `subjectlist`. This is why the data frame has a column `subjectlist` taking string values when we invoke the function:

```
xml_departments(departments).loc[:,['subjectlist']]
```

```
|                                             subjectlist
| id
| CINE                                      CINE/Cinema
| MATH   CS/Computer Science,MATH/Mathematics
```

The `subjectlist` string could later be extended into multiple columns (e.g., `subject1`, `subject2`, etc.), or shifted into a different table as in the database schema from Chap. 12.

We could also read the XML data into two data frames, `departments` and `subjects`, at the same time, and avoid the need to introduce a mashup column. This solution proceeds like the previous solution, but using a second LoD, which we append to where we previously created `subjectlist`. The resulting `subjects` data frame contains subject ids, subject names, and corresponding department ids, following the schema introduced in Chap. 12.

```python
def xml_departments2(departments_root):
    LoD = []       # departments
    LoD2 = []      # subjects
    for dept_node in departments_root:
        rowD = {'id': dept_node.get('id')}
        rowD['name'] = child_value(dept_node, 'name')
        rowD['division'] = child_value(dept_node, 'division')
        chair_node = dept_node.find('chair')
        if chair_node is not None:
            rowD['chair_id'] = chair_node.get('id')
            rowD['chair_last'] = child_value(chair_node,
                                                'lastname')
            rowD['chair_first'] = child_value(chair_node,
                                                'firstname')

        for subject in dept_node.iterchildren(tag='subject'):
            subjD = {'departmentid':dept_node.get('id')}
            subjD['id'] = subject.get('id')
            subjD['name'] = subject.get('name')
            LoD2.append(subjD)
        LoD.append(rowD)
    df = pd.DataFrame(LoD)
    df.set_index(['id'], inplace=True)
    df2 = pd.DataFrame(LoD2)
    df2.set_index(['id'], inplace=True)
    return df,df2
```

We can see the results of the function above with a sample invocation:

```python
departments_df,subjects_df=xml_departments2(departments)
departments_df.head().loc[:, ['division','chair_id']]
```

```
|              division chair_id
| id
| CINE         Fine Arts      9042
| MATH  Natural Sciences      9140
```

```
subjects_df.head()
```

```
|          departmentid                    name
|  id
|  CS              MATH   Computer Science
|  MATH            MATH         Mathematics
```

16.3.1.5   Wrangling Courses

The most complicated child Element of school is courses. Contained within
a course are several attributes of the course (including subject and num), several
children containing data (including title and hours), and several children
representing subtrees, one for each section of the course offered during the year.
We show an example:

```
<course subject="CINE" num="104">
  <title>Film Aesthetics/Analysis</title>
  <hours>4.0</hours>
  <class id="40371">
    <term>FALL</term>
    <section>01</section>
    <meeting>13:30-14:20 MWF</meeting>
    <instructorid>9273.0</instructorid>
  </class>
  <class id="40372">
    <term>FALL</term>
    <section>02</section>
    <meeting>10:30-11:20 MWF</meeting>
    <instructorid>9097.0</instructorid>
  </class>
  <class id="21636">
    <term>SPRING</term>
    <section>01</section>
    <meeting>10:30-11:20 MWF</meeting>
    <instructorid>9097.0</instructorid>
  </class>
</course>
```

We will discuss two ways to extract this data into tabular form. The first approach
mimics what we did with the departments table, and ends up being needlessly
complicated. The second approach is more complicated but results in much better
tabular structure. In the first approach, we read courses into a single data frame,
with a very complicated column, classlist, containing the data of each class
offering. As we did for departments, we use iterchildren for this task.
The column classlist could then be expanded using pandas operations into

a classes table as in the database schema of Chap. 12. The code below carries out this plan, using subject and num as the index for the resulting pandas data frame, and introducing a dividing string | into the column classlist, to make it easier to later extract this data into a classes table. Because some classes are missing data (e.g., meeting time or instructor), we are careful to avoid appending null data into our classlist.

```python
def xml_courses(courses_root):
    LoD = []
    for course_node in courses_root:
        rowD = {'subject': course_node.get('subject'),
                'num': course_node.get('num')}
        rowD['title'] = child_value(course_node, 'title')
        rowD['hours'] = float(child_value(course_node,'hours'))
        classlist = []
        for instance in course_node.iterchildren(tag='class'):
            classlist.append('id/'+instance.get('id'))
            datum = child_value(instance,'term')
            if datum != None:
                classlist.append('term/'+datum)
            datum = child_value(instance,'section')
            if datum != None:
                classlist.append('section/'+datum)
            datum = child_value(instance,'meeting')
            if datum != None:
                classlist.append('meeting/'+datum)
            datum = child_value(instance,'instructorid')
            if datum != None:
                classlist.append('instructorid/'+datum)
            classlist.append("|")

        rowD['classlist'] = ",".join(classlist)
        LoD.append(rowD)
    df = pd.DataFrame(LoD)
    df.set_index(['subject','num'], inplace=True)
    return df
```

The reader may wonder why we chose to use / to represent the key:value mapping in classlist. The reason is that the symbols ':' and '-' already appear inside the data (as part of meeting), and so we needed a different symbol between field name and field value within our classlist string. This decision was made by thinking ahead to the next part of the wrangling process, where we would make a classes table. We omit the invocation of the function above, because classlist is much too messy.

An alternative approach is to read the courses Element into two data frames, courses and classes, following the approach of xml_departments2. We again begin with two LoDs and append to the classes LoD in an inner loop that iterates once for each class section of a given course. As usual, we use get() and child_value() to extract the relevant data in each iteration.

```
def xml_courses2(courses_root):
    LoD = []      # courses
    LoD2 = []     # classes
    for course_node in courses_root:
        rowD = {'subject': course_node.get('subject'),
                'num': course_node.get('num')}
        rowD['title'] = child_value(course_node, 'title')
        rowD['hours'] = float(child_value(course_node,'hours'))

        for instance in course_node.iterchildren(tag='class'):
            D = {'coursesubject':course_node.get('subject'),
                 'coursenum': course_node.get('num')}

            D['classid'] = instance.get('id')
            D['classterm'] = child_value(instance,'term')
            D['classsection'] = child_value(instance,'section')
            D['classmeeting'] = child_value(instance,'meeting')
            D['instructorid'] = child_value(instance,
                                            'instructorid')

            LoD2.append(D)
        LoD.append(rowD)
    df = pd.DataFrame(LoD)
    df.set_index(['subject','num'], inplace=True)
    df2 = pd.DataFrame(LoD2)
    df2.set_index(['classid'],inplace=True)
    return df,df2
```

A sample invocation shows us that the courses data frame matches the schema introduced in Chap. 12, rather than including a messy mashup column.

```
courses_df,classes_df = xml_courses2(courses)
courses_df.head()
```

```
|                                        title   hours
| subject num
| CINE     104   Film Aesthetics/Analysis       4.0
|          219      Elementary Cinema Prod       4.0
|          310   Intermediate Cinema Prod       4.0
|          326          History of Cinema       4.0
|          361             Directed Study       4.0
```

Similarly, the classes table matches the schema of Chap. 12, including course subject and number so that it can be joined back to the courses table if required.

```
classes_df.head().loc[:,['coursesubject','coursenum',
                         'classmeeting']]
```

```
|            coursesubject coursenum      classmeeting
```

```
| classid
| 40371              CINE            104  13:30-14:20 MWF
| 40372              CINE            104  10:30-11:20 MWF
| 21636              CINE            104  10:30-11:20 MWF
| 40373              CINE            219   13:30-16:20 R
| 21638              CINE            219   13:30-14:50 TR
```

16.3.2   Creating XML Data

Previously, we have seen how to create JSON files containing a desired hierarchical structure (e.g., by creating a DoD in memory, and using json.dump()), how to add data to a JSON tree, and how to modify data within a JSON tree. By way of analogy, we now discuss how to create an XML file containing a given tree. We will discuss modifying XML data in the next section.

We assume that the tree comes to us in JSON format, as a DoD. Since an XML file is just a text file, all we have to do is produce a string containing the data in XML format, i.e., with start and end tags, attributes, and text values, all properly nested. We do not need to add whitespace to display the nesting structure, as this can be easily arranged by a custom printing function. As our example, we use the indicators0 data set we know so well. We begin with the data in JSON format:

```
{
  "FRA": {
    "2007": {
      "pop": 64.02,
      "gdp": 2657.21},
    "2017": {
      "pop": 66.87,
      "gdp": 2586.29}
    },
  "GBR": {
    "2007": {
      "pop": 61.32,
      "gdp": 3084.12},
    "2017": {
      "pop": 66.06,
      "gdp": 2637.87}
    },
  "USA": {
    "2007": {
      "pop": 301.23,
      "gdp": 14451.9},
```

```
    "2017": {
      "pop": 325.15,
      "gdp": 19485.4}
    }
}
```

We assume:

- The top level is a dictionary mapping country codes to a country dictionary,
- The country dictionary maps from an integer year to an indicator dictionary, and
- The indicator dictionary maps from indicator name to numeric value.

To convert this data into XML, we need to iterate over the country dictionaries. For each, we need to build the corresponding XML subtree. Doing this will require us to iterate over the indicators dictionary and build the corresponding XML subtree for each. We then need to latch all the XML subtrees together. This will involve several new Element methods:

- For a given string key, the method etree.Element(key) creates an Element whose tag is key.

  – This function automatically creates both the start and end tags.

- For a given string v, and a given node, node.text = v sets the text of node.
- For text tg, att, the method etree.Element(tg, <attribute>=att) creates an Element whose tag is tg and with a given <attribute> set to att.
- For a node p and a node c, the method p.append(c) creates c as a child of p.

  – This automatically nests the start and end tags of c within the start and end tags of p.

We illustrate these functions through the example below. First, we consider the problem of converting from an indicator dictionary into an XML tree. For example, indvars will be {"pop": 64.02, "gdp": 2657.21} the first time we invoke our function below. We assume we already know the year (in this example, "2007"), since by the time we would be building this part of our XML tree, we would have already traversed from the country code to the year and finally to the indicator dictionary. Since we will do this many times in the course of our conversion function, we abstract it as a function:

```
import numpy
def buildTimedata(year, indvars):
    """
    Given a year (as a string or as an int), create an
    Element whose tag is timedata, with attribute of a
    year set to the given year, and with a set of
    children, one for each key-value pair in indvars.
    """
```

```
timedata = etree.Element("timedata", year=str(year))
for key, value in indvars.items():
    if not numpy.isnan(value):
        node = etree.Element(key)
        node.text = str(value)
        timedata.append(node)
return timedata
```

When this function is first invoked, there is no XML Element in sight. The first line creates an Element, corresponding to the year. Then the loop creates several more elements (one per indicator), which are then attached to the year Element via append. In the end, timedate represents the XML subtree. We note that the if-statement ensures that we do not try to build a node with empty text, in case data were to be missing from the JSON file we are reading in. We illustrate with a sample invocation of the function. As the object returned is an Element, we use the function tostring to format the Element as a string we can print. The option pretty_print=True provides newline characters and horizontal whitespace to display the tree in the format we are used to.

```
node = buildTimedata("2007", {"pop": 64.02, "gdp": 2657.21})
util.print_xml(node)
```

```
| <timedata year="2007">
|   <pop>64.02</pop>
|   <gdp>2657.21</gdp>
| </timedata>
```

With this function in hand, we can build the XML tree associated to a country dictionary reasonably easily. We again create an Element to serve as the root (labeled by the code), then attach the subtrees corresponding to each year, which we build using the function above. We again create a functional abstraction that we can call once per country in the JSON file. Here years is the dictionary associated to the given country, as will be clear shortly.

```
def buildCountry(code, years):
    """
    Build a country xml subtree.  Country node is tagged
    'country' and has an attribute of the country's code.
    """
    country_node = etree.Element("country", code=code)
    for year,indvars in years.items():
        timedata = buildTimedata(year, indvars)
        country_node.append(timedata)
    return country_node
```

We again illustrate this function with a sample invocation:

```
data = {"2007": {"pop": 64.02, "gdp": 2657.21},
        "2017": {"pop": 66.87, "gdp": 2586.29}}
node = buildCountry("FRA", data)
util.print_xml(node)
```

```
    | <country code="FRA">
    |   <timedata year="2007">
    |     <pop>64.02</pop>
    |     <gdp>2657.21</gdp>
    |   </timedata>
    |   <timedata year="2017">
    |     <pop>66.87</pop>
    |     <gdp>2586.29</gdp>
    |   </timedata>
    | </country>
```

We now bring the entire solution together, again as a function, which we could use on the full indicators data set.

```
def main(datadir, filename):
    # Load JSON into in-memory dictionary data structure
    path = os.path.join(datadir, filename)
    with open(path) as f:
        ind0Dict = json.load(f)

    # Create root Element and, using add'l function,
      build
    # each of the country subtrees, adding to created
      root.

    # Demonstrates constructor and append method of an
    # Element node
    ind0_root = etree.Element('indicators')
    for code, yearD in ind0Dict.items():
        country_subtree = buildCountry(code, yearD)
        ind0_root.append(country_subtree)

    # Convert tree to a string
  xml_string = etree.tostring(ind0_root).decode('utf-8')

    return xml_string
```

The resulting xml_string represents the XML file, and we could write it to a text file with a .xml format. In practice, our XML files usually have a declaration line at the top, e.g., <?xml version='1.0' encoding='UTF-8'?>. This would be easy to prepend to the xml_string produced above. If the given DoD contained a typo, creating an error in our resulting xml_string, we could fix this using the `node.set()` function (described below) if the typo is in an attribute, or by using `node.clear()` and resetting the node, if the typo is in a tag.

16.3.3 Further Operations

There are additional `Element` operations that one sometimes needs in practice, which we have not needed for our solutions above. For the following, we assume node is an `Element`, and we list further operations that are possible.

- The *i*th child (0 relative) can be indexed, e.g., `node[i]`.
- `node.getparent()` yields the parent of node as an `Element`.
    - The parent of the root is the root itself.
- The right sibling (in document order) can be retrieved with `node.getnext()`.
    - If node is rightmost, this returns `None`.
- The left sibling (in document order) can be retrieved with `node.getprevious()`.
    - This returns `None` if the current `Element` is leftmost.
- `node.iter()` gives an iterator of partial trees left to explore.
    - For example, `for descendent in root.iter()` gives all nodes in the tree, if `root` is the root.
    - We can pass a first argument to `node.iter()` that specifies the string *tag* and iterated nodes are only those that match the given tag.

        This is like `node.iterchildren(tag = <tag>)` but iterating over all descendants, not just children.
- `node.findall(<tag>)` finds a list of `Element` nodes with the given `tag` as children of node.
- In addition to `node.text` there is also `node.tail` if text can follow an `Element` before the next `Element`.
- `node.set(key,value)` creates a `key:value` pair in the attributes of node.

16.3.4 Reading Questions

16.17 Consider the `topnames` data set from the tabular data unit. Please give an example of how this data might logically come to you in XML. Include at least 2 years (so, four "sex" subtrees).

16.18 Describe how you would read the XML-formatted data you just created into `pandas`. You can also answer this question by providing the code. At the very least, be clear about whether this can be done with just one loop, or with doubly nested loops. Justify your answer.

16.19 Below, we provide a snippet of the `topnames` data in XML format. Describe at a high level how you would read this into a `pandas` data frame. Make your solution general, so that it would also work on a larger version of the data. As part of your answer, give examples of the use of the `parse` method in the `etree` module, and give examples of how you would use the `text` and `tag` attributes, and the `get` method of particular Elements you see.

Mary 7065 John 9655

16.20 Write a function (or describe how) to build an XML tree associated with a given subtree of `topnames.json`. Your function may need two parameters, as in the reading. For example, when the dictionary passed to the function is `{"name":"Mary","count":7065}`, then your function will produce the XML Element below:

```
<sex value="Female">
   <name>Mary</name>
   <count>7065</count>
</sex>
```

16.21 Write a function (or describe how) to build an XML tree associated with a year subtree of `topnames.json`. Your function can call the function from the previous reading question. You can assume you are given both the year (as a string) and a dictionary containing the subtree in DoD format.

16.22 Do you think it's possible to read `departments` from XML into `pandas` without creating a mashup column? How about `courses`? Justify your answer.

16.3.5 Exercises

16.23 Write a function:

```
getLocalXML(filename, datadir=".", parser=None)
```

that performs the common steps of creating a path from the given `filename` and `datadir` and parses the XML file, using the passed `parser`, if any, and returns the Element at the root of the tree. If parser is not passed, the standard `XMLParser` should be used.

If the file is not found, or if the parse is unsuccessful (due to XML not being "well-formed"), the function should return `None`. Remember that if a parse is unsuccessful, the `etree` module raises an exception.

As preparation for the following sequence of exercises, please download the file breakfast.xml from the book web page and save it in your data directory. We need to parse the file `breakfast.xml` in the data directory and assign to `broot` the root Element object.

```
broot = getLocalXML("breakfast.xml", datadir)
```

16.24 Using the Element `broot` from above, get the attributes of the first child tagged `'food'`, and store your answer as a dictionary `myAttrib`.

16.25 Using the Element `broot`, find all children with the tag `'food'` and store them in a list of Elements called `foodlist`.

16.26 Using the Element `broot`, find all children with the tag `'description'`, extract the text from each, and store them in a list of strings called `dlist` (one description per food item in `bookstore.xml`). Hint: use loops.

16.27 Write a function

```
findValue(node, tag)
```

that, relative to `node` finds the first subelement matching `tag` and returns the `.text` attribute if found, and None, if no match was found.

Similar to the process employed in this section, the next set of exercises start with a source, `slac.xml`, the course catalog from a small liberal arts college, and have a goal of building a `pandas` data frame from the data set.

In preparation for testing and executing the solutions from these exercises, we set `root` as the root element of the tree:

```
root = getLocalXML("slac.xml", datadir)
```

16.28 Development Step 1: From the tree, build a list of dictionaries of courses with columns (i.e., keys in each of the row-based dictionaries) of `'reg_num'`, `'subj'`, and `'crse'`. Name your list of dictionaries `LoD`.

16.29 Development Step 2: Repeat your code from Step 1 so that the dictionaries in `LoD` include all *leaf* children of each course (i.e., column-keys for everything *other than* `time` and `place`).

16.30 Development Step 3: Finally, use your `LoD` to create a `pandas` `DataFrame`, and set the index to an appropriate column(s) that defines a unique independent variable combination for the data set.

This final exercise of the section is again one without a single correct answer. It shifts the focus from using XML procedural operations to consider, given a data set, *designing* and *writing* an XML tree from the data.

16.31 On the book web page is a file, `breakfast.json`, that contains a list of breakfast menu items. For each, there are fields for `name`, `price`, `description`, and `calories`.

At the top level, we want a function:

```
buildBreakfastTree(path)
```

that, given a `path` to the JSON-formatted file, reads the JSON and traverses the data structure and builds an XML tree, using the structure and tags that you have designed, for holding the data. Your processing should work even if we were to give it a *different* file of information, as long as the structure of the input matched what you see in `breakfast.json`. If no file is found at `path`, or if the file is found but the JSON does not parse, you should return None. Otherwise return the `Element` node for the root of the XML tree you have created.

16.4   XPath

As we have already seen, the ability to specify paths for locating and updating resources is fundamental to the hierarchical data model. For example, we used paths in the previous sections to access and modify data, traversing a path such as ind0 → "FRA" → "2007" → "pop" in a JSON-formatted tree via `ind0["FRA"]["2007"]["pop"]`. However, to do the same in an XML document in the previous section requires either a triply nested loop or three invocations of the `get()` method. XPath is a language that facilitates path-based operations in XML documents, remedying this problem. XPath provides declarative, rather than procedural, solutions to the sorts of problems discussed above [78]. We will often prefer XPath for this reason, as it is less error prone and more compact than XML programming. The name "XPath" comes from "XML Path Language." XPath has a number of important applications:

- Extracting text or attribute data in XML documents.
- Retrieving data stored within HTML files.
- Iterating over web pages or documents in a file structure.

Similarly to regular expressions, XPath is about pattern-matching and extracting data, and is language agnostic [91]. In our discussion to follow, we begin with very specific patterns that identify individual nodes (both leaf nodes and internal nodes), before introducing wildcard characters as a way to match multiple nodes. We then discuss the use of XPath in iteration (e.g., to match all population nodes sequentially in the `indicators` data set) and in functional abstraction (e.g., to allow the country code or year to be a parameter). Just as a regular expression can match many strings, so too, an XPath expression can match many paths in a tree. Throughout, we assume we have already imported `etree` and parsed the given XML file, as in the previous section.

We first introduce an alternative way to represent a snippet of our `indicators` data set without the use of attributes. We call this data set `ind2`.

```
<ind2>
  <FRA>
    <y2007>
      <pop>64.02</pop>
      <gdp>2657.21</gdp>
    </y2007>
    <y2017>
      <pop>66.87</pop>
```

```
      <gdp>2586.29</gdp>
    </y2017>
  </FRA>
  <GBR>
    <y2007>
      <pop>61.32</pop>
      <gdp>3084.12</gdp>
    </y2007>
    <y2017>
      <pop>66.06</pop>
      <gdp>2637.87</gdp>
    </y2017>
  </GBR>
</ind2>
```

Note that ind2 has strictly less information than ind0, i.e., it no longer contains the full names of the countries, and no longer knows that each year Element represents timedata information. However, because it has no attributes, it is simpler for our first examples of XPath. We will use both ind2 and our previous example of ind0 in the discussion to follow. Our year tags start with "y" because, as will be discussed more in Chap. 17, tag names are not allowed to start with digits.

16.4.1 Paths in XML Documents

To fully understand XPath, we must first review paths, which are traversals through a tree. In XPath, nodes are selected by following a path expression, which specifies one or more paths in the tree that match the expression. When the path expression can be interpreted as leading to a particular node, we say the node *matches* the expression. For the purposes of XPath, there are seven *types* of nodes, including element, attribute, and text. In addition to identifying individual nodes, XPath expressions can also return sets of nodes, sets of strings (the text of a node), sets of Booleans, or sets of numbers.

The syntax we use below to represent paths is XPath syntax, but should be familiar to the reader from previous use of URLs and from navigating a file system using ls in a Terminal (on a Mac) or in a Command Prompt (on a PC).

1. A path represents a traversal *from* one node to another in a hierarchical structure.

   • A path can end in any type of XPath node, e.g., an element, attribute, or text.

2. We represent paths as strings that name the sequence of nodes encountered in the traversal.

   • For example "FRA/y2007/pop is a path from the node FRA to the pop leaf node.

3. An absolute path implicitly has the root as the "from" node and the path string starts with a `'/'`.

   • For example `"/ind2/FRA/y2007/pop"` is a path from the root to the pop leaf node.

4. A relative path has, either implicitly or explicitly, a starting point (the "current working directory," or "current node"), and the path string does *not* begin with a `'/'`.

   • For example `"FRA/y2007/pop"` is a path from the node "FRA" to its grandchild "pop" under 2007.

5. We can think of each successive node in the path as a *step*, sometimes also called a *location-step*.

16.4.2   Paths and Expressions in XPath

The syntax of XPath is inspired by the way of thinking of paths described above, as we now explain. First, an XPath expression is a string, representing a sequence of steps in a tree, making up a path from some "initial" node to some "target" node. The initial node is often the root but does not have to be. XPath has:

1. The same notions of absolute and relative (from the "current" initial node) path to specify a traversal.

   • The examples above are correct XPath syntax to identify individual leaf nodes.
   • The path `/ind2/FRA/y2017` identifies an individual internal node.

2. The goal is often like `findall()` to get a *nodeset*, the *set* of nodes (over *multiple* path-traversals) that satisfy a given path *pattern*.

   • Once we have an XPath expression, we give it as a parameter to the `xpath()` method of an `ElementTree` or an `Element`, which returns the nodeset.
   • For example, the paths above yield nodesets of size one.
   • The path `/ind2/FRA/y1950/cell` yields a nodeset of size zero.
   • The path `/indicators/country` yields a nodeset of size three, since there are three nodes with tag `'country'` that are children of `ind0`.

3. A string argument for a *path pattern* defines location-steps with parts that have to "match" as specified, and other parts that are wildcards for one or more nodes or traversals that can match.

   • We have previously seen wildcard characters in regular expressions (where `.` matches all characters) and in SQL (where `%` matches patterns from instances, in the LIKE operator). The logic of wildcard characters in XPath works exactly the same way.
   • For example, `/ind2/FRA/*/pop` matches all grandchildren tagged `'pop'` under the `Element` tagged `'FRA'` in ind2, regardless of the parent of `'pop'`.

- Note: a wildcard matches any node at the given level, but not a path worth of
  nodes. The example above would not match nodes where 'pop' is a great-
  grandchild of 'FRA'.

4. Certain characters/sequences of characters are *expressions* that have *meaning* that
   is not interpreted as tagged element names. These include:

Now that we know how XPath can access attribute data in an XML file, we can
rewrite our XPath expressions from ind2 to work for ind0 (since the two data sets
have the same tree structure). As our discussion progresses, the character sequences
making up our XPath expressions may be longer than the page width, and so we
will "wrap" the displayed expression, often at a / separator, but the value should be
interpreted as a single string with no embedded newlines.

```
/indicators/country[@code='FRA']/timedata[@year
                                ='2007']/pop
```

is equivalent to the path /ind2/FRA/y2007/pop. In doing this, we are using
[@code = 'FRA'] to specify a predicate, which will be satisfied by the
country node whose code is "FRA" but not the other country nodes. If we
wanted the population of France in 2007, we would need the text() of the node
we found:

```
/indicators/country[@code = 'FRA']/timedata[@year
                                    = '2007']
/pop/text()
```

As we work through these examples, the reader is encouraged to use a freely
available XPath tester online, e.g., [14], to test that these expressions really do match
exactly the paths we describe.

If we wanted the population of France and the USA in all years, we would use a
logical *or*, yielding the expression

```
/indicators/country[@code = 'FRA' or @code = "USA"]/
                                    */pop
```

We could also achieve this via the logical or of two complete XPath expressions
by using the | operator.

```
/indicators/country[@code = 'FRA']/*/pop |
/indicators/country[@code = 'USA']/*/pop
```

In the school.xml data, we could use a logical *and*, e.g.,

```
/school/courses/course[@subject='CINE' and
@num='219']/title
```

which locates the title "Elementary Cinema Prod" of the course Cinema 219. We
can also use predicates to test for the presence or absence of nodes with a particular
tag, e.g., in the school data, the expression

Expression	Meaning
/	When the first character, means the traversal starts at the root of the tree. If not the first character, is used to separate location-steps in the set of possible traversals
.	Refers to the current node of the traversal
. .	Means the parent of the current node of the traversal. Every node has a parent, and the parent of the root is the root itself
@	Used to reference/match an *attribute* (instead of an element tag)
[]	Used, relative to the node of the current location-step, to specify a predicate (i.e., something that results in a Boolean true/false, often involving an attribute of the current node)
or, and	Used inside a [] expression to specify logical operators
vert bar	Used between XPaths in the same string to combine the nodeset results from the first XPath with the nodeset results from the second XPath
//	Matches *all* descendant traversals/paths from the node of the current location-step
*	Matches all the element siblings relative to the current location-step level
@*	Matches all the attributes relative to the current location-step level (or predicate, if used with [])
text()	Extracts the text of the current node

```
/school/instructors/instructor[@id]
```

will match all instructors that have an `id` attribute. Similarly

```
/school/instructors/instructor[not(@id)]
```

will match all that are missing the `id` attribute. This has clear applications to the relational data model. We can also match all attributes (rather than element nodes), e.g., the expression

```
/indicators/country/@*
```

first navigates to the three country nodes, then matches all six attributes.

The expression `//` is often a useful start of an XPath expression, e.g., `//course` matches all nodes with tag `"course"` anywhere in the document. Note that this syntax can match nodes at different levels of the tree. We could use this to find all nodes with a specified text, e.g., in `indicators` we could match all nodes with population larger than 300 million:

```
//pop[text() > 300]
```

The expression .. may be familiar from Terminal or Command Prompt. We use it to "back up" one level to the parent of the node that is the current location-step. For instance, we can find all "timedata" nodes where a country had a population greater than 300 million via:

```
//pop[text() > 300]/..
```

This path leads to the "timedata" nodes above every "pop" node where the population is larger than 300 million. Each "timedata" node has an attribute telling which year it represents. We could extract the years themselves from the nodeset above via a path that terminates at the "year" attribute:

```
//pop[text() > 300]/../@year
```

We could extract all year attributes (which, in ind0 is just "2007," "2017," "2007," "2017," "2007," "2017") via //@year, i.e., locating all places in the tree where year appears as an attribute.

As another example, with the school data, if we wanted to find what department offers the course with the title "Elementary Cinema Prod", we could locate that course then back up one level and extract the subject attribute, e.g.:

```
//title[text() = "Elementary Cinema Prod"]/../@subject
```

We provide one more example. In the school data, the following expression finds all nodes with tag "class" with the specified id attribute, then backs up one level and finds the value associated with the attribute subject:

```
//class[@id = 21713]/../@subject
```

The first part of the expression finds exactly one node, identified by a path /school/courses/course/class to the unique class node with the given id (turns out, this node corresponds to the class offering of CS 275: Elementary Graph Theory from the spring semester). When XPath backs up one level, it has the path /school/courses/course going to the course in question. This expression then returns the value associated to the subject attribute, i.e., "CS". We urge the reader to be careful with XPath expressions like //@id, since id attributes can appear on many different types of nodes, with different meanings. For example, in school.xml, the XPath expression departments//@id can be used to find id attributes for departments, whereas //@id yields a jumble of id attributes from both departments and instructors.

Clearly, XPath is designed for finding specific nodes and paths in XML data. Indeed, the find() and findall() functions from XML procedural programming can accept paths as well as node names. For example, if the school XML data is saved in an Element school_root, then school_root.findall(.//title) would find all nodes in the tree, starting from the root (this motivates the use of . in XPath), with the tag title. This includes both the titles of instructor nodes and the titles of course nodes.

16.4.3   XPath Syntax

Having seen many examples of XPath expressions, we now discuss the syntax in general. An XPath expression is built as a sequence of *XPath steps*, each separated by a / character. For example, the expression

```
/indicators/country[@code='FRA']/timedata[@year='2007']/pop
```

consists of three XPath steps: `country[@code = 'FRA']`, followed by `timedata[@year = '2007']`, followed by `pop`.

An XPath step is built from three pieces: an axis, a node-test, and a predicate. In the example above, the node tests look at tags, like `country` (which matches three nodes), `timedata` (which matches two nodes), and `pop`. The predicates restrict the nodeset, so that the first XPath step only matches one node, in this example. We will discuss the "axis" part of an XPath step shortly, but first we give the general syntax for an XPath step:

$$step \models axis::node\text{-}test[predicate]$$

Whitespace does not matter, e.g., `country[@code='FRA']` is the same as `country [ @code= 'FRA' ]`.

The XPath expression above is extremely specific, matching just the population in France in 2007. If we omit the two predicates (since predicates are optional), the XPath expression

```
/indicators/country/timedata/pop
```

will match all six population nodes in the data set, since all three `country` nodes are children of `indicators`, and all six `timedata` nodes are grandchildren of `indicators` via a `country` parent. The result of the XPath expression above will be a list of six `Element` objects. If we go one step further, to the `text()` of each population node, then we will obtain a list of six strings (that could be cast to floats), one for each population in the data:

```
/indicators/country/timedata/pop/text()
```

16.4.4   XPath Axes

An *axis* represents a relationship to the current node, e.g., child, parent, descendant, etc. When we use / in our paths, this is actually a shorthand for the *child* relationship, e.g., `/indicators/country[@code='FRA']/timedata` looks for nodes tagged "timedata" that are children of nodes matching `country[@code='FRA']`. The equivalent non-shorthand version of this same expression, where each step explicitly shows the axis relationship, would be

```
/child::indicators/child::country[@code='FRA']/
child::timedata
```

The first step of this expression has all three parts from the syntax above.

The shorthand `..` is used for searching for the parent. Since there is only one parent of a given node, `..` will match that parent regardless of its tag. Hence an expression like

```
class[@id = 21713]/../@subject
```

is shorthand for

```
class[@id = 21713]/parent::*/@subject
```

The shorthand `//` is used for searching along the axis of descendant-or-self. When this notation is used at the root, it searches the entire tree. However, the notation could also be used deeper in the tree. For instance, if we wanted to find the ids of instructors that are actively teaching sections of classes, we could use the expression

```
/school/courses//instructorid/text()
```

This would find all such instructor ids and the text for those nodes, without getting the instructor ids that are part of the `instructors` branch of the tree. The expanded version of the last two steps of the expression above is

```
child::courses/descendant-or-self::instructorid/text()
```

Lastly, the notation `@` is shorthand for the axis `attribute`, i.e., the expression

```
//course[@num = "101"]/@subject
```

is shorthand for

```
descendant-or-self::course[@num = "101"]/attribute::subject
```

More general axes are summarized in the following table. We recall that a descendant refers to a child, grandchild, etc., and an ancestor refers to the parent, grandparent, etc.

We have already demonstrated most of these in the examples above. In the ind0 data set, the order of the children of `indicators` is FRA, then GBR, then USA. Hence, the expression

```
/indicators/country[@code="GBR"]/following-sibling::country
```

matches the `Element` with start tag `<country code="USA" name="United States">`, as does the expression

```
/indicators/country[@code="GBR"]/following::country
```

because `country` only appears as a sibling of `country` (rather than also appearing as a child, where the distinction between these two axes would matter). Note that the expression

AxisName	Result
self	matches the current node
child	matches all children of the current node
parent	matches the parent of the current node
descendant-or-self	matches all descendants and the node itself
descendant	matches all descendants of the current node
ancestor	matches all ancestors of the current node
ancestor-or-self	matches all ancestors and the current node itself
attribute	matches all attributes of the current node
following	matches all elements in the XML document after the closing tag of the current node
following-sibling	matches all elements that are children of the current node's parent, and come later in the XML document
preceding	matches all elements in the XML document before the current node, excluding ancestors
preceding-sibling	matches all siblings before the current node in the XML document

```
/indicators/country/following::country
```

matches the node corresponding to GBR and also the node corresponding to USA, because both can be seen as "following" a node matched by /indicators/country. The expression

```
/indicators/country[@code='GBR']/following::pop
```

matches the two nodes with tag pop that appears in the XML file after the end tag </country> associated with the start tag <country code="GBR" name="United Kingdom">. In other words, the two populations matched are the population of the USA in 2007 and the population of the USA in 2017. Note that these nodes are not siblings of the GBR country node, so the following matches no results:

```
/indicators/country[@code='GBR']/following-sibling::pop
```

Analogously to the above, the expression

```
/indicators/country[@code='GBR']/preceding-sibling::country
```

matches the country node for France, while the expression

```
/indicators/country/preceding-sibling::country
```

matches both France and the USA, and

```
/indicators/country[@code='GBR']/preceding::pop
```

matches the two population nodes under France, while

```
/indicators/country[@code='GBR']/preceding-sibling::pop
```

matches no results. Lastly, `ancestor` can be useful for finding paths that reach a given resource, e.g., the expression

```
//pop[text() > 300]/ancestor::*
```

first matches the two nodes tagged `pop` where `text() > 300`, then matches all of their ancestors. This includes two `timedata` nodes (with attributes 2007 and 2017), one `country` node (the USA), and the `indicators` node.

16.4.5 XPath Predicates and Built-in Functions

We have seen many examples of predicates above, as well as the use of *and, or,* and *not* within predicates. We have seen that the vertical bar, |, between two XPath expressions means we want to match from either expression. We have seen that we can use = inside a predicate, and we now give more examples of the operators that can appear in a predicate:

Operator	Description
+	Addition
−	Subtraction
*	Multiplication
div	Division
=	Equality test
!=	Non-equality test
<	Less than test
<=	Less than or equal to
>	Greater than test
>=	Greater than or equal to
mod	Modulus (division remainder)

For example, in `topnames.xml`, we could add the male and female births in the year 1883 via:

```
//year[@value='1883']/sex[@value='Male']/count/text()+
//year[@value='1883']/sex[@value='Female']/count/text()
```

We could select only nodes where the count is divisible by five:

```
//count[text() mod 5 = 0]
```

The comparison operator in XPath is able to infer type. For example, if we want the male birth counts for all years in the 1990s, we can use either

```
//year[@value>='1990' and @value<'2000']/sex[@value='Male']/
count/text()
```

which carries out string comparison between the string obtained from `@value` and `"1990"`, or we can use

```
//year[@value>=1990 and @value<2000]/sex[@value='Male']/
count/text()
```

in which XPath infers that `@value` should be interpreted as an integer, and integer comparison should be used.

Many more examples of this nature are possible, but we turn now to examples that allow us to demonstrate some useful XPath built-in functions. As XPath has more than 200 built-in functions, we cannot hope to cover them all.

First, XPath provides indices for results, with index 1 representing the leftmost result (i.e., the first result in the XML document order). For example, in `ind0`, the expression

```
/indicators/country[1]
```

yields the first match, which is France, whereas the expression

```
/indicators/country[2]
```

yields Great Britain. XPath also provides the last match, via the built-in function `last()`, e.g.,

```
/indicators/country[last()]
```

yields the USA `country` node. The second-to-last match is

```
/indicators/country[last()-1]
```

The built-in function `position()` allows one to specify the position of match desired, e.g.,

```
/indicators/country[position() > 1]
```

yields the nodes corresponding to Great Britain and the USA, analogously to slicing in Python. We will make heavy use of the `position()` function in Chap. 22 when we scrape a web page to obtain tabular data. Web pages often come in HTML format, which is a special case of XML. When data is displayed as a table on a web page, the XML file has more structure than we can usually assume. In particular, there is a specified ordering on the children of each node in the part of the XML file representing the table. Hence, the `position()` function allows us to match an entire column. For example, if a `table` node has one child for each row, and every row has one child for each cell, then an XPath expression like `//table/row/cell[position() = 2]` would match exactly the cells of the second column, making it easy to extract a dictionary of lists from a given HTML file. A similar approach is possible procedurally when starting from hierarchical data in JSON list format, where the use of lists allows integer indexing.

The `node()` function allows us to match all nodes along a given path, e.g.,

```
/indicators/country/timedata/node()
```

matches all nodes that fit into path as specified, including both `pop` nodes and `gdp` nodes. Similarly, `//node()` matches all nodes in the document (including both element nodes like `timedata` and text nodes like `"64.02"`), while `//*` matches all element nodes (but not text nodes), and `//@*` matches all attributes (like `"USA"`, `"France"`, and `"2017"`).

XPath can also help us search text nodes, in a way similar to what we saw with regular expressions and in SQL `LIKE` queries. For instance, the `contains()` function allows us to match nodes where the text contains some specified string, e.g., in the `topnames` data, the expression

```
//name[contains(text(),'Jo')]
```

matches all `name` nodes where the `text` contains `'Jo'`. This includes `'John'` and would also contain `'Joanna'` or `'MoJo'` if those were ever top names. Note that this function is case sensitive, so `'Mojo'` would not count. When combined with the `ancestors` axis from above, we can find all paths to the nodes matched. A related function, `starts-with()`, determines whether the text starts with a given string, e.g.,

```
//name[starts-with(text(),'Jo')]
```

matches `'John'` but would not match `'MoJo'`.

16.4.6 Python Programming with XPath

We now show how to use XPath expressions in Python programs. In this section, we assume we have imported `etree` and parsed our set of input files, e.g.,

```python
from lxml import etree
myparser = etree.XMLParser(remove_blank_text=True)

indtree = etree.parse(os.path.join(datadir, 'ind0.xml'),
                      myparser)
indroot = indtree.getroot()

toptree = etree.parse(os.path.join(datadir, 'topnames.xml'),
                      myparser)
toproot = toptree.getroot()

schtree = etree.parse(os.path.join(datadir, 'school.xml'),
                      myparser)
schroot = schtree.getroot()
```

With these root `Element` objects in hand, we can apply XPath expressions via the built-in method `xpath()` that is part of `Element` objects. This function takes as input an XPath expression as a Python string and returns a *list* of nodes that match, which we store as `nodeset`. The items in the list `nodeset` can be strings (if the XPath expression ends at an attribute or `text()` of a node) or `Element` objects. To illustrate, we print the length of the result and the *type* of the first element.

```
nodeset = indroot.xpath("/indicators/country/timedata")
print("Length:",len(nodeset),"Type:",type(nodeset[0]))
```

> | Length: 6 Type: <class 'lxml.etree._Element'>

We can extract information from an `Element` in `nodeset`, as we learned in Sect. 16.3.

```
print(nodeset[0].attrib)
```

> | {'year': '2007'}

We can also print the associated subtree using the utility function for this purpose, which can be helpful for data exploration:

```
util.print_xml(nodeset[0])
```

> | <timedata year="2007">
> | <pop>64.02</pop>
> | <gdp>2657.21</gdp>
> | </timedata>

The ability to match a large number of nodes at once, and to extract their text, allows us to immediately make Python lists containing leaf data of any sort desired. For instance, the following yields a list containing the data from the population leaves (we use `float()` because `text()` yields strings, like `"64.01"`):

```
popListText = indroot.xpath("//pop/text()")
popList = [float(item) for item in popListText]
print(popList)
```

> | [64.02, 66.87, 61.32, 66.06, 301.23, 325.15]

Of course, in practice we might want a list for 2007 and another list for 2017, to compare the 2 years. This is equally easy, and we demonstrate for 2007 below. This time, as an alternative to a list comprehension, we use `map()` to apply the `float()` function to each element in the list returned by `xpath`.

```
path = "//timedata[@year='2007']/pop/text()"
popListText2007 = indroot.xpath(path)
popList2007 = list(map(float, popListText2007))
popList2007
```

> | [64.02, 61.32, 301.23]

It is important to note that the symbol " is being used to start and end the XPath expression we are providing to xpath(). Thus, we must use '2007' instead of "2007" in our code above. We saw the same issue when running SQL queries in Python. Just like SQL queries, we could also use triple quotes for string we pass to xpath(), e.g.:

```
indroot.xpath("""//timedata[@year='2007']/pop/text()""")
```

```
 | ['64.02', '61.32', '301.23']
```

The example above provides a natural place for functional abstraction. The following function uses a format string to create an XPath expression to search for nodes matching a given year. The function then calls xpath() from a given Element parameter, representing a hierarchical data set, like indicators0 or indicators.

```
def getPopList(tree,year):
    expr = """//timedata[@year='{}']/pop/text()""".format(year)
    popListTextY = tree.xpath(expr)
    popListY = [float(item) for item in popListTextY]
    return popListY

print(getPopList(indroot,"2017"))
```

```
 | [66.87, 66.06, 325.15]
```

It is clear how this approach can be used to convert an XML file into a pandas data frame. We simply extract the rows one at a time, and read them into a LoL, which we can then feed to pandas. However, as data we desire is sometimes stored in element attributes (often, metadata), we must illustrate how to extract such data. For instance, we might want to make lists of country codes and names in indicators0. For this, we would want to match all code attributes:

```
codeList = indroot.xpath("//@code")
nameList = indroot.xpath("//@name")
```

As another example, we provide code to find the average number of male births over all years in the topnames data set:

```
expr = "/topnames/*/sex[@value='Male']/count/text()"
countListStr = toproot.xpath(expr)
counts = [int(item) for item in countListStr]
aveCount = sum(counts)/len(counts)
aveCount
```

```
 | 47597.81294964029
```

As previously remarked, we can also find nodes satisfying a given condition, e.g., to find the year, sex, and name where the max count appeared in topnames we can do:

```
str_counts = toproot.xpath("//count/text()")
counts = map(int, str_counts)
m = max(counts)
expr = "//count[text()='{}']/ancestor::*".format(m)
pathToMax = toproot.xpath(expr)
```

The list `pathToMax` contains each node in the path to the maximum count. We can display the relevant information for each node in the path using the following handy function:

```
def results(nodeset):
    print("Length of nodeset result:", len(nodeset))
    for node in nodeset:
        print("Type:", type(node))
        if type(node) == etree._Element:
            print("Tag:", node.tag)
            print("  Text:", node.text)
            print("  Attrib:", node.attrib)
        else:
            print(node)
```

We invoke the function

```
results(pathToMax)
```

```
| Length of nodeset result: 3
| Type: <class 'lxml.etree._Element'>
| Tag: topnames
|   Text: None
|   Attrib: {}
| Type: <class 'lxml.etree._Element'>
| Tag: year
|   Text: None
|   Attrib: {'value': '1947'}
| Type: <class 'lxml.etree._Element'>
| Tag: sex
|   Text: None
|   Attrib: {'value': 'Female'}
```

We see that the max count occurred for the female name in 1947. If we wanted to know the name itself, we could repeat the `xpath()` invocation above with `preceding-sibling` instead of `ancestor`. To get both together, one could use the vertical bar | to combine two XPath expressions (one for the ancestors and one for the sibling).

16.4.7 Case Study Example

We now show how to use declarative XPath to solve the problem, previously solved in a procedural way, of converting the `school` data from XML format into data frames. Using XPath, we can easily match entire lists, e.g., the list of all instructor first names, or the list of all instructor titles. Each such list is a column of the data frame we seek to build. Hence, it is most appropriate to build a Python native dictionary of lists, which can then be fed to the `pandas` constructor. Each mapping in the DoL takes a column name to the list of entries making up that column. The reader is encouraged to compare the declarative solution below with the procedural solution described above.

```
inst_id = "/school/instructors/instructor/@id"
inst_id_list = school0_root.xpath(inst_id)

inst_first = "/school/instructors/instructor/first/text()"
inst_first_list = school0_root.xpath(inst_first)

inst_last = "/school/instructors/instructor/last/text()"
inst_last_list = school0_root.xpath(inst_last)

inst_dept = "/school/instructors/instructor/departmentid/"\
"text()"
inst_dep_list = school0_root.xpath(inst_dept)

inst_city = "/school/instructors/instructor/city/text()"
inst_city_list = school0_root.xpath(inst_city)

inst_state = "/school/instructors/instructor/state/text()"
inst_state_list = school0_root.xpath(inst_state)

inst_state = "/school/instructors/instructor/title/text()"
inst_title_list = school0_root.xpath(inst_state)

DoL = {'instructorid':inst_id_list,
       'instructorfirst':inst_first_list,
       'instructorlast':inst_last_list,
       'departmentid':inst_dep_list,
       'instructorcity':inst_city_list,
       'instructorstate':inst_state_list,
       'instructortitle':inst_title_list
       }

df = pd.DataFrame(DoL)
df.set_index('instructorid',inplace=True)
df.iloc[0:5,0:3]
```

| | instructorfirst instructorlast departmentid |

```
| instructorid
| 9024                            Lori        Williams          MATH
| 9031                          Ankita        Jayaraman         MATH
| 9059                          Arthur        Chapman           MATH
| 9077                          Jordan        Ray               MATH
| 9106                        Kathleen        Campbell          MATH
```

For the case of departments, more care is required, since not all departments have a chair. As with our procedural solution, we will simultaneously extract departments and subjects. For the sake of practice, we show how to extract departments as a list of lists, and how to extract subjects as a list of dictionaries. This involves a blend of XPath expressions and procedural programming. Later, we will show how to build data frames for courses and classes without the need for the procedural functions like get(). The same approach could be used here if desired.

In the solution below, we first make a list of headers, by using a single department, and collecting all its children. We then make lists of department ids, names, and divisions and headers. We iterate over departments, and, where possible, extract lists of chair id, chair last name, and chair first name. We do this using XPath expressions and format strings, to bring in the department and chair name to the XPath expression. For departments without chair data, we instead create the list [None, None, None]. In this way, we ensure that our LoL has a rectangular shape. While we are iterating over departments, we create a row in the subjects table for each subject we encounter. We store these in a list of dictionaries, but could also have used a list of lists.

```python
header_nodes = departments.xpath("./department[@id='CINE']//*")
headers = ['id']+[node.tag for node in header_nodes]

dep_id_list = departments.xpath("./department/@id")
dep_name_list = departments.xpath("./department/name/text()")
division_text = "./department/division/text()"
dep_div_list = departments.xpath(division_text)

LoL = []
LoD = []
for index in range(len(dep_id_list)):
    dep = dep_id_list[index]
    L = [dep_id_list[index],
         dep_name_list[index],
         dep_div_list[index]]
    chair = "./department[@id='{}']/chair/@id".format(dep)
    chair_list = departments.xpath(chair)
    if len(chair_list)==0:
        L = L + [None]*3
    else:
        chair = chair_list[0]
        L.append(chair)
```

```
        expr = "./department[@id='{}']/chair[@id='{}']"\
               "/*/text()".format(dep,chair)
        L = L+departments.xpath(expr)
    LoL.append(L)
    expr = "./department[@id='{}']/subject".format(dep)
    subject_list = departments.xpath(expr)
    for s in subject_list:
        D = {'subjectid':s.get('id'),
             'subjectname':s.get('name')}
        D['departmentid']=dep
        LoD.append(D)
```

```
df = pd.DataFrame(LoL,columns=headers)
df.set_index(['id'],inplace=True)
df.head().loc[:, ['division', 'chair', 'lastname']]
```

```
|                 division chair lastname
| id
| CINE          Fine Arts  9042      Rice
| MATH   Natural Sciences  9140   Bradley
```

```
df2=pd.DataFrame(LoD)
df2.set_index(['subjectid'],inplace=True)
df2.head()
```

```
|                    subjectname departmentid
| subjectid
| CS            Computer Science         MATH
| MATH               Mathematics         MATH
```

Lastly, we show how to build data frames for courses and classes. This time, we build the data frames one at a time. From a developer standpoint, this incremental approach is preferable to the sprawling approach of the previous block of code. As course data is regular, we can build the courses table in the same way we built the instructors table.

```
course_sub_list = courses.xpath("./course/@subject")
course_num_list = courses.xpath("./course/@num")
course_title_list = courses.xpath("./course/title/text()")
course_hours_list = courses.xpath("./course/hours/text()")

DoL = {'subject':course_sub_list,
'number':course_num_list,'coursetitle':course_title_list,
'hours':course_hours_list}
df = pd.DataFrame(DoL)
df.set_index(['subject','number'],inplace=True)
df.head()
```

```
|                                 coursetitle hours
```

```
| subject number
| CINE      104      Film Aesthetics/Analysis    4.0
|           219       Elementary Cinema Prod      4.0
|           310     Intermediate Cinema Prod      4.0
|           326             History of Cinema      4.0
|           361               Directed Study      4.0
```

Since each course can have many associated classes, to build the `classes` table, we will need to iterate over class offerings. We choose to store data in a list of dictionaries. We first construct lists of class ids, terms, and sections. For each class id, we then extract the subject and number, the instructor id if it is present, and the meeting time if present. Unless a class has two instructors, or meets at two times (which should be impossible), the lists `instructor_list` and `meeting_list` below each only have at most one item. The use of a LoD instead of a LoL saves us from having to append a value of `None` in the case of missing data.

```python
LoD = []
class_id_list = courses.xpath(".//class/@id")
class_term_list = courses.xpath(".//class/term/text()")
class_section_list = courses.xpath(".//class/section/text()")

for index in range(len(class_id_list)):
    cid = class_id_list[index]
    D={'classid':cid,
       'classterm':class_term_list[index],
       'section':class_section_list[index]}
    expr = "//class[@id='{}']/../@subject".format(cid)
    sub = courses.xpath(expr)[0]
    expr = "//class[@id='{}']/../@num".format(cid)
    num = courses.xpath(expr)[0]
    D['coursesubject']=sub
    D['coursenum']=num
    expr = "//class[@id='{}']/instructorid/text()".format(cid)
    instructor_list = courses.xpath(expr)
    if len(instructor_list)!=0:
        D['instructorid']=instructor_list[0]

    expr = "//class[@id='{}']/meeting/text()".format(cid)
    meeting_list = courses.xpath(expr)
    if len(meeting_list)!=0:
        D['meeting']=meeting_list[0]
    LoD.append(D)

df2=pd.DataFrame(LoD)
df2.set_index(['classid'],inplace=True)
df2.columns = ['term', 'section', 'subject',
               'num', 'instr', 'meeting']
df2.head()
```

```
|             term section subject  num instr       meeting
| classid
| 40371       FALL      01    CINE  104  9273  13:30-14:20 MWF
| 40372       FALL      02    CINE  104  9097  10:30-11:20 MWF
| 21636     SPRING      01    CINE  104  9097  10:30-11:20 MWF
| 40373       FALL      01    CINE  219  9284   13:30-16:20 R
| 21638     SPRING      01    CINE  219  9284  13:30-14:50 TR
```

16.4.8   Reading Questions

16.32 We have used the analogy to a file system on many occasions. Please explain, for a tree representing file system data, what would be the meaning of:

- internal nodes?
- leaf nodes?
- tag data?
- attribute data?

16.33 Please give an example of an XPath expression that uses . in a non-trivial way, and describe what your expression finds. Our most interesting data set is school.xml, which you can find here:
https://datasystems.denison.edu/data/

16.34 What is the difference between the following two XPath expressions?

```
schroot.xpath("//subject/@id[.='BCHM']")
schroot.xpath("//subject[@id='BCHM']")
```

16.35 Please describe how you might implement an XPath search that uses //. You can assume you have a tree stored in a Python data structure, e.g., as a DoD. You can assume you are searching from the root node and are given some node name to search for. Hint: start by thinking about a concrete example, like ind0, but then also discuss in your answer if you think a general solution is possible (i.e., one that works on any tree parameter). This might also be a good time to look up algorithms like "depth-first search."

16.36 Repeat the problem above, but explain how you could add an extra parameter, startLocation, and only do searches from there, e.g., /indicators/country[@code='USA']//gdp

16.37 Give an example where you would want to match using the axis descendant but not descendant-or-self. Please do not feel restricted to only using the XML files provided. You can think about any tree and any XPath application.

16.38 With reference to the school data, please write an XPath expression to match all attributes with key num anywhere in the document.

16.39 Repeat your solution to the above, but now with a Python line of code that invokes the method `xpath()` on a root node for `school`, and prints the resulting list. To test your code, read in the file (or URL to access the resource given in the link below), then parse it, then extract the root.

https://datasystems.denison.edu/data/

16.40 With reference to the `school` data, please write an XPath expression to match all courses with subject attribute of "CS." In Python, this would yield a list of `Elements` with tag `course`. Hint: you will need a predicate that looks at an attribute (@).

16.41 With reference to the `school` data, please write an XPath expression to match all courses worth 2 or fewer credit hours. In Python, this would yield a list of `Elements` with tag `course`. Hint: this will involve a predicate and backtracking (`..`).

16.42 With reference to the `school` data, please write an XPath expression to match all courses with title text containing "Theory." In Python, this would yield a list of `Elements` with tag `course`. Hint: this is like the above, but uses a built-in function from the reading.

16.4.9   Exercises

The following set of exercises focus on the use of XPath to *declare* the data needed to solve a problem. Your goal, in each, is to determine the XPath string and then to execute the `xpath()` method. You are to avoid procedural solutions unless told otherwise. Make sure the result has the data type specified in the problem, which helps demonstrate that you understand the different types that may result from a particular XPath query.

16.43 Using `countries.xml`, generate a list of all the countries in the `countries.xml` file, then assign the number of countries to the variable `countrycount`.

16.44 Write a function `findPop(root,country)` that finds the population of a given `country` in the data set `countries.xml`. Use an XPath expression and a format string. Return your answer as an integer.

16.45 Create a list named "german50" with the *names* of all countries where over 50% of the population speaks German. This one is more challenging, and may take some additional investigation to do it with just xpath. Hint1: In a conditional predicate (`[ ]`), we can use `text()` to refer to the text of the current element. Hint2: Once we get the right "set" of traversals, we can continue to build our path to get the desired set of information.

16.46 Study the `countries` data carefully. Then use the `position()` function to create a nodeset consisting of, for countries in positions 5–55 inclusive, the population of the second city listed, if there are at least two cities listed. For example, nothing is in the nodeset for Aruba (no cities listed) or Armenia (only Yerevan listed), but Cordoba is in the nodeset thanks to Argentina. Your answer should use a single XPath expression. Please store the results in a list `secondPops` of integers.

16.47 With reference to the `topnames` data set, please find all years where there was a count (either gender) that was strictly larger than 50,000. Please navigate to the appropriate attribute, rather than returning a list of elements.

16.48 With reference to the `topnames` data set, please find all years where the top female name had a count that was strictly larger than 50,000. Please navigate to the appropriate attribute, rather than returning a list of elements.

16.49 With reference to the `topnames` data set, please find all years where the top female name started with an "M" and the count was strictly larger than 50,000. Please navigate to the appropriate attribute, rather than returning a list of elements.

16.50 Get the male counts from all years from 1890–1900 (inclusive). Please use the `position()` function.

16.51 The question above could have also been solved by comparing the value of "year" with 1890 and 1900. However, if you wanted to get the most recent 10 years in the data set, and if you wanted your code to continue to be correct even if more data were added, then you really would need position(). Please write an XPath expression to get the male counts from the last 10 years in the data. Please use the `position()` function.

16.52 Using the `indicators` data and XPath, find the GDP of the USA in 2017, as a float. Use an XPath expression that leads to the `text()` of the node in question.

16.53 Create a list of all female names that appear in `topnames` (duplicates allowed). Note that we want data from every year, so we do NOT use a predicate to restrict the years.

16.54 Create a list of all course titles containing either "Directed" or "Independent." Hint: in your predicate, use the function `contains()`.

16.55 Create a list of all courses in `school` where the course number is 400 or above. This information is stored in an attribute.

16.56 Create a list of courses where the subject is "CS" or there is a class section meeting `11:30-12:20` MWF in the FALL term. This is the kind of query you might do during registration. Hint: break it into two separate queries, and use the | to get all entries matching either of them.

16.57 Use XPath to find the average GDP of all countries in 2017.

Further Reading

There is much more about hierarchical operations than can fit in one chapter. For instance:

- XQuery provides declarative solutions to query problems for hierarchical data, in a way analogous to SQL for relational data [79]. A FLWOR statement is analogous to an SQL SELECT statement with FROM and WHERE clauses. A FLWOR statement has clauses: For, Let, Where, Order by, and Return [38].
- The Document Object Model (DOM) is an interface to XML data that allows programs to access and edit the data. Often, built-in parsers can convert from XML data into an XML DOM object to facilitate this [74].
- jQuery is a JavaScript library that allows for HTML and DOM manipulation in a lightweight way.
- XMLStarlet allows for command-line XML programming [18], and jq allows for command-line JSON programming [25].

Chapter 17
Hierarchical Model: Constraints

Chapter Goals

Upon completion of this chapter, you should understand the following:

- The role of constraints in the communication between data system providers and clients.
- The difference between XML data being well-formed versus satisfying validity of data as encoded by an external constraint file.
- The use of comments in XML files.

Upon completion of this chapter, you should be able to do the following:

- Read and understand the meaning of a DTD file, an XSD file, and a JSON Schema.
- Track down violations of constraints when reading error messages.
- Create simple XML Schema by hand, to enforce constraints desired.

This chapter brings our coverage of data models full circle, by exploring the *constraints* of the hierarchical data model. With the flexibility of hierarchical models comes the danger of modification and/or ambiguity in the structure of a tree containing data or about the valid values for nodes in the tree. Constraints formalize these notions so that structure and relationships of the tree and of the values entailed can be explicitly expressed and then, when data is exchanged between provider and client, that data can be validated for conformance with the constraints.

© Springer Nature Switzerland AG 2020
T. Bressoud, D. White, *Introduction to Data Systems*,
https://doi.org/10.1007/978-3-030-54371-6_17

By way of review, in the tabular model, the structural constraints were the most important, where data must be in row and column form and, further, had the tidy data principles as constraints upon the interpretation of the data—each column must represent exactly one variable and each row must comprise exactly one observation or logical mapping between independent variables and their dependent variables. The weakness of this model is that these constraints were by convention, with no enforcement and non-conformance all too prevalent.

In the relational model, a data set was organized in a set of tables or relations. These, by database design and before any population of data, respect the functional dependencies with, at least, the same underlying principles attributed to tidy data. Further, the valid values in fields could be explicitly specified at table creation time through the schema of the table. Relationships between tables were also explicit through foreign keys. This explicit nature allows enforcement of constraints as tables are populated, so that providers and clients are always interacting through a database that conforms to the constraints.

We will find, through this chapter, that constraints in the hierarchical model follow a middle ground. Constraints need not be explicit nor enforced for a structure to be conceived and populated, and data exchanged, and so constraints may be simply by convention, like in the tabular model. But, through the mechanisms described in this chapter, constraints can, even after conception and population, be *made explicit*. When that occurs, providers and clients can *evaluate* an existing hierarchical data set against that set of constraints and determine if the data set is valid, and to reject a data set that does not conform to the constraints.

17.1 Motivation

The motivation for understanding and using constraints as part of our data models is to allow data systems clients to be designed that work correctly even when a provider supplies new data and updates structures. Consider how we designed our client programs. In most instances, we were given a JSON- or XML-formatted file, which we examined to *infer* the structure, organization, and data types for the data entailed in the file. Many times, we would then write down our assumptions and inferences. For instance, following is an excerpt from what we concluded about `school.xml`:

- `school` is the root, with three children: `departments`, `courses`, and `instructors`.
- Each child of `departments` represents a single department, and can have children such as `name`, `division`, `chair`, `subject`, etc.

- Each department has an id attribute, e.g., `<department id="MATH">`, that we can use to distinguish it from its siblings (which each have a tag of `department`).
- We allow `chair` to have children, for first and last name of the current department chair.

and so forth.

Then, whether procedurally, or in combination with XPath, we designed the operations that *relied on those inferences* so that we could process the data. It was essential to have a detailed understanding of the hierarchical structure before writing code. In practice, JSON and XML data are often obtained by a data system client over a network from a data system provider, as we detailed in Part III of this book. In such cases, it is essential that the client be able to write code that will continue to function even when the provider makes newer data available. That is, the client and provider must have an agreement about the structure, elements, and attributes that will be present in the tree data that the provider will provide, and this agreement facilitates error-free communication.

In the hierarchical model, we can use auxiliary documents to specify the explicit constraints for particular tree structures, so, for instance, there could be a constraint specification for the `school.xml` or for the `school.json` structures, and a different specification for, say, `indicators.xml`. The form of these constraint specifications are still evolving, and, in this chapter, we consider two forms for XML—the Document Type Definition (DTD) [75] standard and the XML Schema [76] standard. For JSON, this problem is solved by a JSON Schema [26].

These constraint specifications are above and beyond the notion introduced in Chap. 15, where we briefly discussed what it means for an XML-formatted file to be *well-formed*, and for a JSON-formatted file to follow the syntactic rules of JSON. These are also a type of constraint that, at the highest level, means that a formatted file defines an unambiguous tree and that the syntactic rules of the format have not been violated. We will give more detail on the well-formedness constraints of XML in Sect. 17.2. For both XML and JSON, defining an unambiguous tree and following the syntactic rules means that processing can successfully proceed through the first step and be *parsed* into an in-memory structure, and we have a correct tree. The software that performs this process is called a *parser*.

The next step is to ensure that the tree behaves as expected from a data perspective. It is this step that evaluates the parsed tree against the explicit constraints (DTD, XML Schema, JSON Schema), and, if successful, we say the data is *valid*. For example, the explicit constraints could specify all the example takeaways inferred for `school.xml` above—that the root is `school` and that `school` has exactly three children, named `departments`, `courses`, and `instructors`, and so forth.

When we study constraints for hierarchical data, they break down into *intra-constraints* (types of data items and/or set of valid values for elements) and *inter-constraints* (relationships between different kinds of elements). We summarize the main types of validity constraints one sees in practice.

Intra-constraints help specify valid values for a particular node in a tree. For JSON, these might be the data type used in a list or for the value-side of a mapping in a dictionary. They could also be an enumerated set of values for an element that represents a categorical variable. The same valid values can apply to XML, with added constraints on attributes, or for the `text` of an element node.

Inter-constraints are about valid edges and paths in the tree. They can entail what elements either *must* or *can* be in a parent–child relationship, like our example of the `school` node being the parent of `departments`, `courses`, and `instructors`, but can also describe sibling relationships. They can also describe valid cardinality between a parent and types of children, like exactly one, zero or one, at most *n*, or, in general, between *n* and *m*.

In the sections below, we discuss what it means for XML to be well-formed, then we discuss validity constraints via document type definitions, XML Schema, and JSON Schema.

17.1.1   Reading Questions

17.1 Give two specific examples of *intra-constraints* for values in the `school` data set, looking specifically at `course` elements.

17.2 Consider the elements that describe a `departments` in the `school` data set. Give two *inter-constraints* that relate a `departments` element to *other* elements in the tree. Be specific.

17.3 Write your own comparison of constraints and their enforcement within the relational model and constraints and their enforcement in the hierarchical model.

17.2   Well-Formed XML

In XML, a set of text, be it a file, or the contents of a network message, that is purported to be XML-formatted is called an *XML document*. We separate the steps of ensuring that an XML document defines a valid tree and conforms to the *syntactic rules* of XML from the (optional) downstream step of checking the tree for conformance to a set of explicit constraints. Satisfying the first step is called the *well-formedness* of an XML document, while satisfying the second step is called the *validity* of the XML document.

An XML document is *well-formed* if it adheres to the following rules:

1. There is *exactly one* root element, containing (i.e., nested within) all other elements. (Note that a *prolog*, which is *not* an element, may precede the single root element.)
2. Elements *must nest properly*, so are opened and closed in the same order.

3. Every tag must close, and, because of case sensitivity, must be spelled and capitalized identically when closed.

   - A self-closing element, terminated with $/>$ is one way to satisfy this requirement.
   - A separate close tag, $</tagname>$ is the only other way to satisfy this requirement.

4. Attributes appear in start tags of elements, and never in end tags.
5. Attribute values have a `name="value"` and the value must be quoted either with matching single or double quotes, but not mixed.
6. Attribute names *cannot* be repeated within a given element.
7. No blank/whitespace between an opening < and the tag name, but whitespace *can* follow the tag name (and will, if any attributes are present).
8. Tag names must:

   - begin with a letter or _,
   - use letters, digits, hyphens, underscores, or periods following the initial character,
   - not include whitespace, a colon, or include the sequence `xml`.

9. The colon is only used for namespace prefixes, whose definition must be in the current element or an ancestor.
10. Within attribute values, the XML-special characters &, <, >, :, and " must be rewritten to avoid misinterpretation.

Namespaces are a way to compartmentalize and differentiate the tag names used in XML. In larger settings, XML can combine information from multiple sources. For instance, local weather observation data may include information from the US Geological Survey [70], or may want to use constraints already defined in some other domain. To uniquely identify tags as belonging to one source or another, XML uses namespaces, where a tag for an element can be prefixed with its namespace followed by a colon. So a tag `quake` in namespace `usgs` would use `usgs:quake` when defining an instance of that element. This would be different from a tag of the same name, like `oh:quake`. We, in general, do not use namespaces in this book, but it is important to know that they exist and are thus part of the well-formedness of an XML document, and can specify their own constraints.

All XML documents we have seen to this point have been well-formed. Example violations of some of the rules above are described in Chap. 15. For the next two sections, we will assume we are working with an XML document that has already been confirmed (e.g., via a parser) to be well-formed. For concreteness, we produce DTD and XSD for the `indicators0.xml` data set introduced in Chap. 15, though our examples also reference `topnames.xml` and `school.xml`.

17.2.1 Reading Questions

17.4 Give a small example of XML that does *not* nest properly. Give a realistic scenario where a well-formedness mistake like this could occur.

17.5 Give a realistic example of an element that is self-closing, making it syntactically correct.

17.6 Give three examples of tag names that violate well-formedness rules.

17.7 Based on reading and understanding the well-formedness rules, is it OK for a start tag that has many attributes be split across multiple lines? Why or why not?

17.3 Document Type Definition

A document type definition is a language/grammar that are *markup declarations* and are used to describe several validity constraints that we lay out below. The definition of DTD is integrated within the definition of XML itself [90], in sections on *element type declarations* and *attribute list declarations*, and *notation declarations*.

Document Type Definition (DTD)
- Defines the structure of the tree

  - Defines the root element.
  - For any element, can define the child tags allowed/required.

- Defines which elements and attributes are allowed in general.
- Defines which are required, and limits on the number allowed.
- Defines the order of children if this is known.
- Defines types to expect for attributes, but a limited set of types.
- Defines which values should be fixed and which can be changed.

Normally, a document type definition consists of a sequence of plain text declarations, following certain syntax rules [75]. We focus first on the case where this text is stored in an external file (with extension .dtd) and later discuss how the DTD constraints can also be included in the XML document in question. In this section, we will build a file, indicators.dtd, that describes the constraints of indicators.xml, and we will also include examples of constraints involving topnames.xml and school.xml.

A DTD is often used when XML data is transmitted between a data system provider and a client. A DTD can also be used to ensure the validity of local files, or for ensuring different XML files are stored with a coherent structure so that they can be processed with the same code. For example, your personal computer probably has DTD files to help it manage its file structure, e.g., on MacOS^{TM}, such files can be found via the path ~/Library/Preferences. Importantly, the use of DTD is native to XML. Each element tag and attribute name in the XML document should be declared as valid in the associated DTD. When this is done, the tree structure can also be specified but does not have to be.

17.3.1 Declaring Elements

Declarations regarding the existence of elements take one of two forms:

- `<!ELEMENT element-name category>`, where *category* refers to what data is allowed, e.g.,

  - `ANY` allows any parsable data,
  - `EMPTY` forces the element text to be empty as one would see with a self-closing tag.

- `<!ELEMENT element-name (element-content)>`, where *element-content* refers to the type of data contained, e.g.,

  - `(#PCDATA)` allows any *parsable character data*; this is appropriate for elements like pop that have one leaf child,
  - `(pop)` declares that there should be one occurrence of a child with tag pop,
  - `(country+)` declares that there must be at least one child tagged country,
  - `(timedata*)` declares that there should be zero or more children tagged timedata,
  - `(cell?)` declares that there should be zero or one child tagged cell,
  - `(pop, life, cell?)` declares that there must be a child pop, there must be a child life, and there can be a child cell, and, furthermore, these three children should appear in the order shown,
  - the vertical bar | represents the "or" of conditions.

    `(#PCDATA|pop|life|gdp|cell)*`

  allows children with any of the four tags shown, with as many replicates as desired, and also allows parsable character data as children of the element in question.

The use of +, *, ?, and | above should be familiar from our treatment of regular expressions in Chap. 4. The reader is invited to think of a DTD as trying to "match" a given XML document, by traversing all paths, analogously to XPath.

17.3.2 Declaring Attributes and Entities

A DTD can also declare attribute lists associated with an element, using the syntax

```
<!ATTLIST element-name attribute-name attribute-type
attribute-value>
```

The `attribute-name` is exactly the name, or key, of the attribute we wish to constrain, associated with the given `element-name`. There are many possible values for `attribute-type` [71]. The main types we will need are:

- `CDATA`, representing a string of character data,
- `ID`, if the type is a unique id,
- `(option1|option2|...|optionN)` if there is an enumerated list of options for the value of the attribute, e.g., for `topnames.xml` our DTD could say `("Female"|"Male")` if we wanted to restrict the attribute on the `sex` element to only take these two values.

The `attribute-value` in a DTD `ATTLIST` declaration can take four possible values: - "value", the default value as a character string, - `#REQUIRED` if the attribute is required, - `#IMPLIED` if the attribute is optional, - `#FIXED` `"value"` if the attribute value is fixed.

We provide examples for each of the four `attribute-value` options (in the same order):

- `<!ATTLIST timedata year CDATA "2020">` says that for each `timedata` node, `year="2020"` is an attribute by default.
- `<!ATTLIST country code CDATA #REQUIRED>` says each `country` element must have an attribute of the form `code="value"` for some string `value`. Since we intend `code` to be part of the index of the associated tabular representation of the data, it makes sense to require that this data be present.
- `<!ATTLIST country name CDATA #IMPLIED>` says each `country` element is allowed to have an attribute of the form `name="value"` but is allowed not to have such an attribute. This allows the XML file to include countries for which the full name is missing.
- `<!ATTLIST course subject CDATA #FIXED "CS">` says each `course` element (in `school.xml`) with attribute `subject="value"` can only have `"CS"` for value. This would make sense if we wanted a subset of course data consisting only of computer science courses.

We provide one more example: `<!ATTLIST class id ID #REQUIRED>` says that each `class` element (in `school.xml`) must have an `id` attribute, and the associated values must be unique over the entire XML document.

In addition to declaring elements and attributes, a DTD can also declare *entities*, i.e., special sequences of characters to be used in the XML document. For example, `<!ENTITY vap "Visiting Assistant Professor">` declares an entity `vap`. Once declared, an entity can be referenced via the syntax `&entity-name;`, e.g., `&vap;`. This could be useful in `school.xml` to be sure that titles of instructors take specified forms, rather than allowing different data curators to type different strings such as "Vis. Assist. Prof." or "Visit. Assistant Prof." Another example the reader may have seen in HTML is the entity `nbsp` that inserts an extra space to be displayed in a document. In an HTML file, this is referenced as ` `. XML also has several built-in entities like `&lt;` for < [75].

17.3.3 Example DTD Declarations

Putting these examples together, a set of DTD declarations for `indicators.xml` is:

```
<!ELEMENT indicators (country+)>
<!ELEMENT country (timedata*)>
<!ELEMENT timedata (pop?,gdp?,life?,cell?,imports?,exports?)>
<!ELEMENT pop (#PCDATA)>
<!ELEMENT gdp (#PCDATA)>
<!ELEMENT life (#PCDATA)>
<!ELEMENT cell (#PCDATA)>
<!ELEMENT imports (#PCDATA)>
<!ELEMENT exports (#PCDATA)>
<!ATTLIST timedata year CDATA #REQUIRED>
<!ATTLIST country code CDATA #REQUIRED>
<!ATTLIST country name CDATA #IMPLIED>
```

These declarations require `indicators` to have at least one `country` child, but allows an arbitrary number of `timedata` children for any `country`, including zero if we only have data about the country's code and name, but no indicators for that country. The `timedata` nodes are tightly constrained. While it is allowed to have any number of indicators (from zero to six), their order is constrained, perhaps because we hope to read the `xml` file into tabular form. A less restrictive constraint would be:

```
<!ELEMENT timedata (pop?|gdp?|life?|cell?|imports?|exports?)>
```

This would allow the indicators to appear in any order. An even less restrictive constraint would be `<!ELEMENT timedata ANY>` which would allow any tree under `timedata`, e.g., allowing XML files with indicators beyond the six that had been included as of the time of writing.

17.3.4 DTD Validation of an XML Document

Now that we understand what DTD declarations look like, we consider our options for validating XML given the declarations to ensure the XML conforms to the constraints. We consider three options:

1. We have an XML file that might not yet explicitly reference the DTD declarations, and have the set of declarations in a separate file. This is a handy mechanism when we are developing the correct declarations.
2. We have an XML file that *references* a DTD document, which could be a local file, or it could be a resource at an external URL. We will illustrate the local file.
3. We can *embed* the DTD declarations within the XML document itself.

In the following we assume we have already imported the etree class from the lxml module. As we proceed through the examples, we will postulate different forms of the indicators.xml data set.

Option 1

In this option, indicators.xml is exactly as specified and used in previous chapters, and has no reference to any DTD declarations.

The etree class has a constructor, DTD, which gives an object that understands a set of DTD declarations and can use them to validate a specific XML tree. For this example, we first constructed a file, indicators_decls.dtd whose contents consist *only* of the 12 DTD declaration given in Sect. 17.3.3. This is assumed to be in the data directory, as is indicators.xml, and we use a utility function that abstracts the usual incantation for creating an XML tree from a file:

```
dtd_path = os.path.join(datadir, "indicators_decls.dtd")
dtd = etree.DTD(file=dtd_path)

xml_path = os.path.join(datadir, "indicators.xml")
assert os.path.isfile(xml_path)
tree = util.getLocalXML(xml_path)
```

Once we have the DTD object and the tree, we use the DTD object's validate() method to perform the actual validation:

```
print("Successful validation:", dtd.validate(tree))
```

```
| Successful validation: True
```

A validation can fail if either the XML document is not conformant, or it could be that our set of declarations is incorrect. Say that we specified that a cell element must have a foobar child: <!ELEMENT exports (foobar)>

```
dtd_path = os.path.join(datadir, "indicators_decls_bug.dtd")
dtd = etree.DTD(file=dtd_path)

xml_path = os.path.join(datadir, "indicators.xml")
assert os.path.isfile(xml_path)
tree = util.getLocalXML(xml_path)

if dtd.validate(tree):
    print("Successful Validation")
else:
    print(dtd.error_log.filter_from_errors()[0])
```

```
| indicators.xml:194:0:ERROR:VALID:DTD_CONTENT_MODEL: Element
| exports content does not follow the DTD, expecting (foobar),
| got (CDATA)
```

The object dtd has an attribute error_log; one of its methods, filter_from_errors() retrieves the list of log messages that are errors, and we look at the first in that list. The error message is as we would expect—when validating the XML, at the first exports element, the text was not a foobar as stated in the (buggy) set of declarations.

Option 2

When we reference DTD declarations from within the XML, we add a line to the prolog of the XML that informs the parser of the expected name of the root and where to find the DTD. In general, the syntax is <!DOCTYPE root-name SYSTEM "filename.dtd">. For our running example, in an updated indicators2.xml, both root-name and file name are indicators, and so we add

```
<!DOCTYPE indicators SYSTEM "indicators_decls.dtd">
```

after the first prolog line.

In the code, in order to request validation at the time of parsing, we construct a parser and pass True as argument to the dtd_validation= named argument. The parse() then happens in the same way as normally:

```
parser = etree.XMLParser(dtd_validation=True)
xml_path = os.path.join(datadir, "indicators2.xml")
tree = etree.parse(xml_path, parser)
```

If the parsing and validation both succeed, execution continues. If not, then the parse() operation will raise an exception, and a try-except block would be needed to allow recovery.

Option 3

Alternatively, the DTD declarations can appear inside the associated XML document, again just before the root start tag. In this case, the syntax is <!DOCTYPE root-name [dtd-text]>, where the square brackets are part of the syntax (not representing an optional piece) and dtd-text is the entire content that would have appeared in the DTD document. As an example, we display an updated indicators3.xml.

```
<?xml version='1.0' encoding='UTF-8'?>
<!DOCTYPE indicators [
  <!ELEMENT indicators (country+)>
  <!ELEMENT country (timedata*)>
  ...
  <!ATTLIST country name CDATA #IMPLIED>
]>
<indicators>
  <country code="ABW" name="Aruba">
  ...
</indicators>
```

In this section, we have seen a number of constraints that can be encoded using DTD syntax. We have discussed the pros and cons of using constraints. In general, constraints should only be used once an XML specification is considered unlikely to undergo large changes in structure. We turn now to XML Schema as an alternative way to encode constraints. Both are common in the practice of data science.

17.3.5   Exercises

17.8 Create a file named `indicators_decls.dtd` in the data directory, and edit it so that it has the contents from the book—the set of constraint declarations as the collection of `!ELEMENT` and `!ATTLIST` declarations for the constraints of the `indicators` data set.

17.9 Write code to explicitly validate `indicators.xml` on the book web page with the `indicators_decls.dtd` in the data directory. Name the DTD object `dtd` and use `tree` for the parsed result from `indicators.xml`.

17.10 On the book web page is the `indicators.xml` file. Copy the file to `indicators2.xml` and then edit `indicators2.xml` to include the `DOCTYPE` specification that declares that the `indicators` constraints can be found in the `indicators.dtd` file. Finally, write the code to create a custom XML parser that will validate DTD, and use this parser to parse and validate an XML tree from `indicators2.xml`.

17.11 Copy the `indicators.xml` file to `indicators3.xml` and then edit `indicators3.xml` to include the `DOCTYPE` specification that includes the DTD declarations for the indicators data set directly in the `indicators3.xml` file. Show that you are successful by repeating your code from above to parse the new file.

The next set of exercises are "pencil and paper" exercises regarding the `school.xml` data set. You may want an example, perhaps of `school0.xml` to remind you of the structure.

17.12 Give an ELEMENT DTD declaration for the root of the tree specifying its children.

17.13 Give an ELEMENT DTD declarations for `departments`, `courses`, and `instructors`.

17.14 Give an ELEMENT DTD declaration for `department`. This should be a single declaration, and not include attributes, which should be in their own declarations. Be careful about the children `chair` and `subject`.

17.15 Give declarations for *all* attributes that are part of the tree. If an attribute is equivalent to a unique identifier, make sure its DTD attribute declaration is restrictive enough.

17.4   XML Schema

XML Schema is a more powerful, but also more complex, alternative to DTD for specifying constraints for an XML document. It has the added benefit of, itself, using XML syntax [87]. XML Schema is supported by the World Wide Web Consortium (W3C), who also hosts web resources that can be referenced from within an XML Schema specification [76]. Below, we list some of the functionality that XML Schema can provide that can go beyond the functionality we have already seen for DTD.

XML Schema Definition (XSD)
- Defines the number and ordering of child elements (more generally than DTD).
- Defines the types allowed (that is, type-interpretation) and value constraints, both for text nodes and for attributes.
- Defines default and fixed values, for text nodes as well as for attributes.
- Can define patterns data must follow, analogously to regular expressions.
- Can define how whitespace should be interpreted.
- Can define elements that extend previous element definitions, or attributes whose constraints reference a previously defined set of constraints.
- Can define allowable substitutions, so, for instance attribute names using different languages can identify the same logical attribute.

XML constraint schema are flexible, and can change. They provide an important tool for validating the correctness of data, ensuring error-free code, and converting data between different data models. Since XML Schema use XML syntax, they are extensible, can be parsed to check well-formedness, and can be manipulated in the same way as XML files, e.g., using the Document Object Model [74]. It is possible to have very different looking XML Schema enforcing the same set of constraints. Since XML constraints are often made with the idea that we want future data to follow the same structure as previous data, constraints on data values should be used cautiously, to avoid valid (if unexpected) future data from breaking existing code during the validation process.

17.4.1   Root of an XML Schema

An XML Schema encodes constraints using tags and attributes. So that an XML file can reference its associated schema, XML Schema are saved as `.xsd` files, even though the content is still in XML format. The root of every XML Schema is an element tagged `<xs:schema>`, and in general all elements in our XSD document will begin with `xs:`, to indicate they are part of the `xs` namespace (see Sect. 17.2). This `xs:schema` element is allowed to contain attributes giving information on the namespace. A common attribute at the root of an XML schema document is `xmlns:xs="http://www.w3.org/2001/XMLSchema"`, which says that the given URL should be referenced for the namespace from which elements and data types with prefix xs: are drawn. In particular, this references an XML Schema document for XSD documents, and constrains the element and attribute names we are allowed to use when creating our XSD document.

An example structure template showing the root is

```
<?xml version="1.0"?>
<xs:schema xmlns:xs="http://www.w3.org/2001/XMLSchema">

<!--Children will define the constraints themselves-->

</xs:schema>
```

Within the body of an XML schema document, we will see tags like xs:element and xs:attribute to declare allowable elements and attributes. When we declare the allowable type of an attribute or text, we will see values like xs:boolean.

In addition to specifying the namespace of the XSD document, the xs:schema root element can specify the namespace for elements without a prefix, again by referencing a uniform resource indicator (URI), e.g., xmlns="a given URI". Any name listed in the namespace defined at the URI will be allowed in the namespace, without prefix. The xs:schema element can also specify a "target namespace" in an analogous way, to store user-created tags. Namespaces can be specified at any element, but are most common at the root. There is much more that can be said about namespaces [63], but we will not require any more than the basics we have just covered.

17.4.2   Declaring Elements and Attributes

As with DTD, the most fundamental constraints declare allowable elements, attributes, tree structure, and types. The tree structure is handled by the nesting of the XML schema document itself, e.g., declaring that there is an element timedata using a start tag, and then declaring that there are elements pop and life before the end tag for the timedata declaration. We summarize the syntax of the building blocks, displaying only start tags, and then display a full example later to clarify the nesting structure:

Declaring Elements
The definition of an element has the following form:

```
<xs:element name="element name" type="xs:type"
            [default|fixed]="value" minOccurs="n"
            maxOccurs="N">
```

This declares an allowable element with the given name, and, optionally, specifies the type of text the element can contain, whether that text has a default or fixed value (just like with DTD), and minimum and maximum numbers of times, inclusive, that this element can occur (the default for both is 1).

- `<xs:element name="city" type="xs:string" default=` `"Granville" maxOccurs="unbounded">` would be a reasonable way to declare elements tagged `city` in `school.xml`, if we wanted instructor city to default to Granville, and if we wanted to allow an unbounded number of city elements.
- `<xs:element name="departmentid" type="xs:string" fixed` `="CHEM" minOccurs="1" maxOccurs="30">` would be a reasonable way to declare `departmentid` elements in `school.xml` if we only wanted to consider instructors in the Chemistry department, and only wanted to allow the possibility of a number of instructors in the data between 1 and 30.

Declaring Attributes

```
<xs:attribute name="attribute name" type="xs:type"
              default|fixed="value" use="required">
```

This declares an attribute, to be part of the element declaration above this line in the XML Schema. The specification of type, default/fixed, and usage are optional. We provide two examples:

- `<xs:attribute name="year" type="xs:integer" default=` `"2020">` could appear in an XSD for `indicators.xml` after the declaration of a `timedata` element.
- `<xs:attribute name="subject" type="xs:string" fixed=` `"CS" use="required">` could appear in an XSD for `school.xml` after the declaration of a `course` element, if we only wanted to allow computer science courses, and wanted to insist that this attribute always appears on `course` elements.

In the partial example below, we demonstrate instances of self-closing `xs:element` tags, for `pop`, `gdp`, and `life`, that, in the example would be subordinate to a `timedata` element that is not shown. In this example, we use the name, type, and minOccurs attributes for the `xs:element`. We give a sense of the nesting with an `xs:element` that will give the structure for `country`; the open tag here need not include attributes, and the close tag will appear in its properly nested location.

```
<?xml version="1.0"?>
<xs:schema xmlns:xs="http://www.w3.org/2001/XMLSchema">

  <!-- Nesting structure to get up to def of country element -->

  <xs:element name="country">

    <!-- Further nesting to get to the children of timedata -->

      <xs:element name="pop" type="xs:float" minOccurs="0"/>
      <xs:element name="gdp" type="xs:float" minOccurs="0"/>
```

```
<xs:element  name="life" type="xs:float" minOccurs="0"/>
<!-- etc. -->

  <!-- Closing of nesting to get back to country level -->

</xs:element>  <!-- close of country element -->

  <!-- Closing of nesting back to root -->
</xs:schema>
```

17.4.3   XSD Types

XML Schema allow many built-in data types, including:

- `xs:boolean`,
- `xs:string`, with related types including options for whitespace interpretation,
- `xs:integer`, with related types for the number of bits allowed,
- `xs:float`, with related types for the number of bits allowed,
- `xs:decimal`, with related types for the number of bits allowed,
- `xs:date`, taking the form "YYYY-MM-DD," for year-month-day,
- `xs:time`, taking the form "hh:mm:ss," for hour-minute-second, and including an option for time zone,
- `xs:dateTime`, merging the two above.

We have already experienced the benefit of declared types, when we used XPath's built-in comparison operator to do integer comparison for data stored in text nodes and attribute values. There are also types for *element nodes*.

- `<xs:simpleType>` declares that the element in question has a text child rather than element children. We will see examples in the next section. This would be used subordinate to the element definition, and that element definition would *not* be self-closing.
- `<xs:complexType>` declares that the element in question is a complex type, i.e., can contain other elements or text. Nested as first child, we typically have one of the following:

  - `<xs:sequence>` declares that the children of the complex type must occur in the order specified by the sequence environment.
  - `<xs:all>` specifies that the child elements can appear in any order and that each child element can occur zero or one time, based on value of `minOccurs` attribute.
  - `<xs:any>` signifies that any type of element is allowed, even if not specified anywhere in the schema. This can be helpful during development of a schema, as it allows testing without requiring *all* elements and attributes to have their full definition.

- <xs:anyAttribute> can also nest as a child of a xs:simpleType or a xs:complexType and it signifies that an attribute is present and any name/value pair of the element will be accepted during validation. The syntax <xs:anyAttribute maxOccurs="unbounded"/> allows any set of attributes for the element in question.

The xs:complexType can have attributes including

- name: attribute that allows other elements to specify that they are of that type.
- mixed="true|false" allows the complex type to contain attributes *and* text, as well as elements.

As an example, consider timedata elements in indicators.xml. Because it has children (like pop, gdp, etc.), it should be defined as xs:complexType. Since its children are all possible with at most one, we use a child of <xs:all> and for the indicators themselves, specify minOccurs="0". The year attribute of timedata is specified as well, using the example from before.

```
<xs:element name="timedata">
    <xs:complexType>
      <xs:all>
        <xs:element name="pop" type="xs:float"
                    minOccurs="0"/>
        <xs:element name="gdp" type="xs:float"
                    minOccurs="0"/>
        <xs:element name="life" type="xs:float"
                    minOccurs="0"/>
        <!-- etc. -->
      </xs:all>
      <xs:attribute name="year" type="xs:integer"
                    default="2020">
    </xs:complexType>
</xs:element>
```

For *using* the above definition of the structure of timedata in the context of an indicators data set, we have two choices:

1. Specify the above *before*, almost like a helper function, and then *reference* this definition later in the schema document. This is the strategy we will use in Sect. 17.4.5. The ability to name and refer to a complex type allows for their reuse in an XSD, e.g., a set of traits we want to associate with every person in our data, whether the person be a student, faculty member, or employee.
2. Place the timedata definition *in its proper location* relative to the other nesting defined by schema. In this case, we would put the above definition *inside* the schema definition of the country element, as shown next:

```
<xs:element name="country">
    <xs:complexType>
```

```
<xs:sequence>
  <xs:element name="timedata" maxOccurs="unbounded"
              minOccurs="0">
    <xs:complexType>
      <xs:all>
        <xs:element  name="pop" type="xs:float"
                     minOccurs="0"/>
        <xs:element  name="gdp" type="xs:float"
                     minOccurs="0"/>

        <!-- etc. -->

      </xs:all>
      <xs:attribute name="year" type="xs:integer"
                    default="2020">
    </xs:complexType>
  </xs:element>
</xs:sequence>
<xs:attribute ref="code" use="required"/>
<xs:attribute type="xs:string" name="name"
              use="optional"/>
  </xs:complexType>
</xs:element>
```

The "inner" definition of timedata is exactly as illustrated before, except for adding the attributes of maxOccurs="unbounded" and minOccurs="0", to allow any number, including zero, of timedata elements for each country. The example also shows how we can specify use="optional" or use="required" when defining the name and code attributes of the country element.

It is also possible to use complex types to constrain an element to be empty (by putting nothing inside the complexType definition) or to build complex types extending other complex types [80], but we will not need this functionality.

17.4.4   XSD Restrictions

We can place restrictions on the values of attributes or text nodes, using the tag <xs:restriction base="xs:type"> and using tags <xs:minInclusive value="min value"> and xs:maxInclusive to set minimum and maximum allowable values.

As an example, suppose we define pop using xs:simpleType instead of as a self-closing xs:element. Then we can restrict that the text node be a non-negative decimal number using the syntax:

```
<xs:element name="pop">
  <xs:simpleType>
    <xs:restriction base="xs:float">
      <xs:minInclusive value="0"/>
    </xs:restriction>
  </xs:simpleType>
</xs:element>
```

Similarly, for string data, we can restrict the length. If we use the tag `<xs:length value="v"/>` as the child of a `xs:restriction` element, where v is an integer (e.g., $v = 3$), then only strings of length v will be allowed. To restrict to strings of length at least v and at most V we would use `<xs:minLength value="v"/>` and `<xs:maxLength value="V"/>`, as children of a `xs:restriction` element.

We can also declare that a value come from a predefined set, by using the tag `xs:enumeration` once for each item in the set of allowable values. This is often done using self-closing tags, which we could have also used above for `xs:minInclusive`. For example, the syntax below declares that `sex` elements in `topnames` have attributes whose values are restricted to `Male` and `Female`. We remind the reader that `sex` elements in `topnames` have tags of the form `<sex value="Male">`, i.e., the name of the attribute is `value` and the value of the attribute is `"Male"` or `"Female"`. We further remind the reader that the `topnames` data comes from the US Social Security Administration, that sex is a required attribute in this data, and that as of now the only allowed values are `Male` and `Female`.

```
<xs:element name="sex">
  <xs:attribute name="value" use="required">
    <xs:simpleType>
      <xs:restriction base="xs:string">
        <xs:enumeration value="Male"/>
        <xs:enumeration value="Female"/>
      </xs:restriction>
    </xs:simpleType>
  </xs:attribute>
</xs:element>
```

Inspection of the similarities between the previous two blocks of XSD syntax makes it clear that text nodes can also be restricted to prescribed sets of values, and attribute values can be restricted to numerical ranges. Another helpful type of restriction is to restrict values to follow a prescribed *pattern*, following the syntax of regular expressions, introduced in Chap. 4. We do this using the XSD self-closing tag `<xs:pattern value="the specified pattern"/>`. For example, to specify that country code in `indicators.xml` must be three uppercase characters, we must specify that the text data match the pattern `[A-Z] [A-Z] [A-Z]` or, equivalently, `[A-Z] {3}`. The following syntax encodes this restriction:

```
<xs:attribute name="code">
  <xs:simpleType>
    <xs:restriction base="xs:string">
        <xs:pattern value="[A-Z]{3}"/>
    </xs:restriction>
  </xs:simpleType>
</xs:attribute>
```

Similarly, we could specify that the value of the `year` attribute on a `timedata` element in `indicators` should match the pattern `[1-2][0-9][0-9][0-9]`. We could also use patterns instead of `xs:enumerate`, e.g., specifying that the value of `sex` must match the pattern `Male|Female`.

Like our definition of `timedata`, the above attribute definition could be placed as a "helper" definition when starting the schema and then referred to later, or it could be placed inline, as part of the definition of `country`.

In general, one should always be wary of overly constraining values. For example, it might be tempting to insist that country names must start with a capital letter, by specifying the pattern `[A-Z]([a-z])*`. However, this pattern would fail to match "South Africa" and any country name containing a space. The result would be a validation error at parsing time.

One last restriction we mention is

```
<xs:whiteSpace value="replace"|"preserve"|"collapse"/>
```

The attribute values, respectively, denote that all whitespace characters (tab, newline, etc.) should be replaced by spaces when the XML file is parsed, or should be preserved, or should be collapsed into a single space whenever whitespace is more. For more value restrictions, we refer the reader to [81].

17.4.5   An XSD Example

We illustrate the preceding discussion with an example of an XML Schema for `indicators.xml`. We use *XML comments* to improve readability. A comment is given with the syntax `<!-- text of the comment -->`. Comment lines are for human readers and are ignored by the XML parser.

In the XML Schema below, we use the strategy of building up our helper definitions first. This gives a *name* (`"code"`, `"year"`, `"timedata"`, and `"country"`) for each helper definition. Then, when we need to *use* the definition previously defined, we use a `ref="prev def name"` instead of the actual definition inline. In the example below, the first place this happens is in the definition of `country`, both where it needs the definition of element `timedata` and where it needs the definition of attribute `code`. The use of the `ref=` attribute recurs when the root element, `indicators` is defined and needs to refer to the definition of `country`.

```xml
<?xml version="1.0"?>
<xs:schema xmlns:xs="http://www.w3.org/2001/XMLSchema">

<!-- Define attributes code and year using regex patterns -->

<xs:attribute name="code">
  <xs:simpleType>
    <xs:restriction base="xs:string">
      <xs:pattern value="[A-Z]{3}"/>
    </xs:restriction>
  </xs:simpleType>
</xs:attribute>

<xs:attribute name="year">
  <xs:simpleType>
    <xs:restriction base="xs:short">
      <xs:pattern value="[1-2][0-9]{3}"/>
    </xs:restriction>
  </xs:simpleType>
</xs:attribute>

<!-- Define timedata element and all its children as
     with 0 occurrences possible -->

<xs:element name="timedata">
  <xs:complexType>
    <xs:all>
      <xs:element name="pop" type="xs:float"
                  minOccurs="0"/>
      <xs:element name="gdp" type="xs:float"
                  minOccurs="0"/>
      <xs:element name="life" type="xs:float"
                  minOccurs="0"/>
      <xs:element name="cell" type="xs:float"
                  minOccurs="0"/>
      <xs:element name="imports" type="xs:float"
                  minOccurs="0"/>
      <xs:element name="exports" type="xs:float"
                  minOccurs="0"/>
    </xs:all>
    <xs:attribute ref="year" use="required"/>
  </xs:complexType>
</xs:element>

<!-- Define country element with unlimited timedata children
     one required attribute of code, optional attr of name -->

<xs:element name="country">
  <xs:complexType>
```

```
    <xs:sequence>
      <xs:element ref="timedata" maxOccurs="unbounded"
                  minOccurs="0"/>
    </xs:sequence>
    <xs:attribute ref="code" use="required"/>
    <xs:attribute type="xs:string" name="name" use="optional"/>
  </xs:complexType>
</xs:element>

<!-- Define indicators as an unlimited sequence of
     country elements -->

<xs:element name="indicators">
  <xs:complexType>
    <xs:sequence>
      <xs:element ref="country" maxOccurs="unbounded"
                  minOccurs="0"/>
    </xs:sequence>
  </xs:complexType>
</xs:element>

</xs:schema>
```

The two strategies of defining helpers up front, versus using nesting and defining in the nested order are only two points in the spectrum, and the same set of constraints can be defined in many different ways.

There is much more that can be said about XML Schema, including how to declare that an element or attribute is unique, how to declare a key, more detailed types and value restrictions, etc. We refer the interested reader to the references to learn more.

17.4.6   Validating an XML Document

As with DTD, we must validate that our XML document satisfies the associated XML Schema. We will consider two options:

1. We have an XML file that might not yet explicitly reference the XML schema, and have the schema in a separate file .xsd file.
2. We wish to embed the location of the .xsd within the XML file, and want the parsing and the validation to be performed together.

In both these cases, we assume there exists an indicators.xsd text file, whose contents are as specified in Sect. 17.4.5. There is not an option comparable to the DTD strategy of *embedding* the schema within the target XML.

Option 1

In this option, indicators.xml is exactly as specified and used in previous chapters, and has no reference to an XML Schema.

Here, we use the XMLSchema constructor of the etree class to build an object to be used for validation. The argument to the constructor is an ElementTree, so we first have to parse the .xsd file. The XMLSchema class has a number of similarities to the DTD class we used in Sect. 17.3.4, in that it has a validate() method and it has an error_log attribute we can use for determining why a validation failed. The argument to the validate() method is the ElementTree of the target XML. The validation, then, is given by the following code:

```
xsd_path = os.path.join(datadir, "indicators.xsd")
xsd_tree = etree.parse(xsd_path)

ind_schema = etree.XMLSchema(xsd_tree)

xml_path = os.path.join(datadir, "indicators.xml")
xml_tree = util.getLocalXML(xml_path)

if ind_schema.validate(xml_tree):
    print("Successful Validation")
else:
    print(ind_schema.error_log.filter_from_errors()[0])
```

```
| Successful Validation
```

Option 2

When we reference XML schema from within the XML, we want to reference the schema location. Since this is, itself, part of XML, we need a specific attribute, which is part of the xsi namespace, whose value gives the location of the schema. We thus need the following two attribute definitions to be added to the *root element* defined in indicators.xml:

- xmlns:xsi="http://www.w3.org/2001/XMLSchema-instance"
- xsi:schemaLocation="indicators.xsd"

The first attribute incorporates the xsi namespace, and the second attribute sets the schema location.

For the purpose of illustration, these changes have been made to a version of indicators.xml that we have named indicators4.sml. Parsing and validation then occur together:

```
xml_path = os.path.join(datadir, "indicators4.xml")
try:
    tree = etree.parse(xml_path)
    print("Successful parse")
```

```
except:
 print("Error in parsing or validating indicators4.xml")

 | Successful parse
```

If the parsing and validation both succeed, execution continues. If not, then the `parse()` operation will raise an exception, and the `try-except` block is present to detect and could be used to allow recovery.

17.4.7 Exercises

17.16 Create a file named `indicators.xsd` in the data directory, and edit it so that it has the contents from the book of the full XSD example.

17.17 Write code to explicitly parse the `indicators.xsd` file, create an `XMLSchema` object from the tree (named `ind_schema`) and to create a parsed tree from `indicators.xml` (named `xml_tree`). The testing of these steps should allow successful execution of

```
ind_schema.validate(xml_tree)
```

17.18 On the book web page is the `indicators.xml` file. Copy the file to `indicators4.xml` and then edit `indicators4.xml` so that the root element has attributes:

- `xmlns:xsi="http://www.w3.org/2001/XMLSchema-instance"`
- `xsi:schemaLocation="indicators.xsd"`

to allow reference to the `xsi` namespace and to then use that namespace to specify a schema location.

Finally, write the code to parse and validate the `indicators4.xml`.

17.19 On the book web page is a relatively simple XML file named `breakfast0.xml`, which has a root named `menu`, whose children are elements named `food`, and each `food` element has an attribute named `price` and text which is the string name of the food item.

Write a complete XML Schema XSD that can validate this XML structure. Name your schema `breakfast0.xsd` and include code to demonstrate your validity checking.

17.20 Create a modified version of `breakfast0.xml` named

```
breakfast0bad.xml
```

that *violates* the constraints of your `breakfast0.xsd`. Provide code that demonstrates that the XML does not pass the validity check, and, using the techniques of the last section, print the error message from the `lxml` error log.

17.21   On the book web page is an XML file named `breakfast1.xml`. This is a
more complex version of the tree providing a breakfast menu.

Write a complete XML Schema XSD that can validate this XML structure. Name
your schema `breakfast1.xsd` and include code to demonstrate your validity
checking.

17.5   JSON Schema

When our tree of data is JSON formatted, we require other mechanisms to define the
constraints on its structure and the constraints on its values. This is an area which
is not as well-developed as validation in XML. Nonetheless, there are research and
standardization efforts underway to define *JSON Schema*, and this JSON-specific
schema language is in *Draft 07* and is in the process of submission to the Internet
Engineering Task Force (IETF) standards organization.

Because there is significant detail in JSON Schema, and it continues to evolve
(but is stabilizing), we will limit our discussion in this section, and show only some
basic examples. Our goal in this section is to enable intelligent *reading* of JSON
Schema and to create basic JSON Schema specifications, which will allow the
constraint of the structure of a tree, but not necessarily to enable the many variations
on constraining values.

A couple of preliminary remarks:

1. Well-formedness is easier to verify for JSON data than for XML data, and this
   is handled automatically by Python when JSON data is loaded (in the same way
   that Python usually checks that brackets are balanced). So the `json.load()`
   or `json.loads()` is, in fact, ensuring the JSON is well-formed.
2. Regarding validation, just as XML Schema are given in XML syntax, so too are
   JSON Schema given in JSON syntax (and saved as `.json` files).
3. When we discuss provider-based APIs for acquiring data in Chap. 23, many
   of these providers return data in JSON format, and must convey to clients
   the structure of what they should expect. While sometimes the structure and
   constraints of their data is conveyed informally, or through example, sometimes
   the documentation may, in fact, use JSON Schema, or may allow the acquisition
   of the JSON Schema along with the data.

JSON Schema

Some of the characteristic abilities of JSON Schema include:

- Using nesting structure in the schema that corresponds to the nesting in the JSON
  data to determine which level of the associated JSON file is being constrained.
- Ability to restrict the *type* at any level of the associated JSON file, including
  object vs. array.
- Ability to restrict the number of items, the individual data types (integer, string,
  etc.), and the allowable values at any level.

- Ability to reference JSON Schema stored at a given uniform resource indicator (URI), including insisting that a given string encodes a valid URI.
- Ability to define new types and to reference them or extend them (e.g., a type for Person that can be extended into either Student or Instructor).
- Using a structure that is both human and machine readable.

17.5.1   Basics of JSON Schema

A JSON Schema contains a JSON *object* that defines a top level set of mappings for the schema. For our purposes, the first two elements in this object/dictionary are:

- `"$id":"a given URI"`, giving a URI that can be used to uniquely identify the JSON Schema, be it a location (like in a file system) or a URL-style identifier.
- `"$schema":"a given URI"`, where the URI identifies the location of a JSON Schema that describes what makes a schema valid. This is analogous to `xs:schema` identifying an XML Schema for XML Schemas.

For instance, the start of the JSON Schema might look like the following:

```
{
    "$id": "https://datasystems.denison.edu/schema/foobar.json",
    "$schema": "http://json-schema.org/draft-07/schema#",
    ...
}
```

The `"$id"` key maps to a URL-style identifier, presumably defining schema for `foobar` JSON data. The `"$schema"` key maps to the current Draft (07) of the JSON Schema standard. This helps define particulars for expectation on properties and validity to be expected by the current schema specification.

A JSON Schema can optionally have top level key:value pairs describing the associated JSON object at a high level: - `"title":"a chosen title as a string"` - `"description":"a chosen description as a string"` - a type of top level object to expect in the associated JSON document (e.g., `"type":"object"` or `"type":"array"`) - `"required":<<an array of keys required at the top level of the JSON>>`

So, for example, in the `school.json`, our top level is an `object` (i.e., dictionary), and the top level children, all required, are `departments`, `courses`, and `instructors`. So the top level of the schema might look like this.

```
{
    "$id": "https://datasystems.denison.edu/schema/school.json",
    "$schema": "http://json-schema.org/draft-07/schema#",
    "title": "School JSON Schema",
    "description": "JSON Schema used as an example",
    "type": "object",
    "required": ["departments", "courses", "instructors"],
    ...
}
```

Since order does not matter in a JSON object, the order of the key-value pairs in the JSON Schema being defined also does not matter, and so the order above could have been switched around arbitrarily.

Finally, and most importantly, a JSON Schema will have a key:value pair where the key is "properties" and the value is a JSON object whose keys match those of the associated JSON object, and whose values provide the schema definition of the respective object. To illustrate, suppose we have a JSON tree:

```
{
    "child1": 42,
    "child2": [4.5, 7.3],
    "child3": { ... },
    "child4": "Hello world"
}
```

So there are four children, where the first and last are scalars, with one an integer and the other a string, child2 is an array of decimal numbers, and child3 is an object with additional structure. An outline of the properties mapping of the JSON Schema might look like this.

```
{
    # mappings for $id, $schema, etc.
    ...
    "properties": {
        "child1": {
            "type": "integer"
        },
        "child2": {
            "type": "array",
            ...
        },
        "child3": {
            "type": "object",
            ...
        },
        "child4": {
            "type": "string"
        }
    }
}
```

Here we see a simple subordinate definition of a schema for child1, and child4, which is an object with a key mapping "type" to one of the defined types for JSON. Even for the schema definition of a scalar, there could be many other mappings in this dictionary, for instance, for a "description", a "minimum", a "maximum", etc.

When we are defining a subordinate schema, like would be needed for
"child2", where the "type" is "array", another mapping in the current
schema definition would map from "items" to a specification for the individual
items to be found in the array. For "child3", where the "type" is "object",
we use a nested schema, and repeat elements like "required", "title", and
then can again use "properties" to define the schema for the elements at the
next level (the children of child3). Continuing our example, if we fill in for
child2 and child3 a little further, we might have:

```
{
  ... # prefix
  "properties": {
    ...
    "child2": {
      "type": "array",
      "items": {
        "type": "decimal"
      }
    },
    "child3": { # nested structure defining schema for child3
      "type": "object",
      "properties": {
        "grand1": {
          "type": "string"
        },
        "grand2": {
          "type": "integer"
        }
      },
      "required": ["grand2"]
    },
    ...
  }
}
```

If we added the initial mappings (for $id, etc.), and the schema in the outer level
properties for child1 and child2 from before, we would have a complete
JSON Schema for the example.

We saw in XML Schema the ability to define helper schema as a way of
compartmentalizing the work and to avoid excessive nesting, and thus increase
readability. The helper schema were defined in one place, and then *referenced*
from the point in a parent schema where they were needed. JSON Schema provide
similar functionality. Here, when we need to reference a schema definition defined
elsewhere, we use an object key "$ref", whose value specifies the location of the
definition. The location for the definition can be:

• a separate local file,

- a URL, perhaps where an organization maintains building block JSON Schema, or
- within the *current* JSON Schema.

We will show an example of the last of these, and the exercises will develop this further. When we want to include schema definitions within the current JSON Schema, they need to be integrated into the tree itself. Typically at the top level object of the JSON Schema, we define a key "definitions" that maps to a value which is itself an object. This object maps from the keys that give a name to our subordinate schema to an object which is the schema definition. Let us illustrate with our running example. Suppose we want to move the definition of the schema associated with child3. As the example below shows, we add the "definitions" key to the outermost object, with an object whose key "child3" has a value of the schema we had for child3 previously. In the properties, we omit the schema for the other child nodes, and, for "child3" we show the reference to the definition. A reference is an object that maps the key "$ref" to the location. In this case, the string starts with #, which indicates an internal location, and the /definitions/child3 defines the path to that location. External references use the same mapping, but use values that give the external location, and would omit the entry in "definitions".

```
{
  "$id": "https://datasystems.denison.edu/schema/example.json",
  "$schema": "http://json-schema.org/draft-07/schema#",
  "title": "Example JSON Schema",
  "description": "JSON Schema used as an example",
  "type": "object",
  "definitions": {
    "child3": {
      "type": "object",
      "properties": {
        "grand1": {
          "type": "string"
        },
        "grand2": {
          "type": "integer"
        }
      },
      "required": ["grand2"]
    }
  },
  "properties": {
    ...
    "child3": {
      "$ref": "#/definitions/child3"
    }
    ...
  }
}
```

As mentioned at the onset, there are many other facilities for specifying constraints on both values and on structure, including things like specifying dependencies, regular expression patterns, cardinalities and lengths, categorical enumerations, uniqueness, and others as well. For the goals of this section, where we have shown constraining data types and have shown building trees through nesting and through separate definition and reference, we hope that we have given a flavor of what can be done, without being overwhelming.

17.5.2 Validating a JSON Document Using a JSON Schema

Once we have a JSON Schema and a JSON document, we can validate the JSON document using the Python package jsonschema [29]. Since both are stored as JSON, validation is a basic three step process:

1. Build the path to the JSON Schema and load() from the open file to a data structure.
2. Build the path to the JSON document and load() from the open file to a separate data structure.
3. Perform the validation.

For the third step, we use the validate() function of the jsonschema module, which takes named parameters, instance= for the JSON document, and schema= for the JSON Schema. If validation is successful, execution proceeds, but if there was an error, which could be a logical error in how the schema was constructed, or could be an actual failure in the validation step, then an exception is raised.

In the following example, we show the process for JSON document school0.json and for JSON Schema school_schema.json. To facilitate the error processing, we reference the ValidationError class of exception from the jsonschema module.

```
import json
import jsonschema

school0_path = os.path.join(datadir, "school0.json")
assert os.path.isfile(school0_path)
with open(school0_path) as f:
    school_json = json.load(f)

school_schema_path = os.path.join(datadir, "school_schema.json")
assert os.path.isfile(school_schema_path)
with open(school_schema_path) as f:
    school_schema = json.load(f)

try:
    jsonschema.validate(instance=school_json,
```

```
                              schemaschool_schema)
except jsonschema.ValidationError as err:
    print("School JSON did not validate: {0}".format(err))
```

17.5.3 Exercises

17.22 Suppose we have designed a JSON for holding a single address, and an example looks like the following:

```
{
    "street": "1851 Longview Dr",
    "city": "Springfield",
    "state": "OH",
    "postal-code": 45504
}
```

Create a *string* that contains a JSON schema supporting this template where city, state, and postal-code *must* be present, and all fields are strings, except postal-code, which is an integer. We give a prefix of the JSON schema:

```
{
    "$id": "https://datasystems.denison.edu/addr.schema.json",
    "$schema": "http://json-schema.org/draft-07/schema#",
    "description": "A schema for an address JSON",
    "type": "object",
    ...
}
```

Your goal is to validate the above example string, and to also create one *bad* example that violates the schema and show that they fail the validation step.

Once you create the schema and examples as strings, write code that uses the json module to instantiate the strings into structures and demonstrate the validation.

In the next set of exercises, you are going to "fill in" JSON Schema to complete a validating schema definition for the school data set, as exemplified in the school0.json file.

Consider the following JSON Schema:

```
{
    "$id": "https://datasystems.denison.edu/schoolschema.json",
    "$schema": "http://json-schema.org/draft-07/schema#",
    "title": "School JSON Schema",
    "description": "JSON Schema used as an example",
    "type": "object",
    "required": ["departments", "courses", "instructors"],
    "definitions": {
```

```
            "department": {
                "type": "object"
            },
            "course": {
                "type": "object"
            },
            "instructor": {
                "type": "object"
            }
        },
    "properties": {
        "departments": {
            "type": "array",
            "items": {"$ref": "#/definitions/department"}
        },
        "courses": {
            "type": "array",
            "items": {"$ref": "#/definitions/course"}
        },
        "instructors": {
            "type": "array",
            "items": {"$ref": "#/definitions/instructor"}
        }
    }
}
```

This is a template that incorporates many of the aspects of the section. The outermost level gives a schema where the root, an object, requires the three main branches, "departments", "courses", and "instructors". Each of these is defined as an array, where the items in the array use a ref to a schema in the definitions of the schema. Each of these definitions should define a schema for a *single* item in the array. At present, the only constraint they enforce is that these elements be an object.

17.23 As a preliminary *setup* question, use the above template to create a file named school_schema.json, and place it in the data directory along with school0.json. Then, follow the three step pattern illustrated in the section to build trees for both JSON-formatted files, and then *validate* school0.json. Because the template is correct, albeit incomplete, the constraints as given are *satisfied* by the target file.

As a bonus in this question, copy school_schema.json to a different file and *intentionally* change the definition so that, when you repeat the steps of validation, you get an error during validation.

17.24 Copy school_schema.json to school_instructors_schema. json. Then, the goal of this question is to complete the schema definition for items in the instructors array—the schema definition of a single instructor object. This should be straightforward, as each instructor has the same set of fields.

Note that the @id field is a *string*, even though the contents of the string are a sequence of digits.

Validate school0.json using the school_instructors_schema.json.

17.25 Copy school_schema.json to school_courses_schema.json. Then, the goal of this question is to complete the schema definition for items in the courses array—the schema definition of a single course object. In addition to scalar children, the schema for a course should optionally have an array of class items, which should also be defined. You can nest this definition, or add it to the definitions and use a reference technique.

Validate school0.json using the school_courses_schema.json.

17.26 Copy school_schema.json to school_departments_schema.json. Then, the goal of this question is to complete the schema definition for items in the departments array—the schema definition of a single department object. In addition to scalar children, the schema for a department should optionally have a chair object and an optional array, subject, with an array of individual subject definitions, which should also be defined. You can nest this definition, or add it to the definitions and use a reference technique.

Validate school0.json using the school_departments_schema.json.

Part III
Data Systems: The Data Sources

Chapter 18
Overview of Data Systems Sources

Chapter Goals

Upon completion of this chapter, you should understand the following:

- the dimension of *sources of data* with respect to the overall goals of understanding data systems,
- the basic characteristics of each of the data sources of:

  - local files,
  - local and remote relational database systems,
  - web servers, and
  - provider APIs.

- the idea of the *forms of data* being realized by specific *formats*, and those formats being the means of providing data from the various *sources*.

The purpose of this chapter is to introduce Part III of the textbook by giving an overview of the *data system sources*. We begin by reviewing the architecture originally described in Chap. 1, which characterizes the client application and the set of data system sources through which the application may obtain data. This review and the relationship between the sources and the data forms and formats

© Springer Nature Switzerland AG 2020
T. Bressoud, D. White, *Introduction to Data Systems*,
https://doi.org/10.1007/978-3-030-54371-6_18

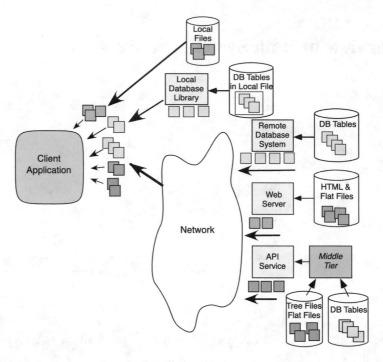

Fig. 18.1 Data sources

are explored in Sect. 18.1. The remainder of the chapter discusses each of the data system source types.

18.1 Architecture

Figure 18.1 presents a broad view of the data systems architecture. On the left of the figure, we depict the client application. A given client may obtain data from any one or multiple of the data sources (providers) shown on the right. The format of the data is dependent on the various data forms, as previously discussed in Part II of this book, through the use of data models.

In the figure, we represent units of data in a particular form/format as a small square and show the various forms being communicated to the client application. In Fig. 18.2, we enumerate the forms of data that a client might obtain from each of the data sources.

When we encapsulate any of our data forms (tabular, relational, or hierarchical) into a file or communicate the form over a network, the problem may be decomposed by thinking in terms of the following layers of functionality:

Data Source/Provider				
Local Files	**Local Database**	**Remote Database**	**Web Server**	**API Service**

Data Models/ Forms

Tabular Hierarchical	*Relational*	*Relational*	*Tabular Hierarchical*	*Tabular Hierarchical*

Fig. 18.2 Data forms for data sources

- *Format*: The translation from a logical structure into a sequence of characters (or bytes) that allows interpretation/parsing, so that one endpoint, the provider, can translate a structure from a data model into a format for receipt and appropriate interpretation back into its structure by a client.
- *Encoding*: The translation from a sequence of characters (provided by the format) into the *bytes* that represent those characters. Bytes are the unit of information storage used within files and carried by network messages.
- *Carrier*: For local files, this is the operating system, which stores the data as named files on a physical storage medium. For providers over the network, the carrier is the *network protocol* used by both ends of the network communication. This could be a specialized protocol, like that used by MySQL or Postgres, or could be a general network protocol, like the HyperText Transfer Protocol (HTTP) used in Internet communication.

We can broadly partition providers into those whose data is provided locally and those whose data is provided over a network. Through the coverage of the data models, we have already seen the data sources of local files and relational database systems, both local and remote. We include them in this overview for completeness, but most of the emphasis in Part III of the book is on data sources over the network, and, in particular, data sources of web servers and API services.

18.2   Data Sources

We start with an overview of the data sources in data systems. Two of our data sources have already received attention—in the tabular and hierarchical models, we have used local files as a convenient source of our data; in the relational model, we used databases managed through a database system as our source of data.

The remaining two data sources, web servers and API services, both involve our clients formulating requests, communicating those requests over a network, and then building corresponding in-memory structure from the network response. For these data sources, the *carrier*, as described above, is a network protocol.

18.2.1 Local Files

Local files may be used to store data for the tabular model and for the hierarchical model.[1] The access to local files is achieved through the operating system. That interaction is covered in Chap. 2, where we discuss direct manipulation of files through open(), close(), and a variety of read() operations. Libraries and packages may be layered above the underlying operating system operations, can parse and interpret a format, and can return a structure based on a path to a local file.

When we use local files for the data of the tabular model, the format involved must support representation of rows and columns. We learned in Sect. 6.2 about CSV and other delimited formats (e.g., tab-delimited files) for doing just that. Our choices for interacting with local files supporting this data model include:

- direct processing with file open(), close(), and read()/readline() operations and controlling the construction of the data structures (e.g., column- and row-centric structures like the dictionary of column lists, or lists of row lists) to represent the two-dimensional data structures,
- use of the Python csv module to facilitate the processing of CSV-formatted files, while maintaining control of the data structures, and
- use of pandas and its read_csv() function, which automates the interaction through the operating system and construction of the DataFrame objects of the module, as described in Sect. 6.3.

For the hierarchical model, our client applications rarely manipulate files directly. The parsing and construction of tree data structures is more complex, and we rely on packages to facilitate our interaction with the local files.

If we have local files that are formatted as XML or HTML, we use the lxml package for reading and parsing the files, and the result is an Element tree. These interactions with local files are described in Chaps. 15 and 16.

When local files are formatted as JSON, we use the json package for reading and parsing the files. The top level local-file-based introduction to this format is given in Sect. 2.4, and more detail on structure and operations is then covered in Sects. 15.3 and 16.2

While using local files as a data source is often convenient, we should also consider some drawbacks:

- The data still must originate from some external source, and in order to make a file local, we typically have to "download" from a source and place the data file(s) in our local file system.

---

[1] In the case of using an SQLite database, a single database is itself is stored in a local file, and so, in that sense, local files also support the relational model.

- Management becomes an issue; the data must be organized and maintained within the file system, and we run the risks of multiple copies of the data, and having different versions of the same data.
- Because of different versions of the same data, and the possibility of local modifications, consistent and reproducible client execution and analysis becomes problematic.
- It can be difficult for multiple people to work on the same data, without introducing local modifications.

18.2.2 Database Systems

As we introduced the relational model, we described the architecture of database systems in Sect. 10.3. From a practical standpoint, we needed to include the possibilities of the database system as a data source in order to be able to learn and work with the operations of the model. The two architectural variants, depicted in Fig. 18.3, unify the earlier coverage with our understanding of data sources in this chapter.

The *connection string* described in Sect. 13.1.1 is our means of understanding the two data source variations. The connection string is a *Uniform Resource Locator* (URL that allows both local and remote resources to by uniquely identified. When the protocol scheme of the URL is, for instance, `sqlite`, the resource is a local one, and its location within the file system identified by a file system path. When the protocol scheme denotes a network protocol, for example, the protocol with prefix `mysql`, the resource is located over the network, and the other elements of the URL provide the necessary information for accessing the remote database system.

Once a connection is established, be it local or remote, the programmatic execution of SQL requests can proceed in similar fashion.

As a data source, a local database has the same drawbacks as local files, with versions and management of the database in the file system presenting problems. The remote database system as a data source can solve many of these problems. In

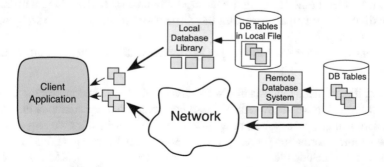

Fig. 18.3 Database source

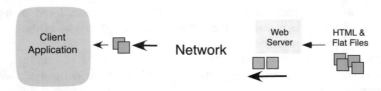

Fig. 18.4 Web server source

this case, the data is shared and centralized, and so there is truly only one resource to be managed. As a result, research and analysis can be more robust and reproducible.

A remote database system as a data source is not without its own drawbacks. These include:

- Need to manage user/password credentials for client users of the database.
- As a shared resource, the database needs management and supervision. Too many client connections and the efficiency of client requests can affect the performance seen by users of the system.
- Database systems and their operations are, by design, strongly consistent when multiple concurrent access occurs. But this consistency comes at a cost, with synchronization communication needed to resolve a set of operations into a consistent sequence. The end result is that database systems have a *scalability problem* as the size of the data, the number of clients, and the frequency of update operations increase.

18.2.3   Web Servers

Our next category of data source is that of the *web server*, pictured in Fig. 18.4. A web server is an application, like Apache [1] or nginx [37], that executes on a remote machine over the network and, in response to requests from clients, sends the requested information back over the network. The vast majority of requests to a web server are made by web browsers, and the vast majority of replies are web pages, formatted in the HyperText Markup Language (HTML), along with images and auxiliary information supporting the display and rendering of the web page(s), like CSS files.

We consider a web server as a source in the data systems context for three primary reasons:

1. The ability to make requests of a web server and receive and interpret the replies is not limited to web browsers. We can construct client applications that "speak the same language" as a web browser and can thus make our own requests.
2. The remote information server by a web server is not limited to HTML and image files but can include files of any type. Thus we can have CSV, XML, JSON, text, and any other files as part of the resource tree provided by the web server.

3. Many web pages in HTML carry useful data in the form of displayed tables of data and lists of items, and our clients may repurpose this data for analysis.

For constructing client applications such as these, we need to understand general topics of networking in the client–server model, and these are addressed in Chap. 19. Building on that understanding of networking, we can then learn details about the primary protocol of the Internet, used by web browsers and web servers, HTTP, in Chap. 20. In the terminology introduced in Sect. 18.1, HTTP is the *carrier*; the *format*, encoded into bytes, is encapsulated in the carrier protocol, and so we show, for our various formats, how to extract data and rebuild structure in Chap. 21.

When the HTML carries useful data, we need to understand the typical ways that data might be represented within a rendering language like HTML. The process of extracting useful data from HTML web pages is known as *web scraping*, and we will look into that further in Chap. 22.

18.2.4 API Service

While web servers can be a data systems source, this may not be their primary functionality. Providers of data can design interfaces that are *intended* as endpoints by which client applications can request collections of data. For a provider, the collection of endpoints for such data access and the specific parameters and rules for making requests are known as an *Application Programming Interface* or API, and thus the name for this category of data source. We depict this type of provider in Fig. 18.5.

In this book, we consider *REpresentational State Transfer* (REST or RESTful) APIs. These have become the most common form of network-based data-providing API. Their popularity comes, in part, from being built upon HTTP, adding conventions for providing arguments serving as the parameters of requests. In Chap. 21 in the introduction, we list some of the many providers in this category.

One of the interesting aspects of an API service as a data source is that it makes opaque the design choices of how to manage the data on the server machines. As shown in the figure, a typical API provider has a *middle tier* that can, in some sense, *translate* between the HTTP based requests coming from the client and then

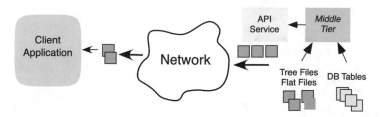

Fig. 18.5 API service source

use back end data sources, like a database system or combinations of tabular and hierarchical data, and can dynamically create responses to those requests.

The coverage of this textbook in the areas of networking, HTTP, and building structure from the HTTP carrier, as covered in Chaps. 19 through 21, is applicable and needed in understanding this data source. We address topics specific to client programming for REST-based API service as a data source in Chap. 23.

18.2.5   Reading Questions

18.1 Please give examples of formats, encodings, and carriers, to illustrate the layering of functionality in a data system architecture.

18.2 There will be an entire chapter on the HyperText Transfer Protocol (HTTP), but to get a sense of some of the subtleties that arise in the "Carrier" part of a data system, please Google "https vs http" and write a sentence to describe the difference. Hopefully by now in your life you have experienced this difference first hand through a web browser. If so, please write another sentence describing a time in your life when the difference mattered.

18.3 Please give an example from your own experience of a time you struggled with a local file. This could have to do with locating the file and writing code to extract data from it, or needing to share the file with a partner, or cleaning the file in some way that was not reproducible (e.g., not documented in your code).

18.4 Please give an example of a real-world situation where a database system might face scalability problems as described in the reading.

18.5 Please give an example of a web browser, and of other ways to obtain data from web pages.

18.6 Please give an example of a web page you know that has a displayed table of data, and then give a different example of a web page hosting a csv file you can download.

18.7 Please give an example of an API and its use. Even better if your example is RESTful.

Chapter 19
Networking and Client–Server

Chapter Goals

Upon completion of this chapter, you should understand the following:

- The basic structure of networks, including

  - host endpoints consisting of network interfaces,
  - routing devices,
  - links connecting endpoints and routing devices, and
  - interconnected local and wide area networks.

- Packets as the unit of network communication and associated ideas of

  - Internet addressing and
  - routing.

- The basic characteristics and use of protocols to solve the network communication problem, culminating in an end result of a reliable byte-stream between endpoints.
- The client–server model of request and reply utilizing the reliable communication of TCP/IP.

In this chapter, we seek to provide the foundation for understanding how data clients can communicate and interact with providers that are located across

© Springer Nature Switzerland AG 2020   591
T. Bressoud, D. White, *Introduction to Data Systems*,
https://doi.org/10.1007/978-3-030-54371-6_19

a network. As described in Chap. 18, this same networking and client–server foundation is an integral part of most of the data sources of interest in this book, from remote relational database systems to web servers to providers accessed through web-based APIs.

The topic of networking is both broad and deep and is, by itself, the subject of full courses and textbooks. By necessity, this chapter only scratches the surface and is meant to provide a basic understanding of the concepts and terminology.

19.1   The Network Architecture

To understand networking, even at a basic level, we must start with the hardware elements that make up the network. A network, as its name implies, can be viewed as a graph with nodes and edges. We originally defined graphs in Chap. 15, and that same definition applies here. Figure 19.1 depicts a network that, in whole, consists of many smaller networks.

In the network graph, the nodes of the graph can be one of the two basic types:

- *Host*: These are the endpoints of communication on a machine. They consist of *each network interface* to a laptop, desktop, smart phone, server machine, or Internet of Things (IoT) device. If a laptop has both WiFi and a wired Ethernet network interface, it consists of *two* host endpoints.

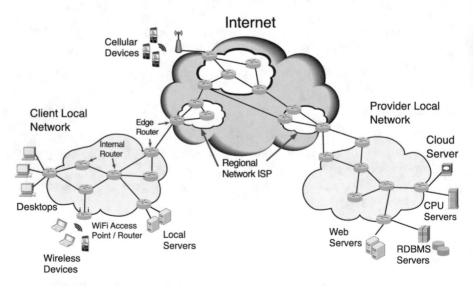

Fig. 19.1 Network architecture

- *Router*: These are the devices that pass bytes of information along as they travel from one host endpoint to another across the network. These are the *internal nodes* in the network graph.[1]

In the figure, we depict, in the lower left, a *Local Area Network* (LAN) for the client. This could be a network for a business, or a non-profit, or a college, and could itself be quite complex. On the other hand, it could be a simple home network consisting of a single WiFi access point. The hosts in the figure consist of desktops, wireless laptops, and smart phones, as well as local servers. Each end host is connected, via a wired or wireless medium, to an *access router*. The access and internal routers are connected via wires, and there are multiple possible paths between the hosts of the LAN. There is also connection to an *edge router* providing connection of the LAN to the rest of the network.

In the center of Fig. 19.1, we show, abstractly, the Internet Service Providers (ISPs) that make up the communication infrastructure of the network. The Internet is, truly, a network of networks, and it employs a loose hierarchical structure. Often, providing access for LANs, like the client's local network and the provider's local network in the figure, there are *regional providers*, each with its own network. For traffic that is not local to the region, these providers may connect to a backbone provider's network. Backbone providers themselves may have connected LANs, or, as illustrated in the top left of the figure, it may connect and provide service through the cellular network.

In a manner analogous to the LAN of the client, any data system provider offering up services on the Internet consists of a provider's local network, as shown in the lower right of Fig. 19.1. This network also has edge routers connecting the network to regional ISPs, has potentially multiple paths for communication traffic from the rest of the Internet, and has access routers by which its own hosts (web servers, database servers, etc.) are connected to the network.

The goal of the network architecture is to support communication between any two *application endpoints* within the architecture. This implies that the solution to this network communication must

- locate, from one application endpoint, both the *host* and a specific *application within the host* as the other application endpoint,
- navigate through the network of routers that define one or more paths between the hosts,
- provide application level communication that is reliable and mimics the stream of bytes that we find when communicating using the operating system to files in a local system.

---

[1]In real networks, internal devices, that we generically here call *routers*, come in many varieties with various names, like *switch*, *bridge*, and *hub*, and can operate at different levels of the protocol stack, but for our purposes, we will categorize them all as routers.

Solving this network communication problem is complex, and the solution must build from the realities of the devices and the architecture that makes up the network. Important aspects of this reality involve *host addressing*, *packet switching*, and *routing*.

19.1.1 Host Addressing

In order to perform the job of *locating* network interfaces of both hosts and routers, all such network interfaces must have an *address* associated with the interface. These are low-level addresses:[2]

- the address consists of a sequence of byte values that, together, make up a multiple byte integer number,
- each address (number value) that is addressable in the public Internet must be globally unique,
- by construction, an address consists of two parts, one that identifies the network where the host resides, and the other that identifies the particular host within that local network.

This low-level numeric address is known as an *IP address*.

Another host addressing concept, familiar to those who set up or manage home networks, is the idea of a *private (IP) address*. A private IP address can *only* exist behind a translation device known as a NAT (Network Address Translation). The NAT still has an outward-facing unique and globally unique IP address, but the addresses within the home/private network need not be globally unique.

19.1.2 Packet Switching and Routing

The Internet was designed for a dynamic, changing nature, with hosts being added and removed, network topology/router placement changes, and new links being established and removed. Management needs to be largely decentralized, so that ISPs can offer varying types and levels of service and could establish relationships with other providers.

This dynamic and changing nature was significant in arriving at the architecture described above. The most important decision in the design of the Internet was to use *packet switching* for data communication. In packet switching, the bytes of data are segmented into maximum sized pieces, called *packets*, and each packet is

---

[2]There are even lower-level addresses, associated with a device hardware interface, and remain static, even as a device might move within or between parts of the network. These are known as MAC or hardware addresses and will not be discussed in this book.

independently sent from a transmitting host, through a sequence of routers, to the destination receiving host. Each packet consists of the data to be communicated plus the destination and source IP addresses along with additional metadata.

Routers accept packets coming in from adjacent routers or from hosts and must forward each packet along a connection that is "best" for getting that packet toward its destination. What is "best," however, can change over time and for a variety of reasons. Some paths or links might be down, new links might be added, or routers along the path could get congested.

Routing is the term used for discovering the network graph topology and conveying to the routers the information they need to make these "best" decisions. Given a packet arriving at a router, the router must use the destination IP address and determine the outgoing link that forwards the packet on toward its destination. The algorithms for routing are continually executing and adapting to the changing topographical and congestion conditions within the Internet. For a system as large as the Internet, and with its continual change, it is not feasible to find optimal paths, and so parts of the routing problem are heuristic and "best effort."

19.1.3   Summary Characteristics of the Network

When we take into account the network architecture and combine this with the packet switching for data communication, performed by routers acting independently on a per-packet basis, we can characterize the reality on the basic facility provided by the network.

1. First of all, routers are devices with finite resources, including memory to store incoming packets. This means that, in times of overload, these resources can be exceeded, and *packets dropped*.
2. To enable reliable communication in the presence of dropped packets, end hosts may retransmit packets. But if the original packet was not dropped, but just taking a longer time or following a congested route, a retransmission can result in a *duplicated packet*.
3. Because packets are routed independently, different packets can take different routes through the network and could be received at a destination host in a *different order* from the source transmission order.
4. A reality of any transmission medium, from transmission through the air, transmission over copper wire, or transmission over fiber optics, is that bits can occasionally be corrupted and 1-bits changed to 0 or vice versa, resulting in *corrupt packets*.

So, in summary, we have a network that can drop, duplicate, and corrupt packets, as well as deliver packets out of order. From this packet-switched network reality, we must solve the network communication problem and deliver a reliable and in-order stream of byte data, in both directions, between two application endpoints. The structure of the solution is the subject of the next section.

19.1.4 Reading Questions

19.1 When modeling a network as a graph, what kinds of considerations go into the decision of whether to make it a "directed graph" (i.e., arrows that point one way between nodes) versus an "undirected graph" (edges can be traversed in both directions)? Which model do you think is more appropriate for the first figure in the reading? Please talk about information flow in your answer.

19.2 In the Network Architecture figure, the "degree" of a node is the number of edges connected to it. Do all routers have degree at least 2? Do all host nodes have degree 1? What does the degree of a node tell us in terms of information flow?

19.3 The reading says a laptop can have two endpoints in the network, if it has both a wireless card and a wired Ethernet interface. Does this mean such a laptop can act like a router and be an internal node in the figure? Can a smart phone ever act like a router? Explain.

19.4 Why is it important that IP addresses be unique?

19.5 IPv4 uses 32 bits per IP address, broken into four blocks of 8 bits each. How many IP addresses are possible on IPv4? Do you think we will run out of IP addresses? Please think about the number of websites and number of hosts in your answer.

19.6 The reading describes how packets can be dropped, in case routers are overwhelmed. But this design decision has many consequences, including the situation of duplicate packets. If we could increase the resources of each router (e.g., doubling their storage capacity via better hardware), would this solve the problem of dropped/duplicate packets? Justify your answer.

19.7 In a world where packets can arrive out of order, how would you structure packets (i.e., what data would you make them contain) to allow them to be stitched together again at their destination in the correct order once they all reach their target?

19.8 In a world where packets can be corrupted due to the physical way they are transmitted, please describe how you would address this problem if you were designing the internet. For instance, could you make the packets contain information that would tell the target node if the packet was corrupted in transit? What if that information gets corrupted? Think through these issues and sketch a plan.

19.9 Consider your favorite web browser and the facility for changing to a *private* browsing mode. Which aspects of the network architecture are changed when you enter this mode? Is it possible for routers to track your IP address and the website you are trying to access?

19.10 Find the IP address for the local machine where you do your computer work. You may have to use a search engine to find possible mechanisms for doing this. You should be able to locate some *online* websites that can reflect your IP address, and there are also ways to use a Terminal or Command Prompt command.

19.11 Have you ever used the same username:password pair on two different websites? If a malicious actor obtained this pair for one website, how might they go about finding if you have used it on another?

19.2 The Network Protocol Stack

Different types of applications have different needs from the network, but the same network architecture, the one described above, must support all application types. For instance, a video streaming application may be able to operate and provide service even with (occasional) packets that are dropped. However, if we are transmitting HTML or a file from a provider to a client, we need *each and every byte, in original order* to be communicated. In data systems applications, unreliable, lossy, or out-of-order data between providers and clients is not acceptable, so the treatment here will focus on that version of the network communication problem.

The goal is to provide a *bidirectional, reliable byte-stream* between two *application endpoints*. Bidirectional means that once communication is established as a *connection*, the applications on both sides may freely transmit and receive data. Reliable implies that all bytes transmitted can be received in the order sent. A *byte-stream* is a model of communication where there are *no boundaries* inherent in the communication itself—the data is a simple sequence of bytes.

Think of a byte-stream like a physical track between two points where marbles may roll down the track. Sometimes a group of marbles may be rolling together. Along the way, they may separate into smaller groups or join together into larger groups. But, in the end, the sequence of marbles is transferred reliably from one point to another. In this analogy, the marbles correspond to individual bytes to be communicated from one endpoint to the other.

Beyond the byte-stream model, if an application needs boundaries, like dividing the bytes of one application message from another, or separating a *line*, defined as a sequence of character bytes terminated by a newline (\n), from the next line, then it is the application that must add these boundary semantics. From the perspective of the network communication, there is nothing special about the sequence; it is just a stream of bytes, and those bytes may be *arbitrarily* broken into the packets that are transmitted over the network. It is a common misconception to think that a line or a message is somehow equal to a packet.

The problem of providing bidirectional reliable byte-stream between application endpoints, when the underlying network provides packet-switched, unreliable, and "best effort" communication between host interface endpoints, is clearly a complex problem. We further need facilities to manage the network, so we can detect when there are connectivity and other problems. Lastly, we need facilities so that client and provider applications and users of the system can be shielded from the low-level concepts like numeric IP addresses and can refer to locations and resources in the network using string-based logical names.

We solve complex problems by decomposing them into smaller problems. In networking, this decomposition takes the form of the *network stack*, a layered set of *network protocols*, each of which logically operates between two communicating parties.

From Merriam-Webster, we present two non-computer definitions of the term protocol:

Definition 19.1 A *protocol* is defined as

- a code prescribing strict adherence to correct etiquette and precedence and
- a detailed plan of a scientific or medical experiment, treatment, or procedure [35].

The first of these is often applied to diplomatic exchange. For instance, there is a protocol associated with having dinner with the Queen of England. The protocol specifies, in a fairly rigorous way, both the *order* of event steps and the specific *content* of each step. Likewise, in the second variation of the definition, a medical treatment would involve a rigorous specification of the *order* of medical steps and the specific *content* that makes up each step.

When applied to data communications, we see these same two aspects of specifying both the allowed ordering of *steps*, in this case, packet messages exchanged between two communicating parties, and the *content* of each packet message, which we refer to as its *format*, specifying various *fields* and their interpretation. This should help to ground our understanding of the Wikipedia definition of a *communication protocol*:

Definition 19.2 A *communication protocol* is defined as a system of rules that allow two or more entities of a communications system to transmit information. The protocol defines the rules, syntax, semantics, and synchronization of communication. Protocols may be implemented by hardware, software, or a combination of both [85].

Figure 19.2 shows the four protocol layers that comprise the network protocol stack used in the Internet.[3] On the left, we show each of the four conceptual protocol layers. The stack starts, at the lowest level, closest to the physical hardware realities of the network with the media access protocol layer, and then proceeding up through the network protocol layer, the transport protocol layer and, at the highest level, the application protocol layer. On the right side of the figure, we enumerate some of the actual protocols that are realizations of the corresponding conceptual layer. The right-hand side highlights one particular *interface* that exists between the layers defining the application protocols and the transport layer of TCP. This is the *socket interface* and is fundamental to all application protocols and will be described further below.

---

[3]Other sources subdivide the lowest layer into a *data link* and a *physical layer*. There is also a seven-layer standard known as the Open Standards Interconnection (OSI) model, which abstracts protocol decomposition based on additional general functionality that may be included.

Fig. 19.2 Network protocol layers

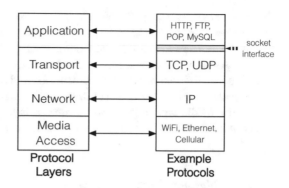

In a layered system such as this, a layer/box in the stack represents an *implementation*, through software and/or hardware, that realizes a particular protocol. Each layer *uses* the services provided by the layer below it. This use comes in the form of the lower layer providing a well-defined interface, which allows an upper layer to "call" on the functionality and to receive data or values returned from the lower layer. In the other direction, the implementation in a layer provides protocol layer services to a layer above it and *presents* a well-defined interface for providing that functionality.

The subsections that follow will start at the bottom, at the media access layer, and work up through the application layer. For each, we will describe the characteristics of the layer and then describe some of the specific protocol instances.

19.2.1   Media Access Protocol Layer

The *media access protocol* layer is responsible for the lowest level of the protocol stack, defining communication between two adjacent nodes. This involves the behavior of the hardware that governs how, for a specific medium, the bits and bytes of packets are physically transmitted and delineated into well-defined frames containing the packet. A given node in the network may need to support multiple specific media access protocols. In a leaf node, like a host, if there are multiple network interfaces, these could be different media access protocols. In an internal node, like a router, different input and output links will often differ in their media access protocol.

Each interface, and thus each instance, of a media access protocol uses *its own network IP address*. In addition, many media network interfaces have their own lower-level device address, known as a Media Access Control (MAC) address.

The three most common media protocol types are WiFi, Ethernet, and Cellular. Within each of these, there are more specific protocol variations. WiFi has a suite of protocols, whose names begin with 802.11 and have variations a, b, g, n, ac, and so forth. The numbering comes from the IEEE Standards Organization. Similarly,

Ethernet is really a range of protocols. These are also standardized by the IEEE and are numbered 802.3 and are known by common names such as "Gigibit Ethernet." Cellular data has evolved through a number of protocol variations from LTE to 4G and 5G. As with all media access protocols, each of these are *different* protocols, and each one has different hardware specifications and requirements, but they all provide an equivalent service for protocol layer above it in the protocol stack. Regardless of specific media type, this *network protocol* layer can call upon the media access protocol layer to transmit and receive packet frames over a single hop in the network.

19.2.2   Network Protocol Layer

The *network protocol* layer is responsible for packet delivery from one machine that is the *source host*, across a network architecture, such as the one presented in Sect. 19.1, to a target machine designated as the *destination host*. To support this functionality, the network layer must support the routing decisions of each router along the path from a source to a destination. Thus, the algorithms for determining topology and supporting the forwarding at each router are both part of the network layer. This also means that beyond the data being carried, the format of a packet must also include the source and destination IP addresses, so that these decisions can be made by each router on a packet-by-packet basis.

A depiction of the network layer, which is implemented on each end host machines as well as on each router along the path, is shown in Fig. 19.3. Starting from the far left, the layer above the network layer, called the *transport protocol* layer, executing on the source host, calls upon the network layer to deliver a packet. This packet delivery request specifies the data and additional transport layer content, along with the IP address of the destination. Each router along the path has already been given information that allows its best-effort routing/forwarding decisions. So, starting at the source host, the network layer determines the outgoing link that will move the packet toward its destination. At each router, the network layer receives the packet and makes its forwarding decision, sending the packet to the next router. This forwarding decision is depicted in the figure by the dashed line in the network layer of each router. The last router will be connected to the local area network of

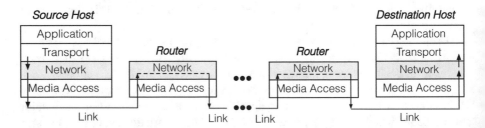

Fig. 19.3 Network layer detail

the destination and will send the packet to the destination host. The network layer on the destination will then pass the received packet up to the transport layer at the destination.

The *Internet Protocol* (IP) is the only network layer protocol used in the Internet.

We can summarize the characteristics of the network protocol layer, and IP, in particular, as follows:

- The unit of communication is a packet, which has a fixed maximum size, typically about 1500 bytes.
- Each public host and router interface has a globally unique IP address.
- The service provided by IP to the layer above is delivery of packets from a source host to a destination host.
- Routing algorithms provide routers with information to determine an output link for a packet based on each packet's destination address.
- Packet delivery across the network has the following characteristics:

  - Packets may be dropped,
  - Packets may experience corruption along the way,
  - Packets may be delivered at the destination in an order different than the source because of multiple paths in the network and independent decisions by routers.

19.2.3   Transport Protocol Layer

The *transport protocol* layer provides communication between a *host application* and a *destination application*. So, this goes beyond the communication between two machines and defines communication between two applications. In data systems, we require a transport service that gives us a reliable bidirectional byte-stream communication between these applications.[4]

The *Transmission Control Protocol* (TCP) is the protocol used in the Internet that provides this service. The services of TCP may be directly accessed by an application, or one or more additional *application protocols* may be layered above TCP, and the application interacts with these higher-level protocols. In the reliable service of a transport protocol, a connection is established from a source host, and the connection specifies both a host and, through an integer (a *port*), a specific application on the destination host. Once a connection is established, either of the hosts involved in the communication can *send* and *receive* sequences of bytes. Once communication is complete, the connection is closed/discontinued by both source and destination hosts.

---

[4]A transport protocol layer can also provide an *unreliable datagram* service between applications, but this does not fulfill our requirements. The unreliable datagram service that is part of the Internet is the User Datagram Protocol (UDP).

The transport protocol of TCP makes use of IP to provide this service. Since IP is unreliable, and packets may be lost or corrupted, TCP must do more than just sending the packets; it must use packet acknowledgments (themselves IP packets) and estimate the time required to traverse the network. If packets do not arrive at the destination in a timely manner, TCP must retransmit packets. To enable distinguishing packets in a stream, TCP adds fields in the packets given to IP that provide identification of the data bytes contained and fields for acknowledging receipt of sequences of data bytes.

The protocol layers of TCP and IP are used by many applications running on a single machine and entail buffering[5] of data between applications and the network interface devices of the machines. For these reasons, TCP/IP is commonly provided as part of the operating system executing on host machines.

19.2.4   The Socket Interface

While not a layer, it is instructive to discuss the *interface* provided by the operating system that allows networking applications as well as higher-level protocols to request services of the transport layer/TCP. This interface is highlighted on the right side of Fig. 19.2 as the line between the HTTP layer and the TCP layer. This interface is known as the *sockets interface*.

When communication is desired, one side of a pair of host applications creates a *socket*. The socket identifies an endpoint of communication and, on a server, will establish the *port* that links a specific application to the logical application destination needed by a client. For instance, the port 80 is logically associated with a web server, and an actual application like Apache [1] or NGINX [37] will create a socket that waits for communication on port 80. A client, when ready to communicate, will create its own socket and request a connection to the destination host/port pair. An application that is waiting on a particular port will allow completion of the connection, and both sides can then perform send() and receive() operations on the socket object, much like a file object provides read() and write() operations.

One other aspect of the network system that is visible at the sockets interface revolves around addressing and naming. The socket interface allows a destination host to be identified by its IP address. But IP addresses are cumbersome to use, given their numeric nature. Client applications should be able to use *logical* names of host machines and perhaps the ports associated with applications. Imagine if the user of a web browser had to remember the IP address of 172.217.4.36 as opposed to www.google.com. The ability to use logical names and to map between logical

---

[5]Buffering is the process of using portions of memory, *data buffers*, so that a slow consumer does not result in a loss of data. The data is buffered (to the limit of the available data buffers), so that it can be used by a consumer when they are ready.

names and IP addresses is a facility provided by the Domain Name System (DNS), and this facility is integrated into the functionality presented by the socket interface.

19.2.5   Application Protocols

Applications can directly use the socket interface to TCP itself, but often application level protocols are layered on top of TCP and provide an even higher level of abstraction to client applications that are based on a "type" of application.

We have already seen one "type" of application—the clients and servers communicating over a network for a relational database system. Each relational database system/vendor is different, and so, for each different system, there is a different application level protocol. There are protocols defined for MySQL, PostreSQL, and Oracle databases. Their goals are the same: to allow an SQL request to be transmitted by a client, received by the DBMS, processed, and then for the resultant table to be transmitted, in whole or in chunks, back to the provider. But each of MySQL, access server, etc. defines its own protocol, which involves the structure and format of messages for requests and replies to be sent along the reliable byte-stream connection and the expected order of all such messages.

By far, the dominant application protocol in use on the Internet is the HyperText Transfer Protocol (HTTP) and its transport-level secure version HTTPS. Like any protocol, HTTP defines expected formats, request and reply messages, and legal ordering. From the perspective of data systems, HTTP is the basis for requests of web servers and for web/API data services, and so we devote Chap. 20 to this topic.

19.2.6   Reading Questions

19.12 Does the byte-stream model also apply to divisions between packets? That is, do network links know when one bit belongs to packet A and one bit to packet B? Or are the "marbles" belonging to two different packets allowed to run together?

19.13 The network layer is running a routing algorithm. Is this algorithm running on routers? On hosts? On some machine that is "running" the internet? Explain.

19.14 The reading briefly describes buffering. What is meant by buffering in the context of internet transmission of data over a packet-switched network? Why is this important?

19.15 Speculate on why we limit the size of a packet transmitted over the network.

19.16 Explain why TCP, the transport layer protocol, is software that is only executed on the host endpoints, and not on the intermediate routers.

19.17 Why would a data scientist need to know about application protocols? Illustrate with a potential real-world application.

19.18 Use Google or other resources, and find the logical application and the application *port number* for the following protocols:

- SSH
- HTTPS
- FTP
- SMTP
- POP3

19.3   Client–Server Model

The term *client–server* is used to describe a computing model for solving computing problems over a network. This model is based on distribution of functions between two types of independent and autonomous processes: the server and the client. A *client* is any process that requests specific services from the server process. A *server* is a process that provides requested services for one or more clients. Client and server processes can reside within the same computer or on different computers linked by a network. The server can provide services for more than one client and, a client can request services from multiple different servers on the network. The network ties the server and client together, providing the medium through which the clients and the server communicate.

One of the most important properties of client–server is the independence of clients from servers. This decoupling allows servers providing the same service to be replicated, allowing scaling of the services. This independence, decoupling, and scaling are the direct results of having a well-defined interface between client and server at the application level protocol. Each side of the client and the server can be developed relative to the application protocol and needs no knowledge of each other.

While client–server is particularly useful in data systems, with data system providers fulfilling the role of server and data manipulation and analysis applications fulfilling the role of client, the client–server model extends to many other application domains. It is used in solutions for wide-ranging problems, like managing print services, distributing CPU intensive computation, and providing shared file servers. Multi-player gaming is another application area that lends itself well to client–server.

Data systems use the client–server model for our data sources that are provided over the network. This includes the ones presented in this book: remote relational database systems, web servers, and API service providers. For some systems, like API service providers, a system that is a *server* for our application client (the middle tier) can also be a *client* for a relational database providing the underlying data. This

Fig. 19.4 Client–server
communication

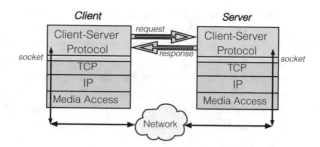

notion of software systems that can serve roles of both server and client allows us to build solutions to larger problems.

We should note that not *all* distributed computing (computing over a network) is structured within the client–server model. There is power in structuring systems with no centralized point of control, in order to adapt to changing conditions and to scale even further. Such decentralized systems often fall into the computing model known as *peer-to-peer*.

In the context of the network protocol stack, we depict a single client, single server scenario in Fig. 19.4.

At a logical level, a client, through an application level client–server protocol, issues a *message* that describes a request for data. On receipt of the request, the server processes the request and generates one or more response messages that represent the data and/or the status from that processing. These responses are also formatted and follow the particular client–server protocol. In implementation, the client and the server, through the application protocol, use the network protocol stack. They establish a connection using the socket interface to TCP. The format defined by the client–server protocol defines the boundaries of what makes up a message, and each message is transmitted/received over the reliable byte-stream channel provided by TCP through the socket interface.

19.3.1   Server Application

The architecture of a server application is generally structured into a client interface component and a processing component. The processing component is given a unit of work from the client, it performs the work, and then generates a result and/or a status indicating the outcome of the work.

The client interface component at the server has one or more threads of execution that repeatedly perform the following steps:

1. *Listen* for and passively await a connection incoming from a client; this could be through direct use of the sockets interface and TCP, or it might be provided by a server-side application library built above the sockets interface.

2. On an incoming connection, spin off work and further communication in a newly created separate connection. This allows the server to also continue to await additional incoming connection requests.
3. Based on the specific processing in application protocol, parse a request and formulate a unit of work for the processing component, and initiate the work.
4. When the work is complete, transmit the results and status back to the client in the form dictated by the protocol. If the protocol allows "chunking," this may involve multiple transmissions over the communication channel, with possible acknowledgments by the client.
5. After the unit of work:

   - If the protocol requires it, close down the connection. Some protocols require one connection per one unit of work. If this is the case, the next request by a client will initiate a new connection.
   - If the protocol allows it, determine whether or not the client has additional requests. If so, repeat steps starting at Step 3. This variation needs an application protocol way of providing graceful termination when there are no more requests.

One other aspect that influences server applications in client–server is whether or not the server *maintains state* about the client. This refers to the collection of information maintained *across* a set of requests/responses with the same client.

A *stateful service* is one that maintains information about all active clients. The information varies by service, but can include, for a given logical client, information about one or more connections, the sequence of requests and responses made so far, and security information starting with authentication information and the set of authorized operations for this particular client.

While being stateful allows some efficiencies and optimizations relative to a client, it can lead to an inconsistent view between clients and servers across outages and interruptions in communication. It also does not scale when the number of clients grows from tens or hundreds of clients to thousands of clients.

A *stateless service* maintains no sustained information at the server about each client. In this case, each connection is afresh, as if the client had never engaged with the server previously. Here, each connection and request must include directly or indirectly *all* the information needed by the server to process the request. It cannot be dependent on the sequence of previous requests.

In this book, we see examples of both stateful and stateless servers in client–server data systems. The remote relational database system and the application level protocols like MySQL and Postres are *stateful*. To support transactions, they maintain information about their clients and the sequence of separate request-response operations. On the other hand, web servers and RESTful API services, implemented over the HTTP application level protocol, are *stateless*. Each request has, within it, all the information needed by the server to respond to the request.

Some compromises are possible. For instance, while HTTP is itself stateless, systems use the so-called *cookies*, whereby information about interaction with a service is stored *at the client*, and may have corresponding client information at the server, to be looked up when needed, but with little to no impact if the client fails and does not return. Thus, this solution does not suffer from the scalability and outage problems associated with stateful services.

19.3.2   Client Application

From the perspective of this book, the client application is the software that we write to achieve our data systems goals. In our context, we almost always build upon one or more packages/libraries that can handle the client–server application level protocol specifics. This higher-level abstraction relieves our clients from the lower-level details of the sockets interface and the send/receive operations on the reliable byte-stream provided by TCP.

The processing at a client mirrors the sequence of steps entailed on the server side. A client performs the following set of steps:

1. The client actively initiates a connection with a particular server. It is expected that a server application is already listening for a connection and is associated with a port linking the logical application with a specific application.
2. Based on client application logic, the client periodically needs to construct and make a request, with steps to:

   - compose the request,
   - transmit the request,
   - await the outcome, which may involve significant time,
   - process the response,
   - if the client–server protocol only allows one unit of work per connection, close the connection,
   - otherwise, continue client application logic until the next request, returning to Step 2.

3. After all units of work have been completed, close and otherwise, clean up the connection.

A response received by the client after a request may indicate that the request could not be completed. On the other hand, the request may result in a significant amount of data, and the response may partition the data response into multiple pieces or chunks and use multiple messages to effect its transfer.

19.3.3   Reading Questions

19.19 Give an example where the client and server processes could reside in the same computer.

19.20 Please give an example where one client could request services from multiple servers.

19.21 The book describes how, for some protocols (and protocol versions), a client–server exchange is limited to exactly one exchange over a socket connection, and for other protocols, there could be *multiple* exchanges over the same connection. Give pros and cons of these two possibilities.

Hint: multiple exchanges over the same connection are called *persistent connections*. A Google search for this term could help solidify your understanding.

19.22 The reading mentions a computational model known as "peer-to-peer." Please read up on this topic online and report what you found. What types of real-world situations would call for a peer-to-peer system?

19.23 What would be the point of a protocol allowing "chunking"?

19.24 Have you ever lost a connection (e.g., internet went out, had to close your laptop to move, and battery died) during a download? If so, what happened when you reconnected? What does that tell you about whether the service was stateful or stateless?

19.25 The reading mentions "denial of service attacks." Please use a search engine to read up on this topic and write a paragraph summarizing what you learned. Then, explain why stateless service is more resistant to these attacks than stateful service.

19.26 What are some pros and cons of the use of *cookies* to save state and relate multiple requests into a *session?

Chapter 20
The HyperText Transfer Protocol

Chapter Goals

Upon completion of this chapter, you should understand the following:

- HTTP as an application level client–server protocol, and one that uses the bidirectional reliable byte-stream of TCP.
- The syntax of the primary messages of HTTP

  - The GET message to request a resource,
  - The POST message to send form data and/or update a resource,
  - Reply messages.

Upon completion of this chapter, you should be able to do the following:

- Construct and send HTTP request messages using Python strings.
- Interpret string-based HTTP reply messages.
- Use the `requests` module to construct and send more complicated HTTP GET and POST request messages.
- Interpret HTTP replies through the `requests` object returned.

The HyperText Transfer Protocol, HTTP, is the lingua franca of the Internet. It is a client–server application level protocol. It is used by clients, which can be web browsers, phone apps, utilities, and other application programs, to request and

© Springer Nature Switzerland AG 2020
T. Bressoud, D. White, *Introduction to Data Systems*,
https://doi.org/10.1007/978-3-030-54371-6_20

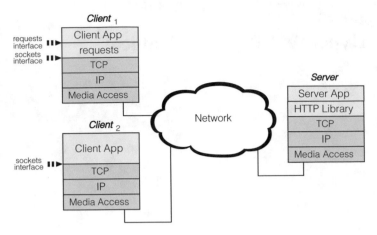

Fig. 20.1 HTTP client–server

update data from applications serving as HTTP providers. These providers can be web servers, providing static and dynamic web pages, and can be other types of data system providers, such as APIs that provide access to data in textual, HTML, CSV, XML, JSON, and other formats. One of the advantages of HTTP is that it provides a common protocol for these requests and responses even though the breadth of applications and use of the protocol is diverse.

As a network protocol, HTTP defines the order and format of the messages that define the communication between a client and a server. But the independence of the server and the set of possible clients, and the fact that HTTP is about the content of the messages, and not about the libraries, packages, or source programming languages, means that clients and servers are free to develop using whatever programming language and library packages they wish.

In Fig. 20.1, we show two client applications and a single server. We have enhanced our depiction of the network protocol stack by showing a layer for the application logic that runs above the stack. $Client_1$ might be an application written in Python, and it is shown as running over the `requests` package, which presents a high level abstraction at the `requests` interface and implements HTTP. $Client_2$, on the other hand, might be written in Java, or perhaps Typescript. It is shown as not using a package that realizes HTTP, but rather uses the socket interface to create connections and use lower-level `send()` and `receive()` functions, and compose its own messages based on the syntactic format of HTTP. On the right side of the figure, the *Server* provides web service to multiple clients. The software of the *Server* could be written in C++, and might use its own library for HTTP.

In this chapter, we focus on the definition of HTTP itself as an application-layer protocol that executes over TCP and uses the `sockets` interface. We then look at a Python package that can greatly simplify writing client software to issue HTTP requests and process replies from providers.

20.1 Identifying Resources with URLs and URIs

Uniform Resource Identifiers (URIs) and Uniform Resource Locators (URLs) define a standard notation for specifying the files, data, and resources of the Internet. URIs are the broader term, and can be used to identify resources *within* a particular provider, using a path notation, and can also include URLs, that, through an explicit protocol scheme and host location along with a resource path, are used to uniquely identify a resource at a specific location in the Internet [19].

20.1.1 Host Locations

The provider of information (aka the *server*) is identified the same way we identify an endpoint for communication—through a host location and a port, and we call this the *location* of the resource. We also need to know the *protocol* (sometimes called the *scheme*) by which the provider will respond to requests for information. In Chap. 13, we used relational database schemes, for instance, those based on MySQL and SQLite. For the remainder of this book, we will use `http` and `https`.

20.1.2 Resource Paths

The actual data/information provided are known abstractly as the *resources* of the provider. Because a provider can have many resources that they provide, and to allow for an organizational structure for these resources, we can think of a provider's resources as forming a *tree*. This is directly analogous to the tree used to represent the *file system* of a local computer system, as seen in Chap. 2. Beyond the path of the tree, resource paths can be further extended to include a *query string* that specifies one or more *query parameters*; following the path, the query string starts with a `'?'` character and includes a sequence of name = value pairs, separated by the & character.

We specify a particular resource in the tree by a string representing its *path*. We start a path with a `'/'` to indicate the root, and then follow the names of the nodes of the tree and end with a leaf child, separating non-leaf children with a path separator of a /. So, for example, `/basic.html` gives the resource path for an HTML file directly under the root at a provider location, and `"/data/ind2016.csv"` gives the resource path for a CSV file positioned under a `data` child under the root. Note the following:

1. The tree could easily have additional levels. The `data` node could have children `foo` and `bar` that could themselves have additional children.

2. There can be many file types/file extensions indicating the content type of the resource. In this example, we have resources that are web pages (`.html`), and resources that are tabular data files (`.csv`), but file-based resources could be of any type.

20.1.3   URL Syntax

Summarizing our requirements for globally identifying resources in the Internet, we need:

Item	Description
protocol	The protocol/scheme identifies the network stack layer above TCP, and which defines the specifics of the form for legal requests made by clients and the form for legal responses returned by servers. We use `http` and `https`
location	Identifies the server/host machine within the Internet
port	Identifies the executing program used for connections. We generally use port 80 for HTTP web server programs and port 143 for HTTPS web server programs
resource-path	Identifies, within the host/port endpoint, a particular resource provided by the server; this component could also include a query string, used for providing parameters to dynamic resources

The general form of a URL, then, includes each of these parts:

protocol : / / *location* [: *port*]*resource-path*

In URLs, if we do not specify a port, then port 80 is assumed when the protocol is HTTP, and port 143 when the protocol is HTTPS. Web browsers and web servers also provide additional shortcuts. So if you do not include a *protocol* : / / in a web browser address line, `http://` is assumed. If no resource path is specified, the web browser provides / (the root) as the resource path. On the server side, if the resource path ends at a non-leaf node of the resource hierarchy, then a leaf resource of `index.html` may be attempted.

A resource path may simply define a path to a resource, but when the URL or URI includes an optional query string, we subdivide the resource path into the part before the query string, called the *endpoint-path*, and the *query-string*, as we show here:

resource-path $\models$ *endpoint-path* [*query-string*]

The query string itself always begins with a question mark character (?). This is followed by one or more name/value pairs, each called a *query parameter*. If there are multiple query parameters, they are separated by an ampersand character (&), and this syntax is presented here:

$$query\text{-}string \models \ ?name=value[\&name=value]^+$$

The *name* and *value* elements of query parameters are significantly constrained in legal characters, and this is true for the rest of the URL or URI as well. For instance, no part of the URL, including the query string, may include embedded spaces, nor may it include a number of other special characters, like :, /, =, or &. The process of converting from forms that have these special characters into ones which do not is called *URL encoding*.

20.1.4   Reading Questions

20.1 Are all URLs also URIs? Are all URIs also URLs? Explain with examples.

20.2 Within the network protocol stack, what are the levels above and below HTTP? What is the interface by which HTTP makes requests of the level below it?

20.3 Is it possible for an application to *not* use a package that implements HTTP? Why or why not?

20.4 Using a web browser, navigate to the public-facing web page for your major within your college. Or, if you are independent of a college or major, search for "Denison Computer Science" to find an equivalent web page. Then, relative to the URL of the page, identify the *protocol*, the *location*, and the *resource path*.

20.5 Consider the following URL:

```
http://httpbin.org/get?foo=bar
```

Identify the *resource path*, and the *query string*. Navigate to the URL, and report the *format* of the returned result, and identify how the query string in the URL is reflected in that result.

20.6 Attempt to navigate to a URL where we try and include a *space* after the = and before the `bar` of the URL:

```
http://httpbin.org/get?foo= bar
```

What happened? In particular, look at the browser's address line; did it change? Look carefully at the returned result. Is it the same or different than in the prior question? Explain *why* you think this happened.

20.2   HTTP Definition

HTTP was created to allow web browsers, as the clients, to request data from web servers. Every time a user in a web browser enters a URL in the address line of the browser, or clicks on a link within a web page, at least one, and often many, HTTP requests are issued by the browser to gather the web page and any other resources, like pictures, video, or other data needed to display the resultant page.

Like other client–server protocols, HTTP is based on request/response exchanges between client and server. As described above, HTTP uses TCP and the sockets interface to establish a reliable byte-stream connection between the client and the server.

1. The executing server program is in an "always ready" state, wherein it has created an unresolved TCP socket endpoint, listening for a connection request from a client targeting the given HTTP port of 80.
2. When ready to make a request, the client first makes a TCP connection to the server endpoint, and at the server, the endpoint is resolved to allow bidirectional communication.
3. The client then constructs an *HTTP request*, consisting of a valid sequence of characters describing the request and conforming to the standard protocol format as given by the HTTP definition and described below.
4. This request is sent over the TCP socket connection to the server.
5. The server receives the request over the TCP socket connection and processes it, constructing an *HTTP response*, consisting of a valid sequence of characters and data bytes giving the result of the request and conforming to the standard protocol format as given by the HTTP definition and described below.
6. This response is sent over the TCP socket connection to the client.
7. The client receives the response and processes it.
8. Both client and server close the TCP socket connection.

Depending on the result, additional request/response exchanges are permitted to occur over the same established connection.[1] This would cause the client and server to iterate Steps 3 through 7 before the closing of the connection in Step 8.

As we will see in the discussion of the specifics that follows, each HTTP request is self-contained. That is, the request provides *all* the information needed by the server to satisfy the request. Further, connections are never *required* to handle more than a single request-response interaction. These two properties are what make HTTP a *stateless* protocol, which simplifies the design of both clients and servers, and makes for more robust networked applications.

---

[1]Multiple request/response exchanges over a single connection were not permitted in HTTP version 1.0, but this is allowed in HTTP 1.1, and may be controlled by parameters that are part of the request.

20.2.1   Message Format

In HTTP, the unit of communication between clients and servers, as a *request* or *response*, is the *HTTP Message* (*message*). The same syntax format is used for both requests and responses:

$$message \models start\text{-}line$$

$$[\,header\text{-}line\,]^{*}$$

$$empty\text{-}line$$

$$[\,body\,]$$

Every message begins with a *start-line*, followed by zero or more *header_lines*, and then a required *empty-line*. Depending on the particular request or response, these may be followed by a message *body*.

Request messages and response messages differ in how they define the *start-line*, with a request message using a *request-line* as the *start-line* and a response message using a *status-line* in that place:

$$start\text{-}line \models request\text{-}line \mid status\text{-}line$$

20.2.2   Request Messages

For requests, HTTP supports multiple *methods*, allowing a client to choose between requests that retrieve information and resources from a server, versus a request that is sending information from the client *to* the server. The former is exemplified by the method named GET, and the latter is exemplified by the method named POST. The syntax of a *request-line* uses the same structure for any of the HTTP methods:

$$request\text{-}line \models method\ \text{SP}\ URI\ \text{SP}\ version\ \text{CRLF}$$

where SP is exactly one space character, and CRLF is the carriage return character ('\r') immediately followed by a newline character ('\n'). The *version* is specified by either 'HTTP/1.0 or 'HTTP/1.1'. *URI* indicates a Uniform Resource Indicator. A URI can either be a URL, which includes protocol and host information (thus the *L*, denoting a *Location*), or, more simply, it can be a *resource-path*, that identifies, from the tree of resources provided by the server, the specific resource being operated upon.

A more complete set of methods possible in an HTTP request message is given in Table 20.1.

Table 20.1 Common HTTP request methods

Method	Description
GET	Fetch the specified resource. The resource path contains all the necessary information the server needs to locate and return the resource. The GET method does not include a body in its HTTP message
POST	Send form data or update a resource based on the information specified through the *resource path* and the *body*. Can also be used to request a dynamic resource when the provider needs additional information. The body of a POST often is formatted in JSON
HEAD	Like GET to request an existing resource, but requests only the "meta"-data. A response will include a status line and the set of header lines for the response, but will not include an actual body of the response. The headers usually include the type of the content and creation and/or last modified times
PUT	Update the specified resource. The body contains the updated data for the resource
DELETE	Delete the specified resource

20.2.3 Connections and Message Exchange

HTTP occurs over TCP and through the *socket* interface. Recall from Sect. 19.2.4 that a socket is an abstraction of the idea of a *reliable byte-stream* connection between two endpoints across a network. In our case, the two endpoints are the client program executing on the client machine and the server program executing on the server machine. The socket is part of the interface, which includes operations like send() and receive() and whose implementation is realized by the Transport Control Protocol (TCP) layer in the network architecture stack.

20.2.3.1 Client-Side HTTP Steps

In its simplest form, a client can make an HTTP request that corresponds to a web browser fetching from a generic URL of the form:

$$protocol : //host[:port]/resource\ path$$

by the following set of algorithmic steps:

1. Establish a TCP socket connection with server machine *host* at port *port*; call it connection.
2. Build a correctly formatted HTTP request string and assign it to a string variable, request_message. This will use an HTTP method (GET) and will include a header Host with value *host* and use *resource path* as the URI in the *request-line*.
3. Perform a send of the string request_message over connection.
4. Perform a receive of the HTTP message response from connection. Assuming a valid response message, this must retrieve the string of characters

up to and including the *empty-line* after the message headers, and then, based on additional information, must retrieve the body of the response.

5. Perform a `close()` operation on `connection`.

To focus on these steps and avoid the substantial detail associated with the actual sockets interface, we have defined a custom wrapper, `mysocket` that is closely allied with our client steps. Complete documentation is in Appendix A, Sect. A.2, but the three functions used in the examples that follow are given in Table 20.2.

20.2.4 Socket Level Programming Examples

20.2.4.1 Example of Socket-Based GET Request

As a first example, suppose we wanted to retrieve the root resource path / from a web server hosted at `httpbin.org` using HTTP version 1.1. If we *just* consider the request-line, and we include the CRLF line termination required by HTTP, we could assign a Python string:

```
request_line = 'GET / HTTP/1.1\r\n'
```

To create a complete request message, we would then include any header lines and then an empty line. The GET method does not use a body, and HTTP 1.1 requires a `Host` header line. Further, we want to specify the control of the connection so that both client and server close the connection after a single exchange. The request could then be constructed as follows:

Table 20.2 Select functions of the `mysocket` Package

Function	Description
`makeConnection(host, port)`	Establish a TCP connection from the client machine to a server at the given machine `host` and listening at the given `port`. This returns the socket connection. This corresponds to Step 1 of the client-side steps
`sendString(conn, s)`	Given an established socket `conn`, take `s`, a string, and send it over the connection. This corresponds to Step 3 of the client-side steps, where `s` would define all the characters making up a complete HTTP request
`receiveTillClose(conn)`	This performs a socket `recv()` from the connection, consuming data until the server closes the connection. This returns the complete HTTP response message. This corresponds to Step 4 of the client-side steps, and assumes that a connection close will define the end of the response message

```
request_line = 'GET / HTTP/1.1\r\n'
host_line = 'Host: httpbin.org\r\n'
one_and_done = 'Connection: close\r\n'
empty_line = '\r\n'

request_message = request_line + host_line + \
                  one_and_done + empty_line
```

If we represent the lines of the HTTP request, recognizing that *both* the invisible characters terminating lines are, in fact, the CRLF ("\r\n") sequence, the complete request is:

```
GET / HTTP/1.1
Host: httpbin.org
Connection: close
```

Given this as a request message, a client would issue this request by

1. making a socket connection to httpbin.org at port 80, the web server port,
2. sending request_message over the socket to the server,
3. receive the response from the socket, and
4. close the socket.

This would complete a single client–server request/response operation.

In the following, we demonstrate using our custom mysocket module. We will discuss reply messages below, but simply show here the retrieval of the stream of data returned from the server into Python variable reply. After the receive, we use index() to seek the first CRLF and thus determine the end of the first line (the *status-line*) and print it out:

```
import mysocket as sock

server = sock.makeConnection("httpbin.org", 80)
sock.sendString(server, request_message)
reply = sock.receiveTillClose(server)
server.close()

status_line = reply[:reply.index('\r\n')]
print(status_line)
```

```
| HTTP/1.1 200 OK
```

20.2.4.2 Example of Socket-Based POST Request

In the `httpbin.org` web server, the `/post`[2] resource path enables HTTP POST requests to that *resource-path*. In the following example, we structure a request message body that consists of a JSON-formatted dictionary. In a request that carries a body, the `Content-length` header is followed by an integer value that captures the number of bytes that the server should expect for the body of the message, and we use `format()` on the `content_length` string to construct that header line:

```
import mysocket as sock

body = """{"a": 1, "b": 2}"""

request_line = 'POST /post HTTP/1.1\r\n'
host_line = 'Host: httpbin.org\r\n'
one_and_done = 'Connection: close\r\n'
content_length = 'Content-length: {}\r\n'
.format(len(body))
empty_line = '\r\n'


request_message = request_line + host_line + one_and_done
+ \                    content_length + empty_line + body

server = sock.makeConnection("httpbin.org", 80)
sock.sendString(server, request_message)
reply = sock.receiveTillClose(server)
server.close()

status_line = reply[:reply.index('\r\n')]
print(status_line)

| HTTP/1.1 200 OK
```

Here, the complete resultant request is:

```
POST /post HTTP/1.1
Host: httpbin.org
Connection: close
Content-length: 16

{"a": 1, "b": 2}
```

---

[2]Note that the `/post` is a *resource path*, and it was chosen by the designers at `httpbin.org` as the resource path where they wish POST requests to go. It is only by this design decision and this particular case that the resource path matches the name of the HTTP method. This will not be the case for other providers.

20.2.5   *Request Header Lines*

Header lines are used in HTTP requests to pass, from the client to the server, additional information about the request and expectations on how the request should be processed. The syntax of a header line is of the form:

$$header\text{-}line \models header\text{-}name: \text{SP } value \text{ CRLF}$$

So, on each header line, we have the name of the header followed by a colon character and a single space character, and then the value to be associated with the *header-name*, with a final line termination of CRLF. The *value* may include embedded spaces and continues up to the line termination. Capitalization is not significant in the *header-name*.

Table 20.3 presents a few of the most common HTTP headers used in request messages or response messages:

Mozilla maintains a rich set of resources including much more detailed documentation on HTTP headers [36].

Table 20.3 Common HTTP header lines

Header name	Description
Host	Name of the server host location. Required for HTTP 1.1
Connection	Specifies whether or not to close the connection after a request/response exchange. Values are close or keep-alive. If the client specifies close, the server will close the connection after a single exchange, but otherwise, in HTTP 1.1, will keep the connection open for subsequent request/response exchanges. If the connection is kept alive, the client must be able (through Content-Length) to determine the length of a response body
Content-Length	Length, in bytes, of the body in this message
Content-Type	Given by a server that understands the underlying type of the data being returned. Values are from the same space as the Accept header, where text/html, text/csv, text/xml, application/json, and application/xml are the most common. The part before the slash is known as the *mime-type* and the part after the slash is the *mime-subtype*
User-Agent	Name of the client application program. User-Agent can be used by servers to customize responses to accommodate differences in HTML support from various web browsers
Accept	Used by a client to specify what formats of reply are acceptable. For instance application/json can be used to request a JSON-formatted reply, or text/html to request an HTML-formatted reply

20.2.6   Response Messages

Response messages, communicated over TCP from the server back to the client, have the same format as request messages, namely a *start-line* followed by zero or more *header-lines* followed by an *empty-line* followed by the *body*. The *start-line* is realized by a *status-line*:

$$status\text{-}line \models version \text{ SP } status\text{-}code \text{ SP } reason \text{ CRLF}$$

The *status-code* field is always exactly three digit characters. The *reason* is allowed to contain embedded spaces and is considered to include all characters up to the line termination of CRLF.

Without the CRLF, the status line from our last example request is:

```
HTTP/1.1 200 OK
```

So, in this case, the server is indicating that it supports HTTP 1.1, and that the request resulted in a *status-code* of 200, and the *reason* is OK. Both the 200 and the OK are indicators of successful processing of the request.

In Table 20.4, we list the categories of defined status codes, which are grouped by values in the 100's. For the full definition of status codes in HTTP 1.1, consult the RFC standard [22].

As another example, suppose we issue a GET request to a server whose root web page has a minimal HTML-formatted message. If we use a web browser and go to URL: http://datasystems.denison.edu/basic.html, we get the result shown in Fig. 20.2.

Using our sockets-based Python string approach and issuing HTTP directly, we have the following example:

Table 20.4 Categories of HTTP status codes

Status code	Reason	Description
100's	Informational	The request was received; continuing to process, but the request is not complete
200's	Success	The request has competed successfully, and if the request entails data to be returned, it may be found in the body of the response. 200 is used for a successful GET, while 201 means a successful creation of a resource
300's	Redirection	The requested resource is no longer at the given resource path. If permanently moved, a 301 is given, while a 302 indicates a temporary relocation. The Location header should be consulted for a new location/path
400's	Client error	The request contains bad syntax or cannot be fulfilled. A syntax error is 400, an unauthorized request is 401, a resource not found is 404
500's	Server error	The server failed to fulfill an apparently valid request

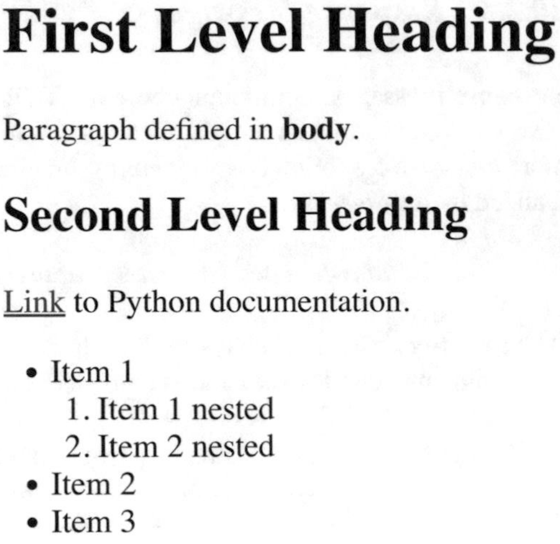

Fig. 20.2 Browser-based
GET

First Level Heading

Paragraph defined in **body**.

Second Level Heading

<u>Link</u> to Python documentation.

- Item 1
    1. Item 1 nested
    2. Item 2 nested
- Item 2
- Item 3

```
request_line = 'GET /basic.html HTTP/1.1\r\n'
host_line = 'Host: datasystems.denison.edu\r\n'
one_and_done = 'Connection: close\r\n'
empty_line = '\r\n'

request_message = request_line + host_line + \
                  one_and_done + empty_line

server = sock.makeConnection("datasystems.denison.edu", 80)
sock.sendString(server, request_message)
reply = sock.receiveTillClose(server)
server.close()
```

Displaying a prefix of the HTTP reply message returned and saved in the code in the variable `reply`, we have:

```
HTTP/1.1 200 OK
Date: Sun, 31 May 2020 14:30:33 GMT
Server: Apache
Accept-Ranges: bytes
Content-Length: 494
Connection: close
Content-Type: text/html; charset=UTF-8

<!DOCTYPE html>
<html lang="en">
  <head>
    <title>Data Systems Basic HTML Page</title>
  </head>
  <body>
    <h1>First Level Heading</h1>
```

```
    ... and so forth
```

While the line termination CRLF is not visible in this displayed result, we can process the reply and use this line separation of CRLF to parse through the status line and the header lines returned by the web server. We know that we have reached the end of these header lines when we encounter the empty line following the last header line (Content-Type in this case). In the following code, we iterate over the reply, searching for the index of the occurrences of CRLF and accumulating the lines into a list. The code then prints out the resultant non-body lines of the response:

```python
reply_lines = []
reached_body = False
start_index = 0
while not reached_body:
    CRLF_pos = reply.index('\r\n', start_index)
    line = reply[start_index:CRLF_pos]
    if line != "":
        reply_lines.append(line)
        start_index = CRLF_pos + 2
    else:
        reached_body = True
        body_start = CRLF_pos + 2

for line in reply_lines:
    print(line)
```

```
| HTTP/1.1 200 OK
| Date: Tue, 30 Jun 2020 21:57:33 GMT
| Server: Apache
| Accept-Ranges: bytes
| Content-Length: 496
| Connection: close
| Content-Type: text/html; charset=UTF-8
```

In the output, we see the status line indicating successful completion of the request. The header lines in the response by the server include Content-Length and Connection, which we have seen before, and also include informative header lines like Date, Server, and Content-Type.

Since the request was for a web page, we expect the *body* of the response to contain a character sequence in the HTML format. The Content-Type header line, with value text/html (along with the body content encoding) in the response confirms that the result is HTML.

In the processing above, when encountering the final *empty-line*, recorded the index of the start of the body within the response as body_start. Simply printing out the body contents from that index for the first ten lines, we see the HTML that corresponds to the displayed page from Fig. 20.2:

```
util.print_text(reply[body_start:], nlines=9)
 | <!DOCTYPE html>
 | <html lang="en">
 |   <head>
 |     <title>Data Systems Basic HTML Page</title>
 |   </head>
 |   <body>
 |     <h1>First Level Heading</h1>
 |
 |     <p>Paragraph defined in <b>body</b>.
```

20.2.7   Redirection

One topic that will come into play in later chapters is that of HTTP *redirection*. Redirection is defined in the HTTP standard and permits resources in the Internet to be moved to new locations, either in the current resource tree or to a different server. When a request is issued to a server and the server determines that the requested resource path is valid, but is for a resource whose location is not as given in the request, but could be available at a different resource path and/or a different host location, the HTTP response message has a status code in the 300's, typically 301 or 302. When this happens, the server, in a header line of the response, will include a header whose name is Location and whose value is a *URI* that gives the redirection information needed to find the resource.

The client, on receiving such a redirect response, can use the Location header value to construct a new request. This redirection happens frequently, but we are rarely aware, because web browsers and libraries issuing HTTP requests process a redirect and, on receipt, automatically generate a follow-up request to the URI specified as the redirect location.

The example code below utilizes a facility of the httpbin.org service that allows a client to build a request that specifies a desired particular return status code. In this case, by using the URI of /status/301, we are requesting httpbin.org to yield a response with a 301 code, indicating a redirection.

After we make the request, we print out the lines of the response.

```
request_line = 'GET /status/301 HTTP/1.1\r\n'
host_line = 'Host: httpbin.org\r\n'
one_and_done = 'Connection: close\r\n'
empty_line = '\r\n'

request_message = request_line + host_line + \
                  one_and_done + empty_line

server = sock.makeConnection("httpbin.org", 80)
```

```
sock.sendString(server, request_message)
reply = sock.receiveTillClose(server)
server.close()

reply_lines = reply.split('\n')
for line in reply_lines:
    print(line.rstrip())
```

```
| HTTP/1.1 301 MOVED PERMANENTLY
| Date: Tue, 30 Jun 2020 21:57:34 GMT
| Content-Length: 0
| Connection: close
| Server: gunicorn/19.9.0
| location: /redirect/1
| Access-Control-Allow-Origin: *
| Access-Control-Allow-Credentials: true
```

We see the 301 status code in the response, along with a *reason* of MOVED PERMANENTLY. The response has a Location of /redirect/1. So a client that handled redirect responses automatically would reissue the request to this URI. If the Location header is not a URL, and so does not include a different host location, the re-issued request should go to the same host server as the original.

This process of getting a redirect response with a Location header could repeat multiple times, and so automatic handling could also repeat until a 200 status code and a successful response allows the process to terminate.

20.2.8   Reading Questions

20.7 The reading mentions the XHTML format. Please read up on this and report what you learned. When would a data scientist need to know about this format?

20.8 Why might you get "many" HTTP requests when you click a link? What does this mean for you when you think about clicking a link in an email from a sender you do not know?

20.9 Why do you think HTTP messages all contain the mandatory empty line?

20.10 Why is it important for HTTP to be able to specify Content-Length in a reply message?

20.11 Why is it important for HTTP to be able to specify an Accept header line in a request?

20.12 Consider the while loop at the end of the section. Why is a loop needed at all? Why do not we just find the first occurrence of \r\n and use that as the break between the start-line and the body?

20.13 Consider the example concluding the section. How does the displayed *body* compare with what you would see by opening "personal.denison.edu" in a web browser then clicking to view the page source using developer tools? Note: this question was written with Google Chrome in mind, but other browsers also let you "view source" for a page.

20.2.9   Exercises

This set of exercises focuses on "raw" HTTP, where requests are built from Python strings and replies will entail the full set of bytes as received over the network from a connected server. For simple interaction with the `sockets` interface, they will presume use of the `mysocket` module described in the book, and available at the author's website for the book. This module should be added to the user's environment so that an `import mysocket` is possible before solving these exercises. See the Appendix A.2 in the book for documentation on the `mysocket` module.

The first set of exercises are about *making requests*.

20.14 Suppose we wish to retrieve (GET) a file via HTTP (so port 80) from `datasystems.denison.edu`. The resource path of the file is `/data/ind0.json`. We wish to use version 1.1 of HTTP and to request that the connection be closed after a single request/reply exchange. We will need a header line to satisfy the HTTP 1.1 requirement of a valid `Host` header. Write a sequence of code to compose a valid HTTP request as a Python string, and assign the result to `message`.

20.15 Write a sequence of code to establish a connection to the host `datasystems.denison.edu` at port 80, to send the string `message` from the previous problem to the host, receive the reply from the host until the server closes the connection, assigning the reply to `reply`, and close the connection. Note: if the request is not completely correct, a network connection can wait forever for a reply that will never come. So if you have difficulty here, double check your answer to the previous problem.

20.16 Suppose we want to generalize the scenario from the first exercise, where the two things that can change are the *host location* and the *resource path*. For example, we might want to change the host to `httpbin.org` and the resource path to `/`, or many other combinations. Write a function

```
buildRequest(location, resource)
```

that constructs and returns a Python string containing a valid HTTP GET request that incorporates the parameters `location` and `resource` into the request at the appropriate places, and includes the appropriate header lines (for the required `Host` and to request the server close the connection after the exchange).

20.17 Write a function

```
makeRequest(location, resource)
```

that first constructs a valid HTTP GET request for `resource` at host `location`, as a Python string, and then performs request-reply steps of making the connection, sending the string request, receiving a reply until the connection closes, and finally closing the client side of the connection. The function should return the reply.

20.18 Write a function

```
makePOSTRequest(location, resource, body)
```

that first constructs a valid HTTP POST request associated with `resource` at host `location`, and where `body` forms the content of the request *following the empty line*, all as a single Python string. The function should then perform the request-reply steps of making the connection, sending the string request, receiving a reply until the connection closes, and finally closing the client side of the connection. The function should return the reply.

The next set of exercises are about parsing through the reply resulting from a request. If we consider an HTTP reply, we can partition it into a status line, the set of headers, and the body. The exercises ask for functions that, parse a given reply and return each of these pieces.

20.19 Write a function

```
parseStatus(reply)
```

that finds and returns a Python string consisting of only the status line of a reply. The returned value should include the line-terminating `"\r\n"`.

20.20 Write a function

```
parseHeaders(reply)
```

that finds and returns a single Python string that starts with the first header in the reply and continues up through the last header in the reply, including the line-terminating `"\r\n"`, but *not* the empty line separating the headers from the body.

20.21 Write a function

```
parseBody(reply)
```

that finds and returns a single Python string that starts with the beginning of the body (i.e., after the empty line of the reply) and continues to the end of the reply.

20.3   Programming HTTP Using `requests`

While the basics of HTTP were simple enough to allow us to construct Python code that used Python strings and programmed to the `sockets` interface, full use of HTTP beyond the basics becomes very complex. We, instead, want a module/package that utilizes the object-oriented facilities of Python and provides an abstraction for the various ways we want to issue requests and process replies.

Such a package can afford the following advantages:

- It can provide Python functions to represent HTTP method invocation.
- It can handle complexities like
    - Line endings,
    - Use of secure HTTP through HTTPS,
    - Binary content types in responses,
    - Use of cookies and multiple operations in a session.

- It can understand and react to HTTP responses like a redirect and, if re-execution to the new location is possible, can automatically handle such cases.
- It can provide object-oriented abstract data types for requests and responses.
- Information within requests and responses, like a set of headers mapping from header-name to value, can be represented in a programmer-friendly structure, like a dictionary.

The `requests` module is a powerful and straightforward implementation of the application protocol layer of HTTP for Python programs. It has become the de-facto standard for Python clients. The module provides an object-oriented abstraction of the HTTP ideas of requests and responses, and has the power to specify header key/value pairs, URL query parameters, and automatically supports redirection. For clients with more sophisticated needs, it can support sessions and cookies.

We import the module in the usual manner:

```
import requests
```

Note that we do not provide an `as` clause alias, so when we refer to components of the module, such references will always be prefixed with `requests`.

In its simplest form, the `requests` module defines, at the top level of the package, a set of functions, one for each of the HTTP method (see Table 20.1).

- `requests.get(url)`
- `requests.post(url)`
- `requests.put(url)`
- `requests.delete(url)`
- `requests.head(url)`

The first argument for each of these functions is a URL, which is a Python string that specifies the protocol, location, and resource path. For these functions, the

URL does *not* typically include query parameters, which can be specified through a separate argument.

In addition to the url first argument, all of these functions support additional named arguments that affect the construction and processing of the request. A few of these are particularly important in the needs we have in this textbook, and are listed and described in Table 20.5.

The requests module uses objects to define abstractions for requests. It uses class Request and PreparedRequest for objects representing the HTTP requests, and which get constructed as part of the process of invoking any of the above functions. It uses class Response for an object representing the result returned from the provider following a request. Each of the get(), post(), and other functions return a Response object, which is used to obtain all information, from status to response headers to the data in the body of a response. In the initial examples below, after using the Response object for getting the *status-code* of a response, we will use the request attribute of the response so that we can examine the specifics of a request resulting from an invocation of get() or post().

In particular, we are interested in looking at the following attributes of a Request/PreparedRequest object listed in Table 20.6.

Table 20.5 Common parameters for HTTP request functions of the requests package

Parameter	Description
headers=	This parameter is used to specify header lines to be included in the request
params=	This parameter is used to specify query parameters. The argument of this parameter is a dictionary mapping
data=	This parameter is used to specify the *body* of the request, appropriate for post() and put(), but not for get() or head(). The argument value references the data bytes to be included in the body and can take any form of binary or text data, but, in addition to a string or bytes parameter, the argument can be a dictionary of key/value pairs, which will be translated and encoded into a *form* in the same style as URL-encoded query parameters
json=	This parameter provides an alternative means of specifying the *body* of the request and the argument references an in-memory data structure, which will be encoded into JSON string form as the bytes of the body
allow_redirects=	Boolean parameter that determines whether or not a request that results in a redirect should automatically issue a subsequent request to the URL specified in the Location header of the redirect. By default, this argument is True

Table 20.6 Class attributes for `Request/PreparedRequest` objects

Attribute	Description
`body`	Body for the request (POST/PUT)
`headers`	Dictionary mapping a header line's *name* to its corresponding *value*
`method`	String specifying the HTTP method for the request (GET/POST, etc.)
`path_url`	The resource path used in the request
`url`	The URL of the request

20.3.1   GET Requests

20.3.1.1   Example 1: GET of HTML

We start with a simple GET to the root of the content served by a web server. The only argument to the `get()` is the target URL.

```
url = "http://datasystems.denison.edu/basic.html"

response = requests.get(url)
request = response.request

print("Response status:", response.status_code)
```

```
| Response status: 200
```

```
print("URL:", request.url)
```

```
| URL: http://datasystems.denison.edu/basic.html
```

For communication between client and server, this request would be translated into the following:

```
GET /basic.html HTTP/1.1
Host: datasystems.denison.edu
User-Agent: python-requests/2.23.0
Accept-Encoding: gzip, deflate
Accept: */*
Connection: keep-alive
```

20.3.1.2   Example 2: GET Specifying Headers for Request

When, as a client, we want to specify our own header lines to be included in a request, or to override some of the headers used by default in the `requests` module, we use the `headers=` named argument and pass a dictionary mapping from header names to header values. This example changes the header lines of a

GET request, but the same technique would be used for any of the other HTTP
methods.

```
url = "http://httpbin.org/get"

headerD = {
    "Accept": "application/json",
    "User-Agent": "datasystems-client"
}
response = requests.get(url, headers=headerD)
request = response.request

print("Response status:", response.status_code)
```

```
| Response status: 200
```

```
util.print_headers(request.headers)
```

```
| {
|     "User-Agent": "datasystems-client",
|     "Accept-Encoding": "gzip, deflate",
|     "Accept": "application/json",
|     "Connection": "keep-alive"
| }
```

So, in this case, the sequence of characters of the HTTP request would be:

```
GET /get HTTP/1.1
Host: httpbin.org
User-Agent: datasystems-client
Accept-Encoding: gzip, deflate
Accept: application/json
Connection: keep-alive
```

20.3.1.3   Example 3: GET with Query Parameters

We mentioned above that, when we need to issue an HTTP request that includes
query parameters we do not, typically, include these as part of the `url` first argument
of one of the request functions of the `requests` module. We instead construct a
dictionary where the mapping defines the relationship between keys and values to
be included as query parameters of the query string, and pass this dictionary as
the argument for the `params=` parameter. Note the embedded space as the value
associated with the desired query parameter whose key is `query` in the example.

```
url = "http://httpbin.org/get"

paramsD = {
    "user": "smith",
    "query": "movies tv"
}
response = requests.get(url, params=paramsD)
request = response.request

print("Response status:", response.status_code)
```

```
    | Response status: 200
```

```
print("Path:", request.path_url)
```

```
    | Path: /get?user=smith&query=movies+tv
```

We see in the output how the `requests` module translated the dictionary of query parameter key/value pairs into the resource path to be included in the request. Recall that the *query string* is the part of the resource path that starts with the ? character. The `requests` module performs all necessary URL encoding; in this case, it translated an embedded space into the legal variant of +. Avoiding having to code URL encoding for ourselves is the reason we much prefer the use of the `params=` named parameter instead of embedding query parameters in the `url` parameter.

Translating the above into the stream of characters that make up the HTTP request, this request becomes:

```
GET /get?user=smith&query=movies+tv HTTP/1.1
Host: httpbin.org
User-Agent: python-requests/2.23.0
Accept-Encoding: gzip, deflate
Accept: */*
Connection: keep-alive
```

20.3.2   POST Requests

20.3.2.1   Example 1: POST with Form Data Body

The HTTP POST method is used to pass information from the client to the server through the *body* of the request. Often a POST occurs when a user, in a web browser, fills in the fields of a web *form*, and the POST is used to pass the user-selected values to the server. Such *form data* is often encoded in the body using the same URL encoding process that is employed for query parameters, where the result is a sequence of *key=value* mappings separated by the & character, and where each *value* has mapped special characters to legal equivalents.

In the following example, we perform a POST to the /post endpoint, and combine a number of the ways of modifying a request, passing query parameters, using client-chosen header lines, and also demonstrating an argument to use as a source for the body of the POST request. When the requests module has a data= named parameter where the argument is a *dictionary*, this is a signal for *form data*, and the dictionary argument has URL encoding performed on it, and the result becomes the body of the request.

```python
url = "http://httpbin.org/post"

paramsD = {
    "user": "jones",
    "query": "TV?episodes"
}
headerD = {
    "Accept": "application/json"
}
body = {"a": 1, "b": 2}

response = requests.post(url, params=paramsD,
                         headers=headerD, data=body)
request = response.request

print("Response status:", response.status_code)
```

```
| Response status: 200
```

```python
print("Request Path:", request.path_url)
```

```
| Request Path: /post?user=jones&query=TV%3Fepisodes
```

```python
print("Request Body:", request.body)
```

```
| Request Body: a=1&b=2
```

Notice, through the output of the example, how the body of the request is the URL-encoded string generated from the dictionary (body) passed as argument for the data= parameter. This is separate from the processing of the params= argument or the headers= argument, each of which has their own effect on the resultant HTTP request:

```
POST /post?user=jones&query=TV%3Fepisodes HTTP/1.1
Host: httpbin.org
User-Agent: python-requests/2.23.0
Accept-Encoding: gzip, deflate
Accept: application/json
Connection: keep-alive

a=1&b=2
```

20.3.2.2   Example 2: POST with JSON Body

In the following example, we illustrate another very common use of POST, wherein the body of the POST captures a more general set of data to be conveyed from the client to the server–the desired body of the POST is a JSON string encoded from a data structure in the client application.

```python
import json

url = "http://httpbin.org/post"

paramsD = {
    "user": "jones",
    "query": "TV"
}
headerD = {
    "Accept": "application/json"
}
json_data = ["foo", "bar", {"a": 1, "b": 2}]

response = requests.post(url, params=paramsD,
                    headers=headerD, json=json_data)
request = response.request

print("Response status:", response.status_code)

   | Response status: 200

print("Request Path:", request.path_url)

   | Request Path: /post?user=jones&query=TV

print("Request Body:", request.body.decode('utf-8'))

   | Request Body: ["foo", "bar", {"a": 1, "b": 2}]
```

The body of the request is the byte encoding of the JSON string derived from the in-memory data structure, json_data in the example.

The request, as communicated from client to server, would look like this:

```
POST /post?user=jones&query=TV HTTP/1.1
Host: httpbin.org
User-Agent: python-requests/2.23.0
Accept-Encoding: gzip, deflate
Accept: application/json
Connection: keep-alive

["foo", "bar", {"a": 1, "b": 2}]
```

20.3.3 Response Attributes

In the examples thus far, we have done very little with the `Response` object returned from each of the HTTP request invocations, other than to show the status code and to obtain the `Request` object so that we could examine the various attributes of the request. In this section, we explore the information of the response in more detail. This will allow processing of the results of the request.

After receiving the response over the TCP socket, the `requests` module constructs an object and uses Python attributes to allow client programmers to access constituent parts of the response, after the module has processed to parse through the response. Some of the most significant Python attributes of the requests `Response` object are presented in Table 20.7.

To facilitate the reflection of the attributes of a response, we write a function to print out some of the most common response attribute values. To allow control of

Table 20.7 Response object attributes

Response attribute	Description
status_code	The three digit integer status code from the status line of the response
content	The raw bytes version of the response body. Useful if the body is binary data, or if it needs to be used as input to a downstream conversion expecting binary data
text	If the response body is textual, this attribute contains the response body decoded into a Python string. The decoding is based on the `requests` module's best understanding of the encoding, and may use response header information to find this encoding specification
headers	A dictionary of the field-name/field-value pairs for the set of headers included in the response
url	The complete url, including any parts based on arguments passed to the request and processed into the URL/URI for the HTTP
request	This is another `requests` module object, representing an abstraction of the request, as prepared by the module. This object has its own attributes, including `headers`, `path-url`, `url`, `method`, and `body`, allowing programmatic inspection of each of these components of the originating HTTP request

the printing of the body of a response, which might be large, we include an optional parameter that places a maximum number of lines of the body to be printed out.

```python
def print_response_info(r, maxlines=None):

    print("Status Code:", r.status_code)
    print("Response Headers:")
    util.print_headers(r.headers)
    if maxlines != None and maxlines == 0:
        return
    print("\nResponse Text Body")
    util.print_text(r.text, nlines=maxlines)


url = "http://datasystems.denison.edu/basic.html"

response = requests.get(url)
print_response_info(response, maxlines=7)
```

```
| Status Code: 200
| Response Headers:
| {
|    "Date": "Tue, 30 Jun 2020 21:57:36 GMT",
|    "Server": "Apache",
|    "Accept-Ranges": "bytes",
|    "Content-Length": "496",
|    "Connection": "close",
|    "Content-Type": "text/html; charset=UTF-8"
| }
|
| Response Text Body
| <!DOCTYPE html>
| <html lang="en">
|    <head>
|       <title>Data Systems Basic HTML Page</title>
|    </head>
|    <body>
|       <h1>First Level Heading</h1>
```

So we pass `response` as argument to our response printing function to associate with the `r` parameter, and, within the body, refer to the attributes of interest: `r.status_code`, `r.headers`, and, for the textual body of the response, `r.text`. Chapter 21 will explore how we can use `Response.text` and `Response.content` for interpreting the data returned into our data systems formats of CSV, JSON, and XML.

20.3.4 Reading Questions

20.22 Create a table that describes a `response` object. List the six most common attributes with a brief description of what information a client gets from each. Then check the online documentation and make a table describing three *methods* that come with a `response` object, what each does, and when you might use it. Template tables for you to follow:

Attribute	Description
First-attribute	What it is good for

Method	Description	Use case
First-method	What it is good for	When to use it

20.23 The reading says we will only consider http and https as protocols by which providers will respond to requests. What are some examples of other protocols you might come across in life? Where would you go to learn about them if you did need to one day?

20.24 The reading discusses thinking about a provider's resources as forming a tree. Some websites are organized following this tree structure. For example, please poke around this website: https://home.apache.org/~taylor/ and see if you can find a path of length three (i.e., with three slashes in the part after "taylor"). What do you notice about the relationship between the URL and the tree structure?

20.25 A powerful tool in the arsenal of any data scientist is the wayback machine: https://archive.org/web/ This allows you to see archived versions of web pages. This can be useful, e.g., if you need access to a data set that was posted and subsequently changed or taken down.
   Please use this tool to find an archived version of

```
personal.denison.edu/~whiteda
```

and then record at least one thing that has changed between the archived version and the present version. Then think up another potential scenario where this tool could be useful.

20.26 When you have multiple web browser windows open, do you have multiple connections? What about if you have multiple tabs? How do you know?

20.27 The reading discusses what happens on the server side when a bad resource path is specified. What does this mean in practice for you when you browse the web.

20.28 Can you think of any examples where the syntax `<username>@<host>` would be used?

20.29 The reading gives detailed Browser Instructions to go with Chrome, and lays out a set of actions you should do. Please describe precisely what happened when you followed those steps.

20.30 Please give examples of experiences you have had with four of the statuses listed in the Status Description table, and whether or not you saw the status code in each case.

20.31 When constructing a `request` using the triple quote approach, do you need the `\n` for the blank line between the header and body, or could you just leave vertical whitespace inside the request string?

20.32 When writing http code, what happens if the port you select when making the connection disagrees with the protocol you select in the request, e.g., if you say `port = 80` but use `https` instead of `http/1.1`?

20.33 When you use http programming to try to `GET` from www.denison.edu using `HTTP/1.1`, you get a 301 response. Explain why.

20.34 How can you use http programming to request a csv file and then save that file on your machine?

20.3.5   Exercises

20.35 The `requests` module uses URLs as the first argument to its HTTP method functions, but we often start with the "piece parts" of the information contained in a URL. Write a function

```
buildURL(resource, location, protocol='http')
```

that returns a string URL based on the three component parts of `protocol`, `location`, and `resource`. Your function should be flexible, so that if a user omits a leading `\` on the resource path, one is prepended. Note that we are specifying a default value for `protocol` so that it will use `http` if `buildURL` is called with just two arguments. The string method `format()` is the right tool for the job here.

20.36 Write a sequence of code that starts with:

```
resource = "/data/ind0.json"
location = "datasystems.denison.edu"
```

and build an appropriate URL, uses `requests` to issue a GET request, and assigns the variables based on the result:

- `status`: has the integer status code,
- `headers`: has a dictionary of headers from the response, and

- `data` has the *parsed* data from the JSON-formatted body

20.37 Suppose you often coded a similar set of steps to make a GET request, where often the body of the result was JSON, in which case you wanted the data parsed, but sometimes the data was *not* JSON, in which case you wanted the data as a string. Write a function

    makeRequest(resource, location, protocol="http")

that makes a GET request to the given `location`, `resource`, and `protocol`. If the request is *not* successful (i.e., not in the 200's), the function should check for this and return `None`. If the request is successful, the function should *use the response headers* and determine whether or not the `Content-Type` header maps to `application/json`. If it is, it should parse the result and return the data structure. If it is not, it should return the string making up the body of the response.

20.38 You have probably had the experience before of trying to open a web page, and having a redirect page pop up, telling you that the page has moved and asking if you want to be redirected. The same thing can happen when we write code to make requests. Write a function:

    getRedirectURL(resource, location)

that begins like your function `makeRequest` but does *not* allow redirects when invoking `get`. This function will return a *url*. If the `get` results in a success status code (one in the 200's), you return the original url (obtained from `buildURL`, with `http` protocol). If you detect that `get` tried to redirect (by looking for a 300, 301, or 302 status code), search within the headers to find the `"Location"` it tried to redirect to, and return that URL instead. If you get any other status code, return `None`.

20.39 The section discussed how to add custom headers along with a get request. This is important in a number of contexts, e.g., when you request a large file, it might make sense to ask it to be compressed for transit. This example motivates our assert statements below. Given parallel lists `headerNameList` and `headerValueList`, you can build a dictionary that maps from header names to their associated values (given by the parallel structure). Write a function

    makeRequestHeader(resource, location, headerNameList,
                                          headerValueList)

that builds a custom header dictionary and then passes it to the `get` method. Your function should call `buildURL` (with protocol `https`) to build the url to pass to `get`. Your function should return the response from the `get` invocation.

20.40 Please write a function

    postData(resource, location, dataToPost)

that uses your `buildURL` function to build a URL (using `https`), then uses the `requests` module to post `dataToPost` to that URL (note: `dataToPost` will be the *body* of the message you send). Please return the `response` returned by the method of `requests` that you invoke. Note: the URL https://httpbin.org/post is set up to allow you to post there.

20.4   Command Line HTTP with `curl`

When working with HTTP, it is helpful to be able to construct requests and try them out before incorporating the equivalent request into client program code. The command-line program, `curl`[6] is a tool to issue many different types of application protocol requests to transfer data from or to a server, and excels at issuing HTTP requests. Because of its ubiquity and generality, many providers, like those explored in Chap. 23, will use `curl` "incantations" to actually help *document* their API. For these reasons, we provide a basic description of `curl` in this section.

The `curl` program is open source, and is available for most operating systems and any POSIX compliant environments, as well as part of package distributions like Anaconda, brew, Cygwin, and others. As a command line program, a Terminal-type program executing a command-line shell, like `bash`, is a prerequisite. The remainder of this section assumes that a Terminal, a shell, and the curl program are available.

20.4.1   Basics

The `curl` command is issued at a command-line prompt. In the examples below, we present the characters making up command invocation, which would be entered after a command prompt, and issued by hitting the `return` key. Like many command-line tools, the syntax for executing the program consists of

- the name of the program,
- zero or more *options* that control how the program performs its function, and
- the *argument* to the program, which gives the URL that is target for the request.

In its simplest form, with no options and a single URL as argument, we can issue

```
curl https://datasystems.denison.edu/basic.html
  % Total    % Received % Xferd  Average Speed   Time    Time
                                 Dload  Upload   Total   Spent
    0     0    0     0    0     0      0      0 --:--:-- -:--:-
  100   494  100   494    0     0    626      0 --:--:-- -:--:-
<!DOCTYPE html>
<html lang="en">
  <head>
    <title>Data Systems Basic HTML Page</title>
  </head>
```

```
<body>
  <h1>First Level Heading</h1>

  <p>Paragraph defined in <b>body</b>.
```

... and so forth

This performs an HTTP GET request to the specified URL, and retrieves the response. Displayed as output from the program are:

- Progress indicators, so that, for large transfers, the user can see work progressing. In this case, the progress begins with `% Total` and ends with `625`.
- The *body* of the response. This starts with `<!DOCTYPE html>` and continues for the full HTML page (abbreviated here).

We use options to control how the program operates, and options come in two varieties, those with *no argument* to the option, and those *with an argument* to the option. Options are indicated with either one or two leading hyphens, (`-` or `--`) with a single hyphen used for a one character version of an option, called *short options*, and a double hyphen used for a multiple character version of the option, called a *long option*. Most options have both short versions, for succinct writing of commands, and long versions, which are more readable.

20.4.1.1   Options Controlling Output

The first two options we want to be able to use control the output. When we use

- `-s` or `--silent`, which takes no argument, we suppress the display of progress output, and
- `-o` or `--output`, which takes an argument that allows us to specify a *file* to be used for the body of the response.

For our examples, we will often use `-s` and `-o`, and use the `head` command after the `curl` if we wish to see a prefix of the result. (The `-n` option to `head` allows an argument indicating how many lines to display.)

```
curl -s -o basic.html https://datasystems.denison.edu/
                        basic.html
head -n 7 basic.html
```

```
| <!DOCTYPE html>
| <html lang="en">
|   <head>
|     <title>Data Systems Basic HTML Page</title>
|   </head>
|   <body>
|     <h1>First Level Heading</h1>
|
```

20.4.1.2   Options to Show Response Metadata

The verbose option (no argument, -v or --verbose) shows significant detail about a request and response. Here we added the -v to the same incantation as before:

```
curl -v -s -o basic.html \
    http://datasystems.denison.edu/basic.html

  *    Trying 140.141.2.184...
  * TCP_NODELAY set
  * Connected to datasystems.denison.edu (140.141.2.184)
  * port 80 (#0)
  > GET /basic.html HTTP/1.1
  > Host: datasystems.denison.edu
  > User-Agent: curl/7.54.0
  > Accept: */*
  >
  < HTTP/1.1 200 OK
  < Date: Fri, 19 Jun 2020 17:00:23 GMT
  < Server: Apache
  < Accept-Ranges: bytes
  < Content-Length: 494
  < Connection: close
  < Content-Type: text/html; charset=UTF-8
  <
  { [494 bytes data]
  * Closing connection 0
```

As a result, we see a line showing establishment of a connection near the beginning of the output, the closing of the connection at the end of the output, and, in between, the set of outgoing headers for the request, and the set of incoming headers for the response. Other than a different User-Agent, these should be familiar from our earlier discussion. If we were to use https instead of http in the target URL, we would see information about the secure connection establishment as well.

To include the response headers *along with* the response body, all as part of the output, you use the -i or --include option, which takes no argument.

```
curl -i -s -o basicplus.txt \
http://datasystems.denison.edu/basic.html
head -n 12 basicplus.txt

  | HTTP/1.1 200 OK
  | Date: Tue, 30 Jun 2020 21:57:36 GMT
  | Server: Apache
  | Accept-Ranges: bytes
  | Content-Length: 496
```

```
| Connection: close
| Content-Type: text/html; charset=UTF-8
|
| <!DOCTYPE html>
| <html lang="en">
|   <head>
|       <title>Data Systems Basic HTML Page</title>
```

To save the response headers to a file, and to keep them separate from the body of the response, use the `-D` or `--dump-headers` option, which takes an argument of the file name for storing those headers.

```
curl -D headers_back.txt -s -o basic.html \
     http://datasystems.denison.edu/basic.html
cat headers_back.txt
```

```
| HTTP/1.1 200 OK
| Date: Tue, 30 Jun 2020 21:57:36 GMT
| Server: Apache
| Accept-Ranges: bytes
| Content-Length: 496
| Connection: close
| Content-Type: text/html; charset=UTF-8
|
```

To perform an HTTP HEAD instead of a GET, we can use the `-I` or `--head` option, which takes no argument. The response of a HEAD omits the body of the response and, in curl, causes the response header lines to be output.

```
curl -I -s https://datasystems.denison.edu/basic.html
```

```
| HTTP/1.1 200 OK
| Date: Tue, 30 Jun 2020 21:57:37 GMT
| Server: Apache
| Accept-Ranges: bytes
| Content-Length: 496
| Connection: close
| Content-Type: text/html; charset=utf-8
|
```

The `-I` could be combined with `-o` in order to save the output of the headers to a file.

Other metadata can be specified to be output by a curl invocation using the `-w` or `--write-out` option, which takes an argument specifying what should be written out. To avoid interpretation by a shell, the argument is specified in double quotes and an item like `%{response_code}` retrieves the three digit response code of the request.

```
curl -w "%{response_code}" -o basic.html -s \
     https://datasystems.denison.edu/basic.html

| 200
```

20.4.2   Sending Custom Request Header Lines

The `-H` or `--header` option can be used multiple times in an invocation and, each
time, specifies exactly one header line to be included in the request. The examples
combine with `-v` to help show the effect.

```
curl -H "Accept: application/json" -H "User-Agent: foobar" -v \
     -o basic.html -s http://datasystems.denison.edu/basic.html

*    Trying 140.141.2.184...
* TCP_NODELAY set
* Connected to datasystems.denison.edu (140.141.2.184)
* port 80 (#0)
> GET /basic.html HTTP/1.1
> Host: datasystems.denison.edu
> Accept: application/json
> User-Agent: foobar
>
< HTTP/1.1 200 OK
< Date: Fri, 19 Jun 2020 17:04:06 GMT
< Server: Apache
< Accept-Ranges: bytes
< Content-Length: 494
< Connection: close
< Content-Type: text/html; charset=UTF-8
<
{ [494 bytes data]
* Closing connection 0
```

The arguments to the `-H` are specified in double quotes to allow embedded
spaces and to avoid interpretation of a single argument as something else. Note the
difference in request headers compared with the default header lines for `Accept`
and for `User-Agent` from the example that used `-v` above.

```
curl -w "%{http_code}" -o basic.html -s \
     https://datasystems.denison.edu/basic.html

| 200
```

20.4.3   Query Parameters

Specifying query parameters through a query string is similar to the `curl` incantations we have shown above. We know the query string is specified as part of the URL. However, because of the special characters of the ? and the & in the syntax of the query string, we must use double quotes as we specify the argument to the `curl` command. To help illustrate, we show a GET request with a query string to the `/get` endpoint at `httpbin.org`. This website is a useful one for understanding and debugging HTTP, and in such a request to `/get`, its response has a *body* that reflects the request being make. In particular, the response is composed of a JSON-formatted dictionary whose first entry is a dictionary showing the received query parameter, and then continues with the request headers and other information gathered at the server.

```
curl -s -o get.json "http://httpbin.org/get?a=1&b=2"
head -n 9 get.json
```

20.4.4   POST Requests

When we want `curl` to use an HTTP method other than GET, we can do so directly by specifying the `-X` or `--request` option, whose argument is the HTTP method.

20.4.4.1   POST with No Body

We start with a POST to URI `/post` at `httpbin.org`, and include query parameters, but have nowhere specified data that should form the *body* of the post:

```
curl -X POST -s -o post.json
                    "http://httpbin.org/post?a=1&b=2"
head -n 12 post.json
```

```
 | {
 |   "args": {
 |     "a": "1",
 |     "b": "2"
 |   },
 |   "data": "",
 |   "files": {},
 |   "form": {},
 |   "headers": {
 |     "Accept": "*/*",
 |     "Host": "httpbin.org",
 |     "User-Agent": "curl/7.68.0",
```

This is successful, and we see the JSON response has, in its top level dictionary, a "data" key, whose value is an empty string, indicating a POST that did not include a body.

20.4.4.2   POST with Form Data

Use the -d/--data option, possibly multiple times, with *name*=*value* specification of key/value pairs. If needed, use quotes around the value if there are embedded special characters. In this example, we include both query parameters as well as the form data, so that, through the result, we can distinguish their different roles.

```
curl -X POST -s -d year=2001 -d newYear=
                              'Get+different+year'\
     -o post.json "http://httpbin.org/post?a=1&b=2"
head -n 11 post.json

  | {
  |    "args": {
  |      "a": "1",
  |      "b": "2"
  |    },
  |    "data": "",
  |    "files": {},
  |    "form": {
  |      "newYear": "Get different year",
  |      "year": "2001"
  |    },
```

Note that, in the JSON returned from httpbin.org, the "form" dictionary has translated back from the URL encoding, which is why that dictionary shows spaces in "Get different year".

20.4.4.3   POST with JSON Data

When the body of a POST has contents that are not a form, and are not URL-encoded, we specify the POST body in a separate file, and then indicate to curl to obtain the body data from the file. Suppose the file postdata.json has the JSON data we wish to POST:

```
cat postdata.json

  | ["foo", "bar", {"a": 1, "b": 2}]
```

The curl command to POST the data uses the -T/--upload-file option, which takes a file specification as its argument, and uses this for the body of the request.

```
curl -X POST -s -o post.json -T postdata.json \
     "http://httpbin.org/post"
head -n 10 post.json
```

```
| {
|    "args": {},
|    "data": "[\"foo\", \"bar\", {\"a\": 1, \"b\":
|                2}]\n",
|    "files": {},
|    "form": {},
|    "headers": {
|      "Accept": "*/*",
|      "Content-Length": "33",
|      "Host": "httpbin.org",
|      "User-Agent": "curl/7.68.0",
```

20.4.5 Exploring Further

Command-line `curl` is a powerful tool, and, for full documentation and usage, the reader should refer to its "man page" [7] and/or an open source book detailing its use [8]. There are advanced options for specifying authentication and certification, using customization files for configuration options, working through a proxy, using cookies, support for more protocols beyond `http` and `https`, and many more.

We summarize most of the options demonstrated in this section in Table 20.8.

Table 20.8 Common options used in `curl`

Long	Short	Arg	Notes
`--request`	`-X`	Yes	GET, POST, PUT, DELETE
`--url`		No	Specify URL for target of HTTP request; can also omit this option and specify as *last* on command line, after all options
`--trace`		Yes	Give file that shows trace information flow
`--verbose`	`-v`	No	Verbose output; can be used to see request headers
`--header`	`-H`	Yes	Specify string to be used as a header line
`--head`	`-I`	No	Send HEAD as HTTP method, to get just headers as return
`--get`	`-G`	No	Select GET as HTTP method
`--include`	`-I`	No	Show response headers as well as body as output
`--data`	`-d`	Yes	Argument can be `'key=value` for specifying one Form element for the POST body
`--silent`	`-s`	No	Hide progress meter
`--output`	`-o`	Yes	Specify file for saving body of the response

20.4.6 *Exercises*

20.41 Write and test a `curl` request that obtains, silently, `/data/ind0.json` from `datasystems.denison.edu`

20.42 Write and test a `curl` request that obtains, silently, `/data/ind0.json` from `datasystems.denison.edu` and saves the output into a file named `ind0.json`.

20.43 Write and test a `curl` request that issues a HEAD request for resource `/data/ind0.json` from `datasystems.denison.edu`.

20.44 Write and test a `curl` request that obtains, `/data/ind0.json` from `datasystems.denison.edu` *and protocol https* and saves the output into a file named `ind0.json`. Request *verbose* output. Repeat with protocol `http` and observe the differences.

20.45 Write and test a `curl` request that obtains, silently, `/data/ind0.json` from `datasystems.denison.edu` and saves the response headers in `ind0_hdr.txt` and the data in `ind0.json`. Also, specify custom header of `Content-Type` mapping to `application/json`. Use `cat` to output the headers.

20.46 Experiment with what happens when you request a non-existent resource, say `/data/foobar.json`. What would be the correct incantation to get the status code? How about to see all the headers, including the 401 Not Found status line? When you request a non-existent resource, what is the *body* of the response message?

20.47 Write and test a `curl` request that performs a POST to resource `/almanac/transportation_data/gasoline/margins/` `index_cms.php` at host `ww2.energy.ca.gov` with form parameters `year` mapping to `2005` and `newYear` mapping to `Get+different+year`. Save the result to `gas2005.html`.

Chapter 21
Interlude: Client Data Acquisition

Chapter Goals

Upon completion of this chapter, you should understand the following:

- The difference between text strings and byte strings, and between text files and binary files.
- The concept of encoding and decoding to get back and forth between text and byte representation.
- The importance of overlaying a file-like view on a set of data acquired over a network.

Upon completion of this chapter, you should be able to do the following:

- Access an HTTP result, both as raw bytes and as a character string.
- Be able to *change* the encoding for interpretation of the raw bytes as a string.
- For these collections of binary data bytes and string-based result data, overlay a file-like interface to enable better access to the data.
- Combine the above to be able to transform the data into the structures for the different formats of data (CSV, JSON, XML) explored in Part II of the book.

© Springer Nature Switzerland AG 2020   649
T. Bressoud, D. White, *Introduction to Data Systems*,
https://doi.org/10.1007/978-3-030-54371-6_21

In Chap. 20, we learned the syntax of HTTP, and how to use the `requests` module to issue GET and POST request messages that allow us to obtain results from a web server. In this chapter, we explore variations of taking the results and transforming the data into an in-memory structure usable in our client programs.

In Chaps. 6 and 15 and in Sect. 2.4, we used local files as our data source to get CSV-, XML-, and JSON-formatted data into `pandas`, into an element tree, and into a dictionary/list composite structure, respectively. Now this same *format* data is being acquired over the network and arrives at our client application in the *body* of an HTTP request. We need to process and obtain the *same* in-memory structure of a `pandas` data frame from CSV-formatted data, an `lxml` element tree from XML-formatted data, or the Python data structure from JSON-formatted data. To do this correctly, we will often require understanding of the encoding, by which the characters of the data at the server are mapped to the bytes transmitted over the network.

21.1   Encoding and Decoding

Recall from Sect. 2.2.2 that the term *encoding* (aka *codec*) defines a translation from a sequence of characters (i.e., a *string* or *text*) to the set of *bytes* that are used to represent that character sequence. Given an encoding, then, for each character, there is a specific translation of that character into its byte representation. Some encodings are limited in the set of characters that they are able to encode, and thus the alphabets and languages they can support. Other encodings allow the full Unicode character set as their input and can adapt to many alphabets and languages. In the reverse direction, given a set of bytes representing a character sequence, along with knowledge of the specific encoding used, we use the term *decoding* for the process of converting a sequence of bytes back into its original character sequence.

In Table 21.1, we list some of the most common encodings. Python documentation provides the set of encodings supported [50].

Suppose we have the character sequence `"All the world's a stage ..."` and are using the specific encoding of `UTF16-BE` for the mapping of characters to bytes. Figure 21.1 illustrates the encoding and decoding process. Given the original string and the encoding of `UTF-16BE`, the *encode operation* translates into a byte sequence. In the figure, we represent the byte sequence using hexadecimal digits (`0-9` plus `a` through `f`), where each pair of hex digits is equivalent to exactly one byte. So the `A` character maps to two bytes, written in hex as `0041`; lower case `l` maps to hex `006c`, and we see these bytes repeated, and so forth. When we are given a collection of bytes, like in the middle of the figure, then the *decode operation*, with knowledge of the correct encoding of `UTF-16BE`, can translate the bytes back into the original character sequence.

Relative to this book, there are three contexts within which we should be aware of encoding. First, within our Python programs, we may have character strings from which we need to explicitly generate a raw bytes representation, or vice versa.

Table 21.1 Common encodings

Encoding	Bytes per Char	Notes
ASCII	Exactly 1	One of the most limited encodings, only supporting the English alphabet, and including A-Z, a-z, 0-9, and basic keyboard special characters
UTF-8	1 to 4, but using 1 whenever possible	Supports full Unicode but is backwardly compatible to ASCII for the one byte characters supported there. This encoding accounts for 95% of the web
UTF-16	2 to 4, with 2 the most dominant	Supports full Unicode and is used more on Windows platforms, where it started as always 2 bytes, until it needed to be expanded. Variations include UTF-16BE and UTF-16LE that make explicit the ordering of multibyte units
ISO-8859-1	Exactly 1	Latin character set, starting from ASCII, but adding common European characters and diacritics. Also known as LATIN_1

Fig. 21.1 Encoding and decoding process

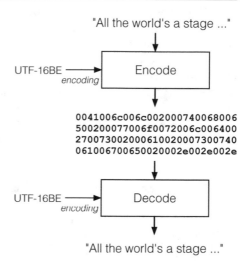

Second, every local file is actually stored as a sequence of bytes. If it is a text file, we need to ensure that, whether reading from a text file or writing to a text file, we are able to specify a desired encoding as appropriate. Third, when we are acquiring data over the network, the body of an HTTP response message is also conveyed over the TCP reliable byte-stream as a sequence of raw bytes. So we want to be able to decode those bytes into their original character sequence when the contents of the message are, in fact, text.

We address the first context, encoding and decoding explicitly in our Python program, here in this section. Encoding and decoding when interacting with files have already been covered in Sect. 2.2.2 and will not be repeated here, but we will give file-based examples in Sects. 21.2 through 21.4. The third context is directly related to the main goals of this chapter and will also be illustrated in Sects. 21.2 to 21.4.

21.1.1   Python Strings and Bytes

The characters of Python strings allow for the full Unicode character set and thus can support a spectrum of alphabets and languages. Further, Python, by default, uses the UTF-8 encoding. These defaults of character set and encoding are often sufficient when our programs are not interacting with text data from outside sources. In this case, we rarely have to do explicit translations, or to specify encodings as we open and use files. But when text data originates from some outside source, and if that outside source might use a different encoding, we must have the tools to interpret the data.

In Python, we represent sequences of characters as a string data type, and the type name is str. Individual characters do not have a separate type and are represented as str whose length is one. Python also has a class bytes that is used to represent a set of raw bytes. This data type can be used both for the result of an encode() operation on a string and for non-string types of binary data. A Python value of the bytes data type can be created using a constant syntax similar to that of strings, but prefixed with a b character. For instance, b'Hello!' defines a bytes value that is the UTF-8 encoding of the string that follows the b character. But this value should not be mistakenly thought of as a str value.

21.1.1.1   The Encode Operation: A String to Bytes

Our first two examples show starting from a Python string, which could be any valid Unicode string in a Python program, and using the encode() method of the string type to perform the encoding operation and yielding a bytes result. We print the type of the original string and the type of the encoded value. The bytes class has a method, hex(), that can display the hex byte sequence for its data, and we use this to help convey the raw data bytes from the encoded value referenced by b16.

```
s = "Hello!"
b16 = s.encode("UTF-16BE")
print(type(s), type(b16), b16.hex())
```

```
| <class 'str'> <class 'bytes'> 00480065006c006c006f0021
```

The next example is functionally the same but instead uses the UTF-8 encoding.

```
s = "Hello!"
b8 = s.encode("UTF-8")
print(type(s), type(b8), b8.hex())
```

```
| <class 'str'> <class 'bytes'> 48656c6c6f21
```

If we are dealing with a single character, Python provides a built-in function, `ord()` for obtaining an integer value corresponding to the encoding of that character based on the default UTF-8 encoding. In the example below, we obtain the encoding of the `'H'` character as well as the Unicode Euro symbol. As shown through the output, the data type of the result is an `int`. We print the integer value of the character's encoding and also use the built-in function `hex()` to show the results in the more familiar hexadecimal byte representation.

```
s1 = 'H'
s2 = 'Ă'
b1 = ord(s1)
b2 = ord(s2)
print(type(b1), b1, hex(b1), type(b2), b2, hex(b2))
```

```
| <class 'int'> 72 0x48 <class 'int'> 8364 0x20ac
```

We can also encode a string s into byte form using `b = bytes(s, encoding = 'ascii')`, which returns a byte version of the string (e.g., if `s = "Ben"` then `b = b"Ben"`).

The important takeaway through all these examples is that *encoding* is operating on a string or a character (type `str`), and the result is one or more bytes (type `bytes`).

21.1.1.2 The Decode Operation: Bytes to a String

The *decode operation* translates a byte sequence back into its original sequence of characters, based on the encoding. As long as the encoding used in the decode operation is the same one used by a prior encode operation, the resulting string will be the same one we started with. In similar fashion to the examples above, we can see decode for a multiple character sequence or for a single character.

From our examples above, b16 is a `bytes` value resulting from a UTF-16BE encode operation, and b8 is a `bytes` value resulting from a UTF-8 encode operation. The bytes class has a `decode()` method, whose argument is the encoding/codec used in the prior encode operation. If no argument is given, the default encoding is used.

```
s16 = b16.decode('UTF-16BE')
s8 = b8.decode()
print("s16:", s16, "s8:", s8)
```

```
| s16: Hello! s8: Hello!
```

We see that although b16 and b8 were clearly not the same `bytes` value, after decoding, s16 and s8 have the same original character sequence.

The Python built-in function chr() performs the single-character reverse of the ord() function. Its argument is an int, and the result is the UTF-8 decode of the value. We demonstrate using b1 and b2, integers obtained using ord() in the above example.

```
s1 = chr(b1)
s2 = chr(b2)
print("s1:", s1, "s2:", s2)
```

```
| s1: H s2: Ă
```

We can also use the conversion capability of the str() function, with a first argument that is the encoded bytes value, and can specify an encoding= named parameter to control the conversion to use the specified encoding. So str(b16, encoding='UTF-16BE') yields the same result as b16.decode('UTF-16BE').

The important takeaway through these examples is that *decoding* is operating on a bytes or int value, and the result is a string (type str).

21.1.2 Prelude to Format Examples

Sections 21.2 through 21.4 will focus on each of the primary formats of CSV, JSON, and XML, with the goal of demonstrating translations of both local files and HTTP response messages into structures usable in our client applications.

In local files, we often know the encoding, but when data is retrieved through the body of an HTTP response message, we cannot assume that the encoding will be ASCII, or UTF-8, or ISO-8559-1. It is important to remember that the bytes of the data files are reaching us via a byte-stream (TCP), which does not mandate textual data, nor require a particular encoding. Furthermore, many different encodings are possible, such as UTF-16BE, which stands for "big endian 16 bits," which results in two raw bytes per character to be encoded. We will discuss this further in the sections below.

For illustration, we will use the following files, both locally, and as retrieved over the network, for our examples.

It will be important to keep in mind the distinction between strings and byte strings, and between text files and binary files, in the discussion to come. We now discuss a sequence of vignettes, with how to acquire data in the formats CSV, JSON, and XML.

The network examples in Sects. 21.2 to 21.4 will all make requests from the book web page, https://datasystems.denison.edu, with resource paths specifying the files in Table 21.2.

As we work with the body of HTTP response messages, it will be important to keep in mind the distinction between strings and binary data as a bytes data

Table 21.2 Example files in various formats and multiple encodings

UTF-8 Encoded	UTF-16BE Encoded	Description
ind2016.csv	ind2016_16.csv	Six country indicators from year 2016 in CSV format
ind0.json	ind0_16.json	Indicator data of pop and gdp for three countries for two different years in JSON format
ind0.xml	ind0_16.xml	Indicator data of pop and gdp for three countries for two different years in XML format

type. The `requests` module gives us two ways to extract data from the body of a response, using the attributes of a `Response` object:

- `Response.content`: the raw bytes version of the data,
- `Response.text`: the decoded translation of the raw bytes into a sequence of characters.

The latter uses an *assumed/inferred* encoding, which can be found through the attribute `Response.encoding`. We will give examples of extracting this information so that it can be used by our code. We will also show how to read both types of data (i.e., either the text data from `Response.text` or the underlying byte data from `Response.content`).

In common amongst the set of network examples, we use our custom `util.buildURL()` function to construct a string `url` prior to each `requests` invocation. This, along with helper functions to print results, is documented in Appendix A, in Sect. A.1. For `buildURL()`, we specify the desired *resource-path* in the first argument and the host location in the second argument.

21.1.3 Reading Questions

21.1 How many characters can be encoded in ASCII, and what are some examples of "basic keyboard special characters"? Use a web search to answer this if you have never seen ASCII before.

21.2 What does it mean that UTF-8 is "backwardly compatible" to ASCII?

21.3 How many characters can be encoded with ISO-8859-1 and is it backwardly compatible with ASCII? You are encouraged to use a web search if this is your first exposure to ISO-8859-1.

21.4 The reading points out that "every local file is actually stored as a sequence of bytes." Have you ever had the experience of trying to open a local file in a program (e.g., a text editor), and seeing something very strange display? This probably had to do with your computer using the wrong decoding scheme. Describe your experience when this happened.

21.5 Is the body of a response to an HTTP GET always text? Justify your answer or give a counterexample.

21.6 What do you notice about the hex byte sequence representations for the two "Hello!" encode() examples? Why is this?

21.7 Conceptually, why do you think there are so many different encoding schemes, and why is it important to keep them compatible?

21.8 As the files ind2016.csv and ind2016_16.csv are stored on the book web page, please go and download them, then try to open them in the most naive program possible (i.e., a simple text editor). Describe what you see.

21.9 Please refer back to an HTTP GET request you made in the previous chapter using the requests module and use the .content and .text attributes to look at the data. Investigate these two quantities using print() and type() and report what you find.

21.10 Please refer back to an HTTP GET request you made in the previous chapter using the requests module and determine the encoding using the .encoding attribute of the response. What did you find?

21.1.4 Exercises

21.11 One of the reasons for the existence of *Unicode* is its ability to use strings that go beyond the limitations of the keyboard. Relative to the discussion in the chapter, *Unicode* is about the *strings* we can use in our programs, and the issue of how they translate/map to a sequence of *bytes* (i.e., their *encoding*) is a separate concept.

When we have the *code point* (generally a hex digit sequence identifying an index into the set of characters) for a Unicode character that is beyond our normal keyboard characters, we can include them in our strings by using the \u escape prefix followed by the hex digits for the code point. Consider the Python string s:

```
s = "Unicode examples: \u2B2C and \u266A and \u1F60 and " \
    "\u265E and \u0394 and \u0402"
```

Write code to print s, then assign to b8 the UTF-8 encoding of s, and b16 the UTF-16BE encoding of s. For each, use the hex() method of the bytes data type to see a hex version of the encoded values. Answer the following questions:

- which of the hex representations is longer?
- give explicit lengths for b8, b16, and for the two hex() transformations.
- how does this compare to the length of s?

21.12 Write a function

```
shiftLetter(letter, n)
```

whose parameter, `letter`, should be a single character. If the character is between `"A"` and `"Z"`, the function returns an uppercase character *n* positions further along, and "wrapping" if the + *n* mapping goes past `"Z"`. Likewise, it should map the lower case characters between `"a"` and `"z"`. If the parameter `letter` is anything else, or not of length 1, the function should return `letter`.

Hint: review functions `ord()` and `chr()` from the section, as well as the modulus operator `%`.

21.13 Building on the previous exercise, write a function

```
encrypt(plaintext, n)
```

that performs a `shiftLetter` for each of the letters in `plaintext` and accumulates and returns the resultant string.

21.14 Write a function

```
singleByteChars(s)
```

that takes its argument, `s`, and determines whether or not all the characters in `s` can be encoded by a single byte. The function should return the Boolean `True` if so, and `False` otherwise.

21.15 Suppose you have, in your Python program, a variable that refers to a `bytes` data type, like `mystery` refers to the `bytes` constant literal as given here:

```
mystery = b'\xc9\xa2\x95}\xa3@\x89\xa3@\x87\x99\x85'\
          b'\x81\xa3@\xa3\x96@\x82\x85@\xa2\x96\x93'\
          b'\xa5\x89\x95\x87@\x97\x99\x96\x82\x93\x85'\
          b'\x94\xa2o@@\xe8\x96\xa4@\x82\x85\xa3Z'
```

Perhaps this value came from a network message, or from a file. But you suspect that it, in fact, holds the bytes for a character string, and you need to figure out how it is encoding. Assume that you have narrowed the encodings down to one of the following:

- "UTF-8,"
- "UTF-16BE,"
- "cp037,"
- "latin_1."

Write code to convert the byte sequence to a character string and determine the correct encoding.

21.2   CSV Data

We have seen that the CSV format is ubiquitous in the world of tidy data. A working data scientist will often work with both local CSV files and CSV files obtained over a network.

21.2.1   CSV from File Data

Suppose we are given a local csv file, ind2016.csv. We have previously seen how to read this file into a pandas data frame:

```
path1 = os.path.join(datadir, "ind2016.csv")
df = pd.read_csv(path1)
df.head()
```

```
|     code          country       pop        gdp    life      cell
| 0   CAN           Canada      36.26    1535.77   82.30     30.75
| 1   CHN            China    1378.66   11199.15   76.25   1364.93
| 2   IND            India    1324.17    2263.79   68.56   1127.81
| 3   RUS           Russia     144.34    1283.16   71.59    229.13
| 4   USA   United States     323.13   18624.47   78.69    395.88
```

Now consider the file ind2016_16.csv. In this file, the same sequence of characters is encoded as UTF16-BE. This is *still* a text file. It just has a different mapping from the characters to the bytes of the file. If we attempt to read into a pandas data frame, the operation seems to complete. However, when we look at the head() of the data, we see that it has been mangled by the read_csv() command:

```
path2 = os.path.join(datadir, "ind2016_16.csv")
df2 = pd.read_csv(path2)
df2.iloc[0:2, 0:4]
```

```
|     Unnamed: 0   Unnamed: 1   Unnamed: 2   Unnamed: 3
| 0          NaN          NaN          NaN          NaN
| 1          NaN          NaN          NaN          NaN
```

Fortunately, the pandas read_csv has a named parameter, encoding=, that we can use to specify the true encoding of the file and thus to get the correct results:

```
df2 = pd.read_csv(path2, encoding="UTF-16BE")
df2.head()
```

```
|     code          country       pop        gdp    life      cell
| 0   CAN           Canada      36.26    1535.77   82.30     30.75
| 1   CHN            China    1378.66   11199.15   76.25   1364.93
```

```
| 2   IND            India  1324.17    2263.79   68.56  1127.81
| 3   RUS           Russia   144.34    1283.16   71.59   229.13
| 4   USA  United States    323.13   18624.47   78.69   395.88
```

We see now that the pandas data frame df2 now correctly contains the data.

21.2.2 CSV from Network Data

It is common to receive CSV files over a network, e.g., when sent as an email attachment, or hosted on a website. We have hosted the data sets mentioned above on the book website and now demonstrate how to retrieve them. We begin with ind2016.csv:

```
csv_url = util.buildURL("/data/ind2016.csv",
                        "datasystems.denison.edu")
response = requests.get(csv_url)
if response.status_code != 200:
    print("Error acquiring file")
```

This file was encoded as UTF-8. We see below that the encoding assumed by the requests module is ISO-8859-1.

```
response.encoding
```

```
| 'ISO-8859-1'
```

This assumed encoding means that it is entirely possible to fail at reading a CSV file into pandas when it is received over the network. Care is required. If we look at response.headers['Content-Type'], we get 'text/csv' and this does *not*, in this case, give more specific information on the encoding.

```
response.headers['Content-Type']
```

```
| 'text/csv'
```

The set of characters actually used in this file are all in the one-byte (0 to 256) range supported by the ASCII encoding. When this is the case, ASCII, UTF-8, and ISO-8859-1 encodings result in the same bytes for this sequence of characters. If we know (or can find out) the encoding, the better thing to do would be to **set** the encoding to the correct one, and then can access the "string" version:

```
response.encoding = 'UTF-8'
util.print_text(response.text)
```

```
| code,country,pop,gdp,life,cell
| CAN,Canada,36.26,1535.77,82.3,30.75
| CHN,China,1378.66,11199.15,76.25,1364.93
```

```
|   IND,India,1324.17,2263.79,68.56,1127.81
|   RUS,Russia,144.34,1283.16,71.59,229.13
|   USA,United States,323.13,18624.47,78.69,395.88
|   VNM,Vietnam,94.57,205.28,76.25,120.6
```

If the data presented above were in a *file*, instead of being in a memory structure of a `Response` object, we could use our file-based techniques from Sect. 3.4 to iterate over the lines and compose the data into a native Python data structure. Otherwise, we would have to manually extract the data. Further, the ability to layer a file-like *view* of a set of data where the bytes or characters reside in memory would allow `pandas`, `lxml`, and `json` to perform parsing and interpretation the same way they do for files.

Fortunately, the Python `io` module has facilities for exactly this purpose: using bytes of data, or bytes making up characters, and constructing a *file-like object*, that presents the same interface and functionality as we get when we perform an `open()` on a file and obtain a *file object*. There are two variations, based on whether the in-memory structure is a `str` or a `bytes` object.

- `io.StringIO()`: takes a string buffer and returns an object that operates in the same way as a file object returned from an `open()` call. Like a file object, this object has a notion of a *current location* that advances as we read (using `read()`, `readline()`, etc.) through the characters of the object.
- `io.BytesIO()\index{BytesIO() constructor}`: takes a bytes buffer and returns an object that operates in the same way as a file object returned from an `open()` call and opened in binary mode. Like a file object, this object has a notion of a *current location* that advances as we perform `read()` operations over the bytes of the object.

We will use these constructors, passing either `Response.text` or `Response.content`, as appropriate, to allow much easier processing of HTTP response message as we consider CSV as well as JSON and XML parsing and interpretation in the sections that follow.

21.2.2.1   Option 1: From String Text

Suppose `response.text` contains the data of `ind2016.csv`, as text data, obtained from `requests.get()` as above. We can use `io.StringIO()` to create a file-like object and explicitly process it into a native Python data structure and then construct a `pandas` data frame.

```
fileLikeObj = io.StringIO(response.text)

headerList = fileLikeObj.readline().strip().split(',')
LoL = []
for line in fileLikeObj:
    rowlist = line.strip().split(',')
```

```
LoL.append(rowlist)

df = pd.DataFrame(LoL, columns=headerList)
df = df.astype({'pop': float, 'gdp': float, 'gdp':float,
                'life': float, 'cell': float})
df
```

	code	country	pop	gdp	life	cell
0	CAN	Canada	36.26	1535.77	82.30	30.75
1	CHN	China	1378.66	11199.15	76.25	1364.93
2	IND	India	1324.17	2263.79	68.56	1127.81
3	RUS	Russia	144.34	1283.16	71.59	229.13
4	USA	United States	323.13	18624.47	78.69	395.88
5	VNM	Vietnam	94.57	205.28	76.25	120.60

This process can be streamlined by use of pandas built-in functions. In the read_csv() data frame constructor, the first argument can be a file object or a file-like object. So we can create the file-like object from the string version of the response and use that as the first argument, with the rest of the benefit in parameter options that come from using read_csv():

```
fileLikeObj = io.StringIO(response.text)
df = pd.read_csv(fileLikeObj, index_col='code')
df
```

	country	pop	gdp	life	cell
code					
CAN	Canada	36.26	1535.77	82.30	30.75
CHN	China	1378.66	11199.15	76.25	1364.93
IND	India	1324.17	2263.79	68.56	1127.81
RUS	Russia	144.34	1283.16	71.59	229.13
USA	United States	323.13	18624.47	78.69	395.88
VNM	Vietnam	94.57	205.28	76.25	120.60

Now let us turn to a case where the encoding is UTF-16BE. Again, the file itself is still a text file.

```
csvurl = util.buildURL("/data/ind2016_16.csv",
                       "datasystems.denison.edu")
response = requests.get(csvurl)
if response.status_code != 200:
    print("Error acquiring file")
```

The object response is very similar to the version of response associated with ind2016.csv. For a web server and the HTTP request, there is little difference between one file and another. So we would not expect the assumed encoding to be correct, and indeed it is not:

```
response.encoding
```

```
| 'ISO-8859-1'
```

If we were to look at the decoded version through `response.text`, we see a nonsense string, exactly because the decoding was incorrect.

```
response.text[:20]
```

```
| '\x00c\x00o\x00d\x00e\x00,\x00c\x00o\x00u\x00n\x00t'
```

To fix this, we set the encoding to the proper value, given our knowledge of how this particular resource was encoded, and we then see an appropriate `response.text`:

```
response.encoding = 'UTF-16BE'
util.print_text(response.text)
```

```
| code,country,pop,gdp,life,cell
| CAN,Canada,36.26,1535.77,82.3,30.75
| CHN,China,1378.66,11199.15,76.25,1364.93
| IND,India,1324.17,2263.79,68.56,1127.81
| RUS,Russia,144.34,1283.16,71.59,229.13
| USA,United States,323.13,18624.47,78.69,395.88
| VNM,Vietnam,94.57,205.28,76.25,120.6
```

If `response.encoding` is correct, then `response.text` will be a correct string containing the textual CSV data. At this point, the *same technique*, where we use the `response.text` string and create a file-like object, can do the same things we did in Chap. 6 and with `pandas`:

```
fileLikeObj = io.StringIO(response.text)
df = pd.read_csv(fileLikeObj, index_col='code')
df
```

	country	pop	gdp	life	cell
code					
CAN	Canada	36.26	1535.77	82.30	30.75
CHN	China	1378.66	11199.15	76.25	1364.93
IND	India	1324.17	2263.79	68.56	1127.81
RUS	Russia	144.34	1283.16	71.59	229.13
USA	United States	323.13	18624.47	78.69	395.88
VNM	Vietnam	94.57	205.28	76.25	120.60

We have seen how to read string text into a `pandas` data frame. We turn now to the case where we use the body from `request.get()` as byte data.

21.2.2.2   Option 2: From Underlying Bytes

In the example above, changes in `response.encoding` and the resultant difference in `response.text` *did not* change the underlying bytes data, available in `response.content`. While it is more complex, particularly across non-standard encoding, to use the bytes data and direct file type operations to construct a data frame, the pandas `read_csv()` can take its input from a file-like object containing bytes data and can perform the decoding itself.

To demonstrate this across our two different encodings, we GET both the UTF-8 encoded CSV file and the UTF-16BE encoded CSV file and use different response objects for the two results:

```
csvurl1 = util.buildURL("/data/ind2016.csv",
                         "datasystems.denison.edu")
response1 = requests.get(csvurl1)
if response1.status_code != 200:
    print("Error acquiring file")

csvurl2 = util.buildURL("/data/ind2016_16.csv",
                         "datasystems.denison.edu")
response2 = requests.get(csvurl2)
if response2.status_code != 200:
    print("Error acquiring file")
```

When we are dealing with the underlying bytes data, and we want/need a file-like object, we use `io.BytesIO()` to construct the file-like object from the bytes in `response1.content` and `response2.content`. We then pass the file-like objects to `read_csv()` and specify the proper encoding:

```
fileLikeObj1 = io.BytesIO(response1.content)
fileLikeObj2 = io.BytesIO(response2.content)
df1 = pd.read_csv(fileLikeObj1, encoding='UTF-8')
df2 = pd.read_csv(fileLikeObj2, encoding='UTF-16BE')
```

We see that this procedure succeeds, demonstrating the power of `read_csv`:

```
df1
```

	code	country	pop	gdp	life	cell
0	CAN	Canada	36.26	1535.77	82.30	30.75
1	CHN	China	1378.66	11199.15	76.25	1364.93
2	IND	India	1324.17	2263.79	68.56	1127.81
3	RUS	Russia	144.34	1283.16	71.59	229.13
4	USA	United States	323.13	18624.47	78.69	395.88
5	VNM	Vietnam	94.57	205.28	76.25	120.60

The value of `df2` is identical. Having demonstrated how to read CSV data, we turn now to reading JSON data.

21.2.3   Reading Questions

21.16 The first example involves specifying the correct encoding for `ind2016_16.csv`. The reading shows what happens if you specify no encoding. What do the results look like if you specify the wrong encoding? Investigate (using the local file `ind2016_16.csv` that you should have downloaded) with at least three encodings.

21.17 Please carry out the `requests.get(csv_url)` block of code and experiment with setting different encodings. Try with both naive encodings like ASCII and also more advanced encodings like UTF-16. Also try with a bytes encoding like UTF-16BE. Describe the results of `util.print_text()` in each case.

21.18 Recall that when you `open()` a file you can choose various modes, e.g., for reading versus writing. What mode would you use to create a bytes file?

21.19 In the code to process `response.text` into a LoL, please explain the purpose of `strip()`, `split()`, and `astype()`. You might want to refer back to earlier chapters.

21.20 Investigate the `read_csv()` method of reading `response.text` into a data frame. Are the entries floating point numbers or strings?

21.21 In the approach to reading from `response.content` into a data frame, did we need to set a value for `response.encoding`? Why or why not?

21.2.4   Exercises

21.22 The purpose of `io.StringIO()` is to create a file-like object from *any* string in a Python program. The object created "acts" just like an open file would.

Consider the following single Python string, `s`, composed over multiple continued lines:

```
s = "Twilight and evening bell,\n" \
    "And after that the dark!\n" \
    "And may there be no sadness of farewell,\n" \
    "When I embark;\n"
```

First, write some code to deal with `s` as a string:

- determine the length of `s`,
- find the start and end indices of the substring `"dark"` within `s`,
- create string `s2` by replacing `"embark"` with `"disembark"`.

Now, create a file-like object from s and perform a first `readline()`, assigning to variable `line1` and then write a `for` loop to use the file-like object as an iterator to accumulate into `lines` a list of the remaining lines, printing each.

21.23 Practice with `io.StringIO()` by using a for loop to print the numbers 1 through 100 into a file-like object, one per line. Then, iterate through this object and confirm that `read()` yields a string representing the entire data, while `readline()` yields one line at a time, and keeps track of the location in the file-like object. Provide your code.

21.24 Repeat the previous problem but with `io.BytesIO()`. Note that you can convert the numbers yielded by your loop into bytes using the `bytes()` function from the previous section. Provide your code.

The next set of exercises involve a file at resource path `/data/mystery3.dat` on host `datasystems.denison.edu`. You can assume the file is textual and is a tab-separated data collection where each line consists of:

```
male_name <tab> male_count <tab> female_name <tab> female_count
```

for the top 10 name applications of each sex to the US Social Security Administration for the year 2015.

21.25 Suppose the encoding of the file is unknown but will be from one of the following:

- "UTF-8,"
- "UTF-16BE,"
- "UTF-16LE,"
- "cp037,"
- "latin_1."

Write code to:

- acquire the file from the web server,
- ensure the status_code is 200,
- assign to `content_type` the *value* of the `Content-Type` header line of the response,
- determine the *correct* encoding and assign to `real_encoding`,
- set the `.encoding` attribute of the response to `real_encoding`,
- assign to `csv_body` the string text for the body of the response.

21.26 In this question, you will start with a *string* and create a *Dictionary of Lists* representation of the data entailed in the string. It is suggested to use the result of the previous problem, `csv_ body`, as the starting point. But to start independently, you can use the following string literal constant assignment to get to the same starting point:

```
csv_body = "Noah\t19635\tEmma\t20455\n" \
           "Liam\t18374\tOlivia\t19691\n" \
           "Mason\t16627\tSophia\t17417\n" \
```

```
"Jacob\t15949\tAva\t16378\n" \
"William\t15909\tIsabella\t15617\n" \
"Ethan\t15077\tMia\t14905\n" \
"James\t14824\tAbigail\t12401\n" \
"Alexander\t14547\tEmily\t11786\n" \
"Michael\t14431\tCharlotte\t11398\n" \
"Benjamin\t13700\tHarper\t10295\n"
```

Construct a file-like object from `csv_body` and then use file object operations to create a dictionary of lists representation of the tab-separated data. Note that there is no header line in the data, so you can name the columns `malename`, `malecount`, `femalename`, and `femalecount`.

21.27 Use `pandas` to obtain a data frame named `df` by using a file-like object based on `csv_body` and use `read_csv()`. Name your resultant data frame `df`. Make sure you have reasonable column names.

Be careful to call `read_csv` so that the separators are tabs, not commas.

21.3   JSON Data

Recall from Chaps. 2 and 15 that JSON is a light-weight format for transmitting simple data types.

21.3.1   JSON from File

When we acquire JSON through a file, we use the `json.load()` function. This function uses an open file object (or file-like object) as an argument. Therefore, to deal with a different encoding, we simply need to specify the encoding as we open the file.

We start by showing what happens when we fail to do this after acquiring data encoded in a non-default `UTF-16BE`:

```
ind0_16_path = os.path.join(datadir, "ind0_16.json")
fh = open(ind0_16_path, mode='r')

try:
    ds = json.load(fh)
except json.JSONDecodeError as e:
    print("Exception encountered in JSON decode")
```

```
| Exception encountered in JSON decode
```

The load() operation raised an exception since it was unable to decode the data. Now we create a file object and specify the correct encoding and voila, things work as we need them to:

```
fh = open(ind0_16_path, mode='r', encoding="UTF-16BE")

try:
    ds = json.load(fh)
except json.JSONDecodeError as e:
    print("Exception encountered in JSON decode")
util.print_data(ds, nlines=10)
```

```
| {
|    "FRA": {
|      "2007": {
|         "pop": 64.02,
|         "gdp": 2657.21
|      },
|      "2017": {
|         "pop": 66.87,
|         "gdp": 2586.29
|      }
```

21.3.2   JSON from Network

In the following examples, we obtain from the web server files with JSON as the body data. In response1, we have UTF-8 encoded data. In response2, we have UTF-16BE encoded data.

```
json_url1 = util.buildURL("/data/ind0.json",
                          "datasystems.denison.edu")
response1 = requests.get(json_url1)
if response1.status_code != 200:
    print("Error acquiring file")

json_url2 = util.buildURL("/data/ind0_16.json",
                          "datasystems.denison.edu")
response2 = requests.get(json_url2)
if response2.status_code != 200:
    print("Error acquiring file")
```

We next show how to get from response1 and response2 to in-memory data structures.

21.3.2.1   JSON from String Data in Response

In common with the examples above, when we want to use the .text (string) version of the response, we *must* get the encoding right. We do this for both response1 and response2, at which point the character string version of the two responses is valid, and we can use a variety of techniques to go from a string into a JSON-based data structure. Since the latter steps are the same after we get the encoding right, we just run through examples using response1.text.

```
response1.encoding = 'UTF-8'
util.print_text(response1.text, nlines=10, json_string=True)
```

```
| {
|    "FRA": {
|      "2007": {
|         "pop": 64.02,
|         "gdp": 2657.21
|      },
|      "2017": {
|         "pop": 66.87,
|         "gdp": 2586.29
|      }
```

```
response2.encoding = 'UTF-16BE'
util.print_text(response2.text, nlines=10, json_string=True)
```

```
| {
|    "FRA": {
|      "2007": {
|         "pop": 64.02,
|         "gdp": 2657.21
|      },
|      "2017": {
|         "pop": 66.87,
|         "gdp": 2586.29
|      }
```

We see that as soon as the correct encoding is specified, the field response.text is legible. We are ready to read the data into memory.

Option 1
Use json.loads(), which takes a string and returns the in-memory data structure.

Given a JSON-formatted string s, the built-in function loads(s), in the json package of Python, returns the data structure encoded. For instance, if s represents a JSON array, then a Python list is loaded, and if s represents a JSON object, then a Python dictionary is loaded. We demonstrate:

```
ds1 = json.loads(response1.text)
util.print_data(ds1, nlines=10)
```

```
|  {
|     "FRA": {
|       "2007": {
|          "pop": 64.02,
|          "gdp": 2657.21
|       },
|       "2017": {
|          "pop": 66.87,
|          "gdp": 2586.29
|       }
```

We achieve a Python dictionary in memory.

Option 2

Create a file-like object, and then use json.load().

Given a JSON file, the load() function returns the data structure encoded. Just as we did with CSV files, we can use StringIO to produce a file-like object, which we can feed to json.load() as follows:

```
fileLikeObj1 = io.StringIO(response1.text)
ds1 = json.load(fileLikeObj1)
util.print_data(ds1, nlines=10)
```

```
|  {
|     "FRA": {
|       "2007": {
|          "pop": 64.02,
|          "gdp": 2657.21
|       },
|       "2017": {
|          "pop": 66.87,
|          "gdp": 2586.29
|       }
```

We achieve a Python dictionary in memory.

Option 3

Use requests.json() method of a response object.

Lastly, the requests module has built-in functionality for JSON files, because of their ubiquity. The following shows how to read directly into a Python dictionary from the HTTP response received.

```
ds1 = response1.json()
util.print_data(ds1, nlines=10)
```

```
|  {
```

```
|    "FRA": {
|      "2007": {
|         "pop": 64.02,
|         "gdp": 2657.21
|      },
|      "2017": {
|         "pop": 66.87,
|         "gdp": 2586.29
|      }
```

In all three of these examples, we have been provided JSON data in string form. We consider now the case of JSON data in byte form.

21.3.2.2   JSON from Bytes Data in Response Body

Because of its alternate encoding resulting in a different set of bytes for the sequence of characters, we use the bytes data of response2 in our examples demonstrating bytes data conversion into JSON-derived data structure.

A Request for Comments (RFC) documents a given Internet Standard. The RFC standard for JSON explicitly allows all three of UTF-8, UTF-16, and UTF-32 to be used in data formatted as JSON. This means that the json module will recognize the bytes data directly, as if it were already a decoded string, greatly simplifying our lives.

Option 1
Use json.loads(), which takes bytes data in UTF-8, UTF-16, or UTF-32 and returns the in-memory data structure.

Analogous to the situation of text data, we can feed the built-in function json.loads() byte data rather than text data:

```
ds2 = json.loads(response2.content)
util.print_data(ds2, nlines=5)
```

```
| {
|    "FRA": {
|      "2007": {
|         "pop": 64.02,
|         "gdp": 2657.21
```

The result is still a Python data structure in memory.

Option 2
Create a bytes file-like object and then use json.load().

Similarly, we can feed the json.load() function a byte file instead of a text file. We use BytesIO to get from the byte data response to a file-like object.

```
fileLikeObj2 = io.BytesIO(response2.content)
ds2 = json.load(fileLikeObj2)
util.print_data(ds2, nlines=5)
    | {
    |     "FRA": {
    |         "2007": {
    |             "pop": 64.02,
    |             "gdp": 2657.21
```

The result is a Python data structure in memory. Having demonstrated how to acquire JSON data, we turn to XML data.

21.3.3   Reading Questions

21.28 Please download ind0_16.json as a local file and experiment with setting different encodings in open(), and the try/except block of code given. Explain what happens.

21.29 In the previous question, we explored errors associated with the JSON load() function when the encoding is wrong. Please do the same now with the loads() function and describe what happens when the encoding is wrong. You may use the code provided to read JSON data from the book web page.

21.30 In JSON Option 2 for response.text, does this assume the encoding has already been specified?

21.31 In JSON Option 3 for response.text, does this assume the encoding has already been specified?

21.32 When extracting JSON from bytes data, do we need to specify response.encoding before applying json.loads() to response.content?

21.33 Why were there three options for extracting in-memory data structures from string data but only two options for bytes data?

21.3.4   Exercises

In many of the following exercises, we will show a curl incantation that obtains JSON-formatted text data from the Internet. Your task will be to translate the incantation into the equivalent requests module programming steps, and to obtain the *parsed* JSON-based data structure from the result, assigning to variable `data`. In some cases, we will ask for a specific method from those demonstrated in the section.

21.34 Using any method, get the JSON data from `school0.json`:

```
curl -s -o school0.json \
      https://datasystems.denison.edu/data/school0.json
```

21.35 Using the bytes data in `.content`, a *file-like object*, and `json.load()`, get the JSON data from `school0.json`.

```
curl -s -o school0.json \
      https://datasystems.denison.edu/data/school0.json
```

21.36 Write a function

```
getJSONdata(resource, location, protocol='http')
```

that makes a request to `location` for `resource` with the specified protocol, then uses the bytes data in the `.content` of the response, with a *file-like object*, and `json.load()`, to get the JSON data. On success, return the data. On failure of either the request or the parse of the data, return `None`.

21.37 The `school0_32.json` resource is encoded with `utf-32`. Use the method of setting the `.encoding` attribute and then accessing the `.text` string body, and get the JSON data.

```
curl -s -o school0_32.json \
      https://datasystems.denison.edu/data/school0_32.json
```

21.38 Repeat acquiring the `school0_32.json` resource, encoded with `utf-32`. This time, use the method of using the bytes data in `.content`, a *file-like object*, and `json.load()`.

```
curl -s -o school0_32.json \
      https://datasystems.denison.edu/data/school0_32.json
```

Where in your code did the encoding of "utf-32" come into play? Can you explain why? What does this mean for the `getJSONdata()` function you wrote previously?

21.39 Use any method you wish to obtain the JSON data associated with the following POST request. Make sure you faithfully translate the `-H` and `-d` options of the `curl` into their `requests` equivalent.

```
curl -X POST -s -o data/reply.json -d field1='value1' \
      -d field2=42 -H "Accept: application/json" \
      "https://httpbin.org/post"
```

21.4 XML Data

Recall from Chap. 15 that XML is a format used for hierarchical data. When an XML file is well formed, we can parse it to map it onto the tree it represents, and can extract the root Element containing the data of the entire tree. This process of turning an XML file into a tree, and finding the root, uses the lxml library and the etree module within it.

In the examples that follow, when we have parsed a tree, we print out the tag of the root Element. Since, for all these examples, the data is the *indicators* data set, and the tree is structured so that the root Element has tag, indicators, we are successful when this is the result we print. Also, the parse() function raises an exception when it encounters a problem, so we place our examples in try-except blocks to help show when such problems occur.

21.4.1 XML from File Data

When we have XML in a file, we have two options for opening and parsing. We can specify the file name, or we can specify a file object. For the former, we first build the relevant path.

```
ind0_path = os.path.join(datadir, "ind0.xml")
ind0_16_path = os.path.join(datadir, "ind0_16.xml")
```

We demonstrate specifying a path to the parse() function:

```
try:
    tree0 = etree.parse(ind0_path)
    root0 = tree0.getroot()
    print(root0.tag)
except:
    print("Exception in parsing XML")

    | indicators
```

Next we demonstrate specifying a file object:

```
fh = open(ind0_path, mode='r', encoding='UTF-8')
try:
    tree0 = etree.parse(fh)
    root0 = tree0.getroot()
    print(root0.tag)
except:
    print("Exception in parsing XML")

    | indicators
```

This can be done for files of any encoding. Again, we first demonstrate specifying a path:

```
try:
    tree1 = etree.parse(ind0_16_path)
    root1 = tree1.getroot()
    print(root1.tag)
except Exception as e:
    print("Exception in parsing XML:", e)

    | indicators
```

Importantly, the parse function is intelligent enough to figure out the encoding, even if we do not specify it when we open the file. Hence, the parse function is actually doing the decoding in the following block of code:

```
fh = open(ind0_16_path)
try:
    tree1 = etree.parse(fh)
    root1 = tree1.getroot()
    print(root1.tag)
except Exception as e:
    print("Exception in parsing XML:", e)

    | indicators
```

Having reviewed how to open and parse XML files locally, we turn to data obtained over the network.

21.4.2   From Network

In common to the following examples, we obtain from the web server files with XML as the body data. In response1, we have UTF-8 encoded data, and in response2, we have UTF-16BE encoded data.

```
xml_url1 = util.buildURL("/data/ind0.xml",
                          "datasystems.denison.edu")
response1 = requests.get(xml_url1)
if response1.status_code != 200:
    print("Error acquiring file")

xml_url2 = util.buildURL("/data/ind0_16.xml",
                          "datasystems.denison.edu")
response2 = requests.get(xml_url2)
if response2.status_code != 200:
    print("Error acquiring file")
```

In both cases, the headers of the response know that the body is in XML format, but do not specify the encoding. The default encoding for a text/xml content type is "UTF-8," as we can see within the text field of the response:

```
print(response1.headers['Content-Type'])
```

> | text/xml

```
util.print_text(response1.text, nlines=10)
```

> | <?xml version='1.0' encoding='UTF-8'?>
> | <indicators>
> | <country code="FRA" name="France">
> | <timedata year="2007">
> | <pop>64.02</pop>
> | <gdp>2657.21</gdp>
> | </timedata>
> | <timedata year="2017">
> | <pop>66.87</pop>
> | <gdp>2586.29</gdp>

For the second encoding type, we must specify the encoding before attempting to retrieve the text of the response. Otherwise, it will not render correctly.

```
print(response2.headers['Content-Type'])
```

> | text/xml

```
response2.encoding = 'UTF-16BE'
util.print_text(response2.text, nlines=10)
```

> | <?xml version='1.0' encoding='utf-16be' standalone='yes'?>
> | <indicators>
> | <country code="FRA" name="France">
> | <timedata year="2007">
> | <pop>64.02</pop>
> | <gdp>2657.21</gdp>
> | </timedata>
> | <timedata year="2017">
> | <pop>66.87</pop>
> | <gdp>2586.29</gdp>

In both cases, we have achieved XML data in the text field of response. We now discuss how to parse this XML data.

21.4.2.1   Using `parse` on Bytes

We consider first the case where the data comes to us in byte form. As usual, we use `io.BytesIO()` to create a file-like object. The `parse()` function is intelligent enough to parse such a file, resulting in a properly formed tree.

```
fileLikeObj1 = io.BytesIO(response1.content)
tree1 = etree.parse(fileLikeObj1)
```

Even when the data is encoded in a way other than UTF-8, we no longer need to specify the encoding, because the `parse()` function is intelligent enough to decode on its own, as we explained above.

```
fileLikeObj2 = io.BytesIO(response2.content)
tree2 = etree.parse(fileLikeObj2)
```

We demonstrate that the two trees returned by `parse` are indeed as expected, i.e., that the root is `indicators` as it should be. This means that the root `Element` will contain all data stored in the XML tree.

```
tree1.getroot().tag
```

```
| 'indicators'
```

```
tree2.getroot().tag
```

```
| 'indicators'
```

We turn now to an alternative way to get the root `Element` that avoids the need for the `parse()` function.

21.4.2.2   Using `fromstring()` with Bytes and Strings

Given a response containing XML data in byte form, we can use the `fromstring()` method, associated with the `etree` type, to extract the root `Element` of the tree represented by the byte data.

```
s_root1 = etree.fromstring(response1.content)
print(s_root1.tag)
```

```
| indicators
```

Like the `parse()` function, we do not need to specify an encoding to the `fromstring()` function, as the following code demonstrates.

```
s_root2 = etree.fromstring(response2.content)
print(s_root2.tag)
```

```
| indicators
```

We turn now to the situation of a response in text form, rather than binary form. In this case, we must read past the head matter of the response.

```
skipheader = lambda s: s[s.index('\n')+1:]
```

We note here that the header we are skipping is a prolog part of the test of the XML data and is unrelated to the headers that come from the `requests` module. The header we are skipping is entirely contained in the body of the `response` we begin with.

We apply this function to both of our text-based XML data responses, to retrieve proper XML strings.

```
xml1 = skipheader(response1.text)
xml2 = skipheader(response2.text)
```

We can feed these XML strings into the `fromstring()` function, in much the way we fed it byte data, and again the function will give us the root of the XML tree.

```
s_root1 = etree.fromstring(xml1)
print(s_root1.tag)

  | indicators

s_root2 = etree.fromstring(xml2)
print(s_root2.tag)

  | indicators
```

In all cases, we are able to retrieve the tree and root `Element` representing the XML data that comes to us either in local form or over the network.

21.4.3 Reading Questions

21.40 The first XML example does not specify an encoding but the second does. Could the second have gotten away without specifying `encoding='UTF-8'`? Justify your answer by actually running the code.

21.41 If the `parse` function is doing the decoding, why do we need to wrap our code in a `try`/`except` block? What could go wrong?

21.42 When reading XML file, how can you print the encoding, to see "UTF-8"? Hint: think back to how we did it for CSV and JSON.

21.43 Please experiment with `response.text` by purposely specifying the wrong encoding and seeing what is printed by `util.print_text()` for one of the XML files accessed over the network. Describe your results.

21.44 The reading shows how to use `parse` on Bytes. Can you also use `parse()` on `response.text`? Do you need to specify the encoding?

21.45 Consider the second block of code that invokes the `fromstring()` method. How does this demonstrate that encoding need not be specified?

21.46 Experiment to find out what happens if you purposely set the wrong encoding, e.g., with `response2.encoding = 'ASCII'` before invoking the `fromstring()` method on `response2.content`. Report your findings.

21.47 Explain the `lambda` function `skipheader` in detail. Why does this skip the header? Why is skipping the header important?

21.48 When feeding the `fromstring()` function text data, do we need to specify the encoding, e.g., with `response1.encoding = 'UTF-8'` before invoking `skipheader(response1.text)`? Investigate by actually running the code, and report what you found.

21.4.4   Exercises

In many of the following exercises, we will show a `curl` incantation that obtains XML-formatted text data from the Internet. Your task will be to translate the incantation into the equivalent `requests` module programming steps and to obtain the *parsed* XML-based `ElementTree` structure from the result, assigning to variable `root` the root of the result. In some cases, we will ask for a specific method from those demonstrated in the section.

21.49 Using any method, get the XML data from `school0.xml`:

```
curl -s -o school0.xml \
      https://datasystems.denison.edu/data/school0.xml
```

21.50 Using the bytes data in `.content`, a *file-like object*, and `etree.parse()`, get the XML data from `school0.xml`.

```
curl -s -o school0.xml \
      https://datasystems.denison.edu/data/school0.xml
```

21.51 Write a function

```
getXMLdata(resource, location, protocol='http')
```

that makes a request to `location` for `resource` with the specified protocol, and then uses the bytes data in the `.content` of the response, with a *file-like object*, and `etree.parse()`, to get the XML data. On success, return the root of the tree. On failure of either the request or the parse of the data, return `None`.

21.52 The `school0_32.xml` resource is encoded with `utf-32be`. Use the method of setting the `.encoding` attribute of the response and then accessing the

.text string body, and using fromstring(). Remember that fromstring() expects to start from an element, not from the header line, so you will need to skip the header to get the string to pass.

```
curl -s -o school0_32.xml \
    https://datasystems.denison.edu/data/school0_32.xml
```

21.53 Repeat acquiring the school0_32.xml resource, encoded with utf-32be. This time, use the method of using the bytes data in .content, a *file-like object*, and etree.parse().

```
curl -s -o school0_32.xml \
    https://datasystems.denison.edu/data/school0_32.xml
```

Where in your code did the encoding of "utf-32be" come into play? Can you explain why? What does this mean for the getXMLdata() function you wrote previously?

21.54 Use any method you wish to obtain the XML data associated with the following GET request. Do **not** simply copy and paste the full url. Translate the set of query parameters into a dictionary to be used in the requests.get() invocation.

```
curl -s -o kivaloans.xml \
    'http://api.kivaws.org/v1/loans/search.xml? \
    sector=agriculture&status=fundraising'
```

Chapter 22
Web Scraping

Chapter Goals

Upon completion of this chapter, you should understand the following:

- HTML as a tree and as an instance of a *markup language*, analagous to XML, but with structure, tags, and attributes intended for the rendering of web pages.
- The particular data set structuring facilities in HTML:
  - the `table` HTML structure,
  - ordered and unordered lists (`ol` and `ul`) as HTML structures for organizing and displaying data.

Upon completion of this chapter, you should be able to do the following:

- Acquire and parse HTML documents from web servers, using:
  - GET for a web page,
  - POST for custom access to a web page based on a form.

- Use combinations of XPath and procedural traversal to construct useable data sets from the most common HTML structures used for rendering data:
  - HTML tables,
  - HTML lists.

© Springer Nature Switzerland AG 2020
T. Bressoud, D. White, *Introduction to Data Systems*,
https://doi.org/10.1007/978-3-030-54371-6_22

The HyperText Markup Language (HTML) is the format used for web pages in the Internet. The format provides for the logical structure of information for navigation, formatting, incorporation of pictures and other media, and structure of web pages, as well as for linking to other web pages and resources both local and remote relative to the web server providing the content. Web pages can display data in various forms, and while the intent of the web page may have been for the *presentation* of data, it can also serve as a data source in the context of our Data Systems.

The term *web scraping* refers to the programmatic access of the HTML of web pages and the extraction of useful information contained therein, even if the intent was for presentation of information and not the providing of data. We should note, at the onset, that data incorporated into web pages is the intellectual property of some owner of the data, often affiliated with the web server provider. As such, there are ethical and legal considerations whenever we decide to acquire data in this manner. The intellectual owner of data has the authority and right to determine the valid use of the data, or its acquisition through these programmatic means. Just because we *can* acquire data through web scraping techniques, does not mean that it is ethical or legal, based on the wishes of the owner of that data.

22.1   HTML Structure and Its Representation of Data Sets

Like XML, HTML can define a tree representing the information for a web page. Unlike XML, HTML is a *specific grammar*. This means that the tags and many of the allowed relationships between parents and children are well defined and have a semantic meaning associated with them.[1] For instance, all HTML documents have a root node of html, and its children can only be head and body. Under head, one of the most common children is title, used in rendering to show the title of the web page, often in a browser tab. Often the children of head include nodes for metadata and scripts, which are code segments to be run within the context of the web browser when the page is loaded. The children of head can also define link elements to tie the page to specific formatting styles.

The body of an HTML document subdivides a web page using div nodes, which are structuring elements. These often have class and id attributes to distinguish their use. Programs that can generate web pages for developers use the structuring mechanism of div nodes to provide, within the child structure of the div, elements for navigation and separation of the objects developers can place into a web page.

The body also often has nodes that represent various levels of headers (in nodes tagged h1 through h6), and these are used to give, in the rendered page, sections, subsections, etc., for the document. We can use the p node to define paragraphs

---

[1] The original HTML had a fixed, defined set of elements/tags. More modern versions of HTML allow for definition of custom elements, but for interpretation by the web browser, these have to be *registered* and given meaning.

within the page, and define figures and captions. Nodes labeled `span` and with `class` attributes are used for grouping of underlying page elements and are often used to impose consistent styling for the rendering of elements.

The text below shows a minimal, but complete, HTML document that is fully capable of being rendered by a web browser. It includes a number of the elements described above.

```
<!DOCTYPE html>
<html>
  <head>
    <title>Title of the Page</title>
  </head>
  <body>
    <h1>First Level Heading</h1>

    <p>Paragraph defined in body.</p>
  </body>
</html>
```

In its original definition, HTML was not as strict as XML and the associated set of rules governing well-formedness that we saw in Chap. 17. For instance, a paragraph element and its start tag, `<p>`, was not constrained to have an end tag `</p>`. There were many such instances that, taken together, means that a single unambiguous tree could not always be parsed and constructed from an HTML source. Specific web browsers, like Chrome and Explorer, would define default heuristics for processing and rendering such cases, and web developers adapted to the rendering choices of the target browsers. As HTML has evolved, newer versions, such as XHTML, have addressed these potential problems by defining a strict version of HTML that adheres to XML well-formedness and is, in fact, defined as an *XML Application* [73].

The reason for pointing out the insufficiency of some HTML to follow a strict tree definition (in part, by omitting some end tags) is a practical one. In the space of performing web scraping, we must take HTML that has been developed by external entities and is hence out of our control, and perhaps not intended for client application processing, and be able to parse it into a tree structure like we did with XML. Depending on how "broken" the HTML may be, we have some programmatic solutions to build a *best-effort* tree from the HTML so that we can proceed with our web scraping task.

Figure 22.1 shows an HTML document structure, including annotation of line numbers. The full text of the HTML has been edited to allow us to more easily see the overall structure. The `html` tree starts on line 2 and ends on the line marked 120. The `head` subtree extends from line 3 to line 11, and we have edited out the contents of a `script` node within the `head` for clarity. Beneath the `body`, which starts at line 12, there are many `div` elements and significant nesting. Lines 18 through 27 contain the HTML for a sidebar used for navigation, with detail omitted at line 23. The main content of the page occurs at the `div` on line 38, which we have collapsed, again to illustrate the overall structure.

```
1  <!DOCTYPE html>
2  <html xmlns="http://www.w3.org/1999/xhtml" xml:lang="en" lang="en">
3   <head>
4    <meta charset="utf-8" />
5    <title>ind2016 | Introduction to Data Systems</title>
6     <script>
7
8     <!-- Script content omitted ... -->
9
10    </script>
11  </head>
12  <body class="sandvox" id="personal_denison_edu_bressoud" >
13   <div id="page-container">
14    <div id="title">
15     <h1 class="title in">
16      <a href="./">Introduction to Data Systems</a></h1>
17    </div><!-- title -->
18    <div id="sitemenu-container">
19     <div id="sitemenu">
20      <h2 class="hidden">Site Navigation</a></h2>
21      <div id="sitemenu-content">
22
23      <!-- Navigation Elements omitted ... -->
24
25      </div> <!-- /sitemenu-content -->
26     </div> <!-- /sitemenu -->
27    </div> <!-- sitemenu-container -->
28   </div> <!-- page-container -->
29   <div id="page-content" class="no-navigation">
30    <div id="main">
31     <div id="main-content">
32      <h2 class="title"><span class="in">ind2016</span></h2>
33      <div class="article">
34       <div class="article-content">
35
36       <!-- Collapse main body content -->
37
38       <div> ... </div>
39
40       </div> <!-- /article-content -->
41      </div> <!-- /article -->
42     </div> <!-- main-content -->
43     <div id="main-bottom"></div>
44    </div> <!-- main -->
45   </div> <!-- page-content -->
46  </body>
47  </html>
```

Fig. 22.1 HTML document structure

The process of web scraping involves, first, one of *discovery*, wherein we use tools to acquire the HTML and then to ascertain its structure and, in particular, determine how the desired data within the page is represented in the HTML, and second, one of *programmatic extraction*, wherein we use our techniques from XML, including both XPath and procedural programming, to process the HTML tree and extract the data, transforming into a usable result, such as one or more data frames.

The remaining subsections describe the two most common ways of representing data sets in an HTML document. Understanding some of the possible HTML representations of data on a web page allows us to proceed through our discovery phase of web scraping. The HTML elements representing a data set would be subordinate in the tree and specific location based on the overall structure. For example, in Fig. 22.1, these data set-containing HTML elements would be present in the subtree rooted at the collapsed `div` at line 38.

22.1.1   HTML Tables

For representing a tabular data set within an HTML page, the most natural structure would be to use an HTML `table` element. This HTML structure can separate the header of the table (using `thead`) from the body of the table (using `tbody`), can define individual rows of the table (using `tr`/*table row*), and, for fields within the row, can define header cells (using `th`/*table header*) and data cells (using `td`/*table data*). The result could look something like the rendered page from a browser as shown in Fig. 22.2.

Fortunately, whether a table is simple and unadorned, as illustrated above, or uses borders and shading and other more complex formatting, the same `table` HTML

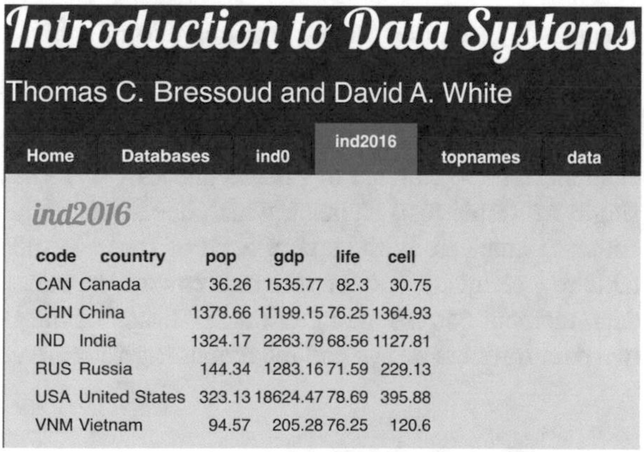

Fig. 22.2 HTML table rendered by browser

Introduction to Data Systems

Thomas C. Bressoud and David A. White

Home	Databases	ind0	ind2016		topnames	data

ind2016

code	country	pop	gdp	life	cell
CAN	Canada	36.26	1535.77	82.3	30.75
CHN	China	1378.66	11199.15	76.25	1364.93
IND	India	1324.17	2263.79	68.56	1127.81
RUS	Russia	144.34	1283.16	71.59	229.13
USA	United States	323.13	18624.47	78.69	395.88
VNM	Vietnam	94.57	205.28	76.25	120.6

Fig. 22.3 HTML for simple
table

```
<table class="table table-bordered">
  <thead>
    <tr>
      <th title="Field #1">code</th>
      <th title="Field #2">country</th>
      <th title="Field #3">pop</th>
      <th title="Field #4">gdp</th>
      <th title="Field #5">life</th>
      <th title="Field #6">cell</th>
    </tr>
  </thead>
  <tbody>
    <tr>
      <td>CAN</td>
      <td>Canada</td>
      <td align="right">36.26</td>
      <td align="right">1535.77</td>
      <td align="right">82.3</td>
      <td align="right">30.75</td>
    </tr>
    <tr>
      <td>CHN</td>
      <td>China</td>
      <td align="right">1378.66</td>
      <td align="right">11199.15</td>
      <td align="right">76.25</td>
      <td align="right">1364.93</td>
    </tr>
    <tr> ... </tr>
    <tr> ... </tr>
    <tr> ... </tr>
    <tr> ... </tr>
  </tbody>
</table>
```

structure is used, the difference coming from style sheets,[2] class attributes, and
possible subordinate structure in the definition of table header (th) and table data
(td) elements.

Figure 22.3 presents the HTML for the table rendered in Fig. 22.2. The table
element has two children of thead and tbody. The child of thead consists of a
single tr (table row) element, which has six table header (th) cell elements. The
column names are in the text of each of these elements. The tbody has multiple
table row elements, one for each data row of the table, and each row has a td, table
data, element corresponding to the six fields of the table. For illustration, the first
two data rows are shown and the remaining four are collapsed.

---

[2]The most common mechanism for defining consistent styling rules is called Cascading Style Sheet
(CSS) and is auxiliary to the web page, much like an XML Schema can be auxiliary to an XML
document.

David White: Publications

Books and Chapters

1. *Introduction to Data Systems: Building from Python,* with **Thomas Bressoud**, Springer, 2020.
2. Statistics for Mathematicians, Chapter in *Data Science for Mathematicians*, edited by **Nathan Carter**, Taylor and Francis, 2020.
3. *Monoidal Bousfield Localization and Algebras over Operads*, Wesleyan University Library, 2014.
4. *Traversals of Infinite Graphs with Random Local Orientations*, Wesleyan University Library, 2012.

Research Publications

20. A Project Based Approach to Statistics and Data Science, *PRIMUS*, Volume 29, Issue 9, pages 997-1038.
19. Curriculum Guidelines for Undergraduate Programs in Data Science, with Richard De Veaux, Thomas Bressoud, et al. *Annual Review of Statistics*, Volume 4, pages 15-30.

Fig. 22.4   HTML list structure displayed

22.1.2   HTML Lists

Another very common structure in HTML for representing collections of data is a *list structure*. In the list, each item could represent a *row* of data. Within the row, there could be a set of HTML structure elements for representing the fields within the row. This is very common in practice. For example, when one searches for top song lists, a site may present the result in list form. The same is true of web pages that visibly look like ordered lists, e.g., displaying recipes (ordered by steps), tourist attractions (ordered by rating), video games (ordered by popularity), lists of publications, etc. An example is shown in Fig. 22.4.

In the discovery step of web scraping, examination of this page reveals that the set of publications are organized as an HTML *ordered list*, which uses the HTML structuring tag ol. The children of ol are *line items*, which use the tag li. Each of these li structures gives a single row of the desired data set. Within each are structuring elements for what we might consider the fields of the row. For instance, we might want fields for the name of the publication, the publisher, the year, the

```
<ol>
 <li>
  <span style="font-style: italic;">
   Introduction to Data Systems: Building from Python,
  </span>
  with <a href="http://personal.denison.edu/%7Ebressoud/">
   Thomas Bressoud</a>, Springer, 2020.
 </li>
 <li>Statistics for Mathematicians, Chapter in
  <span style="font-style: italic;">
   Data Science for Mathematicians</span>,
  edited by
  <a href="https://nathancarter.github.io/">Nathan Carter</a>,
  Taylor and Francis, 2020.
 </li>
 <li>
  <span style="font-style: italic;">
   Monoidal Bousfield Localization and Algebras over Operads,
  </span>
  Wesleyan University Library, 2014.
 </li>
 <li><span style="font-style: italic;">
  Traversals of Infinite Graphs with Random Local Orientations,
  </span>
  Wesleyan University Library, 2012.
 </li>
</ol>
```

Fig. 22.5 HTML list structure to scrape

co-authors, and any associated links. A snippet of the HTML for Fig. 22.4, for the list of Books and Chapters, is displayed in Fig. 22.5. Clearly, regular expressions would be useful for extracting the data desired.

In this particular example, the children of the li tags are a set of span elements. In practice, one often sees div elements as children of li elements as well. One could also imagine, for each li defining a row, a nested list structure with another ordered list (ol) or perhaps an unordered list (ul), e.g., if items in a recipe list had ordered sub-items labeled a, b, c, etc.

As another example, consider the web page displayed in Fig. 22.6. In this case, investigation of the web page reveals a *nested* list structure. In particular, there is an *unordered list*, tagged as ul, whose list elements, li, have children that are themselves ul items, one for each of the years 2007 and 2017. Within these, there is a third level of unordered list ul elements, whose li are the individual indicators of pop and gdp.

If we focus on just the HTML for this nested list structure, we see this nesting in the HTML displayed in Fig. 22.7, which uses indentation and line number annotation to help understand the structure. The "row" of FRA is a li that extends from line 2 through line 17. After the text of li element, the next level ul is defined;

Fig. 22.6 HTML nested lists

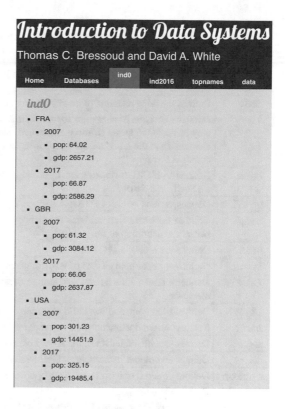

the first child `li` extends from line 4 through 9 and defines the 2007 indicator values, with a subordinate `ul` whose items give individual indicators on lines 6 and 7. The GBR subtree is structured similarly, as is the USA subtree, which is collapsed in this display.

HTML defines many more tags than we have discussed here. In Table 22.1, we present a list of some of the most common tags you are likely to encounter, along with a brief description of their use.

22.1.3   Reading Questions

22.1 Web scraping has received much media attention in recent years. Please share an interesting real-world application of web scraping that you are aware of.

22.2 In the reading, the data sought is located in a table under the `h2` header. Will this always be the case? Either justify a "yes" answer or find an example online to demonstrate a "no" answer.

22.3 The reading discusses the evolution of HTML pages over time. It would be a mistake to only learn the newest form of HTML, e.g., because one day you may

Table 22.1 Common HTML tags

Tag	Description
a	Defines a hyperlink, which is used to link from one page to another, or to a reference point within the current page. The href attribute defines the link itself, while the text gives the displayed version
body	Defines the body contents of an HTML document and is a child of the html root
div	Defines a division or container for an HTML document to allow Structure, and may be nested to group children as needed. The id attribute is used to identify elements, and the class attribute is also used for categorizing containers and for common styling
form	Defines an HTML form for user input. Some possible children include label, select, and input
h1 to h6	Defines various levels of HTML headings
head	Child of the html root, and used to define the metadata for an HTML document before the definition of the body
html	Defines the root of an HTML tree
input	Defines an input field as part of a form, where user can enter data
label	Defines a label within a form for, among others, input and for select elements
li	Defines a list item, and is a child of ordered lists, ol, and unordered lists, ul
link	Relates current document to an external resource, often for style information
meta	Defines specific metadata in the head of an HTML document
ol	Defines an ordered list structure. Children are li elements
option	Defines children of a select as dropdown items in the select list
p	Defines a paragraph
script	Used to embed a client-side script of JavaScript code
select	
span	Defines an inline container used to mark up a part of a document, and is finer grained than a div type of container
table	Defines an HTML table structure, with possible children thead and tbody, or directly to tr rows
tbody	Defines the body of a table and, if present, is a child of table
td	Defines a cell in a table
th	Defines a header cell in a table
thead	Defines the header potion of a table and, if present, is a child of table
title	Defines the title of the document to be shown in the browser's title bar or tab
tr	Defines a row in a table
ul	Defines an unordered list structure. Children are li elements

need to web scrape data on a web page created a long time ago. Indeed, you can even web scrape pages that no longer appear on the Internet, because the "way back machine" stores copies of many web pages. Please explore https://archive.org/web/web.php and think of a real-world example where you would want to access an old copy of a web page, or a web page that was taken down.

Fig. 22.7 HTML nested list
structure

```
 1  <ul>
 2    <li>FRA
 3      <ul>
 4        <li>2007
 5          <ul>
 6            <li>pop: 64.02</li>
 7            <li>gdp: 2657.21</li>
 8          </ul>
 9        </li>
10        <li>2017
11          <ul>
12            <li>pop: 66.87</li>
13            <li>gdp: 2586.29</li>
14          </ul>
15        </li>
16      </ul>
17    </li>
18    <li>GBR
19      <ul>
20        <li>2007
21          <ul>
22            <li>pop: 61.32</li>
23            <li>gdp: 3084.12</li>
24          </ul>
25        </li>
26        <li>2017
27          <ul>
28            <li>pop: 66.06</li>
29            <li>gdp: 2637.87</li>
30          </ul>
31        </li>
32      </ul>
33    </li>
34    <li>USA
35      ...
36    </li>
37  </ul>
```

22.4 Given that web pages can be changed, please discuss best principles for organizations hosting data that a client may want to scrape, and discuss what a client should do before attempting to scrape a page with old code.

22.5 The first figure in the reading shows that in HTML it is possible for two or more tags to appear on the same line, meaning you have to read an HTML document carefully to discover the tree structure. A nice way to see the HTML text of a web page is to open the page in Chrome, go to "View" -> "Developer" -> "View Source." Please follow this procedure with the web page http://personal.denison. edu/~whiteda/math401spring2016.html

Find the table inside the web page and answer the following questions:

1. What tag is used to start and end the table?
2. Does this table have an HTML table header tag?
3. What XPath expression could you use to find the number of rows of the table, and what is the real-world meaning of this number?

22.6 Similar to the question above, please "View Source" on this page: http:// datasystems.denison.edu/data/ and determine the tag name associated to each of the links to data sets listed on the page. How could you determine the number of data sets listed on the page? Is it wise to find and count all elements on the page with the tag you just discovered? Explain.

22.7 The reading mentions "ethical and legal considerations" involved with web scraping. Please think up and describe a situation where, even though a web page is publicly available, it would be unethical to scrape data from the web page.

22.8 The reading mentions "ethical and legal considerations" involved with web scraping. Please describe what you would do in order to determine if it was legal to scrape a web page. Hint: you might consider using a search engine to search for "robots.txt"

22.9 When data is hosted on a publicly available web page, who should be considered the owner of that data? Are you the owner simply because you are looking at the data? This is a difficult question, but please consider it and make an argument in favor of your position. It may help to think about how you would feel if the data consisted of photos of you.

22.10 Review the Terms of Service for Yelp, and additionally search their support pages. Is it legal/permitted to web scrape Yelp pages? Provide specific link references to support your answer.

22.11 Review the Terms of Service for billboard.com and additionally search their support pages. Is it legal/permitted to web scrape, say, the "Hot 100" page? Provide specific link references to support your answer.

22.12 Suppose a web page has information you want to use, but you cannot determine if you are allowed to scrape the web page. Please look up the term "fair use" and describe what you learned and how it might affect you in a situation like this.

22.13 If you blindly web scrape, e.g., using the app SiteSucker to download all files on a given web portal, then you could easily end up possessing files that are illegal to possess (e.g., child pornography). Conversely, it is also possible to view illegal materials without downloading them. Please discuss how you believe the law should cope with these dual situations.

22.2   Web Scraping Examples

We now proceed through a graduated set of examples of web scraping. We start with a simple table, but one that occurs within a realistic web development that uses an application for generating web pages. This is, in fact, the example from Figs. 22.2 and 22.3 above. We then proceed to a table embedded in a publicly available site in Wikipedia. These two examples both use the HTML table structure to represent their data. The exercises will allow the reader to explore web scraping of a nested list structure. Our final example will explore a web page whose content is determined from a simple HTML form, and whose data is acquired through a POST operation.

Each example will describe parts of the discovery process, by which an individual determines the structure used to represent the data on the web page, and the data extraction through the application of the procedural and declarative steps using XML/XPath operations (see Chap. 16).

Before we proceed with the examples, we discuss some considerations that apply to when we retrieve HTML in a request that can make it slightly different than the (mostly similar) techniques of acquiring XML data as explored in Sect. 21.4.

22.2.1 Formulating Requests for HTML

In Sect. 22.1, we described how HTML, as produced by a web developer, might not follow strict tree formatting. Nonetheless, to accomplish web scraping, we need a tree that we can use for both discovery and for programmatic extraction of data. So, instead of using the default ElementTree parser, or specifying a custom XML parser, we define a parser that can take many forms of "broken" HTML and parse into a properly formed ElementTree.

For instance, consider the string version of a "broken" HTML:

```
bad_html = "<html><head><title>test<body><h1>header title</h3>"
```

This HTML lacks the closing tag corresponding to the `html`, `head`, `title`, and body start tags and closes a `h1` with an `h3`. But using the ElementTree constructor for an `HTMLParser` and then using that to parse the above string, converted to a file-like object, we get the following tree as a result:[3]

```
from lxml import etree
htmlparser = etree.HTMLParser()
tree = etree.parse(io.StringIO(bad_html), htmlparser)
util.print_xml(tree.getroot())
```

```
|  <html>
|     <head>
|        <title>test</title>
|     </head>
|     <body>
|        <h1>header title</h1>
|     </body>
|  </html>
```

---

[3]We again invoke our utility function for printing a tree of parsed `Element` structures from a given root. This was first introduced in Chap. 16 and is documented in the Appendix in Sect. A.1.

All unclosed end tags have been closed, and the <h1> was closed with a </h1>. The parser uses known structure and common nesting to make a best-effort construction of a tree from the HTML input.

If, in the course of web scraping, there is HTML input that is even more malformed and unable to be parsed in the above manner, the lxml module has a submodule named html that has its own, even more robust, parser. In addition, the resultant tree has some additional methods and facilities that give additional power when working with HTML. These aspects are beyond the scope of our more basic goals in this chapter, but we show the construction of a tree using the alternative parser in the html submodule:

```
import lxml.html

tree = lxml.html.parse(io.StringIO(bad_html))
util.print_xml(tree.getroot())
```

```
|  <html>
|    <head>
|      <title>test</title>
|    </head>
|    <body>
|      <h1>header title</h1>
|    </body>
|  </html>
```

There are other terrific packages out there that specifically support web scraping. For instance, we highly recommend the package "Beautiful Soup" [61]. Because of the commonality in what we are doing here with our earlier treatment of XML, we chose, in this book, to continue to use lxml.

In the examples that follow, we will use the requests module to make HTTP requests. The first set of examples will use GET requests, and the last will develop the use of a POST request in order to obtain the HTML that will act as a source for our web scraping. We will again use functions in our custom util module for building URLs and printing results.

22.2.2 Simple Table

Our first example entails an HTML table on the page "/ind2016.html" at datasystems.denison.edu. This page is the same one discussed in Sect. 22.1.1 with a straightforward structure, and little that "adorns" the presentation of the table. Our goal is to acquire the page, discover how the data set is represented, and to extract the data into a row and column data frame.

We submit a GET request, passing the url for an HTML resource, and use the raw bytes version of the body of the response to obtain a parse tree to use for subsequent operations. To display the results, we again use util.print_xml() (see Sect. A.1) and use its parameters to limit the depth of the recursion and the maximum number of children at a particular level. This way we can help our discovery and examine subtrees as we explore a given HTML.

```
url = util.buildURL("/ind2016.html",
                        "datasystems.denison.edu")
response = requests.get(url)
assert response.status_code == 200

tree1 = etree.parse(io.BytesIO(response.content), htmlparser)
root1 = tree1.getroot()
util.print_xml(root1, nlines=15, depth=4, nchild=3)
```

```
| <html xmlns='http://www.w3.org/1999/xhtml' xml:lang='en' la
|   <head>
|     <meta charset='utf-8'></meta>
|     <meta http-equiv='X-UA-Compatible' content='IE=edge'></
|     <title>ind2016 | Introduction to Data Systems</title>
|     ...
|   </head>
|   <body class='sandvox has-page-title allow-sidebar no-cust
|     <div id='page-container'>
|       <div id='page'>
|         <div id='page-top' class='no-logo has-title has-tag
|         </div>
|         <<cyfunction Comment at 0x1293d1350>>page-top</<cyf
|         <div class='clear below-page-top'></div>
|           ...
```

In this case, as shown in the prefix printout shown above, and in greater detail in Fig. 22.1, the structure of the HTML has a head and a body, and in the body, we have significant nesting of div nodes.

To be able to process a table for its data content, we must find the HTML table element in question from within the overall tree of HTML. We can start by using XPath and querying for the set of table elements within the body of the document:

```
nodeset = root1.xpath("/html/body/div//table")
print(len(nodeset))
```

```
| 1
```

We find a single table is contained in the body of the HTML, so we assign the first (and only) element in the nodeset and perform an exploratory print of the table:

```
table = nodeset[0]
util.print_xml(table, depth=3, nchild=3)
```

```
| <table class='table table-bordered table-hover table-conden
|   <thead>
```

```
|      <tr>
|        <th title='Field #1'>code</th>
|        <th title='Field #2'>country</th>
|        <th title='Field #3'>pop</th>
|        ...
|      </tr>
|    </thead>
|    <tbody>
|      <tr>
|        <td>CAN</td>
|        <td>Canada</td>
|        <td align='right'>36.26</td>
|        ...
|      </tr>
|      <tr>
|        <td>CHN</td>
|        <td>China</td>
|        <td align='right'>1378.66</td>
|        ...
|      </tr>
|      <tr>
|        <td>IND</td>
|        <td>India</td>
|        <td align='right'>1324.17</td>
|        ...
|      </tr>
|      ...
|    </tbody>
|  </table>
```

We find that the `table` has children of `thead` and `tbody`, that there is a single row in `thead` that has `th` elements for each of the column names, and that the data is contained in the `tr` elements of `tbody`. We can obtain a vector of column names via XPath:

```
column_names = table.xpath("./thead/tr/th/text()")
column_names
```

```
|  ['code', 'country', 'pop', 'gdp', 'life', 'cell']
```

We have two options for acquiring the data itself. Given the simple nature of the table, we see that the field values are uniformly represented in the text of each of the `td` nodes. So we could construct a list of the entire set of field values, and we could then use that single list to construct a list of row lists representation of the data:

```
tdlist = table.xpath("./tbody/tr/td/text()")
LoL = []
fieldcount = 0
for item in tdlist:
```

```
    if fieldcount == 0:
        row = []
    row.append(item)
    if fieldcount < 5:
        fieldcount += 1
    else:
        LoL.append(row)
        fieldcount = 0
util.print_data(LoL, nlines=20)
```

```
| [
|    [
|        "CAN",
|        "Canada",
|        "36.26",
|        "1535.77",
|        "82.3",
|        "30.75"
|    ],
|    [
|        "CHN",
|        "China",
|        "1378.66",
|        "11199.15",
|        "76.25",
|        "1364.93"
|    ],
|    [
|        "IND",
|        "India",
```

Alternatively, we could construct a dictionary of column lists (DoL) representation. Here, we can design an XPath expression that yields a list of data values for a
particular column. This would use a predicate that involves the 1-relative position of
the column (i.e., starting to count at 1 instead of 0). This strategy is then employed
in a loop to create each of the dictionary columns based on the set of column names
obtained above:

```
DoL = {}
for index, column in enumerate(column_names):
    xpath = ".//table//tr/td[{}]/text()".format(index+1)
    DoL[column] = root1.xpath(xpath)
util.print_data(DoL, nlines=20)
```

```
| {
|    "code": [
```

```
|        "CAN",
|        "CHN",
|        "IND",
|        "RUS",
|        "USA",
|        "VNM"
|      ],
|      "country": [
|        "Canada",
|        "China",
|        "India",
|        "Russia",
|        "United States",
|        "Vietnam"
|      ],
|      "pop": [
|        "36.26",
|        "1378.66",
```

Given either representation, it is straightforward to create a pandas data frame for our data. We illustrate with the DoL representation.

```
df = pd.DataFrame(DoL)
df.set_index('code', inplace=True)
df = df.astype({'pop': float, 'gdp': float,
                'life':float, 'cell':float})
df
```

```
|                  country       pop        gdp    life      cell
| code
| CAN              Canada      36.26    1535.77   82.30     30.75
| CHN               China    1378.66   11199.15   76.25   1364.93
| IND               India    1324.17    2263.79   68.56   1127.81
| RUS              Russia     144.34    1283.16   71.59    229.13
| USA       United States     323.13   18624.47   78.69    395.88
| VNM             Vietnam      94.57     205.28   76.25    120.60
```

If a web page were to have multiple tables contained within the body of the HTML, our discovery step would need to be more involved, possibly printing parts of the structure of multiple tables returned from the nodeset = root1.xpath("/html/body/div//table") and determining which one contained the desired data.

Many other variations beyond a simple table like this one are possible. We will see one such variations in our example of Sect. 22.2.3. One of the more common ways in which variations occur is when a data cell (in a td element) has additional structure. For formatting, the td might have a span child. Or part of the data in the cell might be incorporated into a link to another web page or to a relative link

Fig. 22.8 Wikipedia
population table

∧ State rankings

Rank		State	Census population		Change
Current	2010		Estimated July 1, 2019[8]	April 1, 2010[9]	%[c]
1	1	California	39,512,223	37,253,956	6.1%
2	2	Texas	28,995,881	25,145,561	15.3%
3	4	Florida	21,477,737	18,801,310	14.2%
4	3	New York	19,453,561	19,378,102	0.4%
5	6	Pennsylvania	12,801,989	12,702,379	0.8%
6	5	Illinois	12,671,821	12,830,632	–1.2%
7	7	Ohio	11,689,100	11,536,504	1.3%
8	9	Georgia	10,617,423	9,687,653	9.6%
9	10	North Carolina	10,488,084	9,535,483	10.0%
10	8	Michigan	9,986,857	9,883,640	1.0%
11	11	New Jersey	8,882,190	8,791,894	1.0%
12	12	Virginia	8,535,519	8,001,024	6.7%
13	13	Washington	7,614,893	6,724,540	13.2%
14	16	Arizona	7,278,717	6,392,017	13.9%
15	14	Massachusetts	6,892,503	6,547,629	5.3%
16	17	Tennessee	6,829,174	6,346,105	7.6%

within the current web page. In this case, the data might be part of the text of the
link element.

22.2.3   Wikipedia Table

Consider the following common scenario: a client application developer is seeking
data to complement other parts of their application. For instance, they may find
they need data on the latest estimated population of each of the states in the
United States. While the data may be available from a number of open data
sources, (census.gov, for instance), the developer finds the data they want on
Wikipedia: https://en.m.wikipedia.org/wiki/List_of_states_and_territories_of_the_
United_States_by_population. Figure 22.8 shows a screenshot of a subset of this
data from the referenced Wikipedia web page. The subset is in terms of both
dimensions of the columns in the table as well as rows in the table.

22.2.3.1   Goal

We see in the rendered page a table of state populations. Population data and ranks
are relative to the 2010 census and population estimates for 2019. In our case, we
are interested in the most recent data, even if it is an estimate, and so we want to
extract the current rank (as an integer), the string of the name of the state (we do
not care about the state flag picture), and the estimate of the population as of July 1,
2019. These are the first, third, and fourth columns in the table.

Before engaging in web scraping, a developer must look at the acceptable use
policy of a provider and also look at their policy on automatic access to their pages.
In this case, such an investigation revealed that programmatic access should go
through a defined API and that requests can be limited in number and frequency.[4]

22.2.3.2   Discovery

Following Wikipedia policy and using their simple API, where we can build a
URL for a particular desired page by constructing a resource path relative to
`/api/rest_v1/page/html/`, we obtain and parse the tree for the page giving
the population by state data set:

```
pop_page = "List_of_states_and_territories" \
           "_of_the_United_States_by_population"

resourcepath = "/api/rest_v1/page/html/{}".format(pop_page)
url = util.buildURL(resourcepath, "en.wikipedia.org")
response = requests.get(url)
assert response.status_code == 200

tree2 = etree.parse(io.BytesIO(response.content), htmlparser)
root2 = tree2.getroot()
```

In this case, we suspect there are multiple `table` HTML elements in the page,
so we use XPath to obtain the collection and delve into the tables to determine the
correct one:

```
table_list = root2.xpath("/html/body//table")
print(len(table_list))
```

```
| 5
```

If there are multiple tables, we need to discover which table carries the data we
desire. We happen to know that a characteristic of data-carrying Wikipedia tables

---

[4]Policy on automated access is often expressed in a `robots.txt` resource on the web provider.
This file is used by the web server and, in this case, enforces mechanisms to limit or turn away
requests by client applications that are known to violate the provider's policy.

is that they are sortable and carry an xml `class` attribute with a `"wikitable sortable"`[5] value.

```
xpath = "/html/body//table[@class='wikitable sortable']"
table_list = root2.xpath(xpath)
print(len(table_list))
```

> | 2

We look more closely at the first of these two tables:

```
table = table_list[0]
util.print_xml(table, depth=4, nchild=3)
```

```
| <table class='wikitable sortable' style='width:100%; text-a
|    <tbody>
|      <tr style='vertical-align: top;'>
|        <th colspan='2' style='vertical-align: middle'>Rank</
|        <th rowspan='2' style='vertical-align: middle'>State<
|        <th colspan='2' style='vertical-align: middle'>Census
|          ...
|      </tr>
|      <tr>
|        <th>Current</th>
|        <th>2010</th>
|        <th>Estimate,
|          <br></br>
|          <sup ...>
|          </sup>
|        </th>
|          ...
|      </tr>
|      <tr>
|        <td align='center'>
|          <span ...>1</span>
|        </td>
|        <td align='center'>
|          <span ...>1</span>
|        </td>
|        <td style='text-align: left;'>
|          <span ...>
|          </span>
|          <a href='/wiki/California' title='California'>Calif
|        </td>
|          ...
|      </tr>
|        ...
|    </tbody>
| </table>
```

---

[5]Discovery of structure in web scraping can often involve multiple mechanisms and interactive querying of the tree. In this instance, we, in fact, used a combination of printing the various table trees, looking and inferring from the rendered page, and using the developer tools provided by most web browsers to help narrow this search.

We discover that this first table is indeed the table we are looking for. The print of part of the tree above shows us that there is a tbody child of table present, but, in this case, no thead. Looking more closely at the first two tr child nodes of tbody, we see that they are populated with th elements, and these correspond to the two rows of the header of the table seen in Fig. 22.8. The data-carrying rows begin with the third tr child of the tbody. To continue our discovery, we use XPath relative to the table to obtain the third row entry and examine its structure.

```
rowlist = table.xpath(".//tr[3]")
datarow = rowlist[0]
util.print_xml(datarow, depth=3, nchild=5)
```

```
|  <tr>
|    <td align='center'>
|      <span ...>1</span>
|    </td>
|    <td align='center'>
|      <span ...>1</span>
|    </td>
|    <td style='text-align: left;'>
|      <span ...>
|        <noscript>
|        </noscript>
|        <img width='23' height='15' class='thumbborder image-
|      </span>
|      <a href='/wiki/California' title='California'>Californi
|    </td>
|    <td>39,512,223</td>
|    <td>37,253,956</td>
|      ...
|  </tr>
```

The data for this row of the table is contained in the td elements, and this first row corresponds to California, whose first and second columns have value 1, and the third field has a picture and a link whose label is California.

We observe:

1. The current state rank is in first td in row and is the text of a span node under the td Element. It is not always the case that the first visible table entry corresponds to the first td node in a row. Some tables can use additional td elements in the set of rows, used for spacing, borders, and other rendering results.
2. The state rank at the last census in 2010 is in the second td in the row; we will disregard this field based on our goals of wanting the most recent estimate.
3. The third td contains the state information, and we will explore that further below.
4. The fourth td contains the estimated population in 2019, and here the value is directly in the text of the td element.

So we will use positioning to get the columns (as the relative td within the row) that we are interested in.

Focusing on the data cell where we find the state information:

```
util.print_xml(datarow[2], depth=3, nchild=4)
```

```
| <td style='text-align: left;'>
|   <span ...>
|     <noscript>
|       <img alt='' src='//upload.wikimedia.org/wikipedia/com
|     </noscript>
|     <img width='23' height='15' class='thumbborder image-la
|   </span>
|   <a href='/wiki/California' title='California'>California<
| </td>
```

We observe that the name of the state is embedded in a hyperlink, given in the node with tag: a. This node is beneath the td; the name of the state is in the text property of the a node.

22.2.3.3   Data Extraction

Understanding the tree structure of the table, we can now acquire the data for columns 1, 3, and 4 of the table using XPath. We know that the data-carrying rows begin after row position 2, and because the table has a row for the District of Columbia, we want to extract data from 51 rows. We do not want to go further, as these rows contain information on territories and have aggregate data.

Extracting column 1 from rows 3 through 53, and finding the data in the text property of a span underneath the appropriate td, we show the first four entries of the 51 extracted.

```
xpath = ".//tr[position() > 2 and position() < 54]"\
        "/td[1]/span/text()"
rank_column = table.xpath(xpath)
rank_column = [int(s) for s in rank_column]
rank_column[:4]
```

```
| [1, 2, 3, 4]
```

Extracting the state strings from column 3 from rows 3 through 53, by traversing into the a node and extracting the text property, we again show the first four entries of the 51.

```
xpath = ".//tr[position() > 2 and position() < 54]/td[3]"\
        "//a/text()"
state_column = table.xpath(xpath)
state_column[:4]
```

```
| ['California', 'Texas', 'Florida', 'New York']
```

Finally, for the population, we need to do a little work to convert a comma-separated digit string into an integer, so we define a `lambda` function to perform the conversion and then, after we use XPath to extract a vector of strings, use a list comprehension to apply the conversion:

```
convert_pop = lambda p: int(p.replace(",", ""))
xpath = ".//tr[position() > 2 and position() < 54]/td[4]/text()"
pop_column = table.xpath(xpath)
pop_column = [convert_pop(p) for p in pop_column]
pop_column[:4]
```

```
| [39512223, 28995881, 21477737, 19453561]
```

Now that we have the three desired columns, and we can define a dictionary of column list representation and construct the data frame:

```
DoL = {'rank': rank_column,
       'state': state_column,
       'population': pop_column}
df = pd.DataFrame(DoL)
df.set_index('rank', inplace=True)
df.head(10)
```

```
|                        state   population
| rank
| 1              California     39512223
| 2                   Texas     28995881
| 3                 Florida     21477737
| 4                New York     19453561
| 5            Pennsylvania     12801989
| 6                Illinois     12671821
| 7                    Ohio     11689100
| 8                 Georgia     10617423
| 9          North Carolina     10488084
| 10               Michigan      9986857
```

22.2.4   POST to Submit a Form

22.2.4.1   Goal

Consider the website in Fig. 22.9, providing weekly gasoline prices in California for a particular year. This website is available at URL https://ww2.energy.ca.gov/almanac/transportation_data/gasoline/margins/index_cms.php. URLs whose final element of the resource path ends in `.php` are known as PHP files, and these allow a website to create dynamic content. This dynamic content is often the result of a user in a web browser interacting with user interface elements known as *forms*,

Home > Data and Reports > Energy Almanac > Transportation Energy > **Estimated Gasoline Price Breakdown and Margins Details**

Estimated 2020 Gasoline Price Breakdown and Margins Details

Aug 31

	Branded	Unbranded
Distribution Costs, Marketing Costs and Profits	$0.410	$0.550
Crude Oil Costs	$1.040	$1.040
Refinery Cost and Profit	$0.880	$0.730
State Underground Storage Tank Fee	$0.020	$0.020
State and Local Tax	$0.068	$0.068
State Excise Tax	$0.505	$0.505
Federal Excise Tax	$0.184	$0.184
Retail Prices	$3.100	$3.100

Fig. 22.9 CA gas prices web page

where they can enter information, select items from dropdown boxes, and otherwise associate the "answers" to a form with defined fields that are part of the form. When this process is complete, they click an action button that *submits* the form. In HTTP, a form submission uses a POST, where the *body* of the request message encodes the set of form field names and their mapping to the user's selected values. The operation of such a POST was illustrated in Sect. 20.3.2.

By using this mechanism, a web scraping client can design and construct a request and then receive the resultant HTML. With the result, we can then process the constituent `table`, `ul`, or `ol` data and extract it into a usable form. This section will demonstrate this technique for the CA Gas Prices web page.

From a displayed web page perspective, at the bottom of the same page, Figs. 22.10 and 22.11 show the simple forms interface provided on this web page.

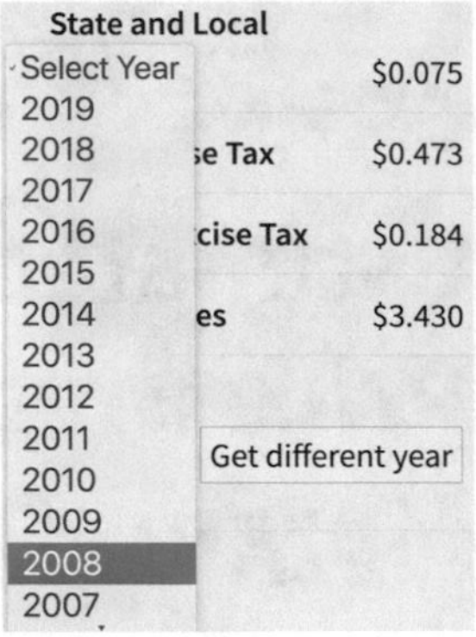

Fig. 22.10 CA gas prices
year selection

State and Local		
Select Year		$0.075
2019		
2018	ie Tax	$0.473
2017		
2016	cise Tax	$0.184
2015		
2014	es	$3.430
2013		
2012		
2011		Get different year
2010		
2009		
2008		
2007		

Fig. 22.11 CA gas prices
form

We see both the dropdown, from which the user can select a particular year, and the selected value and button used to submit the form.

22.2.4.2 Discovery

Given a web page with a form like in our example, our *discovery* must determine the specifics about the form and how we could construct a client-based POST that is able to make a request so that the target web server would respond and so that the client can acquire the HTML-based data.

In this particular example, through examination of the HTML, we can discover the specification of the form on the web page:

```
<form action='index_cms.php' method='post'
      style='margin-left:10px;'>
  <label for='year'>
    <select name='year' id='year'>
        <option value='2020'>Select Year</option>
        <option value='2020'>2020</option>
        <option value='2019'>2019</option>
        <option value='2018'>2018</option>
        <option value='2017'>2017</option>
        ...
        <option value='1999'>1999</option>
    </select>
  </label>
  <input name='newYear' type='submit'
         value='Get different year' />
```

The `<select>` tag defines the dropdown, with `<option>` entries for each of the items displayed. The name and id attributes both have value `year`, and this will be the *field name* used for one of the entries in the form, and the user-selected value is determined by the `value` attribute for a selection.

The `<input>` tag defines the submission button, where the `newYear` will be the second *field name* of the form, and this entry will have the value given by the `value` attribute, so the form entry `newYear` will be associated with `Get different year`.

The `<form>` tag defines the overall operation of the form, using `action` and `method` to indicate that, on a submission, the result should be a POST request to the resource path given by `action`, which here is `index_cms.php`. In this case, this means that the POST of the form is back to the same web page as the original GET.

A discovery process often uses examination of the HTML and couples that with using a web browser and associated *developer tools*, looking at network message interaction, to interact with the form and experimentally determine what happens when a user selects a year and clicks the `Get different year` button.

Summarizing our discovery:

1. The page itself, in HTML, has our desired data by week, and each week is a separate `table` object, with a `tr` (row) for each of the variables and associated values for that week.
2. The form consists of two entries:

   - `year` field maps to the four digit string of the desired year and
   - `newYear`field maps to the constant value `Get different year`

3. The HTTP method is POST, which means that the form entries should be in the *body* of the POST as URL-encoded *field=value* for each field, separated, per URL encoding, with an ampersand (`&`).
4. The resource path/URI of the POST is the same resource path as the original.

22.2.4.3   Request and Data Extraction

In contrast to most earlier examples, we need to change two things in using the `requests` module to make this request:

1. We must get a POST request instead of a GET request.
2. The request must include a body that consists of key-value pairs.

For (1), the `requests` module has a `post` top level function. For (2), we construct a *dictionary* with the desired mappings. We pass that to the `post()` using named parameter `data`. The `requests` module is very flexible in how it interprets an argument provided through `data`. If it is a string, it simply puts the encoded bytes of the string in the body. If it is a dictionary, like in this case, it interprets it and generates a URL-encoded version, as we will see below. Suppose we want to get CA gas price data for the year 2001:

```
page = "/almanac/transportation_data/" \
       "gasoline/margins/index_cms.php"
url = util.buildURL(page, "ww2.energy.ca.gov")

year = 2001

payload = {'year': year, 'newYear': 'Get different year'}
response = requests.post(url, data=payload)
assert response.status_code == 200

request = response.request
print("POST body:", request.body)
```

> | POST body: year=2001&newYear=Get+different+year

The print() helps show the *result* of using the payload dictionary and specifying it as the body of the request by using the data= named parameter to the post(). The requests module translated the dictionary into a URL-encoded set of *field=value* entries separated by &, and with embedded spaces translated into the + character.

```
util.print_headers(request.headers)
```

> | {
> | "User-Agent": "python-requests/2.22.0",
> | "Accept-Encoding": "gzip, deflate",
> | "Accept": "*/*",
> | "Connection": "keep-alive",
> | "Content-Length": "36",
> | "Content-Type": "application/x-www-form-urlencoded"
> | }

Note the specification in the *request headers*, through Content-Type, indicates that the body of the post is a URL-encoded form.

Since our request was successful, we can take the bytes of the response and parse the returned HTML into a tree:

```
cagas_tree = etree.parse(io.BytesIO(response.content),
                         htmlparser)
cagas_root = cagas_tree.getroot()
print(cagas_root.tag)
```

> | html

The tables that we desire all use the table HTML structure and are positioned as direct children of div elements, where the div has a class attribute of "contnr".

```
# Get a list of the weekly tables
table_list = cagas_root.xpath("//div[@class='contnr']/table")
print(len(table_list))
```

```
| 53
```

```
util.print_xml(table_list[0], depth=3, nchild=4)
```

```
| <table>
|   <caption>
|     <h2>Dec 31</h2>
|   </caption>
|   <tr>
|     <td></td>
|     <th scope='col'>Branded</th>
|     <th scope='col'>Unbranded</th>
|   </tr>
|   <tr>
|     <th scope='row' class='tWidth'>Distribution Costs, Mark
|     <td class='numbers'>
|       <span ...>-$0.07</span>
|     </td>
|     <td class='numbers'>
|       <span ...>-$0.04</span>
|     </td>
|   </tr>
|   <tr>
|     <th scope='row'>Crude Oil Costs</th>
|     <td class='numbers'>$0.41</td>
|     <td class='numbers'>$0.41</td>
|   </tr>
|     ...
| </table>
```

Given that each table represents a single week and that the rows in the table represent variables, then each table will give us a single row for a table representing the data of the page. With an eye toward collecting a list of dictionaries for construction of the table, we will develop processing of one table to result in one (row) dictionary.

We can see from the print of the tree that the first piece of data needed, the date, is in a caption child of the table. Let us postulate data columns:

```
['distrib_cost', 'crude_cost', 'refine_cost', 'storage',
 'state_local_tax', 'state_excise_tax', 'fed_excise_tax',
 'retail_price']
```

Assume we just want the branded data.

```
data_cols = ['distrib_cost', 'crude_cost', 'refine_cost',
             'storage', 'state_local_tax', 'state_excise_tax',
             'fed_excise_tax', 'retail_price']
table = table_list[0]
date = table[0][0].text
datastrings = table.xpath("./tr[position()>1]/td[1]/text()")
```

```
datalist = [float(s[1:]) for s in datastrings]
datalist
```

```
 |  [0.41, 0.31, 0.0, 0.08, 0.18, 0.18, 1.1]
```

```
D = {key:value for key, value in zip(data_cols, datalist)}
D['date'] = date
util.print_data(D)
```

```
 |  {
 |      "distrib_cost": 0.41,
 |      "crude_cost": 0.31,
 |      "refine_cost": 0.0,
 |      "storage": 0.08,
 |      "state_local_tax": 0.18,
 |      "state_excise_tax": 0.18,
 |      "fed_excise_tax": 1.1,
 |      "date": "Dec 31"
 |  }
```

In the interest of good functional abstraction, and because we need to repeat the
processing for each of the weekly tables in the web page, we define a function to
perform the work for a single table, and which returns a dictionary representing a
single row of the desired table, with column fields mapping to values obtained as
shown above.

```
def processTable(table):
    """Given an HTML table structure, extract the data values
    for a single row of the CA Gas data set, returning a
    dictionary mapping column fields to extracted values.
    """
    data_cols = ['distrib_cost', 'crude_cost', 'refine_cost',
                 'storage', 'state_local_tax',
                 'state_excise_tax', 'fed_excise_tax',
                 'retail_price']
    date = table[0][0].text
    strings = table.xpath("./tr[position()>1]/td[1]/text()")
    datalist = [float(s[1:]) for s in strings]
    D = {key:value for key, value in zip(data_cols, datalist)}
    D['date'] = date
    return D
```

Now it is just a matter of applying our function to each of the tables. Since we
desire a list result, using a list comprehension makes this easy.

```
LoD = [processTable(table) for table in table_list]
```

Finally, we construct the data frame and set its index.

```
df = pd.DataFrame(LoD)
df.set_index('date', inplace=True)
df.iloc[:6,:4].head(10)
```

```
|             distrib_cost  crude_cost  refine_cost  storage
| date
| Dec 31            0.41        0.31         0.00       0.08
| Dec 24            0.00        0.44         0.23       0.00
| Dec 17            0.08        0.39         0.22       0.00
| Dec 10            0.14        0.38         0.23       0.00
| Dec 03            0.14        0.42         0.23       0.00
| Nov 26            0.17        0.39         0.27       0.00
```

22.2.5 Reading Questions

22.14 Please think up a reason why there might be a lot of broken HTML on the web (i.e., malformed from an XML perspective).

22.15 Please justify the value of the `print_xml()` function to a working data scientist. What would life be like if we did not have this function?

22.16 The reading says "Before engaging in web scraping, a developer must look at the acceptable use policy of a provider and also look at their policy on automatic access to their pages." Please find an example of an "acceptable use policy" and "policy on automatic access" and give your example here, including the link to it. In general, how can you find this kind of information?

22.17 Please study the California gasoline web page used in the POST example and explain what is meant by "Branded" versus "Unbranded" in real-world terms: https://ww2.energy.ca.gov/almanac/transportation_data/gasoline/margins/index_cms.php

22.18 In the first XPath code shown in the reading, explain why we use

```
nodeset = root1.xpath("/html/body/div//table")
```

instead of

```
nodeset = root1.xpath("/html/body/div/table")
```

22.19 With reference to the content printed by `print_xml`, please explain carefully why the XPath expression `./thead/tr/th/text()` yields a list of column names.

22.20 In the `ind2016` example, consider the line

```
tdlist = table.xpath("./tbody/tr/td/text()")
```

Please describe `tdlist` carefully after this line is executed. For instance, how big is this list? What is the *type* of the items in `tdlist`? How does it relate to the table?

22.21 In the code for the LoL solution of extracting `ind2016` data, what is the purpose of `fieldcount`?

22.22 In the DoL approach, what is the point of the `format` string in the following snippet of code?

```
for index, column in enumerate(column_names):
    xpath = ".//table//tr/td[{}]/text()".format(index+1)
```

22.23 What is the point of using the Wikipedia API? For instance, what is the difference between the sources of the two following pages?

- https://en.m.wikipedia.org/wiki/List_of_states_and_territories_of_the_United_ States_by_population
- https://en.wikipedia.org/api/rest_v1/page/html/List_of_states_and_territories_ of_the_United_States_by_population

22.24 The reading invoked `state_column = table.xpath(xs)` for the XPath expression below

xs = ".//tr[position() > 2 and position() < 54]/td[3]//a/text()"

Can you think of a way to rewrite this to avoid the use of the `position` function? You might consider consulting the reading to make sure you understand what this expression is trying to accomplish.

22.25 The reading shows two ways of building a `pandas` data frame from a web page: using either a LoL or DoL. Please write a sentence describing how each approach works, then answer: which do you find more intuitive and why?

22.26 In what way, specifically, does the XPath expression in the reading `"./tr[position()>1]/td[1]/text()"` extract the branded rather than unbranded data?

22.27 Consider the XPath expression used to extract the list of states. Suppose you wanted to extract the list of links (one per state), e.g., for use in a web crawling program. What XPath expression would you use?

22.28 Instead of a list comprehension applying the `lambda` function to create `pop_column`, could we use `map`? Explain.

22.29 When describing the POST, the reading discusses the resource path given by `action`. Can you think of an example where a POST would want to send data to a different resource path than the page the user is currently on?

22.30 Explain the use of the XPath expression

```
"//div[@class='contnr']/table"
```

by referencing the HTML structure of the web page in question.

22.2.6 Exercises

22.31 Write a function

```
getHTMLroot(resource, location, protocol="https", params={})
```

that performs an HTTP GET for the specified `resource` at `location`, using `protocol`, which defaults to `https`. If `params` is specified, these should be used as a dictionary for query parameters for the request. If the GET is successful, the function should verify that the result is HTML. If that is true, parse the HTML and return the root of the tree. If the resource is not found, or if the resource is not HTML, or the resource could not parse, return None.

22.32 Write a function

```
getLinks(resource, location, protocol="https")
```

that retrieves the HTML tree from `resource` at host `location`, using the specified protocol, and then finds all the *external hyperlinks* referenced in the document. Return a list of strings for the URLs. Recall that a hyperlink is referenced with the tag: a, and the `href` attribute within that tag contains the link itself, as opposed to the displayed text. Since many links in a document can be internal, we want to restrict to those that contain a URL, not a URI. Bonus points if the algorithm only returns the *unique* external links.

22.33 Consider the following web page: https://datasystems.denison.edu/databases/index.html. Using a combination of XPath and `util.print_xml()` of the tree or subtrees, answer the following questions:

- What is the *first* ordered or unordered list found in the tree? If you wanted *that specific list*, and wanted to protect against another list being added as the first one, what steps would you take to get the desired list?
- What would the XPath be to get the `div` whose `class` attribute is `"RichTextElement"` that is a *descendent* of the `div` whose `id` attribute is `"main-content"`
- Given the (single) node that results from answering the previous question, write a single XPath expression that would get the set of *strings* naming the available databases.
- Repeat the above, but use *procedural steps* to accumulate the list of databases.
- In English, describe what you learned about the HTML structures involved in this collection of information. Is this the same as one of the structural options considered in the chapter?

22.34 Consider the page: https://datasystems.denison.edu/ind2016_list.html. On that page is a list-structured data that consists, at the outer level, of an ordered list, and at the inner level, an unordered list of indicator values. It also has some adornment in font face, like bolding and italics. Using the techniques of this chapter, write the code to scrape this page and create a pandas data frame containing rows for each of the five countries, and columns for code, name, gdp, pop, and life.

22.35 Consider the page: http://datasystems.denison.edu/ind0.html, which is the page described in the book, which contains triple-nested lists. Note also that all the lists are unordered. Write code to web scrape and create a pandas data frame with three rows, one each for FRA, GBR, and USA, and has a code and four data columns, for each of pop and gdp for years 2007 and 2017. The best solution would use a two-level column index.

The last two exercises ask the reader to scrape tables on Wikipedia. We would ask that you adhere to the acceptable use policy and use the Wikipedia API to access the pages. We also caution that websites not under our control are constantly changing, and while, at the time of this writing, these exercises were feasible, changes could make adaptation necessary.

22.36 At the Wikipedia, page called

```
"List_of_novels_considered_the_greatest"
```

is a table of novels considered great by one or more experts. Using the resource path prefixed with "/api/rest_v1/page/html/", obtain the HTML tree from that page and scrape the page, building a pandas data frame of the results. You are to design the table as you see fit for the columns to be included. Make sure your solution is robust to the case of additional novels being added to this list.

22.37 At the Wikipedia, page called

```
"List_of_Cash_Box_Top_100_number-one_singles_of_1960"
```

is a table containing, by week, the top songs from the year 1960. Using the resource path prefixed with "/api/rest_v1/page/html/", obtain the HTML tree from that page and scrape the page, building a pandas data frame of the results. You are to design the table as you see fit for the columns to be included. This exercise is slightly more difficult than the previous one, because the table has entries that "span" more than one row, when the same song is top for multiple weeks.

Chapter 23
RESTful Application Programming Interfaces

Chapter Goals

Upon completion of this chapter, you should understand the following:

- The foundational ideas of API providers responding to data requests.
- HTTP as the application level protocol for API providers, where a client and a provider are constrained to work within the confines of the protocol.
- REST as a set of principles that broadly govern API providers but allow significant latitude in design.

Upon completion of this chapter, you should be able to do the following:

- Given a provider, be able to issue HTTP requests, both GET and POST, conformant with provider expectations, including conveying parameters as part of the HTTP request in

  - the path part of the resource path,
  - the query string part of the resource path,
  - the headers of the request, and
  - the body of a POST request.

- Process the result, from reacting appropriately to different status codes to parsing and interpreting the data in the body of the result.

© Springer Nature Switzerland AG 2020
T. Bressoud, D. White, *Introduction to Data Systems*,
https://doi.org/10.1007/978-3-030-54371-6_23

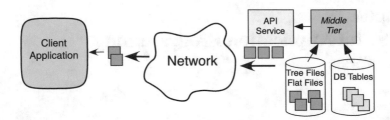

Fig. 23.1 API service provider data source

In Chaps. 1 and 18, we identified a class of providers of data over the network identified as *API service* providers. See Fig. 23.1. These providers desire to make their data available, in programmatic fashion, to data system clients (aka applications, apps). This data source comprises the final data source explored in this book.

23.1   Motivation and Background

Chapter 1 first discussed the characteristics and motivation for the existence of API-based providers. These are organizations and companies that are interested in providing data and to allow the access to that data happen through a client application. The reasons for API providers to exist are diverse and depend, in part, on what the data actually is, and who, if anyone, owns the data (in an intellectual property sense). Some of the more common motivations include

- *Open Data*: Often data is collected directly and indirectly by local, regional, and national governments or other organization, and the data is provided openly as a public service.
- *Service for Customer*: The customer users of a company's products and service expect to be able to use applications running in browsers and on smart devices and to be able to interact with the provider.
- *Service for Profit*: Some providers are purveyors of data, which they gather and curate and then, through charging or ad revenue, establish a business model for their data curation work.
- *Innovation*: For many companies, having client applications that can access data, often on behalf of users, and to innovate with those applications can be a motivation in of itself. The more users tie themselves to providers through the applications they use, the more secure the user–company relationship.

As we consider how API providers, as discussed in this chapter, work, it is helpful to review the three principles involved in data systems, listed here and depicted in Fig. 23.2.

- The *Data System Provider* is the entity providing the interface available over the network and generally is the keeper of the data, hosting it in back-end systems.

Fig. 23.2 Data system principals

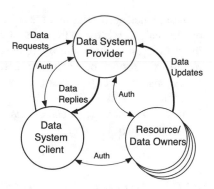

- The *Data System Client*, also known as the *application* or just *App*, is the entity that issues the requests for data and processes the responses. The client app may perform its function to allow for analysis and combination of data from multiple data sources but could also be running at the behest of a human user.
- The *Resource/Data Owners* are the entities with intellectual ownership or stake in the data hosted by the provider. These are sometimes *users* or *customers* of the provider, and the data collected can often be a product of that company–customer relationship.

Chapter 22 on web scraping addressed the acquisition of data over the network from a web server provider. Because of certain commonalities, it is important to distinguish the topic of this chapter from that earlier treatment. Important differences include

1. *Intention*: In web scraping, web pages are designed for presentation for viewing, primarily in browsers, and data extraction is often a repurposing of the data contained therein. In API services, the provider is intending to provide the data itself.
2. *Format*: In web scraping, the format is HTML; while it has a `table` structure and list generation capable of defining trees of information (like `ol` and `ul`), these were not intended for data exchange beyond their capacity for being rendered and seen by users through web browsers. In API services, because of the intention for use by client applications, the format is consistent with the various data models covered in Part II of this book. These formats include CSV, JSON, and XML.
3. *Abstraction*: In API services, the desire is to present a usable interface by which clients can acquire data, but to allow the provider the implementation freedom to structure, store, and query the underlying data in whatever back-end system design is most suitable. In web scraping, the back end is determined by the capabilities of the web server employed by the provider. This is why, in Fig. 23.1, we depict a network-facing component that can service the requests coming over the network, but we abstractly show a *middle tier*, which is responsible for the dynamic query of data from the provider's choice of back end.

Also, when web scraping, the acquired HTML is often static, already consisting of a textual HTML file in the file system of the web server. In API service providers, the data is often dynamic, being generated on the fly based on a request. In this regard, API service providers are more akin to the relational database systems that, in response to an SQL request, dynamically access the back end and respond with the requested data.

Whenever there is a choice, an application should use an API rather than performing web scraping.

23.1.1 General API Characteristics

Given the motivation for providing data over the network through some interface, there are some common requirements and properties of such an API.

- It should follow the basic principles of client–server, with a request followed by a response.
- From a client application standpoint, it should work like a function/method call, where the invocation is passed arguments corresponding to defined parameters, and the function/method, as translated into a client request, returns a data result.
- Following on one of the strengths of the Internet, servers should not be strongly bound to clients, so that, as clients come and go, the servers are not required to maintain state information about the clients and their prior sequence of operation.
- The API should leverage existing application level protocols, so that both servers and clients can be developed independently and do not have to incorporate a novel protocol.

23.1.2 Principles of REpresentational State Transfer (REST)

REST is an acronym for REpresentational State Transfer and is primarily about a set of principles guiding requests for data (state) and accompanying transfer. REST was first proposed by Roy Fielding in a PhD dissertation at UC Irvine [10, 11].

The guiding principles include:

1. *Uniform Interface*—we want servers to "do things" the same way, so that libraries, like our `requests` module, can provide tools to facilitate building requests.

   This principle argues for separation of specifying "what we want" from the representation/format of the data.

2. *Stateless*—most fundamentally, this means that everything needed to successfully execute a request is "packaged together" in a unit, so the functionality can

be provided independent of any prior interaction. Here, state refers to application state, so a server, for instance, should not need to have recorded prior steps related to a client.

3. *Cacheable*—to allow for optimization, and to satisfy requests "more locally," information should be able to be tagged to indicate whether or not it is cacheable.
4. *Client–Server*—Separation of client concerns from the server means that the server can move, or an intermediary can provide information, and the server can implement and/or change the implementation of the service and not affect correct operation of the client.

In practice, many providers have adopted REST principles in defining their API and employ *HTTP* as the application layer protocol that enables many of these principles. While other choices for an application level protocol and other API designs exist, this chapter focuses on what are called *RESTful* services, which all use HTTP.

We have seen in Chaps. 20 and 22 the definition of HTTP and how to use HTTP in a client–server architecture to make requests and get replies and have seen in Chap. 21 how to extract results in our various formats to build in-memory data structures out of the structured data.

What remains is understanding the mechanisms used and specified, in a more formal way, by API providers so that we can understand what data is offered by providers and so that we can formulate the requests, using HTTP, to get the desired data. In relational database systems, our request was formulated as an SQL query that specified the tables, the columns, the filtering of rows, and so forth.

Each provider of an API determines how its own API is going to work, and the specifics, within the overall context of REST principles and by requests sent using HTTP, of how a request must be constructed to be interpreted appropriately by the provider and data returned. There is no standard that is the equivalent of SQL. However, for *all* providers, the client must convey the *parameters* of the request, and, using HTTP, there are a limited number of mechanisms for doing this. Understanding the building blocks of clients specifying request parameters through HTTP is the primary goal of this chapter and the subject of Sect. 23.2. Then, in Sect. 23.3, we bring the building blocks together with a more complete case study.

23.1.3   Reading Questions

23.1 In real-world terms, what is the value of a middle tier for dynamic query of data over a network?

23.2 The reading says "Whenever there is a choice, an application should use an API rather than performing web scraping." How can you find out if there is an API to use?

23.3 Give reasons why a data-providing API should work like a function/method call from the client application standpoint?

23.4 What are the practical consequences of the "general API characteristic" that servers should not be bound strongly to clients?

23.5 Following on the previous question, are there any consequences for your particular computer of the design decision about servers vs. clients?

23.6 Please give a justification in real-world terms of each of the four "guiding principles" of REST. That is, describe, for each principle, what the world would be like if the principle were not followed, and explain why this matters.

23.2   HTTP for REST API Requests

RESTful APIs use the existing application level protocol of HTTP for making their requests and use the HTTP response message for carrying the outcome/status of the request and, if successful, the data resulting from the request. By using an existing protocol, the API must design and decide *how to use* HTTP, as defined, as the vehicle for making requests.

For the examples in this section, we will employ the APIs provided by GitHub [15] and by The Movie Database [64] to demonstrate the various concepts. The GitHub API provides access to public information and data, such as information about organizations and public repositories and events, as well as private information accessed through authentication and authorization on individual users and their private repositories. The Movie Database (TMDB) API allows for searching and discovering information about movies and TV shows, acquiring data about specific shows, actors and actresses, and rating information, and it also has APIs for managing user information and placing and updating ratings for movies and TV shows. The TMDB APIs are not available for anonymous access but require requests to identify themselves as belonging to a particular application that has been previously registered. As a part of their terms of service, TMDB requires attribution for the use of their API and for any images, and we do so here:

For educational purposes and illustration of REST, this book uses the TMDB ⬛⬛⬛ API but is not endorsed or certified by TMDB.

The GitHub API allows a combination of both anonymous and application-/user-specific types of requests to be made. For a request to GitHub, we need to be able to specify things like the organization, the owner, a repository, the content type, etc. For a request to TMBD, we need to be able to specify, for instance, a search term, an application identifier, a movie, or TV show identifier or to convey information like a value for updating a rating. We categorize all such pieces of information that are conveyed from the client to a provider as the *parameters* of the request.

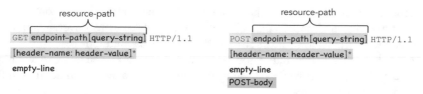

Fig. 23.3 Request parameters in HTTP

The specifics of how to do this have variations as numerous as the number of providers with APIs. But *all* must somehow use HTTP as the basis for the request. In Fig. 23.3, we show the generic syntax of both GET and POST requests.[1] We use shading superimposed on this syntactic representation to indicate *where* parameters of a request may be specified. As used previously, the square brackets ([and]) and the superscript asterisk (∗) are used to indicate optional elements in the syntax and one or more instances of an element, respectively.

- First, highlighted in light green, the HTTP method itself is conveying information about the request, and APIs use GET for a request to read information and a POST to send additional information to a provider and to request an update of some data maintained by the provider.
- Second, highlighted in turquoise, we show an *endpoint-path*, which is part of the *resource-path* specified in the request. APIs use different base paths to distinguish different endpoints and also can encode request parameter items in the elements (location steps) of the path itself.
- Third, highlighted in purple, a *resource-path* can optionally contain a *query-string*, whose start is signified by a ? character, and contains one or more *query-parameters* consisting of key-value pairs.
- Fourth, highlighted in blue, each request contains a set of header lines. These header lines can also be used in an API to define request parameters.
- Finally, in a POST request, there exists a *POST body*, highlighted in orange. This can contain request parameters consisting of key-value pairs (like we saw in Sect. 22.2.4) or, more generally, can consist of a JSON encoding string containing request parameters.

The following subsections discuss how the desired operation and request parameters may be realized with these different elements of an HTTP request, using examples from the GitHub and TMDB APIs. The reader is encouraged to look directly at the API documentation as published by the provider for a number of reasons:

- The specification as given by the provider is the final word on how request parameters are conveyed, the expectations of the provider, required versus

---

[1]For clarity, this syntax depiction omits showing explicitly a representation for the CRLF that terminates each of the start line, the header lines, and the empty line.

optional parameters, any authentication and authorization issues, success and error returns, and the format of any returned data. Any coverage in these examples is meant to be illustrative, and not as documentation.

- APIs change over time, and while these examples all operated as expected at the time of this writing, changes made by providers may change the outcomes of the current set of examples.
- Different providers can use terminology that may be different from each other and could also be different from that used in this book. They may also present information differently. So, it is important to build experience looking at the documentation of different providers to be able to translate and relate to the concepts presented here.

In our examples, we will use a combination of screenshots of the provider's API documentation as well as a `curl` command-line incantation that can issue such a request. When a request is a GET and has no parameters beyond the resource path, we can also issue it through a web browser by entering the corresponding URL in the address line, and the reader is encouraged to deepen his or her understanding by actively executing examples where possible.

In the example code presented below, we assume the usual set of `imports`, and the existence of helper functions documented in Appendix A, such as `util.buildURL()`, `util.print_data()`, and `util.read_creds()`.

23.2.1   *Endpoints*

Generally, API providers, using the resource tree structure of the resource path URI entailed in an HTTP request, define a set of *endpoints* to be used for those requests. Typically, a single endpoint may be organized as having multiple subordinate paths that define different *operations* on that endpoint. One could draw the analogy of an endpoint representing an object *class* in an object-oriented programming language, and the set of subordinate paths as representing the *methods* that may be invoked on the object.

Our first two examples use the GitHub API. In this API, the provider defines, in addition to a *root endpoint* that lies under a resource path element defining the *version* of the API, specific endpoints for interacting with:

- organizations:
  - to list repositories of a particular organization and
  - to get detailed information about an organization.
- repositories:
  - to list public repositories.

- users/owners:
    - to retrieve basic information about the user,
    - to find set of users who are followers of a user,
    - to find set of users this user is following, and
    - to retrieve information about a particular repository of this user,

as well as many other endpoints and subordinate operations.

For the GitHub API, the root endpoint is given, through developer documentation, as https://api.github.com. We present a number of examples illustrating the most common ways of interacting with REST APIs. For each example, we present a command-line method, using `curl` to make the HTTP request, followed by a Python `requests` module programmatic example to accomplish the same end result.

For endpoints without parameters, using an API is no different from requesting a web page or other specific resource in the resource tree.

23.2.1.1   Root Endpoint

GitHub provides a "meta" endpoint at the root of its API resource tree. When accessed, it returns a JSON-formatted result with a key-value dictionary of the top level endpoints supported in the GitHub API.

Curl Incantation

```
curl https://api.github.com/
```

Requests Incantation

```
resourcepath = "/"
location = "api.github.com"
url = util.buildURL(resourcepath, location)

response = requests.get(url)
assert response.status_code == 200

endpoints = response.json()
util.print_data(endpoints, nlines=10)
 | {
 |    "current_user_url": "https://api.github.com/user",
 |    "current_user_authorizations_html_url": "https://github.c
 |    "authorizations_url": "https://api.github.com/authorizati
 |    "code_search_url": "https://api.github.com/search/code?q=
 |    "commit_search_url": "https://api.github.com/search/commi
 |    "emails_url": "https://api.github.com/user/emails",
 |    "emojis_url": "https://api.github.com/emojis",
 |    "events_url": "https://api.github.com/events",
 |    "feeds_url": "https://api.github.com/feeds",
```

We can see from the results that could allow programs, through programmatic means, to discover other endpoints in the API.

23.2.1.2   Non-Root Endpoint

/events is a GitHub API endpoint that, without further parameters, gets the first "page" of the set of current GitHub events: https://developer.github.com/v3/activity/events/. There is a maximum limit of 30 events in a page, so to request subsequent "batches" of 30, a client would have to specify a parameter whose argument gave the desired page.

The JSON result, on success, contains a list of dictionaries, with a dictionary representing each event.

Curl Incantation

```
curl https://api.github.com/events
```

Requests Incantation

```
resourcepath = "/events"
url = util.buildURL(resourcepath, "api.github.com")

response = requests.get(url)
assert response.status_code == 200

eventlist = response.json()
print("Number of events received:", len(eventlist))
```

```
| Number of events received: 30
```

```
util.print_data(eventlist, depth=2, nchild=3, nlines=25)
```

```
| [
|   {
|     "id": "12779420149"
|     "type": "DeleteEvent"
|     "actor":
|     {
|       "id": 270536
|       "login": "kevee"
|       "display_login": "kevee"
|       ...
|     }
|     ...
|   }
|   {
|     "id": "12779420137"
```

```
|          "type": "PushEvent"
|          "actor":
|          {
|            "id": 18057136
|            "login": "peterapps"
|            "display_login": "peterapps"
|            ...
|          }
|          ...
|      }
```

23.2.2   Path Parameters

When an API uses the *endpoint-path* portion of the *resource-path* in a request to encode, as steps in the path, parameters, we call these *path parameters*. Refer back to Fig. 23.3 to reinforce where in a request this information is placed. Say that we want to make a request to get detailed information about a particular organization. The name of the organization is a parameter to the request. In the GitHub API documentation, such a request is shown as

```
GET /orgs/:org
```

What this means is that a request should encode the parameter of the name of the organization into the resource path, at the place indicated by :org. This notation of using a : and then the name of the request parameter is common as a means of specification among many providers, although some providers will use {variable}, or even <<variable>> to mean the same thing. Note that the syntactic decorations (:, { and }, or << and >>) are *not* tokens in the syntax and would not be included in a realization of the path.

Curl Incantation
```
curl https://api.github.com/orgs/denison-cs
```

Requests Incantation]

```
org = 'denison-cs'
path = "/orgs/{}"

resourcepath = path.format(org)
url = util.buildURL(resourcepath, location)

response = requests.get(url)
assert response.status_code == 200
```

```
data = response.json()
util.print_data(data, depth=3, nchild=6)

  | {
  |    "login": "denison-cs"
  |    "id": 28565851
  |    "node_id": "MDEyOk9yZ2FuaXphdGlvbjI4NTY1ODUx"
  |    "url": "https://api.github.com/orgs/denison-cs"
  |    "repos_url": "https://api.github.com/orgs/denison-cs/repo
  |    "events_url": "https://api.github.com/orgs/denison-cs/eve
  |    ...
  | }
```

As a second example of using path parameters, consider the desire to make a request to find the most recent GitHub events associated with a particular repository. But repositories in GitHub are named within either an organization or a GitHub user. So, now we have *two* request parameters to specify—the *owner* of the repository and the *name* of the repository itself.

This is documented in the GitHub API by

```
GET /repos/:owner/:repo/events
```

If the owner has the value `denison-cs` and the repository is named `welcome`, then we can make the request as follows:

Curl Incantation

```
curl https://api.github.com/repos/denison-cs/welcome/events
```

Requests Incantation

```
org = 'denison-cs'
repo = 'welcome'
path = "/repos/{}/{}/events"

resourcepath = path.format(org, repo)
url = util.buildURL(resourcepath, location)

response = requests.get(url)
assert response.status_code == 200

data = response.json()
util.print_data(data, depth=3, nchild=3)

  | [
  |    {
  |       "id": "12091909329"
  |       "type": "CreateEvent"
  |       "actor":
  |       {
```

```
|          "id": 8443005
|          "login": "tcbressoud"
|          "display_login": "tcbressoud"
|          ...
|        }
|        ...
|    }
|    {
|        "id": "12091909183"
|        "type": "CreateEvent"
|        "actor":
|        {
|          "id": 8443005
|          "login": "tcbressoud"
|          "display_login": "tcbressoud"
|          ...
|        }
|        ...
|    }
|  ]
```

23.2.3   Query Parameters

For the next set of examples, we will employ the API provided by The Movie Database (TMDB). While GitHub has endpoints that can illustrate these same concepts, it is instructive to see the topics of Query Parameters, Header Parameters, and POST information from an additional perspective.

The TMDB API has two versions that are active, versions 3 and 4. For purposes of illustration here, we will use version 3, as it has endpoints that are interesting and are good examples of what we are trying to demonstrate. In addition, and in common with a large number of API providers out there, this API requires authentication, by which the client is assured of the identity of the provider and the provider is assured of the identity of the client application. Authenticating the provider to the client happens by virtue of using the `https` protocol in requests. In reference to Fig. 23.2, these two directions of authentication correspond to the bidirectional arrow labeled *"Auth"* between the entities of the provider and the client. This type of authentication is limited to identity of the application, and more sophisticated techniques must be employed to integrate users/data owners. A full discussion of all of these more advanced authentication and authorization techniques is deferred to Chap. 24.

In the TMDB database, all accesses require the request to provide an authenticating *API key*. The API key is obtained by an application developer registering as a TMDB user, complete with a username and password, on the TMDB site, and then, in an additional step, creating the identity for an application by providing detail about the app and its intended use, and, when successful, being given a 32 digit and letter string that *identifies* the application. These steps are taken in the developer's browser. The resultant API key is then integrated into the application, usually by storing the key in a file or database so that the application can securely obtain the key when needed. As a part of authentication, it is assumed that, like a password, only the designated application can possess the provided key, and that if an application makes a request and specifies the key (over an encrypted network connection), then the application has authenticated to the provider, and the request may proceed.

Like we have done previously, we place authentication credential information in a JSON file, `creds.json` that includes information like the following:

```
{
  "tmdb":
    {    "protocol": "https",
         "location": "api.themoviedb.org",
         "apikey":    "64cefb3e82db8d373423a7bd0aa8956d"
    }
}
```

Note that the example given here is NOT a valid API key. To execute the following examples, a reader would need to register for an account and then, in the settings for that account, register their "application," even if the application was for educational purposes. The credentials file should then be updated with the information, and then the following sets up variables for use in the subsequent examples. We use another of our utility functions (Appendix A) that simply finds and parses the JSON-formatted `creds.json`, returns the sub-dictionary for `"tmdb"` information, and sets global variables for future use.

```
tmdb_creds = util.read_creds("tmdb", datadir)
protocol = tmdb_creds['protocol']
location = tmdb_creds['location']
apikey = tmdb_creds['apikey']
```

23.2.3.1   Search for Movies

TMDB has an endpoint `/search` with subordinate operations to search for movies, TV shows, people (actors, actresses, and cast), and more. The endpoint-path of `/search/movie` is the endpoint for searching for movies. Figure 23.4 displays part of the documentation page for this endpoint.

In order to search for a movie, the specification indicates that an HTTP `GET` request to endpoint `search/movie` should be used and that the request parameters that are *required* for such a request are named `api_key` and `query` and are listed under the `Query String` section of the documentation. This conveys that we need to specify two *query parameters* formatted as a query string. The

Search Movies

GET   /search/movie

Search for movies.

Definition	Try it out

Authentication

☑ API Key

Query String

api_key	string	default: <<api_key>>	required
language	string	Pass a ISO 639-1 value to display translated data for the fields that support it.	optional
		minLength: 2	
		pattern: ([a-z]{2})-([A-Z]{2})	
		default: en-US	
query	string	Pass a text query to search. This value should be URI encoded.	required
		minLength: 1	
page	integer	Specify which page to query.	optional
		minimum: 1	
		maximum: 1000	
		default: 1	
include_adult	boolean	Choose whether to inlcude adult (pornography) content in the results.	optional
		default	
region	string	Specify a ISO 3166-1 code to filter release dates. Must be uppercase.	optional
		pattern: ^[A-Z]{2}$	
year	integer		optional
primary_release_year	integer		optional

collapse

Fig. 23.4 TMDB search movie specification

query string is part of the resource path in the HTTP request and is formatted
as a ? followed by a *name=value* for each query parameter, with an & character
separating parameter key-value pairs. If we are constructing a query string by hand,
we substitute + characters, or %20 character sequence for any embedded space in a
value, in accordance with URL encoding rules.

Say that we want to search for movies that match "Star Wars". The curl
incantation of such a request, using the API key example above, follows:[2]

```
curl 'https://api.themoviedb.org/3/search/movie?\
query=Star%20Wars&api_key=64cefb3e82db8d373423a7bd0aa8956d'
```

To make the same query programmatically with the requests module, we
have two choices. First, we could construct a resource path that built and included
the query string. This would involve accumulating the set of query parameters into

---

[2]In this, and the following curl incantations, we show a line break because the line is too long
for display. However, the line break occurs in the middle of a single (path) argument, and so the
syntax would not be correct. If repeating the example, the reader should not break the line in the
middle of a curl option.

key-value pairs separated by & and also knowing, for any values that have spaces (or other characters disallowed in a URI), to perform the appropriate substitutions. Fortunately, the `requests` module can perform this work for us. We simply construct a dictionary containing our query parameters and pass this dictionary to the method call (`get()`, in this case) using the `params=` named parameter. This solution is shown. In the result, before we print the data of the result, we use the `request` object referred to by the `response` and print part of the `path_url` to see how `requests` performed the translation from dictionary to query string.

```python
resourcepath = "/3/search/movie"

url = util.buildURL(resourcepath, "api.themoviedb.org")
queryParams = {"query": "Star Wars", "api_key": apikey}

response = requests.get(url, params=queryParams)
assert response.status_code == 200

search_results = response.json()
print(response.request.path_url[:46] + '...')
```

```
| /3/search/movie?query=Star+Wars&api_key=e36d72...
```

```python
util.print_data(search_results, depth=2, nchild=9, nlines=30)
```

```
| {
|   "page": 1
|   "total_results": 160
|   "total_pages": 8
|   "results":
|   [
|     {
|       "popularity": 84.602
|       "vote_count": 13870
|       "video": False
|       "poster_path": "/6FfCtAuVAW8XJjZ7eWeLibRLWTw.jpg"
|       "id": 11
|       "adult": False
|       "backdrop_path": "/zqkmTXzjkAgXmEWLRsY4UpTWCeo.jpg"
|       "original_language": "en"
|       "original_title": "Star Wars"
|       ...
|     }
|     {
|       "popularity": 155.674
|       "vote_count": 5072
|       "video": False
|       "poster_path": "/db32LaOibwEliAmSL2jjDF6oDdj.jpg"
|       "id": 181812
|       "adult": False
```

```
|        "backdrop_path": "/jOzrELAzFxtMx2I4uDGHOotdfsS.jpg"
|        "original_language": "en"
|        "original_title": "Star Wars: The Rise of Skywalker"
|        ...
|    }
```

Using the documentation page for this API endpoint, we see that we could employ additional query parameters to, for instance, search for movies whose language is other than en-US, or, for results with multiple pages, request a different page of the results. Note how the JSON dictionary result informs the requestor of the total number of results and pages, so that programmatic control to get multiple pages is possible.

23.2.4   Header Parameters

We have already discussed the use of headers in Chap. 20, and APIs often require headers for certain types of requests. For example, even though TMDB always returns data in JSON format, many APIs allow the client to specify the format desired, with a header of the form "Accept": "application/json" or "Accept": "text/xml". As usual, if this is an option, it will be specified by the API documentation. We will see an example shortly where headers are required for TMDB. Inspection of the TMDB documentation will show that headers are required for POST requests, DELETE requests, and GET requests related to a specific account (e.g., the endpoint /account/{account_id}/lists), where the headers must contain authentication information as will be discussed further in Chap. 24. Even though headers are not required for the GET requests discussed above, we can still create a header dictionary and pass it to the requests module for our requests above. Note that the invocation of get() now takes three arguments.

```
resourcepath = "/3/search/movie"

url = util.buildURL(resourcepath, "api.themoviedb.org")
queryParams = {"query": "Star Wars", "api_key": apikey}
headerParams = {"Content-Type":"application/json",
                "Accept-Encoding":"gzip, deflate"}

response = requests.get(url, params=queryParams,
                        headers=headerParams)
assert response.status_code == 200

request = response.request
util.print_headers(request.headers)
    | {
```

```
|     "User-Agent": "python-requests/2.22.0",
|     "Accept-Encoding": "gzip, deflate",
|     "Accept": "*/*",
|     "Connection": "keep-alive",
|     "Content-Type": "application/json"
| }
```

We see that this request is successful, and that our chosen headers are reflected in the `response` object, even though the headers themselves do not appear in the URI. Of course, we could create a functional abstraction for the code above, so that the content type and/or accept-encoding are parameters, and any header dictionary as in Chap. 20 could be passed to the `requests` module. However, TMDB will only provide data in JSON format, regardless of the value of `"Content-Type"`. Since APIs are much more restrictive than general web resources, it is unwise to attempt to use headers beyond those required or supported by the API documentation. For example, if we added `"Transfer-Encoding":` `"chunked"` to `headerParams` in the code above, it would result in a status code of 501, since the TMDB API is unable to provide data in a chunked encoding.

23.2.5   POST and POST Body

In our final example on constructing requests to satisfy the specification for an API request, we consider one where many of the mechanisms for conveying request parameters are exemplified:

- POST request
- Path parameter
- Header parameter
- Query parameter
- Authentication through query parameter
- POST body.

The provider's specification for our desired operation, submitting a rating for a movie, is given in Fig. 23.5. Within the figure, look closely at the specification under the heading `Request Body`. This is specifying the expectation that the *body* of the POST request uses a JSON-formatted string, whose contents are a JSON `object` (i.e., Python dictionary), and that there is one key, `"value"`, that maps to the number representing the rating to be set.

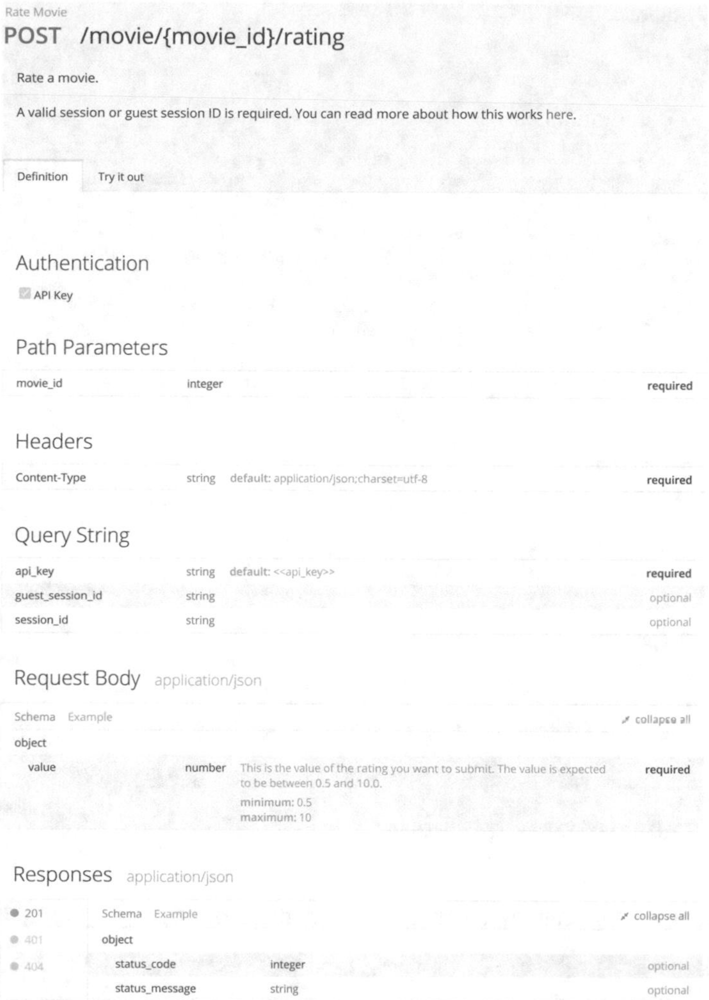

Rate Movie

POST   /movie/{movie_id}/rating

Rate a movie.

A valid session or guest session ID is required. You can read more about how this works here.

Definition   Try it out

Authentication

☑ API Key

Path Parameters

movie_id	integer	required

Headers

Content-Type	string default: application/json;charset=utf-8	required

Query String

api_key	string default: <<api_key>>	required
guest_session_id	string	optional
session_id	string	optional

Request Body   application/json

Schema   Example   ✎ collapse all

object

value	number	This is the value of the rating you want to submit. The value is expected to be between 0.5 and 10.0. minimum: 0.5 maximum: 10	required

Responses   application/json

● 201   Schema   Example   ✎ collapse all
● 401   object
● 404   status_code   integer   optional
               status_message   string   optional

Fig. 23.5 TMDB rate movie specification

From the figure, we also discover that the query parameters include both a
$session\_id$ and a $guest\_session\_id$, which are both optional. But to post
a rating, there must exist some session associated with the user wishing to rate the
movie. So, to be able to make the rate movie request, we must at least have a guest
session. So, we start by showing that preliminary step (Fig. 23.6).

Following the documentation, the curl incantation uses the specified endpoint
and includes an api_key query parameter.

```
curl 'https://api.themoviedb.org/3/authentication/\
guest_session/new?api_key=64cefb3e82db8d373423a7bd0aa8956d'
```

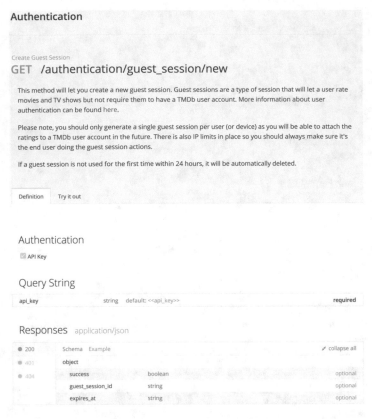

Fig. 23.6 TMDB guest session specification

In our client application, the request would look like this:

```
resourcepath = "/3/authentication/guest_session/new"

queryParams = {"api_key": apikey}

url = util.buildURL(resourcepath, "api.themoviedb.org")
response = requests.get(url, params=queryParams)
assert response.status_code == 200

results = response.json()
util.print_data(results, depth=2, nchild=3)

   | {
   |    "success": True
   |    "guest_session_id": "eac354d08266e75fb4eba8f4dcdd35ab"
   |    "expires_at": "2020-07-01 21:57:58 UTC"
   | }
```

```
sessionid = results['guest_session_id']
```

At the conclusion of this code, the variable `sessionid` refers to the desired value, and we can use this in the POST to follow.

With a valid session id, the `curl` incantation must specify that the request is *not* a GET, by using the `--request` option, and, for the POST, specifies the body of the request by using the `--data` option.

```
curl --request POST --data '{"value": 9.5}' \
-H 'Content-Type: application/json;charset=utf-8' \
'https://api.themoviedb.org/3/movie/181812/rating?\
api_key=64cefb3e82db8d373423a7bd0aa8956d&\
guest_session_id=7ca08f87015c752f9c2811fb61be0018'
```

The translation of the request to a programmatic sequence demonstrates the use of a *path parameter*, to specify the movie id (obtained from the earlier search, with value `181812`), *query parameters* for the API key and for the guest session id, a *header parameter* for the content type, and, finally, use of the POST *body* to send the rating information as a JSON-formatted string. Note the use of `json.dumps()` in the code.

```
movieid = 181812
path = "/3/movie/{}/rating"
resourcepath = path.format(movieid)

# Use the session id acquired in the previous step
queryParams = {"api_key": apikey,
               "guest_session_id": sessionid}
headerParams = {"Content-Type":
                "application/json;charset=utf-8"}

body = {"value": 9.5}
bodystr = json.dumps(body)
url = util.buildURL(resourcepath, "api.themoviedb.org")
response = requests.post(url, params=queryParams,
                         headers=headerParams,
                         data=bodystr)
assert response.status_code == 201

util.print_data(response.json())

    {
        "status_code": 1,
        "status_message": "Success."
    }
```

```
print(bodystr)
```

| {"value": 9.5}

```
print(response.request.body)
```

| {"value": 9.5}

In formulating a request like this, there are a number of relatively common mistakes to be on the lookout for:

1. This request is authenticating a *client*, not a *user*. An API key is not the same as a session id, which is associated with the specific user wishing to rate the movie. This is why we need a guest session id. If we do indeed want to associate the rating with a specific user, we would need to obtain a session id, which requires a different form of authentication. OAuth is used for this purpose and is discussed in the next chapter.
2. The body of the message *has to be JSON*. If you instead pass a dictionary as the `data` parameter to `requests`, it assumes you want the data body to be URL-encoded as a form. We needed the `dumps` to convert from JSON to the string needed for the body and to keep `requests` from doing the URI-encoding processing.
3. For this POST, success is a 201, not a 200. A 201 is also a successful response, and some providers use this status code for a successful POST. Other providers could choose to use a 200, so the documentation must be consulted.

It is very important to look at any non-201 response, and, in particular, to look at the response body to get a more detailed message. For instance, it would not be appropriate to generate a guest session each time, since an update like this has its real use in being associated with a user. But a guest session has a limited duration, so a guest session id could expire, and an operation that "used to work" could start not working. And without attention to the status code and message, it would appear as "buggy," when it is not.

23.2.6   Reading Questions

23.7 Please give a real-world justification for the design decision to make the TMDB API not available for anonymous access.

23.8 Please give concrete examples to illustrate the kinds of things we need to be able to specify for a request to GitHub. Then, do the same for a request to TMDB.

23.9 Please open the following link in your favorite web browser and follow the directions there for creating an account and requesting an API key:

- https://developers.themoviedb.org/3/getting-started/introduction

You can fill in any application name and URL that you like, but in the `Application Summary` field, you need to write a sentence or two. Feel free to simply describe that you are going to use the API key to learn about APIs for a course you are taking. When you finish, please write down the first four digits of the API key you received in the cell below. Keep the rest of your API key secret, and keep it in a place you will never forget it.

23.10 As part of the process in getting your API key above, you had accept some terms. Since you are going to use this API key to develop software, it is important to *actually read and understand* those terms of use, to be sure you are not violating them. In case you did not save a copy, you can see the terms of use you accepted at this link:

* https://www.themoviedb.org/documentation/api/terms-of-use

Please write down two terms of use you would expect, and one that surprised you, plus why. If any of the terms of use confused you, please write down your confusion below, so you can clarify it later.

23.11 The reading gives an example of an endpoint-path to the root for the GitHub API. Please formulate an endpoint-path to the root for TMDB and then describe what happens when you navigate to that link in your web browser.

23.12 The reading gives an example of an endpoint-path to a non-root endpoint. Please give an analogous example for TMDB and describe what happens *and why* when you navigate to that path in a browser.

23.13 We have seen query strings in our study of HTTP. Please study the GitHub API documentation and write down an example where query strings are used in a resource path:

* https://developer.github.com/v3/#parameters

23.14 For the curl example of searching TMDB for "Star Wars," are the parameters shown query parameters or path parameters and how do you know based on the URI?

23.15 The reading mentions TMDB query parameters to search for different pages and languages. Please investigate the documentation and provide example URIs to justify this statement.

23.16 The "TMDB Search Movie Specification" figure is based on the following link:

* https://developers.themoviedb.org/3/search/search-movies

Please read the information at that website carefully and then answer the following questions:

1. On the left sidebar, please give an example of a different kind of SEARCH you can do with the TMDB API. Then, please find the goal of a POST message under MOVIES on the left sidebar.
2. In the Responses table, what example of a status code 401 is provided?
3. In the schema for a successful `application/json` response, what path would lead to `vote_average`?
4. At the top of the page, in "Try it out," please put in the required api_key (you can use the one from the reading), a search for "citizen kane," not including adult films, and only including films released in 1941. Please write down the resulting URL and describe what happens when you click "Send Request." Is it the same thing that happens when you copy and paste the URL into your web browser?
5. Experiment with different settings in the "Try it out" menu. What happens if you leave out the API key? Does capitalization matter? Can you search for a film with the symbol + in the title (e.g., the movie +1 released in 2013 about three college friends)? How are spaces and pluses converted into URLs, and why?

23.17 Suppose we created a URL as follows:

```
s = "https://api.github.com/repos/{}/welcome/events" \
    "?type=CreateEvent"
url = s.format("denison-cs")
```

Please identify all parameters and their type (that is, query parameter, path parameter, header parameter, etc.).

23.2.7   Exercises

The first set of exercises will work with the TMDB developer API, version 3, documented here: https://developers.themoviedb.org/3/getting-started/introduction. Part of the point of these exercises is to *find* the documentation for obtaining various types of information and then being able to *apply* the techniques of the section to actually make the request.

For these exercises, we presuppose that you have registered and have an API key to use for making requests.

Many of the exercises will ask that you construct an answer to the question by using both the command-line and the `curl` and that you write code to do it programmatically, with `requests`.

23.18 Suppose we want to get detail from the TMDB API about actor Samuel L. Jackson. Before we can get detail information, we need to do a search to find the unique ID associated with the actor.

Find the documentation for searching for an actor, and, following the model in the book, and possibly combining with the "Try It Out" option in the documentation, construct a command-line as well as a `requests`-based search.

23.19 In the previous question, you might have been specific and included first and last names, in which case you may have obtained a single result. But you may have used, for instance, just the last name, or could have specified last name followed by first name, which would give you a different total number of results. Suppose your goal is to continue searching until you get down to a single result. Write code that interacts with a user to get a search term and then uses a `while` loop to make a request and then process the reply. If the result has a single result, print information about the result and terminate the loop. If the result has multiple results, print the total number of results, ask for a new search term, and continue the loop.

23.20 As a result of the previous questions, you should have the TMDB id associated with Samuel L. Jackson. Explore the documentation and find the API associated with getting the *movie credits* for a particular person. Write the curl and the `requests` programming to get the first page of movie credits for Samuel L. Jackson.

23.21 Write a function

```
getMovieCredits(personid, apikey, page=1)
```

that uses `apikey` to retrieve the movie credits associated with the person identified by `personid` and fetch the given `page`. The return value should be a data structure interpreted from the JSON-formatted body of the response. Return None if the request was not successful.

23.22 Suppose you want to rate a particular TV show, for instance, the show "Silicon Valley." Works the techniques you have learned as well as the model shown in the section, to do the following:

1. Find the id associated with the TV show.
2. Obtain a guest session id.
3. Formulate the POST request to perform the actual rating.

23.23 Please study the documentation for deleting a movie rating and then write the code that uses the `requests.delete()` method to carry out such a request, including the required header dictionary. You may assume that variables containing the API key, guest session id, and movieid have already been defined. The documentation may be found at the following link:

- https://developers.themoviedb.org/3/movies/delete-movie-rating

The next set of exercises involves the API associated the World Bank. This API does *not* require an API key, and openly provides their data, but does require attribution. The "top" page for the documentation is at

- https://datahelpdesk.worldbank.org/knowledgebase/topics/125589-developer-information.

Below that page, the reader is well advised to read the basics of API call structure. Note, in particular, how to specify multiple indicators, requested output format, and date ranges.

- https://datahelpdesk.worldbank.org/knowledgebase/articles/898581-api-basic-call-structures

And the API endpoints associated with countries and with indicators:

- https://datahelpdesk.worldbank.org/knowledgebase/articles/898590-country-api-queries
- https://datahelpdesk.worldbank.org/knowledgebase/articles/898599-indicator-api-queries

And to combine and aggregate:

- https://datahelpdesk.worldbank.org/knowledgebase/articles/898614-aggregate-api-queries

Relating these exercises to our use of the indicators data set in early chapters, we are interested in countries whose three character code is CAN, CHN, GBR, RUS, USA, and VNM. In the API, and as maintained by the World Bank, the indicators that they use map to the more intuitive names used in this book as follows:

Textbook indicator	World bank indicator
pop	SP.POP.TOTL
gdp	NY.GDP.MKTP.CD
life	SP.DYN.LE00.IN
imports	BM.GSR.GNFS.CD
exports	BX.GSR.GNFS.CD

Formulate API requests for each of the following scenarios:

23.24 Retrieve detailed information about VNM. Note the information available about the country.

23.25 In a single query, retrieve information about USA, CHN, and IND in the JSON format.

23.26 Retrieve information about the imports indicator.

23.27 Retrieve imports of the USA over all years.

23.28 Retrieve imports of the USA in 2011.

23.29 Retrieve the GDP of CHN, IND, and USA for years 2015–2019.

23.3   Case Study

In this section, we present a case study demonstrating the process common in data systems clients to use a REST API as a data source, acquire data, and put it into a useful tabular form and corresponding pandas data frames. We use the requests module to obtain data from the TMDB API about popular movies and cast members in those movies. We divide our work into two phases. In the first phase, we focus on constructing a table of popular movies and demonstrate the need, in the design, for using more than a single request to achieve a result. In the second phase, we use the popular movies table as *input* to the process of generating a table of top cast members from those movies. This involves the input driving *multiple requests* to obtain the needed data, another common pattern.

First, we define local variables from our creds.json file, including the API key that we will need to interface with TMDB.

```
tmdb_creds = util.read_creds("tmdb", datadir)
apikey     = tmdb_creds['apikey']
```

23.3.1   Phase 1: Build a Table of Popular Movies

We focus on using the /movie/popular endpoint as a relatively simple way to get set of movies that "rise to the top" in terms of being interesting. Other options for getting an interesting set of movies might use other API endpoints, like

- /movie/now_playing: Gets list of movies in theaters;
- /movie/top_rated: Gets the top rated movies in TMDB;
- /movie/upcoming: Gets a list of upcoming movies in theaters;
- /discover/movie: Gets a flexible list of movies based on many different criteria and provides sort options. Some of the criteria include
  - release year minimum and maximum
  - vote average minimum and maximum
  - vote count minimum and maximum
  - with particular people in cast or crew
- /search/movie: General search function equivalent to entering a search string in the interactive version of the TMDB site.

There is also the ability to use movie or TV identifiers from other databases, like IMDB, FreeBase, Instagram, Twitter, and others, and get corresponding TMDB movie information. Before we can proceed, it is essential to carefully study the documentation, reproduced below (Fig. 23.7).

We will solve this phase of generating a pandas table of popular movies in two parts:

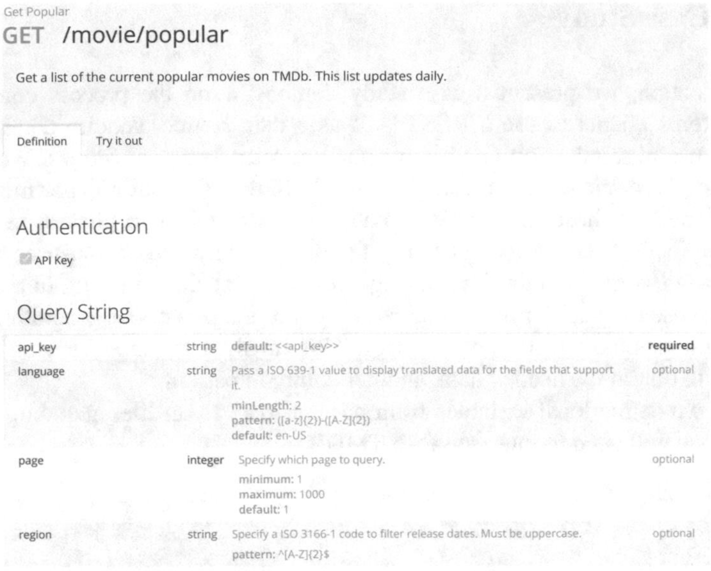

Get Popular
GET /movie/popular

Get a list of the current popular movies on TMDb. This list updates daily.

Definition   Try it out

Authentication
☑ API Key

Query String

api_key	string	default: <<api_key>>	required
language	string	Pass a ISO 639-1 value to display translated data for the fields that support it.	optional
		minLength: 2 pattern: ([a-z]{2})-([A-Z]{2}) default: en-US	
page	integer	Specify which page to query.	optional
		minimum: 1 maximum: 1000 default: 1	
region	string	Specify a ISO 3166-1 code to filter release dates. Must be uppercase.	optional
		pattern: ^[A-Z]{2}$	

Fig. 23.7 TMDB popular movie API

- retrieving a page of results through the API, and
- using the results to build a native Python data structure of tabular information.

These parts can then be combined and repeated to retrieve multiple pages and aggregate the native Python data structure, which will then allow construction of the data frame.

23.3.1.1 Design a Function to Issue Request

With a mind toward developing clean and reusable software, we start with a building block function that can be used to issue a single request of the API. We illustrate how to do this using `curl` and then how to do it using the `requests` module. In the design of our function, we want a parameter for the page, so that we can easily request different pages of results.

```
curl 'https://api.themoviedb.org/3/movie/popular?page=1&\
   api_key=75df0c4f93ec9e484534b8ce1db9a67e'
```

```
def GetPopularMoviePage(apikey, page=1):
    """
    Use popular movies endpoint to request a page of the
    movies currently deemed popular by some metric.

    Parameters:
    apikey: previously obtained key for client from TMDB
```

```
    page: which page of data to request, default 1

    Return: data structure from JSON-formatted results
    """
    host = "api.themoviedb.org"
    resourcepath = "/3/movie/popular"

    url = util.buildURL(resourcepath, host, protocol='https')
    queryParams = {"api_key": apikey, "page": page}

    response = requests.get(url, params=queryParams)
    assert response.status_code == 200

    return response.json()
```

We display the results we have just obtained, which we will next need to convert into a pandas data frame.

```
popmovies_page = GetPopularMoviePage(apikey)
util.print_data(popmovies_page, nchild=7, nlines=30)
```

```
| {
|    "page": 1
|    "total_results": 10000
|    "total_pages": 500
|    "results":
|    [
|      {
|        "popularity": 207.629
|        "vote_count": 3815
|        "video": False
|        "poster_path": "/xBIIvZcjRiWyobQ9kxBhO6B2dtRI.jpg"
|        "id": 419704
|        "adult": False
|        "backdrop_path": "/5BwqwxMEjeFtdknRV792Svo0K1v.jpg"
|        ...
|      }
|      {
|        "popularity": 160.436
|        "vote_count": 146
|        "video": False
|        "poster_path": "/9zrbgYyFvwH8sy5mv9eT25xsAzL.jpg"
|        "id": 531454
|        "adult": False
|        "backdrop_path": "/jMO1icztaUUEUApdAQx0cZOt7b8.jpg"
|        ...
|      }
|      {
|        "popularity": 155.674
|        "vote_count": 5072
```

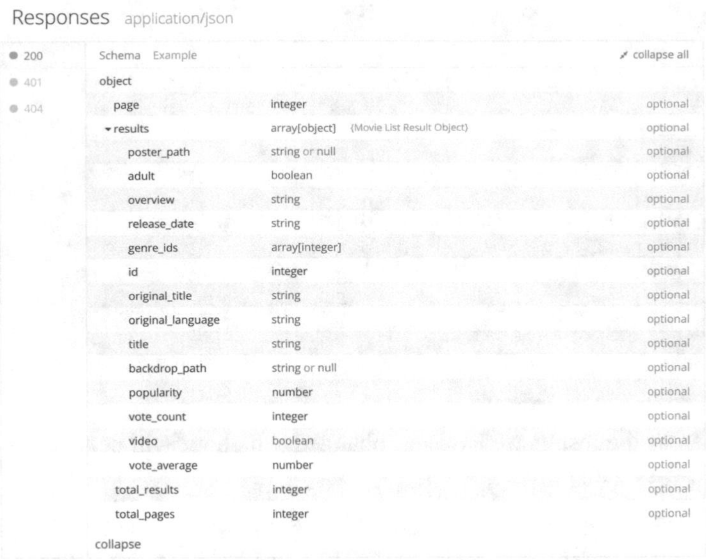

Responses application/json

● 200	Schema Example		⚲ collapse all
● 401	object		
● 404	page	integer	optional
	▾ results	array[object] (Movie List Result Object)	optional
	poster_path	string or null	optional
	adult	boolean	optional
	overview	string	optional
	release_date	string	optional
	genre_ids	array[integer]	optional
	id	integer	optional
	original_title	string	optional
	original_language	string	optional
	title	string	optional
	backdrop_path	string or null	optional
	popularity	number	optional
	vote_count	integer	optional
	video	boolean	optional
	vote_average	number	optional
	total_results	integer	optional
	total_pages	integer	optional
	collapse		

Fig. 23.8 TMDB popular movie results

```
"video": False
```

23.3.1.2   Understand Results

To help understand the structure of the results, Fig. 23.8 shows the TMDB documentation in the form of a schema for the JSON-formatted reply.

Result Observations
From the documentation and also relating that to our first result, we see the following:

1. The root is a dictionary (JSON object) with keys `page`, `total_results`, `total_pages`, in addition to `results`.
2. The primary information is in `results`, which is of type `array[object]`, which is the JSON way of saying a `list` of `dict` (dictionary) objects.
3. Each individual dictionary in the `results` list gives information about exactly one movie.
4. The full set of popular movie information extends significantly beyond the results in this one request/response exchange. The result tells us what page we have and also tells us the total number of pages and the total number of results. Depending on the goal, we may need to issue multiple requests to obtain the full set of results.

23.3.1.3   Design Movie Table

To build a tabular representation of the movies, we need to decide on the fields. For simplicity, we will select just a subset of the possible information and will include one non-tidy field in our movie table, as noted in our target list of fields:

Field	Python type	Notes
`id`	int	The TMDB identifier for the movie
`title`	str	The TMDB title
`genres`	str	Encode list of integers from result into comma-separated list as string; not tidy, but convenient for our current needs
`popularity`	float	Metric computed by TMDB for the popularity of a movie
`vote_count`	int	Number of rating votes for this movie at TMDB
`vote_average`	float	TMDB computer average of votes
`release_date`	str	Release date of the movie

With this design, we can then write a function that produces a list of row dictionaries (LoD) representation from the JSON-parsed data structure of a request. We choose the LoD representation because of its ease in later constructing a data frame and its row-centric nature, which matches our processing of the request results.

```
def MovieResult2LoD(result, maxelements=None,
                    include_adult=False):
    """
    Process the JSON-based result of a popular movie query
    and, for each element of the result, create a row
    in a LoD representation of the movie fields.

    Parameters:
    result: JSON-based data structure from popular movie
            query
    maxelements: limit on the number of results to
            process.  If None, process all
    include_adult: boolean indicating whether or not
            results tagged as adult should be included

    Return: LoD with rows of movies
```

```
    """
    assert isinstance(result, dict)
    assert 'results' in result

    LoD = []
    count = 0
    for movie in result['results']:
        if maxelements != None and count >= maxelements:
            break
        if movie['adult'] and not include_adult:
            break
        D = {}
        D['id'] = movie['id']
        D['title'] = movie['title']
        genre_list = movie['genre_ids']
        genres = ",".join([str(g) for g in genre_list])
        D['genres'] = genres
        D['popularity'] = movie['popularity']
        D['vote_count'] = movie['vote_count']
        D['vote_average'] = movie['vote_average']
        D['release_date'] = movie['release_date']
        LoD.append(D)
        count += 1
    return LoD
```

We illustrate the function above and print the first row it yields.

```
LoD = MovieResult2LoD(popmovies_page)
util.print_data(LoD[0])
```

```
| {
|    "id": 419704,
|    "title": "Ad Astra",
|    "genres": "18,878",
|    "popularity": 207.629,
|    "vote_count": 3815,
|    "vote_average": 6.1,
|    "release_date": "2019-09-17"
| }
```

23.3.1.4   Handle Multiple Pages

Very often, an API service provider throttle results in various ways. If the number of results from a request can be large, the results are typically divided into *chunks* or *pages*, and a request indicates the total number of pages and/or results as part of each response. It is up to the client to navigate this and, if necessary, issue additional requests until all the desired data is acquired.

If all we want is the first page, we could just take our LoD and construct a pandas data frame, and we are done with Phase 1. However, suppose we want more than just the 20 results that a single query offers. We want a higher-level function that can call these two functions for as many times as we need and composes the results together, with a final composite LoD that we then turn into a DataFrame. Although we select 20 as a default value for the parameter num_movies, other values may be passed to the function.

```python
def GetPopularMovies(apikey, num_movies=20):
    """
    Use the building block functions so that, based on
    num_movies, make a series of requests and process
    responses, aggregating into a composite LoD structure
    and finally converting to a data frame.

    Parameters:
    apikey: previously obtained key for client from TMDB
    num_movies: total number of movies to retrieve and
        process

    Return: DataFrame structure
    """

    Composite_LoD = []

    page = 1
    movies_left = num_movies
    more_pages = True

    while more_pages and movies_left > 0:
        movie_page = GetPopularMoviePage(apikey, page)
        if movie_page['page'] == movie_page['total_pages']:
            more_pages = False

        page_LoD = MovieResult2LoD(movie_page)
        Composite_LoD.extend(page_LoD)

        movies_left -= len(page_LoD)
        page += 1

    df = pd.DataFrame(Composite_LoD)
    return df
```

We illustrate our function with a sample call, requesting 40 movies this time (which will require results beyond those present on page 1). Because titles exceed the length of the displayed line, we just project ids, the genre list, and the popularity.

```
movies = GetPopularMovies(apikey, num_movies=40)
print("Number of Movies in DataFrame:",len(movies))

 | Number of Movies in DataFrame: 40

movies.iloc[:5, [0,2,3]]

 |           id             genres   popularity
 | 0    419704             18,878      207.629
 | 1    531454          35,10402      160.436
 | 2    181812          28,12,878     155.674
 | 3    475430   12,14,878,10751      131.333
 | 4    496243          35,18,53      124.070
```

23.3.2   Phase 2: Build Table of Top Cast Given Movie IDs

Given the set of movie IDs from Phase 1, the second phase of our case study was to build a table of *top cast* from that set of movies. That will involve multiple requests, one per movie, to obtain the cast list for each movie. Then, for each of these cast lists, we compose a cast table. We use a threshold so that we only include a limited number of cast members per movie. Finally, we remove duplicates in the cast table.

We subdivide our work into

- understanding the relevant API,
- designing a table for the cast members and a function to build a native Python data structure for the cast of a given movie, and
- designing a function to perform this repeatedly over a set of movies.

23.3.2.1   Understand Movie Credits API

Just as before, we must begin by studying the API documentation. Figure 23.9 shows the API that, given a movie id, retrieves the credit information about that movie. The pattern for making a request here is similar to others we have seen, where the *movie id* is included as a path parameter, and the API key is a query parameter.

The schema for the response is shown in the figure as well. We summarize our conclusions about the response as follows:

1. The root of the return is a JSON object
2. The root has three children: the id of the movie and children for cast and crew.
3. The cast child is a JSON array of JSON objects.
4. Among the fields for a cast member, the name, id, and gender of the cast member appear to be independent of the particular movie.

Get Credits

GET /movie/{movie_id}/credits

Get the cast and crew for a movie.

Definition	Try it out

Authentication

☑ API Key

Path Parameters

movie_id	integer	required

Query String

api_key	string default: <<api_key>>	required

Responses application/json

● 200	Schema Example			⤢ collapse all
● 401	object			
● 404	id	integer		optional
	▾ cast	array[object]		optional
	cast_id	integer		optional
	character	string		optional
	credit_id	string		optional
	gender	integer or null		optional
	id	integer		optional
	name	string		optional
	order	integer		optional
	profile_path	string or null		optional
	▾ crew	array[object]		optional
	credit_id	string		optional
	department	string		optional
	gender	integer or null		optional
	id	integer		optional
	job	string		optional
	name	string		optional
	profile_path	string or null		optional
	collapse			

Fig. 23.9 TMDB movie credits API

5. The `order` field is an integer that gives the relative importance of this cast member.

If we seek to limit to the "top" cast, we can use a threshold to filter by the cast whose `order` value is less than or equal to a given threshold.

We illustrate a request to the API using `curl`.

```
curl 'https://api.themoviedb.org/3/movie/181812/credits?\
api_key=75df0c4f93ec9e484534b8ce1db9a67e'
```

23.3.2.2   Goal: Design Cast Table

To build a tabular representation of the movies, we need to decide on the fields. For simplicity, we will just select three fields, already available in the results of the credits from a movie. This could be enhanced by requests to get detailed information about each of the selected persons in the cast. This is our target list of fields:

Field	Python type	Notes
id	int	The TMDB identifier for the person
name	str	The String name of the person
gender	int	Encoding as an integer for the gender of the person and may be null

We are now ready to solve the top cast problem, in two steps. First, we write a function that uses the `requests` module to get the cast data associated with a given movie, which we wrangle into a list of dictionaries.

```python
def GetMovieTopCast(apikey, movieid, threshold=None):
    """
    Given an apikey and a movieid, make a request for the
    movie credits for the given movie.  Build a list of
    row dictionaries with the result for those cast whose
    order field is below threshold.

    Return: the LoD structure on success, None on failure
    """
    host = "api.themoviedb.org"
    path = "/3/movie/{}/credits"

    rpath = path.format(movieid)
    url = util.buildURL(rpath, host, protocol='https')
    queryParams = {"api_key": apikey}

    response = requests.get(url, params=queryParams)
    if response.status_code == 401:
        print("Failed: {} with Invalid API".format(rpath))
        return None
    if response.status_code == 404:
        print("Failed: {} with Not Found error".format(rpath))
        return None
    assert response.status_code == 200

    result = response.json()
    assert result['id'] == movieid
    assert 'cast' in result

    LoD = []
    for person in result['cast']:
```

```
      D = {}
      if threshold == None or person['order'] <= threshold:
          D['id'] = person.get('id')
          D['name'] = person.get('name')
          D['gender'] = person.get('gender')
          LoD.append(D)

  return LoD
```

We illustrate this function with a sample call.

```
topcast = GetMovieTopCast(apikey, 181812, threshold=6)
util.print_data(topcast, depth=2, nchild=3)
```

```
| [
|    {
|       "id": 4
|       "name": "Carrie Fisher"
|       "gender": 1
|    }
|    {
|       "id": 2
|       "name": "Mark Hamill"
|       "gender": 2
|    }
|    {
|       "id": 1315036
|       "name": "Daisy Ridley"
|       "gender": 1
|    }
|    ...
| ]
```

With the function GetMovieTopCast() in hand, we carry out the second step of our solution, which involves calling GetMovieTopCast() once for each movie and collecting the results in a pandas data frame.

```
def GetTopCast(apikey, movie_list, threshold=10):
    """
    Given a list of movies, repeatedly request the top
    cast from that movie, subject to an order threshold,
    and aggregate them into a list of the union of the
    top cast members.

    Parameters:
    apikey: API key for client to TMDb
    movie_list: integer list of TMDb movie ids
    threshold: integer for definition of "top" in cast
```

```
    Return: data frame of unique cast members
    """

Composite_LoD = []

for movieid in movie_list:
    MovieLoD = GetMovieTopCast(apikey, movieid,
                                          threshold)
    Composite_LoD.extend(MovieLoD)

df = pd.DataFrame(Composite_LoD)
df.drop_duplicates('id', inplace=True)
return df
```

We illustrate this function with a sample invocation.

```
movie_list = list(movies['id'])
cast = GetTopCast(apikey, movie_list, threshold=8)
cast.head()
```

	id	name	gender
0	287	Brad Pitt	2
1	2176	Tommy Lee Jones	2
2	17018	Ruth Negga	1
3	40543	John Ortiz	2
4	882	Liv Tyler	1

And we see a successful outcome with a top cast data frame.

23.3.3   Summary Comments

In this chapter, we have learned how to interface with APIs, how to formulate requests that satisfy the API documentation, how to read the documentation to learn what kind of response we will receive, and how to wrangle the response into a pandas data frame. These skills are tremendously useful in data science.

When taking what we have learned into practice, we will see that there is no standardization, and that there are as many different ways for a provider to specify an API as there are providers out there. One of the more recent directions one can observe in the landscape of API services is for providers to *reduce* the number of endpoints, in the limit providing just a *single* endpoint, but to *increase* the expressive power of what can be requested. In this type of API service, the client builds a *tree structure* that describes the information they desire using a *Graph Query Language* or *GraphQL* [17]. This, in essence, extends the ideas presented in this chapter and combines them with parts of what we learned in Chaps. 15 through 17 in the hierarchical model (and hence the term *graph* in the name of this query language).

We turn next to issues of authentication and authorization, so that our client apps can obtain sensitive and/or protected user data from API providers.

23.3.4 Reading Questions

23.30 If `array[object]` is the JSON way of saying a list of dict (dictionary) objects, what do you think is the JSON way of saying a list of lists?

23.31 Please explore the documentation for TMDB API and find an example where the responses give you a different JSON type (other than `array[object]`) and report your answer here. Study the figures in the chapter carefully to see where the type is specified in the Responses table.

23.32 What would happen if we did not use the parameter `threshold` in our functions?

23.33 What is the connection between the URI sent by `requests.get()` in `GetPopularMoviePage` and the URL in the `curl` command above the function? Explain your answer.

23.34 Please justify the book's reason for using a LoD for the desired movie table. Please refer back to earlier chapters for the pros and cons of the various ways to represent two-dimensional data in Python.

23.35 Carefully study the function `GetPopularMovies()` and then explain the logic behind the `while` loop. Specifically, when does the code escape the loop and what does that mean in the context of the problem?

23.36 Why is it important to remove duplicates from the Top Cast table? Please reference the tidy data assumptions.

23.37 Study the code for `GetMovieTopCast()` and `GetTopCast()`. In the latter, why do we use `Composite_LoD.extend()` instead of `.append()`?

23.38 The Internet is full of APIs that can provide data on a wide variety of subjects. Please locate an API on a subject you are interested in, read the documentation, and describe a data analysis project you could complete using this API. Please make sure to provide sufficient detail to justify your plan. At the time of writing, a search engine for APIs was available at http://apis.io/

23.3.5 Exercises

23.39 Please run the code from the reading, which carries out a POST to TMDB, but use the *session id* from the book, without getting a new guest session id of your own. The session id will be expired, so you should not get a 201 status code. Write code to demonstrate the extraction information from the response to figure out the problem with your request. (Printing the status code is not sufficient.)

23.40 The reading discusses several "common mistakes" including attempting to POST a body that is not a JSON-formatted string. In this exercise, you will write code intentionally to one of these mistakes and observe the behavior.

1. Start by replicating and successfully executing the code to obtain a guest session id, and name it `guest_session_id`.
2. Repeat the code from the book to perform the POST but do NOT convert the dictionary with `value` mapping to a number into a JSON-formatted string.

   - Run this code and print out the status code and the text body of the response.
   - From the `request` object (obtained from `response.request`), print out the body of the request.

23.41 The reading mentioned the TMDB endpoint `/discover/movie`. Many examples of what can be done with this endpoint are provided at the following link:

```
https://www.themoviedb.org/documentation/api/discover
```

However, as always with APIs, it is unwise to rely solely on examples. Rather, you must read and understand the API documentation. Navigate the TMDB API documentation and find the URL that explains how to use the endpoint `/discover/movie` (the place with the "Try it out" tab). Record that URL as the answer to this question.

23.42 Write a function

```
getGenreDict(my_api_key)
```

that returns a Python dictionary mapping genre names to id numbers, e.g., `{'Action':28, 'Adventure':12,...}`. Your function should only make a single call to `requests.get()`, with the appropriate endpoint-path for TMDB and with the given API key. As always, return `None` if something goes wrong.

23.43 Write a function

```
searchPersonId(api_key, name)
```

that uses the `/search/person` endpoint to conduct a search for the given `name` and returns *just the* `id` of the first entry in the `results` list. If you solved `searchPerson()` in the last exercise set, this only needs to go a bit further to retrieve the first element and the id from that first element.

As always, return `None` if something goes wrong, but return `-1` if there were no results.

23.44 Suppose you want to generate a list of movies featuring your favorite actors/actresses and in the genres you like. Suppose `actorList` and `genreList` are lists of strings, like "Humphrey Bogart" or "Adventure." You can use your functions `searchPersonId` and `getGenreDict` to find the ids associated with these strings.

After studying the API documentation and examples for the `/discover/movie` endpoint, please write a function

```
makeBigMovieLoD(api_key, actorList, genreList)
```

that returns a LoD with one dictionary per movie, with columns `title` (that is, "original title"), `id`, `popularity`, `language`, and `overview`. For full credit, your function should use a single call to `requests.get()` and should include movies with *all* of the actors on `actorList` and with *any* of the genres on `genreList`. Be careful with your ANDs and ORs to solve this problem as stated. If you come across a non-existent actor or genre (e.g., "adventure"), do not include it in the search. Please sort your results by popularity, from most popular to least popular.

23.45 Write a function

```
makeBigMovieDF(api_key, rating, year)
```

that returns a data frame consisting of *all* US movies whose rating is less than or equal to the given rating (e.g., G, PG-13, R, etc.) in the given year, sorted by revenue, from largest revenue to smallest. Your columns should be `title`, `release_date`, `popularity`, `vote_average`, and `overview`.

Be careful, as your initial invocation of `requests.get()` might only return page 1 out of some large number of pages. You should be sure to get the information from all the pages. A helper function would be wise. The filter on year should use `year`, not `release_date`.

Hint: study the documentation for the `/discover/movie` endpoint carefully. What we have called rating they call something else.

Chapter 24
Authentication and Authorization

Chapter Goals

Upon completion of this chapter, you should understand the following:

- The differences between authentication and authorization.
- The roles of provider, client, and resource owner within the context of security.
- How authentication is achieved between provider and client and client and provider.
- The concept of delegated authority and the differences from impersonation in providing clients access to protected data.
- The process involved in OAuth2, the framework in the Internet used for delegated authority.

Upon completion of this chapter, you should be able to do the following:

- Construct a valid "basic authentication" as defined by HTTP for use in secure authentication over an encrypted communication stream.
- Program the client steps for negotiating delegated authority between provider, client, and resource owner.
- Use the access token achieved in the above for subsequent data requests from an API provider.

© Springer Nature Switzerland AG 2020 757
T. Bressoud, D. White, *Introduction to Data Systems*,
https://doi.org/10.1007/978-3-030-54371-6_24

We conclude Part III of this book, our exploration of data sources and data acquisition, with coverage on the aspects of security that impact the writing of data systems clients that must access data that is, in some form, *protected* at the data systems provider.

In this chapter, the goal is to understand the separation of the entities of the data system client application and the provider from the resource owner, to understand the security concepts of authentication and authorization, and to see how a client application can securely perform authorized operations on behalf of one or more resource owners.

This chapter builds to the use of OAuth2 [24, 23], an Internet standard framework that encompasses delegated authority, whereby a resource owner can convey permission for a limited set of operations to be performed by a client application. The use of OAuth2 assumes understanding of client application interaction with providers using APIs via HTTP, including specifying parameters using the various mechanisms of resource path elements, query parameters, header field-values, and data in a POST body. Coverage of these topics may be found in Chaps. 20 and 23.

24.1 Background

24.1.1 Principals

At the onset of this book, in Chap. 1, and reiterated in Chap. 18, we discussed the *principals* as the entities involved in client–server based data systems. The principals include the data system provider (provider), presenting an interface and allowing access to the data resources it provides, the data system client (client application), responsible for acquiring the data from various providers and consolidating and transforming the data in preparation for analysis, and the resource owners, representing the users or organizations that are the source/origination of the data, and whose ownership and privacy must be protected by unwanted access.

Figure 24.1 depicts these entities and their interactions. As we have seen through earlier chapters, we have client applications that interact with a spectrum of providers, from web servers to database management systems to API service providers; the client application makes data requests and receives data replies from the provider. The resource owners, while they can also make requests and receive replies, are also the entities that are the source of the data and any changes or appending of data on their behalf. Pairwise between all three category of principals are edges labeled *Auth*. These represent the communication interactions to provide security in the form of both *authentication* and *authorization*.

As an illustrative example, suppose we have a provider that is a vendor for wearable fitness devices. The provider collects data updates from its customers, the resource owners, with fitness data like steps taken over time, measures of heart rate, etc. These data are private to each resource owner and constitute the

Fig. 24.1 Data systems
principals

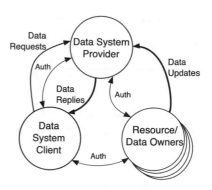

protected resource. To upload such data, authentication identifies the particular
resource owner, and based on that identity, authorizes the appending of data to that
resource owner's private data. External to the resource owners, a client application
might be created that, with permission of resource owners, compares and contrasts
fitness metrics across one or more subsets of resource owners based on categories
like age or gender or other metric. Of interest to us in this chapter is how the resource
owners can interact with client application and the provider to control what data can
be accessed and/or limited, and how the client application can then interact with the
provider and, in a secure and authorized way, obtain the protected data.

24.1.2   Authentication and Authorization Concepts

We define more precisely our fundamental security terms, starting with authentication:

Definition 24.1 *Authentication* is an act, process, or method of showing something
to be real, true, or genuine [32].

In the data systems context, we wish to authenticate (i.e., show to be genuine
or prove the identity of) one or more principals in a system. Authentication can
be considered in both directions between all pairs of principals. Providers need to
authenticate resource owners before they are authorized to change or append to their
protected data. Resource owners and client applications both need to authenticate
providers to ensure that private data is not being exchanged with a malicious third
party, pretending to be the provider. And providers need to authenticate client
applications as part of their obligation to protect the data of their resource owners.
This last can involve authentication of the *developer*, the person or organization
responsible for designing and coding the client software, and also can involve
authentication of an executing instance of the client application itself.

Authentication involves the entity being authenticated proving their identity by:

- something they (and only they) *know*, like a password or other secret,
- something they (and only they) *have*; such as a bio-metric, like a fingerprint or unique facial structure, or by a device like a smart phone.

Multi-factor authentication is simply the layered use of multiple authentication mechanisms. For instance, a developer might prove who they are through the use of both a password and by having, and responding to, a notification on a smart phone.

Separate from authentication is the foundational security concept of authorization:

Definition 24.2 *Authorization* is the function of specifying access rights/privileges to resources [83].

Resources can be files, or can be data sets or subsets of data. The term, authorization, in common usage, can be both a verb (the function or process) and a noun (the "right" conveyed). An authorization defines an access policy: what entity is permitted to access a given resource, and through what operation(s).

Authorization is relative to a particular identified principal, and so authentication must precede authorization so that a principal is known and verified before access to resources is granted. For instance, a resource owner, once identified, might have implicit authorization to both read and write all of their own private data.

On the other hand, a third party, like a client application, must be conveyed *delegated authorization* if they wish to act on behalf of a resource owner and be granted *limited* access rights for the private data of any given resource owner. This type of authorization is explicit, because it involves the permission of the resource owner in a multi-way exchange with both the provider and the client application. Further, the permissions granted should have limits, and should be restricted to certain data and/or operations, like read versus write authority. Permissions can also be for a limited time period, or until revoked by the resource owner. The extent of permissions and/or the set of permissions granted are called the *scope* of the delegated authorization.

24.1.3   Impersonation

If authentication of a resource owner by the provider is by something the user/resource owner *knows*, like a password, then an alternative solution for client access to protected resources would be *impersonation*. In this case, the resource owner would convey their credentials (e.g., user name and password) for a particular provider directly to a client application. The client application could then authenticate *as if the client application was, in fact, the resource owner.*[1]

---

[1]The reader should note that conveying a password to another party violates the assumption that only one principal knows the password, and subverts authentication.

Impersonation has significant disadvantages. First and foremost, the resource owner is essentially giving to the client application full access to their data. A malicious or hacked client application could then delete or modify the data of a resource owner, instead of just having limited access. The resource owner, in order to rescind access, has no choice but to change the password. Further, once possessed by the client application, the credentials of the resource owner might be obtained from the client application by a malicious third party. Also, if the resource owner uses the same user name and password to other providers, those other providers could be compromised as well. For all of these reasons, impersonation in not viable for our goals, and will not be considered further here.

24.1.4 Encryption, Keys, and Signatures

Three other concepts that are helpful in understanding the security mechanisms we use are *encryption*, *keys*, and *signatures*. While a full and detailed understanding of these ideas are beyond the scope of this book, building a conceptual/intuitive understanding will enable better understanding of how the security techniques in the remainder of the chapter can perform their function.

Encryption is an algorithmic process that takes the bytes of a communication that could otherwise be visible to an onlooker/eavesdropper, like a text sequence or set of numeric values, and transforms them into a cipher (also correctly spelled cipher) byte sequence that cannot be understood/interpreted by an onlooker. We learned in Chap. 19 that eavesdropping on communication links, particularly on shared media like WiFi or cellular, is easily accomplished. *Decryption* performs the reverse algorithm,[2] taking the bytes of a cipher byte sequence and transforming them back into their original form. In modern cryptography, the algorithms themselves are *not secret*, as algorithms can be reverse-engineered and would not stay secret for long. Encryption is fundamental to providing the security aspect of *privacy*, ensuring that communication between two principals remains confidential to just those two parties.

A *key*, or more precisely a *cryptographic key*, is a value that is input to either an encryption or decryption algorithm. An encryption algorithm uses both a key and the communication bytes as input and generates the cipher byte sequence as output. A decryption algorithm uses a key and the cipher byte sequence as input and generates the original bytes as output.

The simplest and most efficient encryption/decryption algorithms use the *same* key as input to both algorithms. In this case, the key is said to be *shared* and called a *secret* that must only be known to the two principals involved. Because of the use of the same key, this form of encryption is known as *symmetric* encryption. This type

---

[2]The encryption and decryption algorithms must be compatible with each other, but are not necessarily identical.

of encryption, while efficient, has drawbacks that require us to use other additional methods to accomplish our goals.

In *asymmetric* encryption, *different* keys are used for encryption and decryption. A principal (typically the provider, in our context[3]) has both a *public key* and a *private key*. The public key, as its name indicates, is *not* a secret, and can be well known, distributed, or posted, such that any party can obtain the public key. The private key, generated at the same time as the public key, is a secret that is *only known* to a single principal, the entity that generated the key pair. The key pair can be used such that data encrypted with one of the two keys can only be decrypted with the other key in the pair. So data encrypted with the public key can only be decrypted with the private key, and data encrypted with the private key can only be decrypted with the public key.

Interestingly, if communication data is encrypted with the public key of a principal, and if an onlooker can obtain the cipher data, the possession of the public key and the cipher data does *not* allow a transformation to the unencrypted data. Only if an onlooker could guess the unencrypted data and encrypt the guess with the public key and then see if the resultant cipher data matches could the process be broken.

An entity can also encrypt a communication with their own *private key* to obtain cipher data. Any other entity could use the corresponding public key to decrypt the cipher data and, if the transformation yields an interpretable result, then the data could *only* have been generated by the entity that holds the private key. This, then, gives us one mechanism for authentication.

There are times when, as opposed to encrypting data, we need to take a set of data and ensure that its contents have not been altered and that the contents are attested to by a known entity. This concept, like its analog in the non-digital world, is known as a *signature*, or more precisely in this context, a *digital signature*. A digital signature encompasses a collection of text and/or set of numeric values (i.e., *what is being signed*) and, by incorporating an aspect of the contents of the data, and by using properties of asymmetric encryption, allows the data and digital signature to be communicated to a recipient, such that the recipient can:

- detect if the data has been altered in any way,
- verify that the digital signature has not been forged,
- verify the identity of the signer.

Note that the mechanism of a digital signature does *not* provide for privacy of the data itself.

In general, our data systems use a *combination* of mechanisms and techniques to perform required aspects of security, including the mechanisms for authentication, for maintenance of privacy during communication, and mechanisms to convey delegated authority between the three principals. In the following sections, we focus

---

[3] We will see in Sect. 24.2.1 that *certificate authorities* are another example of a principal that generates and uses asymmetric key pairs.

on the specifics of how these security mechanisms are employed together to solve the data systems problem of delegated authority and secure and private acquisition of protected resources from providers by client applications.

24.1.5 Reading Questions

24.1 The reading discusses the example of wearable fitness devices. Please demonstrate your knowledge of the terms "authentication" and "authorization" via a different real-world example, where data privacy is a serious concern, and where there are resource owners, client applications, and providers.

24.2 Please give a real-world example (perhaps from your own experience) of impersonation as defined in the reading, and how it could go badly.

24.3 If you wanted to set up a secure encryption/decryption algorithm using a shared key, between yourself and some other person we will call X, could you email the key to X? Why or why not?

24.4 The Caesar cipher, used by Roman general Julius Caesar, involves encrypting a string by shifting all letters by some fixed number. For example, you could encrypt "sentence" to "vhqwhqfh" by shifting all letters to the right by three (since v is three letters after s). If a letter is at the end of the alphabet, the shift takes you to the beginning, e.g., y is sent to b. In this example, please identify the key. Then describe a decryption algorithm and explain whether or not it uses the same key.

24.5 The reading describes a signature as a way to ensure that a document (or data) has not been altered, and that its contents are attested to by a known entity. In the real world, when you sign a document does your signature have this effect? Explain.

24.6 Please use a search engine to read about the "man in the middle attack" and report what you find. Then describe how signatures help prevent this attack and keep our data secure from it.

24.7 Please hypothesize (or use a search engine to read about) how public and private keys can be used to encrypt information in such a way that it cannot be decrypted (even if the encrypted text and the setup communications are sent over an open channel) by an eavesdropper without the private key. You would learn more about this topic in a full course on cryptography.

24.8 Please hypothesize (or use a search engine to read about) how computers can generate digital signatures that cannot be forged by a malicious entity. You would learn more about this topic in a full course on cryptography.

24.2   Authentication and Privacy

A number of our security requirements may be satisfied through the use of `https` as a protocol/scheme for communication between two data systems endpoints.

In particular, the `https` scheme can provide

- authentication of the *provider* to another principal,
- privacy of the contents of communication (data and any further authentication/authorization information) between the provider and another principal.

The protocol scheme of `https` is realized by the operation of HTTPS and is covered in Sect. 24.2.1. HTTPS gives privacy and authentication of the provider to the client, and so, notably missing is authentication in the other direction, from the client application to the provider. This often uses an authentication method defined as part of HTTP, not HTTPS, and will be discussed in Sect. 24.2.2.

24.2.1   HTTPS

The goal of HTTPS is to provide the full set of regular HTTP request/response messages (as covered in Chap. 20) over an *encrypted*, and therefore private, reliable byte-stream. This is accomplished by adding a security layer *below* HTTP and *above* the socket interface to TCP in the network protocol stack, as depicted in Fig. 24.2.

The operation of HTTPS can be partitioned into four phases of communication exchange. The first three phases (A, B, and C) provide for authentication of the provider/server as well as the establishment of privacy through an encrypted reliable byte-stream. These are part of a protocol specified by the Secure Socket Layer

Fig. 24.2 HTTPS in data systems

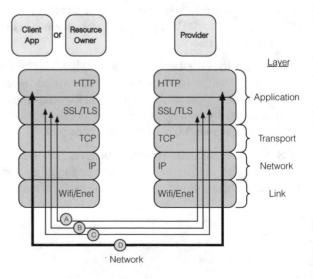

(SSL), which supersedes its predecessor, a protocol known as Transport Layer Security (TLS). The fourth phase (D) is simply standard HTTP.

Phase A: Setup

An HTTPS session begins at the socket interface to TCP, establishing a conventional TCP connection, but, typically, to port 443 instead of port 80. The connection is initiated by the client side, and in this case our client side might be the client application, or it might be the resource owner. The logical host name of the provider is obtained from the target URL of the HTTPS request, and is translated into an IP address and packets are exchanged to establish the TCP connection, like any regular socket setup. This part has no security aspects involved. After TCP connection, the setup is furthered by the client side sending a message requesting an SSL/TLS connection and conveying the client-side cryptographic capabilities. The server completes the interaction of this phase by accepting (if it chooses to) and conveying its own cryptographic capabilities. An eavesdropper could see and interpret all of the communication in this phase.

Phase B: Certificate Exchange

The next goal, once both parties are intent on establishing a secure connection, is to obtain a *certificate* from the server side.[4] A certificate is a set of text bytes in a message that authenticates the identity of the sender, even though the message is sent in the clear, and could be observed by an eavesdropper. This is accomplished by trust and digital signatures. As explained earlier, a digital signature is a cryptographic mechanism that allows the signer to be verified, even though the signature itself is not private, and uses the public/private asymmetric keys of an entity known as a *certification authority* to do its job. Recall that these digital signatures cannot be forged, nor the message information changed without detection. If the client side receives a certificate that is digitally signed by an entity that it *trusts*, like one of the few global and well-known certification authorities, then it can trust the information. The information in a certificate (beyond the signature) establishes the validity of the destination address of the provider, and also includes the *public key* of the provider, to be used in Phase C.

Phase C: Key Exchange

By the end of Phase B, the client side is assured of the identity of the provider and has the public key of the provider. Based on the highest level of capabilities supported by both communication endpoints (from Phase A), the client generates a random secret key to be shared and used for encryption of the data in the duration of the communication session denoted by Phase D. This secret key is encrypted with the public key of the provider and sent across the communication channel. Data encrypted by a public key cannot be decrypted by any entity but the provider, which uses its private key to decrypt and obtain the shared secret key. Alternative to a unilateral client-side generation of a shared secret key, for additional security, the

---

[4]The specification of SSL allows for a client side to provide its own certificate, but this is, in practice, rarely utilized.

client side and provider can cooperate using a protocol known as Diffie-Hellman Exchange to exchange and determine a shared key. In either case, at this point, both parties have the same secret key, to be used for encryption of the data payload of all further communication in the connection session.

Phase D: HTTP Session

Once Phase C is complete, the two endpoints have a secure means of communication (i.e. an encrypted reliable byte stream) and the provider has been authenticated. For the duration of the persistent connection, messages are exchanged using standard HTTP. This means that *everything* other than the IP address of the two endpoints is encrypted and private. This includes

- any user id and password, even if it appeared as part of the original URL,
- the HTTP method/verb,
- the HTTP status reply,
- the requested resource path,
- any query parameters,
- all header fields in the request and the response, including cookies,
- POST or PUT body data,
- response body data.

Clearly, given the overhead work involved in Phases A through C, it is very desirable, from a performance standpoint, to use a single SSL/TLS connection for multiple HTTP request/response exchanges, and to thus support and use persistent connections.[5]

It is tempting to think of HTTPS as having the ability to solve many of the security issues we face in data systems. So it is important to well understand, based on the capabilities discussed above, what HTTPS does *not* do.

In particular, HTTPS does not:

- hide or obscure the identities of the endpoints of communication,
- hide the timing or frequency of communication between endpoints,
- keep data private once it has made it up the protocol stack on the sending or the receiving machine; data is only private within the protocol levels in the operating system and below.

This last point sheds some light on why certain facilities, like the developer tools in most browsers, are able to explore the HTTP content of an HTTPS channel.

---

[5]Recall from Chap. 20 that *persistent connections* refer to the use of a single socket connection to be used for more than one request-reply exchange, and are achieved by specifying a "keep-alive" header line in the request.

24.2.2 HTTP Authentication

As explored above, HTTPS provides authentication of a server, specified as the endpoint of an `https`-scheme URL. The server, for us, is most often the provider. For a full solution of delegated authority with separate resource owners in a full data system, we must also consider the other authentication possibilities:

1. A developer of a client application needs to authenticate to the provider and set up an identity so that the provider is aware of the client application and its needs.
2. Each resource owner must authenticate to the provider, both when managing their own resources, and also when giving approval for a client application to access certain data on their behalf.
3. Each resource owner may need to authenticate to the client application, when setting up the service provided by the client application, or when participating in the exchange that gives delegated authority from the provider to the client application.
4. A client application may need to authenticate to the resource owner, so that the resource owner knows they are engaging the correct endpoint for the client application, and not some malicious impostor.
5. An executing client application must authenticate to the provider.

Most of these authentication contexts and their solution are outside the scope of our goals here. Many may be provided for through a combination of web servers working with the interactive features of a web browser (or *web agent*[6]). Our focus, in the remainder of this section, is on possibility 5. We want to understand how the clients that we write might need to use mechanisms provided for by HTTP to allow the client to authenticate to the provider.

The specifics are mandated by the provider, but the most common solution is the one discussed in the remainder of this section, and employs an authentication mechanism defined by HTTP. The standard for HTTP authentication comes in two flavors—"Basic" and "Digest", and uses *headers* of HTTP requests and replies to provide information in both directions. For instance, a client might issue a request for a protected resource and be denied, but a response header could convey the provider's expectation for authentication. If communication is occurring over HTTPS, which provides privacy of the transmitted and received headers, the simpler "Basic" flavor of authentication is adequate, and is used in most instances, and will be covered here.

Providers can also specify, through their API, other ways of conveying authentication credentials. For instance, as one of the steps in OAuth2 (see Sect. 24.3), a provider may allow or require an application to send a client identifier and a client secret that are specified through URL query parameters sent with an HTTP request.

---

[6]The vast majority of the time, a web agent is a browser, such as Internet Explorer or Safari or Google Chrome, but could also be realized by utilities such as `curl` or `wget`, or be programmatically implemented in a library.

If communication is over HTTPS, this is just as secure as using HTTP request and response headers.

24.2.2.1   Basic Authentication

A client side providing authentication information via HTTP to a server side could be in response to an HTTP request to a protected resource without appropriate authentication. In this case, a request is made where the resource path specifies a protected resource.

The response is an HTTP response with a status code of (typically) `401 Unauthorized`. This response should be detected and treated as a *challenge* for the necessary authorization. Such a response should have a response header field with the following syntax:

WWW-Authenticate : Basic realm = "*realm-area*"

This challenge is specifying the type of authentication required by the server. Here the challenge is for the `Basic` flavor, along with a *realm* specifier. The *realm* should be thought of as an area of the resource tree to be governed by this authentication challenge. When a client provides authentication, either because of a challenge or as a proactive measure because it is part of a framework scheme like `OAuth2`, the authentication is conveyed as part of an HTTP request, and is specified in a header field of the request. The general syntax of the authentication header reflects the type of authentication (and should correspond to the challenge) along with the credential information itself:

Authorization : Basic *base64-credentials*

Credentials are composed of two parts, a user/client id and a password/secret. These two components, as strings, are colon separated, string encoded into a byte sequence, and the byte sequence then encoded into a form known as *base64*.

If we call the two pieces of information *id* and *secret*, the base64 credentials in basic authentication are given by:

$$base64\text{-}credentials \models base64\text{-}encode(char\text{-}encode(id \,||\,':'\,||\, secret))$$

where *id* and *secret* are strings and || denotes string concatenation.

The following example shows how, using credential information in Python string variables `userid` and `secret`, we can construct the header line needed for HTTP Basic authentication, assuming the server expects UTF-8 encoding of credential strings. We use the standard Python library `base64` to provide the algorithm for generating the base64 encoding:

```
import base64

userid = 'jdoe'
secret = 'mypassword'

string_creds = "{}:{}".format(userid, secret)
bytes_creds = string_creds.encode('utf-8')

base64_creds = base64.standard_b64encode(bytes_creds)

str_creds = base64_creds.decode('utf-8')
header_line = "Authorization: Basic {}".format(str_creds)
header_line
```

| 'Authorization: Basic amRvZTpteXBhc3N3b3Jk'

Note well that the base64 encoding of credentials is not, by itself, secure. The process can be reversed, and, given the base64, any malicious user could obtain the id and the secret. The following code starts with base64_creds, whose value from the above code segment is b'amRvZTpteXBhc3N3b3Jk', and can easily obtain the userid and secret:

```
base64_creds
```

| b'amRvZTpteXBhc3N3b3Jk'

```
encoded_bytes = base64.standard_b64decode(base64_creds)
string_creds = encoded_bytes.decode('utf-8')
string_creds
```

| 'jdoe:mypassword'

This is why Basic authentication *must* occur over a secure connection (i.e., HTTPS) to provide the privacy of the header, and thus the authentication information.

24.2.3   Authentication Considerations

For many providers and for other contexts, authentication may be solved in other ways beyond the HTTPS-based and HTTP Basic authentication covered in this section. Consider the following scenarios.

- A provider may decide to allow a client access to resources by using the http scheme (and foregoing privacy), but still want authentication. In this case, the provider may issue a challenge for "Digest" authentication. Digest authentication entails additional complexity, for instance, the use of *cryptographic hashes*, which are beyond the scope of the current presentation.

- Client and provider are not using `http/https` at all. For instance, the application protocol might be the `MySQL` protocol. For other protocols, these authentication mechanisms must also be addressed, but are again beyond the scope of our current goals.

Another consideration that often arises, when designing and building client applications, involves decisions on what libraries and packages to employ. Most of our work in Part III of this book has limited the use of libraries and packages to a minimal set and selecting the most common. But in the area of security, authentication, and authorization, there are many packages that can assist in the process, and can present an object-oriented abstraction for various security concepts. In a real development setting, these may be very worthwhile to explore and to use.

24.2.4   Reading Questions

24.9 When you use your personal computer to access a website from a browser, are you on the client, the resource owner, or the provider? What about when you are running a program or notebook that accesses an API by a provider?

24.10 What does TCP stand for, and why does the HTTPS session begin there?

24.11 The reading mentions "globally known certification authorities." Please use a search engine to learn a bit more about this topic and report what you found. In particular, how/when does the host of a website get a certificate? Do they do it every time a client navigates to their web page? Under what circumstances would a certificate be denied?

24.12 The reading points out that the client side receives a certificate from the server side. Does the server side also require a certificate from the client? Why or why not? If it did, would that solve the problem of the client authenticating to the provider?

24.13 During the key exchange, the client generates a secret key, encrypts it using the public key of the provider, and sends it to the provider. If an onlooker sees this transmission, can the onlooker determine the secret key? Explain.

24.14 Why is it that, in Phase D, an onlooker cannot steal a user id and password typed by the client? If the client had used `http` instead of `https`, would this still be true? Explain.

24.15 The reading highlights that, even with `https`, we cannot "hide or obscure the identities of the endpoints of communication." Is security for that kind of information important?

24.16 The reading highlights that `https` does not keep data private "once it has made it up the protocol stack." Why does this mean you can see HTTP content using developer tools? What does it mean about the security of your data on the provider's servers? Can you think of a real-world example where these considerations have mattered?

24.17 Please give an example from your own experience of a time you got a 401 authorization error.

24.2.5 Exercises

24.18 Suppose you were able to observe, over WiFi, the following HTTP header line in a TCP packet:

```
Authorization: Basic YmlsbGJpeGJ5OnNoaGGhfdmVyeV9zZWNyZXQ=
```

Write a sequence of code to determine and assign to variable `user` and `password` the values encoded as part if thus base64 encoding of credentials. You may assume the character-to-bytes encoding is UTF-8.

24.19 Write a function

```
parseBase64Creds(headerline)
```

that, given a header line (from an HTTP request message) that includes the *entire* Basic HTTP authorization header, parses it and returns a tuple of the user id and password.

24.20 Write a function

```
base64Creds(user, password)
```

that returns a *dictionary* that, in the manner expected by the `requests` module for specifying header lines on a request, creates a new dictionary with a mapping appropriate for HTTP Basic authentication.

24.3 Authorization

As defined in Definition 24.2, authorization governs the access rights of a principal to particular resources, and comes into play after authentication of the principal that wishes to access and to operate on a particular set of resources.

As noted in Sect. 24.1.2, if we are in a situation where the *owner* is the principal that has been authenticated, authorization can be *implicit*. By virtue of the authentication, the owner can automatically have rights for operations on their own data. Our focus in this section, relative to the principals shown in Fig. 24.1, is where the entities are separate, and it is the client that desires access to resources of the owner.

In this case, there are just two possibilities. Either the resource owner conveys their user id and password to the client, and enables impersonation, or else the client must obtain *explicit rights*, known as delegated authority, for some subset of the owner's resources, and for some subset of the possible operations on those resources.

We dismiss impersonation as a solution for its previously discussed limitations:

- it gives too much power to the client,
- there is no ability for the resource owner to revoke the permission, without a full password reset or resource owner account deletion, and
- the client holding of user ids and passwords leads to the possibility of compromising the resource owner's password and identity.

The concept of delegated authority should, in fact, be familiar to most readers. Whenever we interact with an app on a mobile device or on a streaming device and are sent to a provider to give permission in a dialog to access information from Facebook or Google or Dropbox or iCloud, we are acting as a resource owner giving delegated authority and we are using OAuth2. We now proceed by describing and exploring the framework known as OAuth2, the most commonly employed mechanisms for delegated authority in the Internet and used by API service providers.

24.3.1   OAuth2 Background

A data system provider has an obligation to its resource owners to protect their owned resources and to only allow properly identified and authorized clients appropriate access. At the same time, for many providers, it is in their best interest to promote a rich set of client applications as both value to their customers, the resource owners, as well as building sustained demand for their service.

Toward that end, providers have come to understand the value of delegated authority and the need for a standard framework to accomplish such an exchange between the three principals in a secure way. This framework should allow flexibility for various types of providers, client applications, and resource owners.

Within the auspices of OAuth2, there are a number of related processes (flows) of the framework that can be used to achieve different goals, and are based on different underlying assumptions. The objective of a flow is for an application to acquire an *access token*, which represents an authorization for some resource(s) and some operation(s) on those resources. In Table 24.1, we summarize these various flows, known as *grant types*, before we move into the details of the grant type/grant flow most applicable to our use in building client applications in our setting of data systems.

Table 24.1   OAuth2 grant flows

Grant flow/type	Summary description
Authorization code	Full delegated authority with separation of resource owner from client application, and execution environment of the client permits "keeping a secret." Used by *web server apps* and the so-called *native apps*, and, at its core, it first acquires an authorization code which it then exchanges for an access token
PKCE	Proof Key for Code Exchange (PKCE) is an extension of the Authorization Code Flow to perform the exchange step from public clients, thereby allowing for mobile and JavaScript type applications
Client credentials	This is the OAuth equivalent of an API key seen in the last chapter, where the access token is *not* for a particular user, but applies to the application instead
Device code	This allows authorization by obtaining a device code and directing a user to enter the code at a known location. Often used in streaming devices to allow easier setup
Implicit flow	Simplified flow that obtains an access token without requiring an exchange, for simpler (but less secure) implementations for mobile and JavaScript applications. This flow is no longer recommended, and an Authorization Code extended with PKCE should be used instead
Password grant	This flow is also not recommended, as it distills down to impersonation

24.3.2   Delegated Authority: Authorization Code Grant Flow

In OAuth2, the process for the full implementation of delegated authority between separate entities of resource owner, provider, and client application is defined by the steps of OAuth2's *Authorization Code Grant Flow*.

The overall goal in this process, defined for *each* resource owner and a given client application, is to convey an explicit permission for approved operations delegated by the resource owner, called a *token*, to a client application. Because this involves three distinct principals, only two of which are communicating at any point in time, and each two-way communication must also involve authentication, the process proceeds in stages. To preclude a malicious entity from observing any of the exchanged information, we presume the process occurs over HTTPS to ensure privacy, as well as provide authentication of the server-side endpoint in such communication.

We can think of the steps involved as being grouped into three stages. In the first stage, the resource owner is central, interacting with the client application, getting and then conveying information to the provider, interacting with the provider to approve of the delegated authority by the client app, and finally conveying the granted permission back to the client application. The granted permission is represented by an object called a *code*.

The code is not enough to allow a holder of the code to access protected resources. As we will see shortly, at the point where the code represents the wishes and permission of the resource owner and is securely conveyed from the provider to the resource owner, there could be a security problem. The web agent of the resource owner could be compromised, the code observed, and the code used by some other party to request protected data. Or, if a malicious impostor impersonated the client app and received the code from the resource owner, that impostor could then, without authentication to the resource owner nor to the provider, use the code to access protected data.

To complete the security between the principals, the code must first be securely conveyed to an authenticated client app, and that client app must authenticate to the provider and pass along the code in order to receive the data request token. This occurs in Stage 2. The end goal of Stage 2 is for the client application to acquire the token. The holding of the token ("bearing" the token) is then enough for a data request to be validated. The token, in some sense, carries both the authority granted along with the verified identity of the client app to request data.

Stage 3 can involve many request/response pairs and operates according to the data request specifications of the provider. This is analogous to our use of HTTP in Chap. 23 where requests must specify all the needed parameters and replies, when successful, result in the requested data. The difference here is that the data may consist of protected resources, and that, in the request, the token obtained from Stage 2 is included as one of the parameters of the request.

Precondition

This is the most complete and secure of the possible flows defined by the OAuth2 framework, and listed in Table 24.1. But, in the model and pattern described here, it is only appropriate in settings where the data system client and its software can be securely run inside a *trust perimeter* and can provide authentication to the authorization server of the data system provider. In particular, this is *not* the case if:

- The software/code of the data system client is executed on a mobile device (smart phone or tablet, for instance), typically of the resource owner.
- The software/code of the data system client is executed as part of the environment of the resource owner (or other outside entity). For instance, if the client software is part of a JavaScript code base that gets downloaded during a web page visit and runs as part of the web browser/agent of the resource owner.

In both of these cases, the data system client code must somehow include the authentication information (the so-called *client secret*), and in these environments where that code is outside the trust perimeter, the code is said to be unable to "keep the secret." Even when using SSL (https) between the location of the client code and the data system provider, the secret is in the clear before any encryption, and may be obtained through reverse engineering of the code and/or files of the data system client. The PKCE flow, which is an extension to the authorization code flow, should be used to handle these cases.

24.3.2.1 Pre-Stage: Application Registration with Provider

Stage Notes
- Happens only once per Client Application

To support the OAuth2 framework, a provider must be able to identify each and every client application. This involves client developers creating an identity with the provider and, for each client application, creating a unique client identifier and a "secret," essentially a password, that can be used at the appropriate time to authenticate that particular client. The id and secret are conveyed from provider to the developer of the client app.

Eventually, a resource owner, typically through a web agent, will be presented by the provider with a screen asking for permission to allow the client application to perform certain operations with their protected data. To support informing the resource owner at the time of this permission request, the client is typically required to provide supporting information, like a logo and web page for the application to allow a resource owner to "vet" a request. When registering an application, the client is also required to supply information that is used to bridge from a resource owner accepting a delegated authority request to conveying that authorization back to the client.

Figure 24.3 depicts the information transfer that must occur prior to the steps that occur for a particular resource owner. On the left side of the figure, we show, for the Data System Client principal, that we have a developer that is using a *web agent*, which might be a standard browser like Google Chrome, Microsoft Internet Explorer, etc. The developer, as a human user, creates login credentials with the provider. On the right of the figure, we show the provider as being composed of at least two separable parts, the *Authorization Server*, which is responsible for all security aspects like authentication, communication privacy, and authorization, and the *Resource Server*, which is responsible for satisfying resource request to obtain protected data. These two entities often have their own logical and IP addresses in the network, and we use *DSA* to represent the address of the Data System Authorization server, and *DSR* to represent the address of the Data System Resource

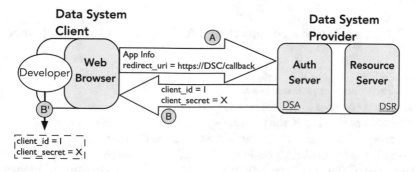

Fig. 24.3 Authorization pre-stage

server. When needed later, we will use *DSC* to represent the address of the Data System Client.

Step *A* represents the developer, through their web agent, supplying information about the client application. This might include a logo for the client organization, a website for the application, prose to be presented to the resource owner at the time they are asked for their permission, and other "soft" information that will inform the resource owner and allows them assurance of the validity of a request for permission/authority. These may be supplied through dialogs or forms as presented by the authorization server.

A critical piece of information supplied by the developer at this pre-stage is known as the *redirect-uri*. This piece of information will be used in Stage 1 of the steps of the OAuth2 framework and will be detailed there, but we note, from the figure example, a couple of salient points about the value of the redirect-uri:

- In the most typical case, the redirect-uri is a URL, and it most often uses an `https` protocol/scheme, so that interactions using the URL in Stage 1, including resource paths, query parameters, and all other HTTP information will be protected from a privacy standpoint.
- The URL has a web server location that uses DSC, our notation for the logical or IP address of the Data System Client. This implies that, when this URL is used, it expects an HTTP service (like a web server) to be executing at the client location, DSC.
- The URL has a resource path, generically denoted here as `/callback` that should correspond to the HTTP service processing needed by OAuth2 during Stage 1 at the client. The actual resource path depends on the client app specifics for how it handles HTTPS requests as part of OAuth2 Stage 1 processing.

If the authorization server at the provider is satisfied with the client application information provided by the developer in this pre-stage, it will generate and give to the developer a *client-id* (I) and a *client-secret* (X), shown as Step *B* in the figure. These often simply appear as fields in a rendered page of the developer's web agent. The developer needs to use these values in subsequent OAuth stages, and so the figure depicts, as Step *B'*, the storage of these values in a file or database repository maintained for the client application.

24.3.2.2   Stage 1: Client Obtains Code with Cooperating Resource Owner

Stage Notes
- Happens once per *resource owner*.
- Acquired *code* has to be kept secure until Stage 2.

The goal of Stage 1 is for the data system client to receive a *code*, an object, as described above, that represents the acceptance of the resource owner for the identified client to perform a designated set of operations on the resource owner's protected data. This set of interactions is depicted in Fig. 24.4 as Steps *A* through *G'*

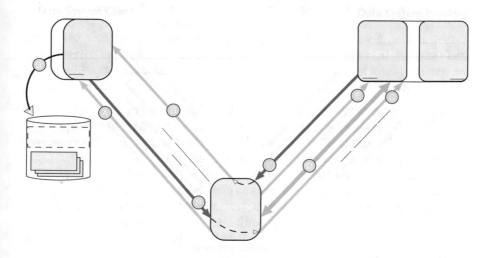

Fig. 24.4 Stage 1: obtain code

and involves all three principals. This is the most complex of the stages in OAuth2 and, depending on the setting, some steps may be aggregated, and some steps may not be directly visible.

We assume in this discussion that there is a web server, or more generically an HTTP service, executing on behalf of the data system client,[7] and that the location/IP address of that web server is denoted by address DSC. We further assume, in accordance with the precondition of using the Authorization Code Grant Flow, that the web server executing at DSC is in a secure/trusted environment, and can thus keep the secret of the client's authentication information.

The resource owner composed of a user running a web agent (i.e., browser) initiates the process in Step *A* through an HTTP request to the client application web server at DSC. This could be initiated by clicking a link in a web page, or clicking a "Sign Me Up" kind of button in a form. The flow must now transition to the Authorization Server at DSA, and so the result of this HTTP request needs to translate into a subsequent request at DSA. As discussed in Chap. 20, HTTP supports a request translating into another request with a different location and/or a different resource path with an HTTP response status of a *redirect*. A redirect response includes a header field, the redirect Location that specifies the information needed for the redirected request. In this case the redirect is shown as Step *B* and the redirect location is a URL whose host address is DSA and whose

---

[7]Note that a web server at the *client* is new for us. Up to now, web servers have been run only at the provider.

resource path gives the authorization endpoint at the Authorization server. The client application acquires these two pieces of information either through developer documentation from the provider, or through the Pre-Stage. In addition to location and endpoint, the redirect location also includes a number of other parameters to be conveyed from the client application to the provider, and are encoded as HTTP query parameters of this URL:

- `redirect_uri`: this is the *same* redirect_uri registered by the developer with the provider in the pre-stage, and will be used in Step F, later in this stage,
- `client_id`: this is the unique identifier for this client application as received from the provider by the client developer in the pre-stage,
- `state`: this is non-guessable random value generated by the client application that is used to align later OAuth steps as part of the same OAuth sequence,
- `scope`: this is a value that conveys the particular operations the client application wishes to have delegated to it; the particulars of the one or more values that are valid for scope are defined and determined by the provider and may typically be found in the provider's developer documentation.

Construction of this URL by the client will be demonstrated in Sect. 24.3.3.1.

If the resource owner's role in this sequence is being driven by a browser as the web agent, the browser will automatically handle the redirect response and initiate Step C, the corresponding HTTP request. Many packages and libraries supporting HTTP will also handle the redirect response and automatically initiate the Step C request. In either case, Step C is an HTTP request directed at the authorization server and containing the path and parameters described above.

As an alternative to a redirect in response to an HTTP request of Step A, the client application web Server at DSC could also generate the full URL used in Step C and provide this link directly for the resource owner to click on. This alternative would avoid the redirect, and effectively combine steps A through C into a single step.

Once the HTTP request in Step C is received at the authorization server, a sequence of HTTP requests and responses occur, shown generically as bidirectional Step D in the figure. Here, the provider is interacting with the resource owner to specify the client, the supporting information of the client, and the set of operations being requested, as determined by the scope. The resource owner, by virtue of HTTPS, is assured the identity of the provider. The resource owner authenticates to the provider in the same manner as usual, using a login form, cached credentials in the browser or the user's operating system, or other mechanism. If the scope has a collection of values for multiple types of operation being requested, the resource owner may select some of the operations but not others. If the resource owner is satisfied with the scope and with the supporting information regarding the client, they come to a point of clicking an "Accept" button, thereby granting permission authorizing the operations specified by the scope. This acceptance is another HTTP request to the authorization server at DSA and is denoted Step E.[8]

With acceptance, the provider generates the code, an object capturing all the necessary information of client, resource owner, and scope of the accepted authority. Per the goal of this stage, the code needs to be conveyed to the client application. This is accomplished by using the resource owner as a kind of "messenger" and using the mechanism of an HTTP redirect. The HTTP request of Step E generates a result of an HTTP redirect as the response from the authorization server of the provider. The redirect location, in the header field of the response denoted as Step F, contains a URL where the location of the URL is the client at address DSC and this location and the resource path are simply the redirect_uri given in the pre-stage and validated by including them as part of the request of Step C. The redirect location in Step F also includes the important information of the code itself, along with the validating information state. The state *must* be the same as generated in Step B.

Again, as a redirect, the processing of response Step F by the web agent of the resource owner into a subsequent HTTP request using the redirect location happens automatically, and results in Step G as an HTTP request to the web server running at the data system client at location DSC.

The client application needs to use the `code` and the `state` from Stage 1 in subsequent OAuth stages, and so the figure depicts, as Step G', the storage of these values in a file or database repository maintained by the client application. Note that these values are maintained *for each* resource owner that goes through this delegated authority process. This part of the OAuth flow is reinforced in Sect. 24.3.3.2.

24.3.2.3   Stage 2: Client Exchanges Code for Bearer Token

Stage Notes
* Happens once per resource owner.

  – Code from stage 1 can only be used once.

* Exchange is *time sensitive*—the code can have a limited lifetime, and must be exchanged for a token within that lifetime.

The missing piece, to this point, is the authentication of the client to the provider, coupled with the code obtained by the client application in Stage 1. This is realized as a simple HTTP request and response between the client application and the authorization server of the provider. The end result, once the client application is authenticated and proves to be in possession of the code, is the sending of the token from the provider to the client. The interaction of this stage is shown in Fig. 24.5, and

---

[8]This is the step that should be familiar to anyone who has signed up and for a third party service and enables login using Facebook or Google, or uses an app on a mobile device, and gives permission to access iCloud, Dropbox, or other providers of protected resources.

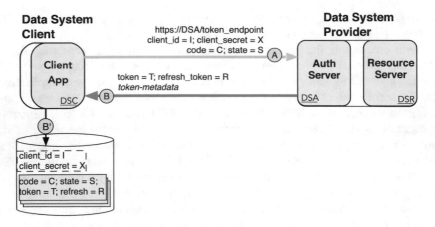

Fig. 24.5 Stage 2: obtain token

also appears in our walk-through in Sect. 24.3.3.3 showing actual code to illustrate the authentication and exchange.

Step *A* is an HTTP request. It uses https to authenticate the provider and uses a resource path to route to the token-processing functionality of the authorization server, and is denoted `token_endpoint` in this illustration. The parameters of the request include the request type of an `authorization_code` along with parameters of the client_id, the client_secret, the code, and the state. The mechanisms for passing parameters vary with the provider, and range from query parameters, header fields, and POST body data (URL-encoded). It is important to understand a particular provider's requirements.

The standard for conveying authentication information is to use HTTP Basic authentication, discussed in Sect. 24.2.2, where the client_id and client_secret are used like a login and password, are `' : '` concatenated together, are encoded using base64, and are placed as the part of the value in a request header field whose header field name is `Authorization`.

If successful, meaning that the code is not expired, the code is valid for the client, the resource owner, and the scope, and the client id and client secret properly authenticate the client application, the authorization server, through its token endpoint, yields an HTTP response with the desired token. Beyond the token, the response can include an object known as a *refresh token* as well as some metadata about the provided token, like its type (`bearer` in our flow), and issue time and/or validity duration. The response may also include the `scope`, particularly if the scope granted by the resource owner is different than the scope requested by the client application in Stage 1. These result parameters are sent as JSON within the body of the response.

It is expected that the exchange of a code for a token happens quickly after the code is acquired by the client in Stage 1. A code could expire in a matter of seconds or minutes and, if expired, the Stage 1 steps must occur all over again. Further, a code is only valid for exchange into a token *one time*. After being exchanged the first time, an authorization server will reject any subsequent request that uses the same code to request a token.

24.3.2.4 Stage 3: Client Acquires Data Using Token

Stage Notes
- Happens multiple times, once per data request for a given resource owner.
- Interaction in Stage 3 is just with the Resource Server, not the Authorization Server.
- Data requests must be monitored for success, and if not successful, a client must determine what the situation might be from multiple possibilities.

  – The token expires and no refresh token was given in Stage 2; here, no recovery is possible, and the flow must begin again.
  – The token expires, and a refresh token was given in Stage 2; here, recovery may be possible, exchanging a refresh token for a new token, as defined in Stage 4.
  – The ongoing data requests exceeded a rate limit or data limit; it may be possible to recover or fix:

   . recovery: wait till enough time elapses to enable resumption of request stream,
   . fix: use meta-information to programmatically take remaining time/request count into account and slow request rate to match the rate limit and *avoid* the error through programmatic means.

With a valid (bearer) token, the data system client application is able to make one or more requests for protected data of the resource owner through the data system provider (Fig. 24.6). Each request, labeled A_i for the ith request, is an HTTP request, whose location is the *resource server* of the data system provider, and whose various data endpoints, HTTP verb, and parameters, are as defined in the developer documentation of the provider. This proceeds in exactly the same way as data requests of a RESTful API with the following extension: each request must contain, as a parameter, the token value obtained during Stage 2. The token parameter, depending on acceptable methods defined by the provider, might be conveyed as a query parameter or as a header field. If as a query parameter, the parameter name is `access_token`. If as a header field, the field name is `Authorization` and the value is `Bearer` followed by the token itself. A provider could also require the token be placed in a POST body in a form/urlencode format.

The token is checked by the resource server and, if valid, the server will respond as appropriate with a response B_i. If a token is not of unlimited lifetime, it will eventually expire. When it does, in request A_{i+1} of the illustration, the HTTP response will have an invalid token error. The client may have to examine more than the response code to distinguish a token expiration error from an error due

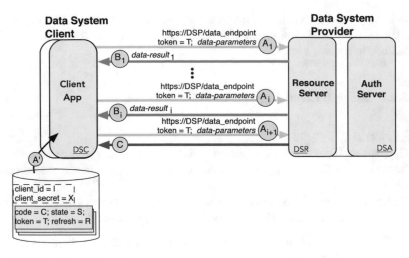

Fig. 24.6 Stage 3: acquire data

to other causes. This could involve examination of the response body. If the latest request failed due to token expiration, we proceed and execute the steps of Stage 4, if applicable, before we can resume data requests anew.

24.3.2.5   Stage 4: Client Exchanges Refresh Token for New Token

Stage Notes
- Occurs after a data request fails, and the reason for the failure is an expired token, and the client had previously received and stored a so-called *refresh-token*.
- There can be other reasons for a data request failing, even after there have been one or more successful data requests. For instance, exceeding some data or request limit imposed by the provider, making a request that exceeds the authority given in the scope, or a resource owner revoking the authority.
    Stage 4 looks much like Stage 2, only this time, instead of exchanging a code for a token, we exchange a *refresh_token* for a new token.

Step *A* is an HTTP request. It uses HTTPS to authenticate the provider and uses a resource path to route to the token-processing functionality of the authorization server, and is denoted `token_endpoint` in this illustration (Fig. 24.7).
    The following are the parameters for this HTTP request:

- `grant_type` of `refresh_token`,
- `refresh_token` with value taken from the client storage associated with this resource owner from the last successful acquisition of a token and refresh token in a Stage 2 or a prior Stage 4,
- `scope` the scope originally granted, or a subset thereof; not required,
- `Authorization: Basic` with client id and client secret.

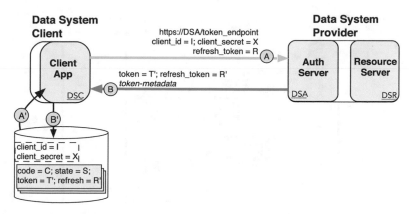

Fig. 24.7   Stage 4: refresh token

The standard specifies the first three as part of a POST body with indicative header fields in the request, and the last is specified as a header field.

Result, shown as Step *B* will, if successful, convey a token and possible refresh token along with token metadata, just like a response in Stage 2. The new information should be stored in the file or database maintained by the client application, as illustrated in Step *B'*.

24.3.3   OAuth Dance Walkthrough

It is helpful to look at the OAuth2 process for delegated authority and the stages of the authorization code flow in its entirety. It can also be instructive to see some code examples that illustrate the key steps. Because of the steps back and forth between the principals, the sequence is often referred to as the *OAuth Dance*, and we do so here as well, with Fig. 24.8 outlining the steps of the dance.

The labels at the top of the figure depict each of the principals involved in the exchange. From left to right we have the *Client Application*, the *Resource Owner*, and then the *Provider*, subdivided into the *Authorization Server* and the *Resource Server*. Time progresses from top to bottom along the vertical lines in the diagram. Black dashed and solid lines are HTTP messages between principals, with the dashed lines representing a *redirect* message, while a solid line represents a regular HTTP message. The double-headed wide gray lines denote multi-step interaction that could involve graphical interface dialogs and displayed web pages as well as underlying HTTP messages. Where a *key* symbol is shown on an arrow, it represents an authentication secret shared between two principals. Numbers in circles on the lines are used to reference the steps in the dance.

We assume, by the start of the summary actions in this figure, that the pre-stage discussed in Sect. 24.3.2.1 has already occurred, and so we begin at the client

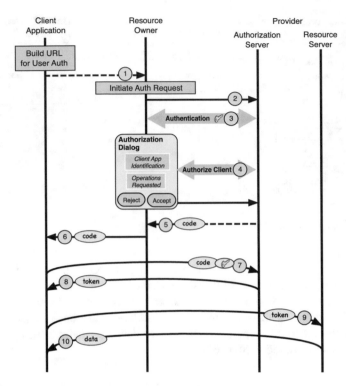

Fig. 24.8 OAuth2 dance for the authorization code flow

and its operation to build a URL that, when issued as an HTTP request by the user/resource owner, will allow that user to both authenticate with the provider and give permission for the delegated authority.

The examples shown below are typical for a provider supporting OAuth2 delegated authority through the *Authorization Code Flow*. While an actual provider would incorporate the same logical set of steps, some aspects of the particulars will vary from provider to provider. Providers supporting an API and OAuth2 authority mechanisms will document their own process in detail, and so the reader should always refer to a desired target provider's documentation and not rely solely on the examples shown here.

To support the code examples, we assume a JSON file named creds.json whose contents will contain a dictionary with a key mapping to a dictionary with the information needed about a particular provider to use in the steps of the example. We also have utility functions, documented in appendix Sect. A.1, for reading (util.read_creds()) and updating (util.update_creds()) a particular entry in the creds file, so that we can use it for recording additional information acquired through the steps, and to enable code that might reside in different parts of the client to be able to access the information.[9]

```
{
  "provider": {
    "client_id":       "304b9ef49da340d8",
    "client_secret": "994ee891af846424715382",
    "redirect_uri":  "https://client.org/code_callback",
    "provider_auth":
        "https://provider.com/oauth/authorize",
    "provider_token":
        "https://provider.com/oauth/access_token",
    "scope": "group_read user_readwrite"
  }
}
```

As can be seen from these file contents, our hypothetical client has a location identified by client.org, and our hypothetical provider has a location identified by provider.com. The provider's documentation will specify the endpoints for authorization and for token support, and here we use illustrative resource paths of /oauth/authorize, and /oauth/access_token, respectively. For our hypothetical provider, we show scope as a string of space-separated values, with the first conveying a right to *read* some group level information, and the second conveying a right to *read* and *write* user level information. For real providers, these scopes may be relatively simple strings, but can also be quite complex in their construction. Again, see the provider's documentation for details on the possible scopes allowed through their API.

```
provider_creds = read_creds("provider", folder=datadir,
                            file="creds.json")
```

24.3.3.1 Build User Auth URL

Starting at the upper left of the summary figure, the client begins by constructing a URL that is incorporated into Steps 1 and 2 of the OAuth dance. The URL, through query parameters, contains information on the identity of the client, the scope of the operations to be performed on behalf of a resource owner, the location and endpoint back at the client used when a resource owner grants delegated authority, and state information that can be used across a sequence of the steps in the dance to thwart cross-site scripting types of malicious attacks.

---

[9]In a real system supporting a client application, some form of database would support the recording and access of this information, along with records for each resource owner and other client information to be shared between the client's web server and the client application modules that need to acquire data.

In the code below, the location and endpoint of the provider to be used as the destination of the built URL are obtained from `provider_creds`, as are the `client_id`, `redirect_uri`, and `scope`. The state to be used for this sequence of steps of the dance is generated as a random string, and the credentials are updated with its value.

```
url = provider_creds["provider_auth"]

state = util.random_string()
provider_creds["state"] = state
update_creds("provider", provider_creds, folder=datadir,
            file="creds.json")

paramsD = {
  "client_id": provider_creds["client_id"],
  "redirect_uri": provider_creds["redirect_uri"],
  "scope": provider_creds["scope"],
  "state": state
}

request = requests.Request("GET", url, params=paramsD)
prepared = request.prepare()
user_auth_url = prepared.url
```

Because we only wish to *generate* the URL, but not to actually issue an HTTP request, we use the `Request` constructor of the `requests` module where we can convey the base URL and the query parameters and build a Request object. This object is then *prepared*, where the `requests` module performs the URL encoding of the query parameters. We obtain the desired generated URL (as `user_auth_url`) from the `url` attribute of the Prepared Request object.

Note that there is nothing about the URL we just built that is specific to a particular resource owner. So a URL could be constructed once and that same URL be used for *multiple* resource owners. Each resource owner would be taking part in their own instance of the OAuth dance.

The URL, `user_auth_url`, comes into play twice in the steps of the OAuth2 dance of the figure. First, it is the value of the `Location` header line Step 1 of the dance, where, on a user action, through their web browser with the web server at the client application, the client application responds with an HTTP redirect. Second, because of the redirect, it becomes the URL for a GET request by the resource owner's browser in Step 2 of the dance, and because of the value of `provider_auth` in `provider_creds`, takes the web browser to the Provider at the Authentication Server.

24.3.3.2   Delegation by Resource Owner

The GET of the built URL from the resource owner to the provider initiates a set of steps between these two principals. First the provider, through HTTPS authenticates to the resource owner, and the resource owner, typically through a username and password, authenticates to the provider. This is Step 3 of the dance shown in Fig. 24.8. The authorization server at the provider, through an interface provided in the browser of the client, then presents what is depicted as the `Authorization Dialog`. This can take different forms, depending on the provider, but allows the resource owner to identify the client application and to examine the set of operations that are being requested for delegated authorization. This is depicted as Step 4 of the dance, and is finalized with the resource owner either rejecting or accepting the authorization for the client to act on its behalf. If the authorization dialog were an HTTP Form, the acceptance would result in an HTTP POST from the resource owner's web browser to the authorization server, shown as the unlabeled solid line from the resource owner to the authorization server at the end of Step 4.

The finalization of Step 4 is responded to by the authorization server by means of an HTTP redirect, shown as Step 5 of the dance. This redirect has a `Location` header line whose value is composed in part by the URL location value included in `user_auth_url` as the `redirect_uri`, and also includes query parameters that repeat the `state` and include the real objective of Steps 1 through 6: the `code` that encodes the delegated authority of the client application to act on behalf of the resource owner. The web browser of the resource owner responds to this redirect by performing a GET request to this URL, which, because of `redirect_uri`, is back to the client application. This is shown as Step 6 of the dance. To recap, Steps 1 through 6 of the summary figure comprise Stage 1 in Sect. 24.3.2.2, whose goal was for the client to obtain this code of delegated authority.

24.3.3.3   Exchange Code for Token by Client

A code obtained through steps 1 through 6 has no meaning unto itself. It is a string, often comprised of decimal digits and letters from the set a to f (i.e., hexadecimal digits), but can be made of other characters too. It is unique to the authorization server and can be used by the authorization server to look up and uniquely identify the three important components of this transaction, the client application, the particular resource owner, and the set of allowed operations of the scope.

To allow our code examples to proceed, we simply show a Python variable, `code`, being set to what would arrive at the client application as the value of `code` in the *query string* of the HTTP GET request from Step 6. In the code below, we add this key-value pair to `provider_creds` and update the pseudo-database in `creds.json`.

```
code = "69ed5cc592908474c37c"
provider_creds["code"] = code
```

```
update_creds("github", provider_creds,
             folder=datadir, file="creds.json")
```

Corresponding to Step 7 of the dance, the client constructs an HTTP request that will serve to provide the code and also authentication of the client and expect an HTTP response that conveys a token, shown as Step 8 of the dance. These two steps correspond to Stage 2 of Sect. 24.3.2.3.

Actual OAuth API providers must specify a location and endpoint to be used for this exchange of a code for a token, and use query parameters to convey most of the arguments needed in the request of Step 7. Often, the client can additionally specify the form of the reply, and both this capability and how it is specified will vary from provider to provider. The HTTP response might contain the token and other returned values as header lines of the response. Or the body of the response may be formatted as JSON or XML and contain the returned values. The developer of an application would need to carefully consult a provider's documentation to determine the particulars for that provider.

In Table 24.2, we summarize the set of information needed in this request and how it may be incorporated in the request as illustrated in our example. For use in our code example, most information is obtained from our `provider_creds` pseudo-database.

For our hypothetical provider, we assume that the provider has specified a POST operation, and that the provider allows an `Accept` header wherein we can request the response format be in JSON. In our example provider, we assume that the request parameters are conveyed as query parameters. Alternatively, a provider could require that a client structure these as form-style key-value pairs, and place them in a POST body.

Care should be taken in checking the success of a POST request, as, for some providers, success is indicated by a 200 status code, while others return a 201 status code.

The following example, based on these parameters and assumptions, makes the POST request and obtains the result. Since, on success, the result is in the JSON format, we parse the result and place the three result values of `access_token`, `refresh_token`, and `expires_in` into our stored credentials.

```
url = provider_creds["provider_token"]

paramsD = {
    "grant_type": "authorization_code",
    "client_id": provider_creds["client_id"],
    "client_secret": provider_creds["client_secret"],
    "redirect_uri": provider_creds["redirect_uri"],
    "code": provider_creds["code"],
    "state": provider_creds["state"]
}
headerD = {"Accept": "application/json"}
```

Table 24.2 Exchange code for token request parameters

Parameter	Description	Incorporation
HTTP method	Request may be either a GET or a POST and is specified in the provider documentation	This becomes the function invoked from the `requests` module
endpoint	The base URL for token exchange, as given in provider documentation	This is used as the `url` in the request invocation
`grant_type`	Used when a provider can accept additional OAuth flow variations. For delegated authority and a secure client, we use `authorization_code`	Query parameter
`client_id`	The unique identifier for this application, generated by the provider in the pre-stage	Query parameter
`client_secret`	The authentication shared secret, generated by by the provider in the pre-stage, and known only to the provider and this client	Query parameter
`redirect_uri`	The URL in the client application where users are sent after authorization. May be used by some providers as an additional check, but may be optional or not present for other providers	Query parameter
`code`	The code query parameter conveyed to the client in Step 6 of the OAuth dance	Query parameter
`state`	The state string generated by the client in the build of the URL and used in Step 2 of the OAuth dance	Query parameter
response format	Options that may be given by some providers on the format of the information returned by the request. A response typically uses the body to convey information like a token, refresh token, and expiration, and may be as URL-encoded key-value data format, as JSON, or as XML	If not specified, providers often have a default format, but if specified, is typically through the `Accept` header line

```
response = requests.post(url, headers=headerD,
                         params=paramsD)
if response.status_code != 200:
    print("POST error:", response.status_code)
    print(response.text)
    exit()

result = response.json()
provider_creds['access_token'] = result['access_token']
provider_creds['refresh_token'] = result['refresh_token']
provider_creds['expires_in'] = result['expires_in']
update_creds("provider", provider_creds, folder=datadir,
             file="creds.json")
```

The above code makes the assumption that the client is acting on behalf of a single user/resource owner. In the case of multiple resource owners, the process, after building the URL, would be repeated per user, and the recording of the token shown in this last example would have to involve more sophisticated mechanisms, as the client would need to have records of the access token for *each* user, and, when making requests, look up the access token appropriate to the request.

24.3.3.4   Data Requests

Once the client has an access token recorded for a resource owner, data requests can then be made to the provider, and, for each request, the client must convey the token as a part of the request. This is shown as Step 7 of the OAuth dance in Fig. 24.8, and the reply from the provider is shown as Step 8. These steps are exchanges between the client and the *Resource Server* of the provider as principals. Also, within the client, the software for making data requests may be handled by an entirely separate part of the client application as that responsible for obtaining the token in the first place. Steps 7 and 8 may be repeated for as many data requests are needed by the client, at least up and until the expiration of the token.

For some providers, the access token is specified as a query parameter of the data request, while for other providers, the access token is specified as header line of the request, where the name of the header is `Authorization` and the value concatenates either the string `Bearer` or the string `token` and the value of the access token. We illustrate both variations in the two examples below.

Suppose our provider, at `provider.com`, supports an endpoint served by the resource server at `/data/v1/me` that obtains user information via delegated authority for the resource owner associated with our access token. If the provider requires the access token be provided as a query parameter named `token`, the HTTP request could be made as follows:

```
provider = read_creds("provider", folder=datadir,
                               file="creds.json")

location = "provider.com"
resourcepath = "/data/v1/me"
url = buildURL(resourcepath, location)

paramsD = {
  "token": provider['access_token']
}
response = requests.get(url, params=paramsD)
assert response.status_code == 200
```

If, on the other hand, the provider specifies that the access token be conveyed through an `Authorization` header naming a `Bearer` token, the same HTTP request could be made as follows:

```
provider = read_creds("provider", folder=datadir,
                              file="creds.json")

location = "provider.com"
resourcepath = "/data/v1/me"
url = buildURL(resourcepath, location)

authHeader = "Bearer {}".format(provider['access_token'])
headerD = {
  "Authorization": authHeader
}
response = requests.get(url, headers=headerD)
assert response.status_code == 200
```

Once the provider-specifics of conveying an access token have been handled, most often in one of the two ways illustrated here, and for as long as the access token is valid (has not expired), other aspects of requesting data and interpreting results from a REST-based API provider proceeds as covered previously, in Chaps. 23 and 21, and using hierarchical formats, as covered in Chap. 16.

24.3.4 Reading Questions

24.21 Please give a real-world example where a Provider would want to give a client a certain type of access but not another type of access. Please be specific.

24.22 Please give a real-world use of the word "code" that aligns with the data systems use of the word. For example, this is not referring to code you type in Python, right?

24.23 Stage 2 describes the need for a token, how a client application can get a token, and issues that would arise if we allowed anyone with the code to access the data. But why are these issues resolved with a token? In other words, how do we know a malicious entity cannot intercept the token when it is being used in stage 3, just as a malicious entity could have intercepted the code at the end of stage 1?

24.24 The reading mentions a trust perimeter and gives two examples that do not count. Please provide one example that *is* a trust perimeter.

24.25 Please give an example from your own experience that illustrates the user experience of the pre-stage. Please be specific.

24.26 The reading points out that the redirect-uri expects a web server to be executing at the client location. When you use Internet at home, do you have a web server? How about when you use Internet on campus? How do you know?

Q If you have experience in web development, or if not, by searching for answers on the Internet, postulate at least two ways that you could run a web server on a personal machine, like a laptop, but without the large-scale web servers like Apache or NGINX.

24.27 Consider the figure illustrating Stage 1, where the client obtains a code. Please give an example from your own experience of a time you have received a redirect-uri. Then describe whether or not your browser automatically initiated the corresponding HTTP request, or whether you were provided a link to click. Please be specific.

24.28 The reading mentions the use of "cached credentials in the browser" as a way to streamline the authentication process when a resource owner is directed to the provider for the delegation step. What happens (and why) if you attempt to use "private browsing mode" to obtain a code from a data system provider, even if you have previously authenticated?

24.29 Consider step E of Stage 1. The reading says if the resource owner (that is, you, using your web browser) is satisfied with the supporting information regarding a client, from an authorization point of view, the resource owner can click "accept." Using what you know about how humans behave, describe if there are security vulnerabilities in this process from a user interface point of view. Illustrate your point with a real-world example.

24.30 Have you ever had an experience with an expired code or token or other "permissions" issued for the purposes of authentication? If so, please describe it.

24.31 Stage 4 lists several reasons why a data request could fail. Please write a justification for why we should design data systems with such failures in mind. For example, setting a cap on the total number of requests helps protect us against a denial of service attack. What about the other reasons for failures?

24.3.5   Exercises

The goal of the exercises in this section is to give real-world experience based on the OAuth2 Authorization Code Flow for delegated authority. A few words of caution are in order:

1. The exercises will use real API providers that support OAuth2 and the flow defined by the Authorization Code Grant.
2. Use of these particular providers is not an endorsement of their services. We do, however, thank them for providing examples and making public their API and documenting their requirements for OAuth2.
3. Providers are out of our control, and so instructions, steps, and exercises that were validated at the time of this writing could *change*, and the reader will need to adapt.

4. Dealing with the real world can be messy, and this can be reflected in these exercises.
5. A reader/student solving these problems will be playing the role of *both* the *client* (or developer) and the *resource owner*. During the solving of the problems, it would be wise to refer back to the figures to help reinforce the role at any particular point in time.

Consistent with the presentation in the section, we will *not* be running a web server or PHP server, that would be used in a real system to handle the HTTP request by the client at the *redirect-uri*. This will expose some additional "messiness" in conveying a code (and state), after delegation, back to the client application.

The following set of exercises use Spotify (tm) as the API provider. We list their developer home page, a link to their documentation for their Web API, and a link to their Authorization Guide.

- https://developer.spotify.com/
- https://developer.spotify.com/documentation/web-api/
- https://developer.spotify.com/documentation/general/guides/authorization-guide/

24.32 The goals of this first exercise are primarily setup and include:

- Create a Spotify account that will be associated with you as a *developer* of a client (if you are an existing user, that is fine to use as well),
- Register your client application,
- Build a `creds.json` with needed information associated with your application.

Once you have a Spotify account, the following summarizes the steps to register an application. We are indebted to our former student Caileigh Marshall for the tool used below.

1. Go to https://developer.spotify.com/dashboard/login
2. On the Dashboard, use the `Create A Client ID` or, equivalently, click the `My New App` template box. 3.In Step 1 of 3

   - Name your app something like: `datasystems-<login>`, using a login of your choosing.
   - Add whatever description you like.
   - Read and accept terms.
   - Click `Create`.
   - You should see a message that your application has been created, and be on a page dedicated to that application.

3. Click the `Edit Settings` button and fill in the form as follows:

   - For `Website`, you can use https://datasystems.denison.edu, or can use some other website that you are affiliated with. For some API providers, a check is performed to see if the website exists.

- For `Redirect URIs`, enter

    - https://caileighmarshall.github.io/cs181project/ (including the trailing slash), and click `Add`,
    - https://localhost/callback/, and click `Add`

- Click `Save` to save your application settings edits.

4. On your application page, click `Show Client Secret`, and leave the page up for the next set of steps.

The Quick Start also has instructions which may be more up-to-date than the specific steps given here.

- https://developer.spotify.com/documentation/web-api/quick-start/

Finally, you will need a creds.json. You are welcome to start by editing a text file and specify an outer dictionary with an initial mapping from `"spotify"` to an empty dictionary.
Assuming `creds.json` is in the data directory, then, start with the line:

```
spotify = util.read_creds("spotify", datadir)
```

Write code and update the `spotify` dictionary above with:

- "client_id" mapping to the string of the client id given in your Spotify Application
- "client_secret" mapping to the string of the client id given in your Spotify Application
- "redirect_uri" mapping to the first of the redirect-uris you added to your application: `"https://caileighmarshall.github.io/cs181project/"`

When complete, your code should write the dictionary back out to your creds file, nominally by

```
util.update_creds("spotify", spotify, datadir)
```

24.33 Suppose our client application will want to, on behalf of a user:

- Create public or private playlists
- Add items to a public or private playlist
- Remove items from a public or private playlist

    Using the documentation at

- https://developer.spotify.com/documentation/general/guides/scopes/

    determine the needed scope(s) and then create a global variable:
`app_scopes` that is a Python string with the set of scopes, and using a single space to separate for multiple scopes. Add to your dictionary with `"scope"` mapping to the value of `app_scopes`, and update the creds file.

24.34 Write code, modeled on the book example, that creates an Auth URL to be given to a resource owner. The URL should have a resource path of `"/authorize"` and the host location should be `accounts.spotify.com`. Based on the OAuth2 requirements, the URL should have query parameters that include:

- `client_id`
- `response_type`
- `redirect_uri`
- `state`
- `scope`

For Spotify, also include a query parameter `"show_dialog"` that maps to the string `"true"`. Most of these can be found in the creds dictionary. For `state`, use `util.random_string()` to generate a random state string, and you should add it to the credentials and update the credentials.

Ultimately, the code should assign the built URL to `user_url`, and to print its value.

24.35 The goal of this exercise is for the *resource owner* to give permission for delegated authority for playlist manipulation, to get an authorization `code` from the provider, and to "give" it to the client application by writing code to set a `code` variable in the client program and update the credentials file.

For a regular application, the `redirect_uri` would take the resource owner to a web server under the control of the client app, which would take the `code` provided, and put it in a database.

We simulate that with copy-and-paste.

1. Copy the URL built in the last step and paste it into the address line of your favorite web browser.
2. As a *resource owner*, you will authenticate and read the authorization dialog and, if you approve, click on the "AGREE" button.
3. The redirect-uri (entered for your app, and repeated in the URL you built) takes you to a website that displays `Copy this code:` with the authorization code generated by the provider. Follow the instructions and select and copy the code.
4. Before leaving the website, look carefully at the address line of your browser (as resource owner) and write down:

    - the host location
    - the endpoint-path
    - the name and value for *each* query parameter

5. Write code that

    - assigns to variable `code` a string with the copied code,
    - updates the credentials dictionary with a mapping from `"code"` to the variable `code`,
    - updates the credentials file.

24.36 Bonus questions. Repeat the build of the URL, but for the `redirect_uri`, specify the `https://localhost/callback/` that was also added to the application on the Spotify Dashboard.

Answer the following:

1. Note what, if any, differences you encounter when you copy this new URL into a browser and in the delegation process.
2. When you "AGREE," what happens during the redirect? DO NOT CLOSE the window.
3. For the web page, focus again on the *address line* in the browser, and write down:

   - the host location
   - the endpoint-path
   - the name and value for *each* query parameter

Explain *why* you saw what you saw, and argue whether or not you could proceed with the setting of the `code` as was done in the previous problem.

24.37 Write code, modeled from the book example, that exchanges an authorization code for an access token. Based on the OAuth2 requirements and Spotify documentation, the request should be a POST, and the POST body should be a *form* with mappings for the following:

- `grant_type`
- `code`
- `redirect_uri`
- `client_id`
- `client_secret`

Based on Spotify documentation, the resource path is `"/api/token"` and the host location is `accounts.spotify.com`.

Verify the response is successful, and add *all* name-value mappings in the JSON-returned data into the credentials dictionary and update the dictionary.

24.38 Given the access token, now in the credentials file, write code that can perform one or more delegated operations on behalf of the resource owner. Possibilities include:

- Create a new public or private playlist
- Add one or more items to a playlist
- Read items from an existing playlist

This is intentionally an open-ended question. Now that the earlier questions have obtained an access token, interaction with this provider, but now on behalf of a particular resource owner, can proceed in the same manner as pursued in Chap. 23.

Appendix A
Custom Software

In support of the examples and exercises in the book, we provide two Python modules, `util`, and `mysocket`. The source code for these two modules is available at the book website: https://datasystems.denison.edu. To use these modules, they must be available in the Python path searched when an `import` occurs. The reader/user has some options:

1. Python distributions have a directory, `site-packages`, that exists to allow custom software installation, and this directory is part of the Python path, so placing these two Python files in this directory will make them available.
2. These two files could be placed in the same directory as the files that use them. This becomes less than ideal if the reader has many different folders being used for exercises.
3. The files could be placed in a `modules` directory that is a *sibling* to the set of directories used for solving exercises and homeworks. Then, code like the following could be inserted in a Python script or notebook to add the `modules` directory to the search path and import either of these modules:

```
moduledir = "../modules"
path = os.path.abspath(moduledir)
if not path in sys.path:
    sys.path.append(path)

import util
import mysocket
```

4. Not install at all. The primary benefit of the `util` module comes from the printing functions, which can be achieved through other means. The `mysocket` module simplifies direct socket-based programming for HTTP, but using the `requests` module alone may work fine, depending on the reader and/or instructor's goals.

© Springer Nature Switzerland AG 2020   797
T. Bressoud, D. White, *Introduction to Data Systems*,
https://doi.org/10.1007/978-3-030-54371-6

A.1   The util Module

The purpose of the util module is to provide utility functions that, while not overly complex themselves, can accelerate the process of working with websites and with data in various forms.

A.1.1 *buildURL*

Signature

```
buildURL(resource_path, host, protocol="https",
         extension=None, port=None)
```

Description

Construct a full URL from its constituent parts, given as parameters.

Parameters

Name	Type	Req'd	Description
resource_path	str	Y	Path to a resource at a server; if a query string is needed, it should be included here. If path does not lead with "/," it is prepended
host	str	Y	The "location" in the URL. Can be a DNS name or an IP address
protocol	str	N	Protocol scheme for the URL. The default is "https," and "http" is the other most common
extension	str	N	If present, used to append "." and a file extension at the end of the resource path. The default is None
port	int	N	If present, includes ":" as part of the URL. The default is None

Type	Description
str	Constructed string URL

Return

A.1.2 *random_string*

Signature

```
random_string(length=8)
```

Description

Generate a random string, built from upper case characters and decimal digits, with length given by the argument.

Parameters

Name	Type	Req'd	Description
length	int	N	Number of characters to be included in the random string. The default is 8

Return

Type	Description
str	Generated random string

A.1.3 *getLocalXML*

Signature

```
getLocalXML(filename, datadir=".", parser=None)
```

Description

Given a source XML file, located in datadir, read and parse the file using XMLParser if another parser is not specified and return the root Element.

Parameters

Name	Type	Req'd	Description
filename	str	Y	Name of the XML file for input
datadir	str	N	Allows specification of the directory folder where the file can be found. The default is "."
parser	etree parser	N	Parser to use, if specified. The default is None

Return

Type	Description
Element	Root of the parsed ElementTree if successful. None is returned if the file is not found or if the parse is unsuccessful

A.1.4 *read_creds*

Signature

```
read_creds(key, folder=".", file="creds.json")
```

Description

Reads and returns a sub-dictionary from a JSON-formatted file, based on the passed key. Assumes IDS convention of credentials, typically in a "creds.json" file being a JSON object with keys giving provider credentials and mapping to an object whose keys give the information for that provider.

Parameters

Name	Type	Req'd	Description
key	str	Y	The key for the sub-dictionary, typically values like "mysql," or "sqlite," or "spotify," designating a provider
folder	str	N	Allows specification of the directory folder where the credentials file can be found. The default is "."
file	str	N	Allows for a credentials file named something other than "creds.json." The default is "creds.json"

Return

Type	Description
dict	On success, returns the provider sub-dictionary; returns None if the file is not found, or if the given key is not present, or if there is a problem parsing the JSON-formatted file

A.1.5 `update_creds`

Signature

```
update_creds(key, keycreds, folder=".", file="creds.json")
```

Description

Given a provider key, and a dictionary of keys mapping to values for that provider, updates a credentials file that follows the IDS convention of credentials stored in a JSON-formatted file whose root is a JSON object with keys giving provider credentials and mapping to an object whose keys give the information for that provider. The provider's credentials are read in before then being updated, in case external changes have occurred. Uses dumps before write to get "pretty" indented JSON output.

Parameters

Name	Type	Req'd	Description
key	str	Y	The key for the sub-dictionary, typically values like "mysql," or "sqlite," or "spotify," designating a provider
keycreds	dict	Y	Dictionary for provider-specific credential information. These entries, and just these, are updated
folder	str	N	Allows specification of the directory folder where the credentials file can be found. The default is "."
file	str	N	Allows for a credentials file named something other than "creds.json." The default is "creds.json"

Return

Type	Description
None	Raises an exception if the file is not found, if the file is not parsable, if the given key is not present, if the new credentials are not JSON-formattable, or if there is a problem writing to the file

A.1.6 `print_text`

Signature

```
print_text(s, nlines=None, width=59, truncate=True,
           wrap=False, wrap_space = False, suffix="",
           prefix="", json_string=False)
```

Description

Print/output a multi-line text string, providing limits on the number of lines to print as well as width of the output, and control truncation and wrapping. Input can be a JSON-formatted string, and, if so, can be reformatted with indenting for a pretty-print capability.

Parameters

Name	Type	Req'd	Description
s	str	Y	Multi-line text to be printed
nlines	int	N	Maximum number of lines to be printed. The default is None, in which case all lines are printed
width	int	N	Maximum width of a processed line. The default is 59
truncate	bool	N	Determine whether to truncate lines that exceed width. The default is True
wrap	bool	N	Determine if the line should be wrapped. The default is False
wrap_space	bool	N	Determine, if wrapping, if breaks should be limited to where spaces occur. The default is False
suffix	str	N	Used when truncating to add characters to help indicate the truncation. The default is ""
prefix	str	N	Used when wrapping to use as prefix for non-whitespace characters to help indicate the continuation lines. The default is ""
json_string	bool	N	If True, reformat string by loading as JSON and then dumping as JSON, but with tree-based indentation. The default is False

Return

Type	Description
None	Would raise an Exception as a byproduct of an error in parsing when doing JSON processing

A.1.7 print_data

Signature

```
print_data(data, nlines=None, width=59, truncate=True,
           wrap=False, wrap_space = False, suffix="",
           prefix="", depth=None, nchild=None)
```

Description

Print/output a multi-line tree data structure, assumed to be JSON-format compliant, providing limits on the number of lines to print as well as width of the output, with control of truncation and wrapping. If either of depth or nchild is specified, it uses recursion controlled by the depth and nchild parameters to allow tree-based control of the output beyond just the number of lines. Otherwise, simply does a JSON dump of the structure with indentation.

Parameters

Name	Type	Req'd	Description
data	dict or list	Y	Data structure to be printed, assumed to be JSON-format compliant
nlines	int	N	Maximum number of lines to be printed. The default is None, in which case all lines are printed
width	int	N	Maximum width of a processed line. The default is 59
truncate	bool	N	Determine whether to truncate lines that exceed width. The default is True
wrap	bool	N	Determine if the line should be wrapped. The default is False
wrap_space	bool	N	Determine, if wrapping, if breaks should be limited to where spaces occur. The default is False
suffix	str	N	Used when truncating to add characters to help indicate the truncation. The default is ""
prefix	str	N	Used when wrapping to use as prefix for non-whitespace characters to help indicate the continuation lines. The default is ""
depth	int	N	Number of tree levels (depth) to be printed. The default is None
nchild	int	N	Controls how many siblings are printed at a given level. The default is None

Return

Type	Description
None	No return value

A.1.8 *print_xml*

Signature

```
print_xml(node, nlines=None, width=59, truncate=True,
              wrap=False, wrap_space=True, suffix="",
              prefix="", depth=None, nchild=None)
```

Description

Print/output a multi-line XML Element data structure, rooted at the specified node, providing limits on the number of lines to print as well as width of the output, with control of truncation and wrapping. If either of depth or nchild is specified, it uses recursion controlled by the depth and nchild parameters to allow tree-based control of the output beyond just the number of lines. Otherwise, uses tostring/fromstring and "pretty printing" to get formatted output.

Parameters

Name	Type	Req'd	Description
node	etree Element	Y	Data structure to be printed, assumed to be JSON-format compliant
nlines	int	N	Maximum number of lines to be printed. The default is None, in which case all lines are printed
width	int	N	Maximum width of a processed line. The default is 59
truncate	bool	N	Determine whether to truncate lines that exceed width. The default is True
wrap	bool	N	Determine if the line should be wrapped. The default is False
wrap_space	bool	N	Determine, if wrapping, if breaks should be limited to where spaces occur. The default is False
suffix	str	N	Used when truncating to add characters to help indicate the truncation. The default is ""
prefix	str	N	Used when wrapping to use as prefix for non-whitespace characters to help indicate the continuation lines. The default is ""
depth	int	N	Number of tree levels (depth) to be printed. The default is None
nchild	int	N	Controls how many siblings are printed at a given level. The default is None

Return

Type	Description
None	Exception could occur as a result of parsing

A.1.9 *print_headers*

Signature

```
print_headers(data, nlines=None, width=59, truncate=True,
              wrap=False, wrap_space = False, suffix="",
              prefix="")
```

Description

Print/output a dictionary data structure, providing limits on the number of lines to print as well as width of the output, with control of truncation and wrapping.

Parameters

Name	Type	Req'd	Description
data	dict	Y	Data structure to be printed, assumed to be JSON-format compliant
nlines	int	N	Maximum number of lines to be printed. The default is None, in which case all lines are printed
width	int	N	Maximum width of a processed line. The default is 59
truncate	bool	N	Determine whether to truncate lines that exceed width. The default is True
wrap	bool	N	Determine if the line should be wrapped. The default is False
wrap_space	bool	N	Determine, if wrapping, if breaks should be limited to where spaces occur. The default is False
suffix	str	N	Used when truncating to add characters to help indicate the truncation. The default is ""
prefix	str	N	Used when wrapping to use as prefix for non-whitespace characters to help indicate the continuation lines. The default is ""

Return

Type	Description
None	No return value

A.2   The mysocket Module

The mysocket module provides useful functions that simplify programming with the sockets interface. For readers with limited network programming experience, it allows network programming without all the customary detail.

A.2.1 `makeConnection`

Signature

```
makeConnection(location, port=80)
```

Description

Establish a TCP/socket connection between the current host, as the client, and the host specified by location as the passive server, which should be awaiting a connection.

Parameters

Name	Type	Req'd	Description
location	str	Y	Host location of the server. Can be a DNS name or an IP address, formatted as a string
port	int	N	Application port for the attempted connection. The default is 80

Return

Type	Description
socket connection	Established, connected socket object if successful, and None if a connection could not be established

A.2.2 `sendString`

Signature

```
sendString(connection, text, encoding="utf-8")
```

Description

Send the given string out the given connection. No interpretation is performed on the string. Call blocks until all bytes have been sent.

Parameters

Name	Type	Req'd	Description
connection	socket object	Y	Open, established connection to network endpoint
text	str	Y	String to be sent
encoding	str	N	Encoding to be used for translating the characters of text into the bytes to be sent over the network. The default is "utf-8"

Return

Type	Description
None	No return value

A.2.3   *receiveTillClose*

Signature

```
receiveTillClose(connection, encoding="utf-8", eol=False)
```

Description

Receive data over the given connection up through the point where the connection has been closed by the other side. Bytes of the input data are decoded into characters, and, optionally, CRLF sequences can be translated into newlines ('\n'). This last option should never be enabled on data that might be binary.

Parameters

Name	Type	Req'd	Description
connection	socket object	Y	Open, established connection to network endpoint
encoding	str	N	Encoding to be used for translating the characters of text into the bytes to be sent over the network. The default is "utf-8"
eol	bool	N	Determine whether to convert instances of CRLF into newline characters. The default is False

Return

Type	Description
None	No return value

A.2.4 *sendBytes*

Signature

```
sendBytes(connection, data)
```

Description

Send the (raw) bytes in data out the given connection. Call blocks until all bytes have been sent.

Parameters

Name	Type	Req'd	Description
connection	socket object	Y	Open, established connection to network endpoint
data	bytes	Y	Data to be sent

Return

Type	Description
None	No return value

A.2.5 *receiveTillSentinel*

Signature

```
receiveTillSentinel(connection, sentinel, encoding="utf-8")
```

Description

Receive data on the given connection, accumulating received data until the sentinel is received (or the connection is closed by the other side). Sentinel must be a single character; decoding is performed based on the given encoding.

This function can give a network analog to a readline(), if the sentinel is a ' \n '.

This function is inefficient, and a receiveTillClose() or a receiveBySize() is preferred when applicable.

Parameters

Name	Type	Req'd	Description
connection	socket object	Y	Open, established connection to network endpoint
sentinel	str	Y	Sought character to terminate a receive
encoding	str	N	Encoding to be used for translating the raw bytes received into characters of text. The default is "utf-8"

Return

Type	Description
str	Accumulated characters received, up to and including the sentinel

A.2.6 *receiveBySize*

Signature

```
receiveBySize(connection, size, encoding=None, eol=False)
```

Description

Attempt to receive size bytes of data from the given connection. Can result in fewer bytes if the other side closes the connection. If encoding is specified, decodes the bytes into characters based on the encoding. Optionally, CRLF sequences can be translated into newlines (`'\n'`). If eol is specified, so also must encoding.

This function can be called when binary data is expected, but encoding must *not* be specified, and eol must be False for correct operation.

If the application protocol above is HTTP (1.1), this function can be used to receive the *body* of an HTTP message, provided parsing of header lines has determined the Content-Length.

Parameters

Name	Type	Req'd	Description
connection	socket object	Y	Open, established connection to network endpoint
size	int	Y	Number of bytes to receive
encoding	str	N	Encoding to be used for translating the raw bytes received into characters of text. The default is None, in which case no decode is performed
eol	bool	N	Determine whether to translate CRLF into newline. Only applicable if encoding is specified. The default is False

Return

Type	Description
str or bytes	Received data, which will be a string if encoding is specified and bytes data otherwise

A.2.7   *sendCRLF*

Signature

```
sendCRLF(connection)
```

Description

Send a CRLF sequence out the given connection.

Parameters

Name	Type	Req'd	Description
connection	socket object	Y	Open, established connection to network endpoint

Return

Type	Description
None	No return value

A.2.8 sendCRLFLines

Signature

```
sendCRLFLines(connection, text, encoding="utf-8")
```

Description

Convenience function to send a Python line-ending block of text out the given connection. Text is generally assumed to be `'\n'` terminated and the function "converts" `'\n'` into CRLF as it performs the send operations over the connection. If text does not end in a LF or CRLF, one is appended.

Parameters

Name	Type	Req'd	Description
connection	socket object	Y	Open, established connection to network endpoint
text	str	Y	String to be sent
encoding	str	N	Encoding to be used for translating the characters of text into the bytes to be sent over the network. The default is "utf-8"

Return

Type	Description
None	No return value

References

1. Apache: HTTP server project (1995). URL https://httpd.apache.org/. Accessed: 2020-05-25
2. Apple: Terminal user guide (2020). URL https://support.apple.com/guide/terminal/welcome/mac. Accessed: 2020-06-23
3. Bayer, M.: Sqlalchemy. In: A. Brown, G. Wilson (eds.) The Architecture of Open Source Applications Volume II: Structure, Scale, and a Few More Fearless Hacks. aosabook.org (2012). URL http://aosabook.org/en/sqlalchemy.html
4. Codd, E.F.: A relational model of data for large shared data banks. Communications of the ACM **13**(6), 377–387 (1970)
5. Code Beautify: XML viewer (2020). URL https://codebeautify.org/xmlviewer. Accessed: 2020-06-26
6. curl.haxx.se: curl homepage (2020). URL https://curl.haxx.se. Accessed: 2020-06-06
7. curl.haxx.se: curl manpage (2020). URL https://curl.haxx.se/docs/manpage.html. Accessed: 2020-06-06
8. curl.haxx.se: Everything curl – the book (2020). URL https://curl.haxx.se/book.html. Accessed: 2020-06-06
9. ecma International: The JSON Data Interchange Format (2013). URL https://www.ecma-international.org/publications/files/ECMA-ST/ECMA-404.pdf. Accessed: 2020-06-11
10. Fielding, R.: Architectural styles and the design of network-based software architectures. Ph.D. thesis, University Of California, Irvine (2000)
11. Fielding, R.: Chapter 5: Representational State Transfer (REST) (2000). URL https://www.ics.uci.edu/~fielding/pubs/dissertation/rest_arch_style.htm. Accessed: 2020-06-25
12. freeformatter.com: JSON formatter (2020). URL https://www.freeformatter.com/json-formatter.html. Accessed: 2020-06-26
13. freeformatter.com: XML formatter (2020). URL https://www.freeformatter.com/xml-formatter.html. Accessed: 2020-06-26
14. freeformatter.com: XPath tester (2020). URL https://www.freeformatter.com/xpath-tester.html. Accessed: 2020-05-31
15. GitHub: GitHub developer (2020). URL https://developer.github.com/v3/. Accessed: 2020-05-31
16. Google: Google Python style guide (2020). URL https://google.github.io/styleguide/pyguide.html. Accessed: 2020-06-23
17. graphql.org: GraphQL homepage (2020). URL https://graphql.org/. Accessed: 2020-06-25
18. Grushinskiy, M.: XMLStarlet command line XML toolkit (2020). URL http://xmlstar.sourceforge.net/. Accessed: 2020-05-31

© Springer Nature Switzerland AG 2020

T. Bressoud, D. White, *Introduction to Data Systems*,

https://doi.org/10.1007/978-3-030-54371-6

19. Internet Engineering Task Force (IETF): RFC 3896 (2005). URL https://tools.ietf.org/html/rfc3986. Accessed: 2020-05-16
20. Internet Engineering Task Force (IETF): RFC 4180 (2005). URL https://tools.ietf.org/html/rfc4180. Accessed: 2019-08-26
21. Internet Engineering Task Force (IETF): RFC 4627 (2006). URL https://tools.ietf.org/html/rfc4627. Accessed: 2020-06-11
22. Internet Engineering Task Force (IETF): RFC 7231: Section 6 (2015). URL https://tools.ietf.org/html/rfc7231#section-6. Accessed: 2020-06-04
23. Internet Engineering Task Force (IETF): RFC 6749 (2019). URL https://tools.ietf.org/html/rfc6749. Accessed: 2019-11-22
24. Internet Engineering Task Force OAuth Working Group: OAuth 2.0 (2019). URL https://oauth.net/2/. Accessed: 2019-11-22
25. jquery.com: jQuery (2020). URL https://jquery.com/. Accessed: 2020-05-31
26. json-schema.org: JSON schema (2020). URL https://json-schema.org. Accessed: 2020-05-31
27. jsonformatter.org: JSON formatter (2020). URL https://jsonformatter.org/. Accessed: 2020-06-26
28. json.org: Introducing JSON (2020). URL https://www.json.org/json-en.html. Accessed: 2020-06-11
29. jsonschema: jsonschema (2020). URL https://python-jsonschema.readthedocs.io/en/stable/. Accessed: 2020-05-31
30. Kerrisk, M.: grep(1) — Linux manual page (2020). URL https://man7.org/linux/man-pages/man1/grep.1.html. Accessed: 2020-06-23
31. lxml.de: lxml – XML and HTML with Python (2020). URL https://lxml.de. Accessed: 2020-05-31
32. Merriam-Webster: authentication (2019). URL https://www.merriam-webster.com/dictionary/authentication. Accessed: 2019-11-13
33. Merriam-Webster: database (2019). URL https://www.merriam-webster.com/dictionary/database. Accessed: 2019-09-27
34. Merriam-Webster: system (2019). URL https://www.merriam-webster.com/dictionary/system. Accessed: 2019-06-26
35. Merriam-Webster: protocol (2020). URL https://www.merriam-webster.com/dictionary/protocol. Accessed: 2020-03-26
36. Mozilla: HTTP headers (2020). URL https://developer.mozilla.org/en-US/docs/Web/HTTP/Headers. Accessed: 2020-06-19
37. NGINX/F5: NGINX (2010). URL https://www.nginx.com/. Accessed: 2020-05-25
38. Progress: Restructuring data: FLWOR expressions (2020). URL https://www.progress.com/xquery/resources/tutorials/using-xquery/flwor-expressions. Accessed: 2020-05-31
39. pydata.org: pandas documentation (2019). URL https://pandas.pydata.org/pandas-docs/stable/. Accessed: 2019-08-30
40. pydata.org: pandas.read_csv (2019). URL https://pandas.pydata.org/pandas-docs/stable/reference/api/pandas.read_csv.html#pandas.read_csv. Accessed: 2019-08-30
41. pydata.org: pandas.DataFrame.melt (2020). URL https://pandas.pydata.org/pandas-docs/stable/reference/api/pandas.DataFrame.melt.html. Accessed: 2020-02-13
42. pydata.org: pandas.DataFrame.pivot (2020). URL https://pandas.pydata.org/pandas-docs/stable/reference/api/pandas.DataFrame.pivot.html. Accessed: 2020-02-13
43. pydata.org: pandas.DataFrame.pivot_table (2020). URL https://pandas.pydata.org/pandas-docs/stable/reference/api/pandas.DataFrame.pivot_table.html. Accessed: 2020-02-13
44. pydata.org: pandas.DataFrame.stack (2020). URL https://pandas.pydata.org/pandas-docs/stable/reference/api/pandas.DataFrame.stack.html. Accessed: 2020-02-13
45. pydata.org: Reshaping and pivot tables (2020). URL https://pandas.pydata.org/pandas-docs/stable/user_guide/reshaping.html. Accessed: 2020-02-13
46. Python Software Foundation: csv – CSV file reading and writing (2019). URL https://docs.python.org/3/library/csv.html. Accessed: 2019-08-30

47. Python Software Foundation: The assert statement (2020). URL https://docs.python.org/3/reference/simple_stmts.html#the-assert-statement. Accessed: 2020-06-23
48. Python Software Foundation: Built-in types: Mapping types – dict (2020). URL https://docs.python.org/3/library/stdtypes.html#typesmapping. Accessed: 2020-06-26
49. Python Software Foundation: Built-in types: String methods: str.split() (2020). URL https://docs.python.org/3/library/stdtypes.html#str.split. Accessed: 2020-06-23
50. Python Software Foundation: codecs – codec registry and base classes: Standard encodings (2020). URL https://docs.python.org/3/library/codecs.html#standard-encodings. Accessed: 2020-05-29
51. Python Software Foundation: Compound statements: Function definitions (2020). URL https://docs.python.org/3/reference/compound_stmts.html#function-definitions. Accessed: 2020-06-26
52. Python Software Foundation: Data structures: Dictionaries (2020). URL https://docs.python.org/3/tutorial/datastructures.html#dictionaries. Accessed: 2020-06-26
53. Python Software Foundation: Errors and exceptions (2020). URL https://docs.python.org/3/tutorial/errors.html. Accessed: 2020-06-23
54. Python Software Foundation: Input and output: Methods of file objects (2020). URL https://docs.python.org/3/tutorial/inputoutput.html#methods-of-file-objects. Accessed: 2020-06-23
55. Python Software Foundation: json – JSON encoder and decoder (2020). URL https://docs.python.org/3/library/json.html. Accessed: 2020-05-31
56. Python Software Foundation: os – miscellaneous operating system interfaces (2020). URL https://docs.python.org/3/library/os.html. Accessed: 2020-06-23
57. Python Software Foundation: os.path – common pathname manipulations (2020). URL https://docs.python.org/3/library/os.path.html. Accessed: 2020-06-23
58. Python Software Foundation: re – regular expression operations (2020). URL https://docs.python.org/3/library/re.html. Accessed: 2020-06-26
59. Python Software Foundation: Regular expression HOWTO (2020). URL https://docs.python.org/3/howto/regex.html. Accessed: 2020-06-26
60. Rhino: Python basic syntax: End of statements (2020). URL https://developer.rhino3d.com/guides/rhinopython/python-statements/. Accessed: 2020-06-23
61. Richardson, L.: Beautiful soup documentation (2020). URL https://www.crummy.com/software/BeautifulSoup/bs4/doc/. Accessed: 2020-06-23
62. SQLAlchemy: SQLAlchemy 1.3 documentation: Engine configuration (2019). URL https://docs.sqlalchemy.org/en/latest/core/engines.html. Accessed: 2019-10-10
63. Srivastava, R.: XML schema: Understanding namespaces (2020). URL https://www.oracle.com/technical-resources/articles/srivastava-namespaces.html. Accessed: 2020-05-31
64. The Movie Database (TMDb): TMDb: API overview (2020). URL https://www.themoviedb.org/documentation/api. Accessed: 2020-05-31
65. tutorialspoint: Learn JSON (2020). URL https://www.tutorialspoint.com/json/index.htm. Accessed: 2020-05-31
66. tutorialspoint: Learn XML (2020). URL https://www.tutorialspoint.com/xml/index.htm. Accessed: 2020-05-31
67. Ullman, J.D., Widom, J.: A First Course in Database Systems, 3rd edn., pp. 17–18. Pearson Prentice Hall, Upper Saddle River, NJ (2008). ISBN 978-0-13-600637-4
68. US Social Security Admin: Get ready for baby (2019). URL https://www.ssa.gov/oact/babynames/. Accessed: 2019-07-19
69. USGS: Earthquake events (2020). URL https://earthquake.usgs.gov/fdsnws/event/1/query?format=xml&starttime=2019-04-01&endtime=2019-05-31&minmagnitude=6. Accessed: 2020-05-31
70. USGS: U.S. geological survey (2020). URL https://www.usgs.gov/. Accessed: 2020-06-16
71. w3schools.com: DTD – attributes (2020). URL https://www.w3schools.com/xml/xml_dtd_attributes.asp. Accessed: 2020-05-31
72. w3schools.com: HTML tutorial (2020). URL https://www.w3schools.com/html/. Accessed: 2020-06-23

73. w3schools.com: HTML versus XHTML (2020). URL https://www.w3schools.com/html/html_xhtml.asp. Accessed: 2020-05-31

74. w3schools.com: XML DOM tutorial (2020). URL https://www.w3schools.com/xml/dom_intro.asp. Accessed: 2020-05-31

75. w3schools.com: XML DTD (2020). URL https://www.w3schools.com/xml/xml_dtd.asp. Accessed: 2020-05-31

76. w3schools.com: XML Schema (2020). URL https://www.w3schools.com/xml/xml_schema.asp. Accessed: 2020-05-31

77. w3schools.com: XML tutorial (2020). URL https://www.w3schools.com/xml/default.asp. Accessed: 2020-05-31

78. w3schools.com: XPath tutorial (2020). URL https://www.w3schools.com/xml/xpath_intro.asp. Accessed: 2020-05-31

79. w3schools.com: XQuery tutorial (2020). URL https://www.w3schools.com/xml/xquery_intro.asp. Accessed: 2020-05-31

80. w3schools.com: XSD complex elements (2020). URL https://www.w3schools.com/xml/schema_complex.asp. Accessed: 2020-05-31

81. w3schools.com: XSD restrictions/facets (2020). URL https://www.w3schools.com/xml/schema_facets.asp. Accessed: 2020-05-31

82. Wickham, H.: Tidy data. Journal of Statistical Software, Articles **59**(10), 1–23 (2014). DOI 10.18637/jss.v059.i10. URL https://www.jstatsoft.org/v059/i10

83. Wikipedia: Authorization (2019). URL https://en.wikipedia.org/wiki/Authorization. Accessed: 2019-11-13

84. Wikipedia: Comma-separated values (2019). URL https://en.wikipedia.org/wiki/Comma-separated_values. Accessed: 2019-08-26

85. Wikipedia: Communication protocol (2020). URL https://en.wikipedia.org/wiki/Communication_protocol. Accessed: 2020-03-26

86. Wikipedia: Relational database (2020). URL https://en.wikipedia.org/wiki/Relational_database. Accessed: 2020-05-31

87. Wikipedia: XML Schema (2020). URL https://en.wikipedia.org/wiki/XML_schema. Accessed: 2020-05-31

88. Windows: Windows Server: Windows commands (2020). URL https://docs.microsoft.com/en-us/windows-server/administration/windows-commands/windows-commands. Accessed: 2020-06-23

89. World Bank: World Bank open data: Free and open access to global development data (2019). URL https://data.worldbank.org/. Accessed: 2019-07-19

90. World Wide Web Consortium (W3C): Extensible markup language (XML) 1.0 (fifth edition) (2008). URL https://www.w3.org/TR/xml/. Accessed: 2020-06-12

91. World Wide Web Consortium (W3C): XML Path language (XPath) 3.1 (2017). URL https://www.w3.org/TR/2017/REC-xpath-31-20170321/. Accessed: 2020-06-29

Index

© Springer Nature Switzerland AG 2020
T. Bressoud, D. White, *Introduction to Data Systems*,
https://doi.org/10.1007/978-3-030-54371-6

Printed in the United States
by Baker & Taylor Publisher Services